SUSTAINABLE CATALYSIS

SUSTAINABLE CATALYSIS

Challenges and Practices for the Pharmaceutical and Fine Chemical Industries

Edited by

PETER J. DUNN
Pfizer Green Chemistry Lead
Sandwich, Kent, United Kingdom

K. K. (MIMI) HII
Imperial College London
South Kensington, London, United Kingdom

MICHAEL J. KRISCHE
University of Texas at Austin
Austin, Texas, United States of America

MICHAEL T. WILLIAMS
CMC Consultant
Deal, Kent, United Kingdom

WILEY

Published by John Wiley & Sons, Inc., Hoboken, New Jersey.
Published simultaneously in Canada.

Library of Congress Cataloging-in-Publication Data:

Sustainable catalysis : challenges and practices for the pharmaceutical and fine chemical industries / edited by Peter J. Dunn, Pfizer Green Chemistry Lead, Sandwich, Kent, United Kingdom, K.K. (Mimi) Hii, Imperial College London, South Kensington, London, United Kingdom, Michael J. Krische, University of Texas at Austin, Austin, Texas, United States of America, Michael T. Williams, CMC Consultant, Deal, Kent, United Kingdom.
pages cm
Includes index.
ISBN 978-1-118-15542-4 (cloth)
1. Environmental chemistry–Industrial applications. 2. Chemical engineering. 3. Catalysts. 4. Pharmaceutical industry–Waste minimization. I. Dunn, Peter J. (Peter James) editor of compilation. II. Hii, K. K., 1969- editor of compilation. III. Krische, Michael J., editor of compilation. IV. Williams, Michael T. (Michael Trevelyan) editor of compilation.
TP155.2.E58S86 2013
541′.395–dc23

2012040248

10 9 8 7 6 5 4 3 2 1

CONTENTS

FOREWORD

It is our pleasure to introduce this book on the application of catalysis to the manufacture of pharmaceuticals and fine chemicals.

Many scientists study catalysis for the thrill of discovering new knowledge, whereas the applied scientist has the additional motivation of seeking to do something useful with that knowledge. The science of catalysis, in particular, is transformed by the discipline of targeted outcomes. There are an infinite number of combinations of reactions and catalyst formulations, but only a small fraction will ever be useful in some way for mankind.

Until recently catalysis has played a modest role in the pharmaceutical and fine chemical sector, which is concerned with the manufacture of small volumes of large, and often complex, organic molecules by multi-step synthetic routes. The affordable cost of reagents, relative to the high value of the products, meant that there was little incentive to develop individual catalytic steps. This situation began to change with the growing social and industrial interest in "greener," safer manufacturing processes, which generate less waste and avoid hazardous reagents. Economics was a driver due to the increasing cost of environmental protection and waste treatment. The potential for new catalytic methods to create new "chemical space" was a parallel attraction. Catalysis was now part of the solution, with many opportunities for innovation. In 2005, the ACS Green Chemistry Institute together with leading pharmaceutical corporations, set up the Pharmaceutical Roundtable. In a landmark study, this body developed a list of 12 key research areas for green chemistry research, including 10 types of synthetic reaction [1].

If the matching of industrial need with scientific discovery is the beginning of the story, the next stage is the achievement of efficiency and selectivity in the research laboratory. However, even then there is still much to be done. Many issues arise when a process is scaled up for commercial production, and so the successful development of new catalytic processes also needs the complementary skills of industrial application.

This was the vision for a dedicated symposium on the theme of "*Challenges in Catalysis for Pharmaceuticals and Fine Chemicals*," which was jointly organized by the Applied

Catalysis Group (ACG)[1] of the Royal Society of Chemistry and the Fine Chemicals Group (FCG)[2] of the Society of Chemical Industry. The intention from the outset was broad participation and ownership. Having canvassed opinions among our members, we set about finding authoritative speakers from industry who could describe the challenges for commercial application, and from academia who could tell us how to meet them, so combining the industrial perspective with academic reports on the scientific "state of the art." The first meeting in 2007 was a resounding success, and has since been followed by "*Challenges II*" and "*Challenges III*" in 2009 and 2011.

In line with the aims of the "*Challenges*" meetings, the contents of this book have been selected to represent topical areas of catalytic synthetic chemistry, including several on the original "*Challenges*" list. In order to encourage a greater degree of realism in research, most subjects have been covered initially from an academic angle and then from an industrial angle.

We hope that this book will be both enjoyable and stimulating for those who are interested in this exciting field. Most of all, we hope that it will inspire both more academic discovery and more industrial application of catalysis for pharmaceuticals and fine chemicals.

John Birtill
Highcliffe Catalysis Ltd. and University of Glasgow, RSC Applied Catalysis Group
Alan Pettman
Pfizer Ltd., SCI Fine Chemicals Group and RSC Applied Catalysis Group

REFERENCE

1. Constable DJC, Dunn PJ, Hayler JD, Humphrey GR, Leazer, Jr., JL, Linderman RJ, Lorenz K, Manley J, Pearlman BA, Wells A, Zaks A, Zhang TY (2007). Key green chemistry research areas—a perspective from pharmaceutical manufacturers. *Green Chem.* 9, 411–420.

[1] www.rsc.org/appliedcatalysis
[2] www.soci.org

PREFACE

"There must be a better way to make things we want, a way that doesn't spoil the sky, or the rain or the land."

—Sir Paul McCartney

There has been an increasing awareness within the fine chemicals and pharmaceutical industry of the need to improve the environmental and production costs of synthesis, driven largely by both the perceived need to improve society's image of the industry and the tightening regulatory controls over the release of waste products and toxins into the environment. The replacement of stoichiometric reagents for synthetic transformations by catalytic routes is playing a major role in this drive toward "greener," safer, and more economic chemical processes. The development of scalable catalytic methodologies suitable for relatively complex pharmaceutical intermediates, which often contain multiple H-bond donors and acceptors, is a significant synthetic chemical challenge. However, robust catalytic processes are increasingly emerging and have begun to make a significant impact upon the "greening" of pharmaceutical processes. The scene is thus set for an exciting period of further growth for the discovery and development of "green" catalytic processes, which will remain an important technology for the foreseeable future.

The content of the book is carefully chosen to represent key areas that are particularly important for the fine and pharmaceutical industries, including C–H, C–N, and C–C bond forming reactions, featuring chemo-, bio-, and organocatalytic approaches. It has been our aim to provide examples of the more recently discovered catalytic methodologies, particularly those that are featured on the list of reactions identified by the GCI Pharmaceutical Roundtable as "most important" or "aspirational," as well as topical areas of catalytic synthetic chemistry that were highlighted in the "Challenges" meetings, such as the catalytic reduction of amides and esters, biocatalysis, amide formation, addressing concerns with the use of genotoxic intermediates for nucleophilic substitution, and C–H activation of aromatics.

We have enlisted an illustrious team of academic and industrial experts and leaders as contributors. In seven of the chosen topics, an academic overview of the current innovations is followed by an industrial case study at the process scale, with the aim of providing valuable insights into a catalytic methodology, from proof of concept (mg scale) to eventual application on the synthesis of organic molecules (kg to multi-tonne scale). The remits of academic/industrial research are thus united by a common theme, providing a balanced perspective on the current limitations and future challenges.

We hope that this approach will highlight the technology gap between "blue-sky" and "applied" research that will translate curiosity-driven research to the industrial manufacture of high-value chemical products that will sustain and improve quality of life, without exerting unnecessary demands on our environment and the needs of future generations.

We hope that this book provides a useful resource for both academic and industrial readers, and helps foster growing awareness of the challenges involved in this exciting and rapidly developing area. Last but not least, we thank all our authors for the high quality of their contributions, and for their patience with all our demands and deadlines.

Peter J. Dunn
K. K. (Mimi) Hii
Michael J. Krische
Michael T. Williams

CONTRIBUTORS

Joanne E. Anderson, GlaxoSmithKline Inc., Research Triangle Park, NC, USA

Martin A. Berliner, Chemical Research and Development, Pfizer Inc., Groton, CT, USA

Johann Chan, Chemical Process R&D, Amgen Inc., Thousand Oaks, CA, USA; and Chemical Development, Gilead Sciences, Foster City, CA, USA

Jannine Cobb, GlaxoSmithKline Inc., Research Triangle Park, NC, USA

David J. Cole-Hamilton, School of Chemistry, University of St. Andrews, North Haugh, Fife, Scotland, UK

Justyna Czaban, Institute of Organic Chemistry, Polish Academy of Sciences, Warsaw, Poland

Roman Davis, GlaxoSmithKline Inc., Research Triangle Park, NC, USA

Deborah L. Dodds, School of Chemistry, University of St. Andrews, North Haugh, Fife, Scotland, UK; and Johnson Matthey plc, Billingham, UK

Peter J. Dunn, Pfizer Global Supply, Pfizer Ltd, Sandwich, Kent, UK

Russ N. Fitzgerald, GlaxoSmithKline Inc., Research Triangle Park, NC, USA

Hiroaki Gotoh, Department of Applied Chemistry, Graduate School of Engineering, Yokohama National University, Hodogaya-ku, Yokohama, Japan

Karol Grela, Institute of Organic Chemistry, Polish Academy of Sciences, Warsaw, Poland; and Department of Chemistry, Warsaw University, Warsaw, Poland

Yujiro Hayashi, Department Chemistry, Graduate School of Science, Tohoku University, Aoba-ku, Sendai, Japan

Koji Hirano, Division of Applied Chemistry, Graduate School of Engineering, Osaka University, Suita, Osaka, Japan

Jinkun Huang, Chemical Process R&D, Amgen Inc., Thousand Oaks, CA, USA; and Chengdu Suncadia Pharmaceutical Co., Ltd., A Subsidiary of Hengrui Medicine Co., Ltd., China

Jacob M. Janey, Department of Process Research, Merck Research Laboratories, Merck & Co Inc., Rahway; and Chemical Development, Bristol-Myers Squibb, New Brunswick, NJ, USA

Michael J. Krische, Department of Chemistry and Biochemistry, University of Texas at Austin, Austin, TX, USA

Stephen P. Marsden, School of Chemistry, University of Leeds, Leeds, UK

Masahiro Miura, Division of Applied Chemistry, Graduate School of Engineering, Osaka University, Suita, Osaka, Japan

Benjamin M. Monks, Department of Chemistry, Durham University, Durham, UK

Joseph Moran, Department of Chemistry and Biochemistry, University of Texas at Austin, Austin, TX, USA; and ISIS, University of Strasbourg, Strasbourg, France

Alan J. Pettman, Chemical Research and Development, Pfizer Ltd, Sandwich, Kent, UK

Lionel A. Saudan, Corporate R&D Division, Firmenich SA, Geneva, Switzerland

Chris Senanayake, Chemical Development, Boehringer Ingelheim Pharmaceutical, Inc., Ridgefield, CT, USA

Christian Torborg, Department of Chemistry, Warsaw University, Warsaw, Poland

Matthew D. Truppo, Merck Research Laboratories, Rahway, NJ, USA

Nicholas J. Turner, Manchester Institute for Biotechnology, School of Chemistry, University of Manchester, Manchester, UK

Xiang Wang, Chemical Process R&D, Amgen Inc., Thousand Oaks, CA, USA; and Chemical Development, Gilead Sciences, Foster City, CA, USA

Xudong Wei, Chemical Development, Boehringer Ingelheim Pharmaceutical, Inc., Ridgefield, CT, USA

Andrew Whiting, Department of Chemistry, Durham University, Durham, UK

Jonathan M.J. Williams, Department of Chemistry, University of Bath, Claverton Down, Bath, UK

Feng Xu, Department of Process Research, Merck Research Laboratories, Rahway, NJ, USA

Nathan Yee, Chemical Development, Boehringer Ingelheim Pharmaceutical, Inc., Ridgefield, CT, USA

ABBREVIATIONS

Abbreviations	Full Name
3,5-*t*-Bu-4-MeO-MeO-BIPHEP	2,2′-Bis[di(3,5-di-*tert*-butyl-4-methoxyphenyl)phosphino]-6,6′-dimethoxy-1,1′-biphenyl
Ac	Acetyl
acac	Acetylacetonate
Ad	1-Adamantyl
Alkyl groups	Me, Et, *n*-Pr, *i*-Pr, *sec*-Bu, Pent, Hex, Hep, Oct
aq.	Aqueous solution
BARF	Tetrakis(3,5-bis(trifluoromethyl)phenyl)borate
BINAP	2,2′-Bis(diphenylphosphino)-1,1′-binaphthyl
BINOL	1,1′-Bi-2-naphthol
BIPHEP	2,2′-Bis(diphenylphosphino)-1,1′-biphenyl
Bn	Benzyl
Boc	*tert*-Butoxycarbonyl
b.p.	Boiling point
bpy	Bipyridine/bipyridyl
Bz	Benzoyl (PhCO)
CBz	Benzyloxycarbonyl
Cl,MeO-BIPHEP	5,5′-Dichloro-6,6′-dimethoxy-2,2′-bis(diphenylphosphino)-1,1′-biphenyl
COD	1,5-Cyclooctadiene
COE	Cyclooctene
Cp	Cyclopentadienyl/cyclopentadiene
CTH-P-PHOS	2,2′,6,6′-Tetramethoxy-4,4′-bis(diphenylphosphino)-3,3′-bipyridine
Cy	Cyclohexyl
dba	Dibenzylideneacetone

DBU	1,8-Diazabicyclo-[5.4.0]-undec-7-ene
DCE	1,2-Dichloroethane
de	Diastereomeric excess
DEAD	Diethyl azodicarboxylate
DIAD	Diisopropyl azodicarboxylate
DIPPF	1,1′-Bis(diisopropylphosphino)ferrocene
DMAc	*N,N*-Dimethylacetamide
DMAP	*N,N*-Dimethylaminopyridine
DME	1,2-Dimethoxyethane
DMF	*N,N*-Dimethylformamide
DMPU	*N,N*′-Dimethyl propylene urea
DM-SEGPHOS	5,5′-Bis[di(3,5-xylyl)phosphino]-4,4′-bi-1,3-benzodioxole
DMSO/dmso	Dimethylsulfoxide
DPEphos	Bis(2-diphenylphosphinophenyl)ether
DPPB	1,4-Bis(diphenylphosphino)butane
DPPE	1,2-Bis(diphenylphosphino)ethane
DPPF	1,1′-Bis(diphenylphosphino)ferrocene
DPPP	1,3-Bis(diphenylphosphino)propane
dr	Diastereomeric ratio
EDCI	[3-(Dimethylamino)propyl]ethylcarbodiimide
ee	Enantiomeric excess
equiv.	Molar equivalent(s)
er	Enantiomeric ratio
ETP	2-Bis(diphenylphosphinoethyl)phenylphosphine
HBTU	*O*-Benzotriazole-*N,N,N*′,*N*′-tetramethyl uronium hexafluorophosphate
HOBt	1-Hydroxybenzotriazole
IMes	1,3-Bis(2,4,6-trimethylphenyl)imidazol-2-ylidene
IPr	1,3-Bis(2,6-diisopropylphenyl)imidazol-2-ylidene
JohnPhos	2-Di(*tert*-butyl)phosphinobiphenyl
Josiphos	1-[2-(Diphenylphosphino)ferrocenyl] ethyldicyclohexylphosphine
MeO-BIPHEP	2,2′-Bis(diphenylphosphino)-6,6′-dimethoxy-1,1′-biphenyl
MIBK	Methyl isobutyl ketone
mol	Moles
Ms	Methanesulfonyl
MS	Molecular sieves
MTBE	Methyl *tert*-butyl ether
NBD	Norbornadiene
NMP	*N*-Methyl pyrrolidone
Ns	*p*-Nitrophenylsulfonyl
PEG	Poly(ethylene glycol)
phen	Phenanthroline
Phth	Phthaloyl
Piv	Pivaloyl (*t*-BuCO)
Py	Pyridyl/pyridine
rr	Regioisomeric ratio
rt	Room (ambient) temperature

S/C	Substrate-to-catalyst ratio
SEGPHOS	5,5′-Bis(diphenylphosphino)-4,4′-bi-1,3-benzodioxole
SIMes	1,3-Bis(2,4,6-trimethylphenyl)-4,5-dihydroimidazol-2-ylidene
TADDOL	α,α,α,α-Tetraaryl-1,3-dioxolane-4,5-dimethanol
TBAB	Tetra-*n*-butylammonium bromide
TBAF	Tetra-*n*-butylammonium fluoride
TBAI	Tetra-*n*-butylammonium iodide
TBDPS	*tert*-Butyldiphenylsilyl
TBS	*tert*-Butyldimethylsilyl
TBTU	*O*-(Benzotriazol-1-yl)-*N*,*N*,*N*′,*N*′-tetramethyluronium tetrafluoroborate
Tetraphos	1,2-Bis((2-(diphenylphosphino)ethyl)(phenyl)phosphino) ethane
Tf	Trifluoromethanesulfonyl
TFA	Trifluoroacetate or trifluoroacetic acid
THF	Tetrahydrofuran
TIPS	Triisopropylsilyl
TMBTP	2,2′,5,5′-Tetramethyl-3,3′-bis(diphenylphosphine)-4,4′-bithiophene
TMEDA	*N*,*N*,*N*′,*N*′-Tetramethylethylenediamine
TMS	Trimethylsilyl
TOF	Turnover frequency
TolBINAP	2,2′-Bis(di-*p*-tolylphosphino)-1,1′-binaphthyl
TON	Turnover number
Triphos	1,1,1-Tris(diphenylphosphinomethyl)ethane
Ts	*p*-Toluenesulfonyl
WALPHOS	1-[(*R*)-2-(2′-Diphenylphosphinophenyl)ferrocenyl] ethyldiphenylphosphine
Xantphos	4,5-Bis(diphenylphosphino)-9,9-dimethylxanthene
XPhos	2-Dicyclohexylphosphino-2′,4′,6′-triisopropylbiphenyl
Xyl-BINAP	2,2′-Bis[di(3,5-xylyl)phosphino]-1,1′-binaphthyl
XylylWALPHOS	1-[-2-(2′-Di-3,5-xylylphosphinophenyl)ferrocenyl]ethyldi-3,5-xylylphosphine
μ-wave	Microwave

1

CATALYTIC REDUCTION OF AMIDES AVOIDING $LiAlH_4$ OR B_2H_6

DEBORAH L. DODDS[1,2] AND DAVID J. COLE-HAMILTON[1]

[1]*School of Chemistry, University of St. Andrews, North Haugh, Fife, Scotland, UK*
[2]*Johnson Matthey plc, Billingham, UK*

1.1 INTRODUCTION

Amines are key components in a variety of pharmaceutical compounds, chemical intermediates, and commodity chemicals. A detailed review by Jung and coworkers describes the synthesis of secondary alkyl and aryl amines [1]. The synthesis of amines by metal-catalyzed reactions generally falls into one of two categories: (i) reduction of an unsaturated nitrogen-containing species or (ii) tandem reactions involving amination and reduction steps. Synthetic routes to primary amines include the reduction of nitro arenes, nitriles, or amides; amination of alcohols; and hydroaminomethylation of alkenes (Scheme 1-1).

Routes to secondary and tertiary amines are more limited, but they can generally be made via amide reduction, amination of alcohols, and alkene hydroaminomethylation (Scheme 1-2).

This chapter focuses primarily on the synthesis of amines via amide hydrogenation. Particular aspects considered are the atom economy (AE) of the reactions, the operating conditions, and the safety of the reagents/processes. These catalyzed processes are then compared with stoichiometric metal hydride reagents.

1.2 AMIDES

Amides are particularly challenging substrates for hydrogenation reactions, which is a consequence of their stable resonance structure. The conjugation of the nitrogen's electron

Sustainable Catalysis: Challenges and Practices for the Pharmaceutical and Fine Chemical Industries, First Edition. Edited by Peter J. Dunn, K. K. (Mimi) Hii, Michael J. Krische, and Michael T. Williams.

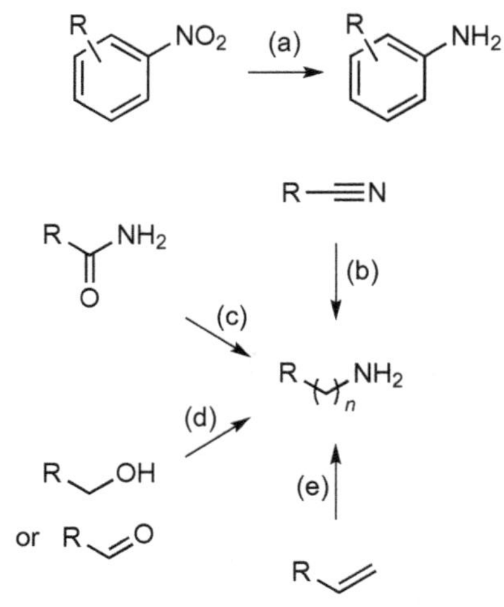

SCHEME 1-1. Homogeneously catalyzed routes to primary amines: (a) hydrogenation of nitro arenes (H_2); (b) hydrogenation of nitriles (H_2, $n = 1$); (c) hydrogenation of amides (H_2, $n = 1$); (d) amination of alcohols (H_2, NH_3, $n = 1$); (e) hydroaminomethylation of alkenes (CO/H_2, NH_3, $n = 3$).

lone pair with the π-bond of the carbonyl is so effective that the double-bond character is shared across both the C–O and C–N bonds, leading to planarity within the molecule. The delocalization extends to the first carbon of a substituent attached to the carbonyl or nitrogen, such that there is no longer free rotation about the C–N bond, an effect that is readily observed by ^1H NMR spectroscopy. This resonance adds stability to the amide functionality, making them significantly harder to reduce than other carbonyl groups, such as ketones.

In addition, the reaction is less favorable at higher temperatures as a result of a negative ΔS of hydrogenation, which in turn leads to a more positive ΔG. Despite a lower, more favorable ΔG value observed at lower temperatures, the reaction has a high kinetic barrier that requires high temperatures for the reaction to proceed. This is why heterogeneous amide hydrogenations traditionally require extremely forcing reaction conditions.

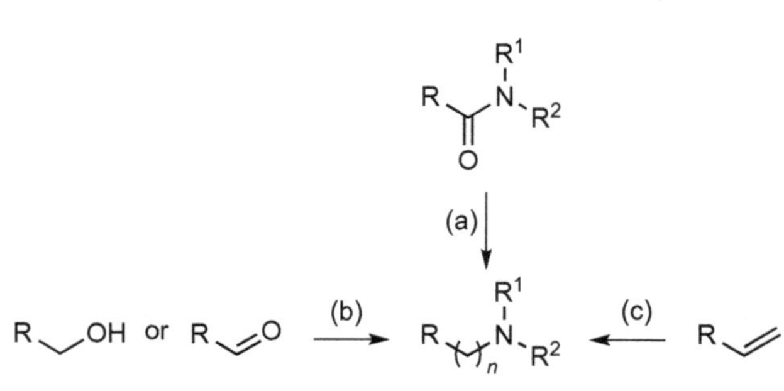

SCHEME 1-2. Homogeneously catalyzed routes to secondary ($R^1 = H$) and tertiary amines: (a) hydrogenation of amides (H_2, $n = 1$); (b) amination of alcohols (H_2, HNR^1R^2, $n = 1$); (c) hydroaminomethylation of alkenes (CO/H_2, HNR^1R^2, $n = 3$).

1.3 IMPORTANCE OF AMIDE REDUCTIONS IN PHARMACEUTICAL SYNTHESIS

Amide formation followed by reduction to the amine is a common route to C—N bonds as they are very reliable, yet versatile. The reduction step is, more often than not, carried out with a stoichiometric amount of a metal hydride reducing agent such as lithium aluminum hydride ($LiAlH_4$) or borane (B_2H_6); however, these types of reagents have a number of inherent problems associated with their use, particularly on a large scale. First, they are difficult and potentially hazardous to handle and have complex workup procedures. Second, there is a large amount of waste generated as a by-product, such as mixed metal hydroxides or boric acid, which must be disposed of in a responsible manner—this is both an environmental and an economic drawback. As a result, amide reduction avoiding the use of $LiAlH_4$ and B_2H_6 has been identified as a key area of development by the ACS Green Chemistry Institute and members of the pharmaceutical round table [2].

In 2006, a study of the synthesis of 128 drug candidates carried out in the process chemistry departments of GlaxoSmithKline, AstraZeneca, and Pfizer highlighted the popularity of metal hydride reducing agents [3]. Of the 94 reduction reactions in the study, 44% were heterogeneous hydrogenations, 41% were metal hydride/borane reductions, and only 4% represented homogeneous hydrogenations. In fact, no carboxylic acid derivatives were reduced using homogeneous methods. Although this is not the whole picture, it does give a reflection of the trends that are present in industrial process chemistry.

Stoichiometric amide reductions are commonplace in the pharmaceutical industry. In this section, examples are chosen to highlight the various challenges faced by the synthetic chemists in the reduction of a molecule with multiple functional groups. The synthesis of paroxetine, a selective serotonin reuptake inhibitor used to treat depression, involves the reduction of an imide intermediate **1** that incorporates an ester side chain (Scheme 1-3). The global reduction of all three C=O units is carried out in one step (90%) using 5 equiv. of $LiAlH_4$ as the reducing agent to give the cyclic amine **2** [4]. Clearly, the AE and safety of this reaction could be significantly improved with a homogeneous catalytic hydrogenation using molecular hydrogen, as water would be the only by-product.

Another example of $LiAlH_4$ amide reduction can be found in the synthesis of tolterodine (Scheme 1-4), an anticholinergic used to treat urinary incontinence. One step in its preparation involves the reduction of an amide **3** prior to the final debenzylation

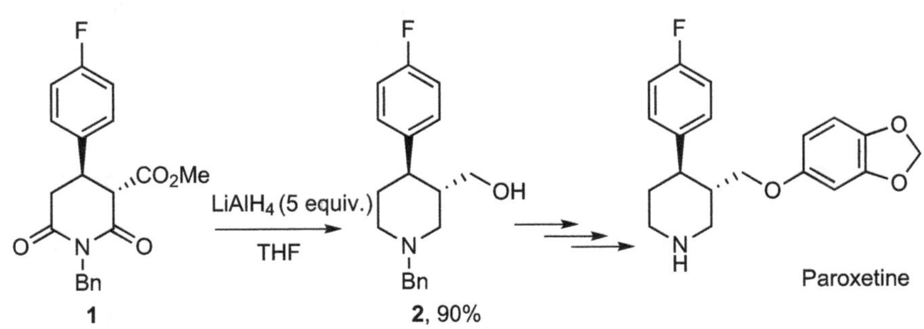

SCHEME 1-3. Imide/ester reduction step in the synthesis of paroxetine.

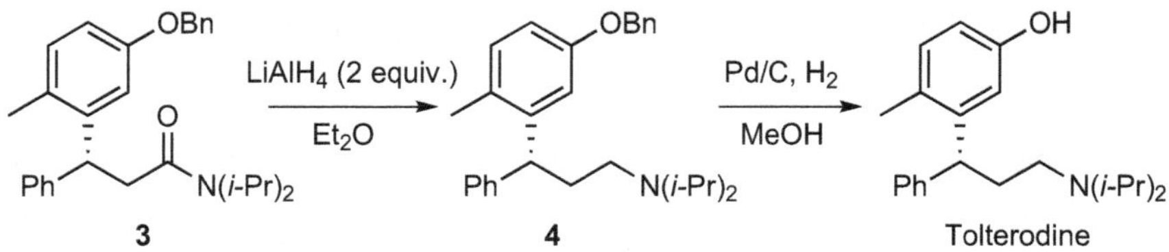

SCHEME 1-4. Amide reduction step in the synthesis of tolterodine.

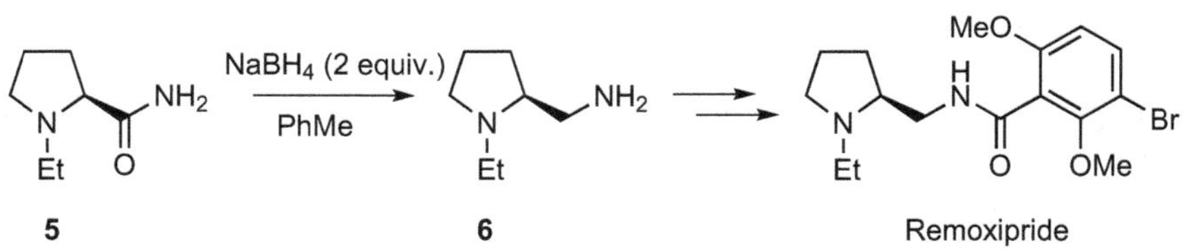

SCHEME 1-5. Amide reduction step in the synthesis of remoxipride.

step [5]. This proceeds with an overall yield of tolterodine of 74%. A combined amide reduction/debenzylation would improve AE and remove the need to workup and isolate the intermediate, which has significant cost and time implications.

Remoxipride is an atypical antipsychotic drug that has been used to treat schizophrenia. The amine intermediate **6** was prepared via a sodium borohydride reduction of the primary amide **5** to the primary amine (Scheme 1-5), which proceeded in 54% yield (crude) [6]. As is the case for paroxetine (Scheme 1-3), the reduction occurs adjacent to a stereogenic center, which must not racemize during the reaction.

One of the late-stage transformations in the synthesis of sibenadet, which is used to treat chronic obstructive pulmonary disease [7], is a borane reduction of the secondary amide **7** to a secondary amine, which is then isolated as the hydrochloride salt (Scheme 1-6). The overall yield over these two steps was only 20%, a result of competitive reduction of the benzothiazolone. The impurities were not only difficult to separate and remove; they also appeared to hamper the crystallization of the product.

Selective reactions are particularly important in pharmaceutical processes, as the final molecule often has more than one functional group. Verapamil, a calcium channel blocker used in the treatment of cardiovascular ailments, provides a good example of this [8], where the last step of the synthesis is the borane reduction of tertiary amide **8** in the presence of a nitrile group (Scheme 1-7), which proceeds in 60–73% yield.

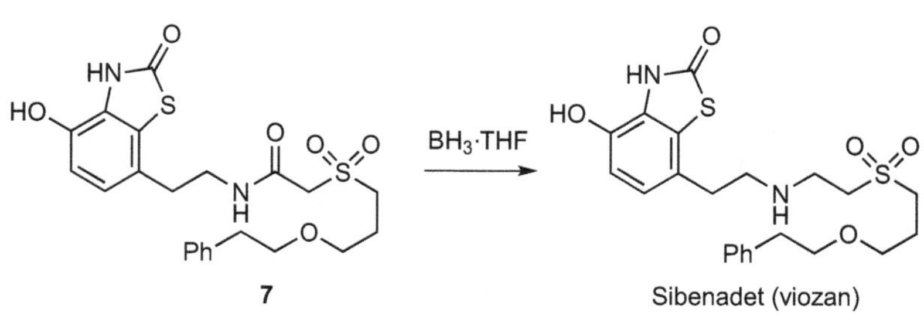

SCHEME 1-6. Amide reduction step in the synthesis of sibenadet.

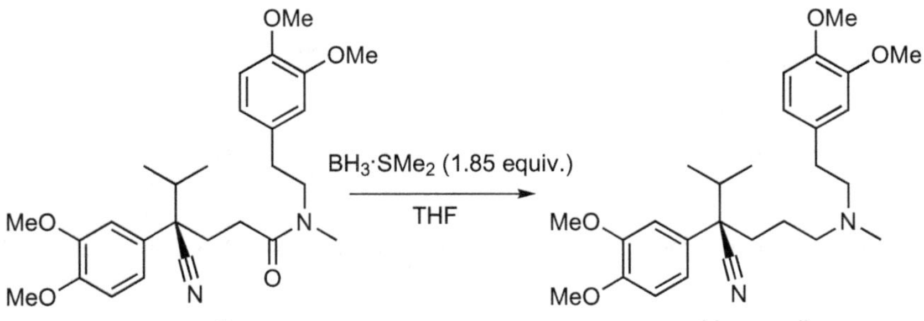

SCHEME 1-7. Amide reduction step in the synthesis of verapamil.

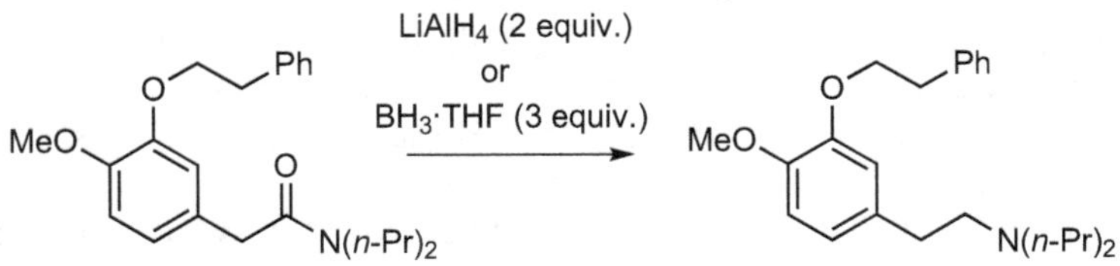

SCHEME 1-8. Amide reduction step in the synthesis of NE-100.

Finally, an example of a tertiary amide (**9**) reduction by either $LiAlH_4$ or BH_3 is provided in the case of NE-100, a σ receptor antagonist with potent antipsychotic effects (Scheme 1-8) [9].

From these examples, it is clear that metal hydride and borane reductions of amides represent important and widely used reactions in the pharmaceutical industry, and improvements need to be made to obtain a safer, greener, and more efficient transformation. Catalytic methods may fulfill these requirements, although steps need to be taken to ensure that procedures can be carried out with high selectivity under relatively mild conditions, preferably without the need for specialist equipment.

1.4 HETEROGENEOUS AMIDE HYDROGENATION

Catalytic hydrogenation of amides was first reported by Adkins and Wojcik in 1934 [10], which was achieved by using heterogeneous copper chromite catalysts under extremely forcing reaction conditions (300 bar, 250 °C), under which the reactions were prone to side reactions, such as further alkylation of the product (primary amides) and C–N bond cleavage (mainly secondary and tertiary amides) [10]. Improvements to the copper chromite method were reported in 1984 by King, of the Procter & Gamble Company, where the introduction of zeolite resulted in milder reaction conditions of 140 bar and 287 °C [11]. This allowed the reduction of *N,N*-dimethyldodecanamide (**10**) in 1 h, with a conversion of 92% and 81% selectivity to **11** (Scheme 1-9). The reaction without zeolite under the same conditions only gave 47% conversion and 47% selectivity.

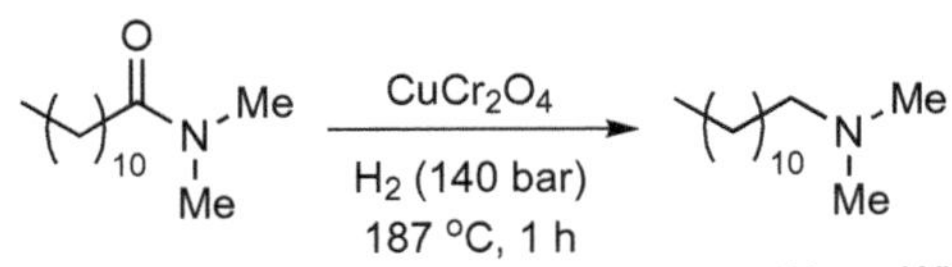

10 11 With zeolite 92% Without zeolite 47%

SCHEME 1-9. Hydrogenation of *N,N*-dimethyldodecanamide **10** to *N,N*-dimethyldodecylamine **11**.

Obviously, these extreme conditions are incompatible with pharmaceutical and fine chemical synthesis, where compounds may contain many thermally sensitive functional groups. However, recent advances in heterogeneous bimetallic catalyst systems have allowed drastically improved conditions to be developed. For example, Fuchikami and coworkers [12] reported the use of bimetallic catalysts comprising rhodium and rhenium carbonyl species, capable of reducing primary, secondary, or tertiary amides under milder conditions (typically 160–180 °C and 100 bar). However, the reaction is hampered by overreduction, including of phenyl groups to cyclohexyl groups.

An extensive patent published in 2005 by Smith et al. [13] at Avantium International B. V. describes the screening of bi- and trimetallic catalysts for amide reduction, using the reduction of *N*-acetylpyrrolidine (**12**) as a test substrate. Typical tests were carried out at 10 bar and temperatures of 70–160 °C, screening hundreds of catalysts (Scheme 1-10). Combinations of Pt, Rh, or Ir with Re, Mo, or V provided the most active catalysts, achieving yields in excess of 80% at 130 °C.

Recently, Whyman and coworkers reported a similar series of bimetallic heterogeneous catalysts using combinations of Rh/Mo [14], Ru/Mo [15], Rh/Re, and Ru/Re [16]. A detailed study was carried out on each of these systems employing the primary amide, cyclohexane carboxamide **14**, as the test substrate (Scheme 1-11) to give cyclohexylmethanamine **15** in good yields. Minimum operating conditions were found to be either 100 bar and 130 °C, or 50 bar and 160 °C in the case of Rh/Mo. At lower temperatures and pressures, lower conversions, higher amounts of alcohol, and unwanted amine products were observed.

Using the Ru/Mo catalyst system at 100 bar and 160 °C, primary amides were readily hydrogenated to the desired primary amines. Although benzamide gave 83% primary amine (accompanied by16% of the alcohol), the phenyl ring was also reduced. In comparison, the hydrogenation of butanamide and 2,2-dimethypropanamide gave 77% and 40% primary amine, respectively, with the remainder attributable to alcohol. Conversely, the two secondary amides tested, *N*-methyl benzamide and *N*-methyl cyclohexamide, were only hydrogenated to the corresponding amines in trace amounts. In contrast, reductions of tertiary aliphatic amides proceeded much more smoothly, with up to

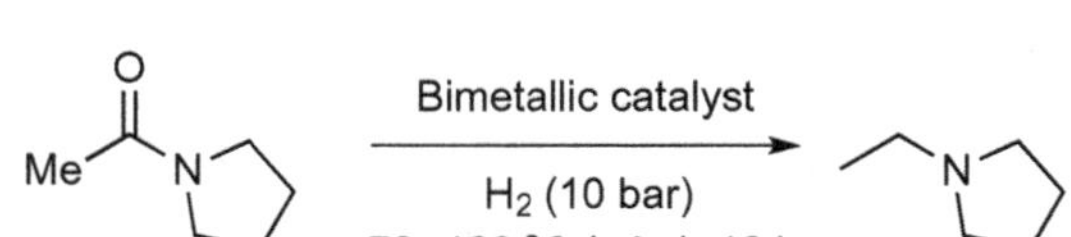

12 13

SCHEME 1-10. Hydrogenation of *N*-acetylpyrrolidine **12** to *N*-ethylpyrrolidine **13**.

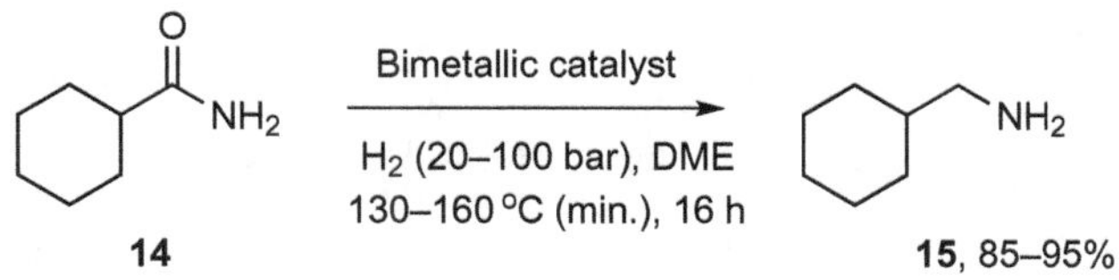

SCHEME 1-11. Hydrogenation of cyclohexane carboxamide **14** to cyclohexylmethanamine **15**.

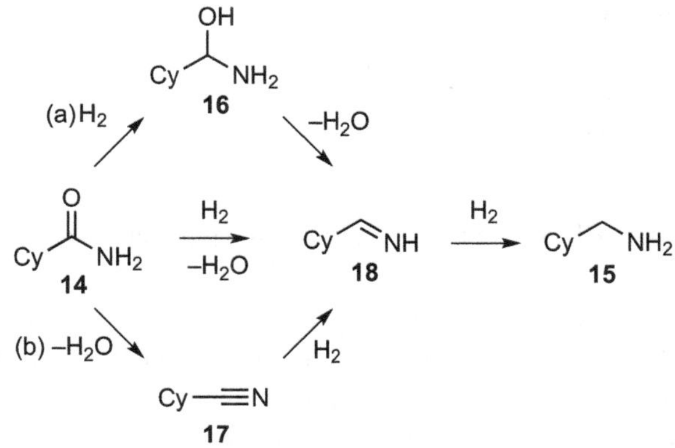

SCHEME 1-12. Potential amide hydrogenation pathways: (a) proceeds via the hemiaminal **16**; (b) proceeds via the nitrile **17**.

100% conversion for *N,N*-diethylpropanamide. Higher conversions and selectivities were also achieved with the Re-based catalysts, although the operating temperatures were also higher.

The same authors also conducted a study of the mechanism by examining thermochemical data for the hydrogenation of **14**. They proposed that the amide hydrogenation could proceed via two pathways: the first is through the hemiaminal **16** followed by a second hydrogenation, with a concerted loss of water (Scheme 1-12, route a). The second pathway could proceed, in the case of primary amides, through the nitrile **17** (dehydration), which is then hydrogenated to give the amine (Scheme 1-12, route b).

The calculated free energy $\Delta G^{\circ}_{298.15}$ of the formation of the hemiaminal is much greater than that of the dehydration reaction (104.8 kJ mol^{-1} vs. 26.5 kJ mol^{-1}, respectively) [16], suggesting that the formation of the nitrile intermediate may be more favorable. Pathway (b) should also be more selective for the formation of the amine, as water is eliminated, reducing the likelihood of alcohol formation (from **16**). The authors proposed that nitrile formation is rate limiting, and under the adopted reaction conditions, the two routes may be competitive processes, accounting for the difference in observed reactivity (primary $>$ tertiary $\gg$ secondary).

1.5 HOMOGENEOUS AMIDE HYDROGENATION

The first report of a homogeneous catalytic amide reduction was described in a patent by Crabtree and coworkers at Davy Process Technology, using a triphosphine ligand,

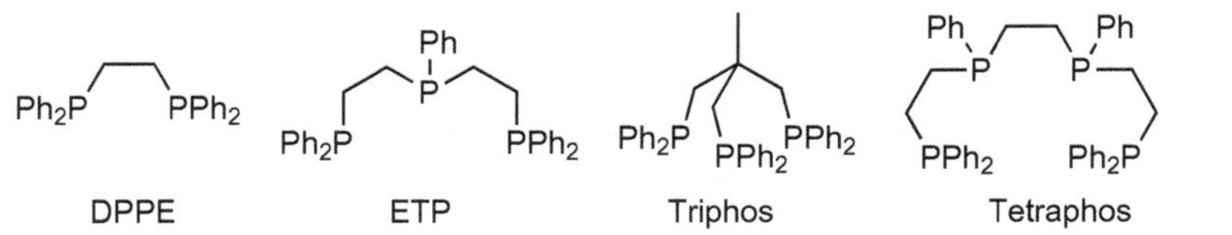

FIGURE 1-1. Selection of the ligands tested in hydrogenation of dimethyl oxalate [18].

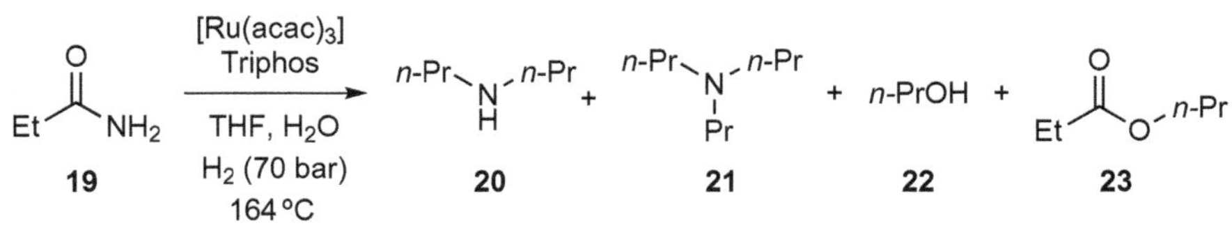

SCHEME 1-13. Hydrogenation of propanamide **19**.

1,1,1-tris(diphenylphosphinomethyl)ethane (Triphos, Figure 1-1), in a ruthenium-catalyzed reaction [17]. Examination of product mixtures revealed that the hydrogenation of propanamide **19** did not result in the expected propylamine, but a mixture of dipropylamine **20**, tripropylamine **21**, propanol **22**, and propyl propanoate **23** (Scheme 1-13).

The use of Triphos was not unfounded, as it had previously been found to be a useful ligand for the hydrogenation of carboxylic acid derivatives by Elsevier and coworkers in 1997 [18,19]. Used in conjunction with [Ru(acac)$_3$], the hydrogenation of dimethyl oxalate **24** proceeded smoothly to ethylene glycol **26** (Scheme 1-14). The addition of Zn as a cocatalyst was found to increase the yield of ethylene glycol—it is thought to have a dual role in the process: (a) acts as a reducing agent for the Ru(III) precatalyst and (b) the resultant Zn(II) acts as a Lewis acid to activate the ester group toward attack by the Ru catalyst.

In a later study by the same authors, a series of ligands was screened, including mono-, bi-, tri-, and tetradentate phosphines (Figure 1-1), as well as arsines and amines. Of those tested, PPh_3, DPPE, ETP, and Tetraphos showed conversion to **21** in 36, 11, 67, and 85% yields, respectively. Among these, Triphos was the only ligand that can effect the second reduction of **25** to give the diol **26** [20]. The TON for Triphos was also high (160, four times greater than that afforded by ETP and Tetraphos).

The success of the Triphos ligand is attributed to its ability to only adopt a *facial* (*fac*-) geometry around the metal center, which is catalytically more active than the other tridentate ligand, ETP, which can form facial and meridional (*mer*-) isomers (Figure 1-2). A similar effect is observed in the hydrogenation of 2-cyclohexen-1-one, where Triphos

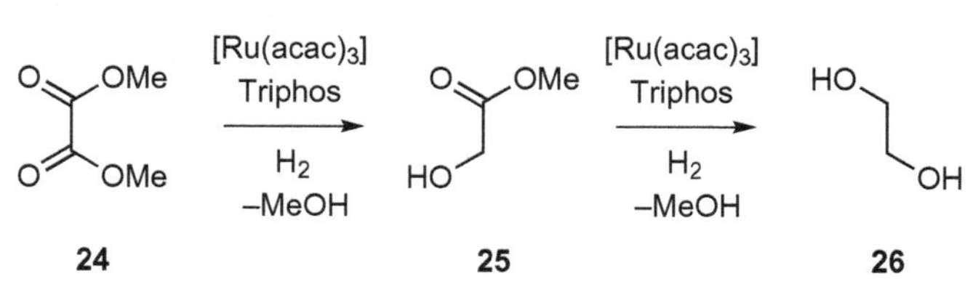

SCHEME 1-14. Hydrogenation of dimethyl oxalate **24** to ethylene glycol **26** via methyl glycolate **25**. Conditions: MeOH, H_2 (80 bar), 120 °C, 16 h, Zn (0.3 mol%).

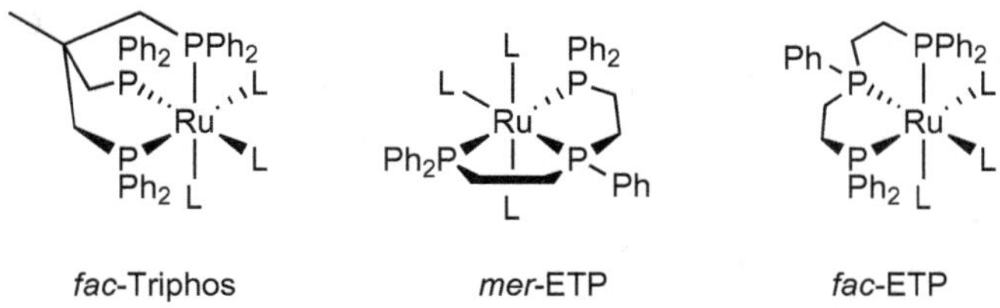

FIGURE 1-2. Coordination geometries of tridentate ligands.

reacts twice as fast as ETP [21]. This could also explain the reduced performance of the tetradentate ligand, Tetraphos, as this can also form a number of geometric isomers that can have different catalytic activities [22]. Another distinct advantage of the Triphos ligand over other phosphine ligands is the fact that it is an air-stable solid.

Since this initial study, a variety of other homogeneous catalysts has been applied to the hydrogenation of esters, and these will be discussed in the following chapter.

1.5.1 Hydrogenation of Primary Amides

Following the work of Crabtree and coworkers [17], Cole-Hamilton and coworkers [23] reported their initial results on some hydrogenation studies, where 100% conversion of butanamide **27** to dibutylamine **35** and tributylamine **36** can be achieved in ca. 50:50 ratio, with no observed formation of butylamine **29** (Table 1-1, entries 1 and 2, and Scheme 1-15). In order to obtain **29**, butanamide **27** must first undergo hydrogenation with the loss of water to give the imine **28**; this is then hydrogenated to give the desired primary amine. However, the reaction does not stop here, and **29** can undergo transamidation with the amide **27** to afford secondary amide **34**, or it can form an imine **33** with the aldehyde **31** (generated from **27**). Both of these observed intermediates are then readily hydrogenated to the secondary amine **35**. This cycle can then be repeated to give the tertiary amine **36**.

TABLE 1-1. Hydrogenation of Butanamide 27[a] (Scheme 1-15) [23]

Entry	H_2O: THF[b]	NH_3(aq): THF[b]	NH_3(l): THF[b]	1° Amine **29** (%)	Alcohol **32** (%)	2° Amide **34** (%)	2° Amine **35** (%)	3° Amine **36** (%)	Conversion (%)
1[c]	0.1	–	–	0	Trace	Trace	46	53	100
2[c]	0.01	–	–	0	Trace	Trace	48	51	100
3	0.1	–	0.5	44	8	10	38	0	100
4	0.1	–	1	36	3	14	6	0	59
5	–	0.3	–	78	12	10	0	0	100
6	–	0.5	–	85	15	0	0	0	100
7	–	0.7	–	85	15	0	0	0	100
8	–	1	–	73	25	2	0	0	100
9[d]	–	1	–	75	25	0	0	0	100

[a]Conditions (unless otherwise indicated): Butanamide **27** (1 g, 11.4 mmol), $[Ru_2(Triphos)_2Cl_3]Cl$ (91 mg, 0.05 mmol), 164°C (external), 220°C (internal), H_2 (40 bar), 14 h, THF (10 ml), Hastelloy autoclave.
[b]v/v.
[c]$[Ru(acac)_3]$ (45 mg, 0.1 mmol) and Triphos (142 mg, 0.22 mmol) were used instead of $[Ru_2(Triphos)_2Cl_3]Cl$.
[d]NH_3 (4 bar).

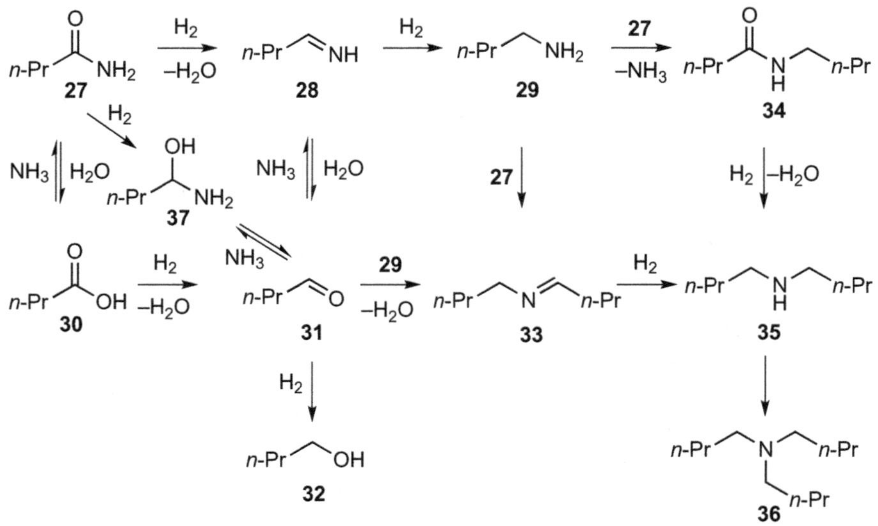

SCHEME 1-15. Proposed mechanism of amide reduction indicating possible intermediates and routes to side products [23].

The initial processes involve the formation of many unwanted side reactions, that is, hydrolysis of amide **27** and imine **28**, liberation of amine from the aminal **37**, and transamidation of **29** with **27**, all proceeding with the liberation of ammonia. However, the proposed mechanism suggests that these processes may be reversed or suppressed by working in the presence of ammonia. Indeed, the introduction of liquid ammonia increased the selectivity for primary amine **29** to 44%, while the formation of the tertiary amine **36** was sequestered (entry 3). A higher concentration of liquid ammonia increased the selectivity of the primary amine to 61% (entry 4), although this somewhat suppressed the yield to 59%. The use of aqueous ammonia was more fruitful, and a selectivity of 85% toward primary amine could be achieved while complete conversion was maintained (entries 6–7). The downside to using aqueous ammonia is the inevitable accumulation of water in the reaction, which leads to the formation of a higher amount of alcohol **32** (entry 8). By the same token, a combination of aqueous ammonia and ammonia gas also did not lead to any improvement. Nevertheless, this reaction represents the first example of the homogeneously catalyzed hydrogenation of a primary amide to a primary amine using only molecular hydrogen, with a high level of selectivity.

The protocol may be adapted for the hydrogenation of nonanoic acid **38**, which proceeds in the presence of ammonia to produce nonylamine **39** with 49% selectivity (the other products obtained are shown in Scheme 1-16) [23].

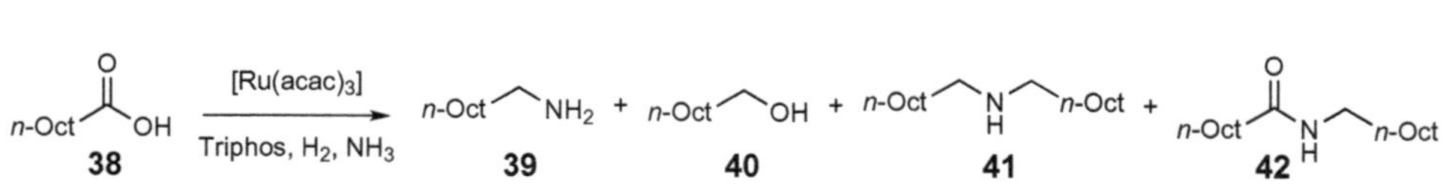

SCHEME 1-16. The production of nonylamine **39** by the hydrogenation of nonanoic acid **38** in the presence of ammonia [23].

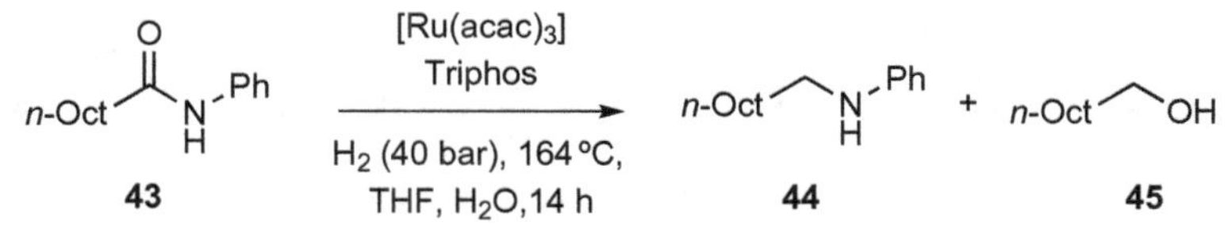

SCHEME 1-17. Hydrogenation of *N*-phenylnonanamide [23].

1.5.2 Hydrogenation of Secondary Amides

Secondary amides are challenging substrates as they may potentially undergo further reaction to give tertiary amines, rather than the desired secondary amines. To date, the only example of homogeneous hydrogenation of secondary amides was reported by Cole-Hamilton and coworkers. In the original communication on amide hydrogenation [23], the reaction temperature was "set" at 164 °C using collar-type heaters used for heating the autoclaves. Subsequently, by using an autoclave fitted with a thermocouple pocket, the internal temperatures were in fact found to be some 60 °C higher (the temperatures quoted in the current chapter are actual reaction temperatures).

Choosing *N*-phenylnonanamide **43** as a test substrate, the reduction furnished a mixture of the corresponding secondary amine, *N*-phenylnonylamine **44** and nonanol **45**, where the selectivities are dependent upon the reaction conditions employed. The alcohol is thought to originate either from the hydrolysis of the amide to the acid or from the hydrolysis of the imine to the aldehyde. Subsequent hydrogenation of these intermediates leads to the alcohol (see Scheme 1-17).

The optimum reaction conditions for the hydrogenation of **43** were reported to be 220 °C at 40 bar hydrogen pressure in THF for 14 h. The reaction requires both [Ru(acac)$_3$] and Triphos in order to proceed (Table 1-2, entries 1–3). In the absence of Triphos, a lower conversion is observed (entry 2). The addition of water appears to have a detrimental effect on the selectivity (entries 4 and 5). However, water is thought to also have a stabilizing effect on the catalyst. The reaction still gives full conversion at 220 °C, with only a slight loss in selectivity. Below this temperature, the conversion drops dramatically with a significant loss in selectivity. In fact, only alcohol is observed at 160 °C, as a result of C–N cleavage, which was also reported by Milstein and coworkers [24].

TABLE 1-2. Hydrogenation of *N*-Phenylnonanamide 43[a] (Scheme 1-17) [23]

Entry	[Ru(acac)$_3$] (%)	Triphos (%)	T (°C)[b]	Amine **44** (%)	**45** (%)	Conversion (%)
1	–	–	220	0	0	0
2	1	–	220	57	4	61
3	–	2	220	0	0	0
4	1	2	220	93	7	100
5[c]	1	2	220	99	1	100
6	1	2	200	91	9	100
7	1	2	180	48	32	80
8	1	2	160	0	40	40

[a]Conditions (unless otherwise indicated): *N*-phenylnonanamide **43** (4.3 mmol), H_2 (40 bar), 14 h, THF (10 ml), H_2O (1 ml), Hastelloy autoclave.

[b]Internal temperature, external autoclave temperature actually around 60 °C lower.

[c]No water added.

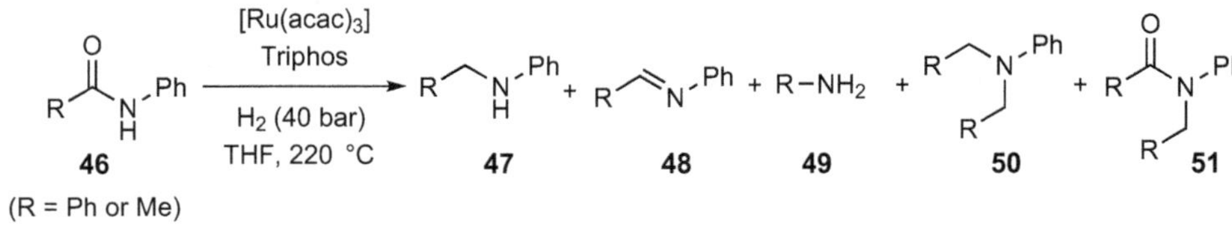

SCHEME 1-18. Hydrogenation of benzanilide (R = Ph) and acetanilide (R = Me), showing the products and side products obtained.

By studying a variety of substrates, it was found that the presence of an aryl group on the nitrogen atom is a key requirement for the amide substrate. Conversely, the reaction was less sensitive to changes of substituents on the C=O, where both aromatic and aliphatic groups are tolerated. Thus, benzanilide and acetanilide (**46**, R = Ph and Me, respectively) were chosen as model substrates for further optimization (Scheme 1-18).

The first area of optimization was the pressure, which, at 40 bar, was too high for widespread application in the pharmaceutical industry (Table 1-3). Subsequently, it was found that the reaction could be performed at 10 bar with no loss of conversion, and a rather unexpected improvement in selectivity (entries 1 and 2). Lowering the pressure to 5 bar (entry 3) further improved the selectivity, but a concurrent loss of conversion was also observed. By extending the reaction time, it was possible to obtain 89% conversion, but with reduced selectivity (entry 4).

Previously, the hydrogenation reactions could be run at 200 °C without any detrimental effects on conversion or selectivity. In the present system, a decrease in both was observed by lowering the temperature to 200 °C (entry 5). Further decrease to 180 °C led to the formation of only a trace of the desired product, the main product being aniline. By running the reaction at lower temperatures and pressures simultaneously (10 bar, 200 °C), 80% conversion can be achieved after an extended reaction period of 63 h.

Shortly after its publication, several research groups reported problems with reproducing the results reported in the original paper (M. Beller and coworkers, private communication). By a process of elimination, the purity of the ligand was found to exert an important effect on the reaction outcome. Several batches were tested, alongside purified samples stored under an inert atmosphere, but each showed a much reduced activity. It was

TABLE 1-3. Optimization of the Hydrogenation of Benzanilide (46, R = Ph)[a]

Entry	pH_2	T (°C)	*t* (h)	Amide **46** (%)	2° Amine **47** (%)	Imine **48** (%)	Aniline **49** (%)	3° Amine **50** (%)	Conversion (%)	Selectivity (%)
1	40	220	16	8	62	2	26	2	92	67
2	10	220	16	8	71	1	17	2	92	78
3	5	220	16	29	64	3	3	1	71	90
4	5	220	66	11	58	2	26	3	89	66
5	40	200	16	27	38	1	34	0	73	52
6	40	180	16	42	<1	0	58	0	58	<1
7	10	200	16	51	33	1	15	<1	49	67
8	10	200	63	20	55	1	22	1	80	69

[a]Conditions: Benzanilide (5 mmol), $[Ru(acac)_3]$ (1 mol%), Triphos (2 mol%), THF (10 ml), Hastelloy autoclave. Product distribution calculated based on GC-FID.

TABLE 1-4. Hydrogenations of Benzanilide (46, R = Ph) Employing Catalytic Amounts of MSA[a]

Entry	pH_2 (bar)	MSA (%)	t (h)	2° Amine **47** (%)	Imine **48** (%)	Aniline **49** (%)	3° Amine **50** (%)	Other (%)	Conversion (%)	Selectivity (%)
1	10	1	16	84	3	3	10		100	84
2	10	0.5	16	91	4	0	5		100	91
3	10	10	16	12	<1	0	28	53[b]	100	12
4	5	0.5	62	76	6	5	13		100	76
6[c]	10	1	16	Ring hydrogenation occurred						
7[d]	10	1	16	89	3	0	7		100	89
8	10	1	8	92	4	0	4	0	100	92

[a]Conditions (unless otherwise indicated): Benzanilide (5 mmol), $[Ru(acac)_3]$ (1 mol%), Triphos (2 mol%), 220 °C, THF (10 ml), Hastelloy autoclave, product distribution calculated based on GC-FID.
[b]*N*-phenylpyrrolidine, tertiary amide, and two other impurities.
[c]Stainless steel autoclave.
[d]Aniline (1%) added.

subsequently discovered that the catalytic activity and the selectivity of the reaction could be restored by adding a catalytic amount of acid, specifically, methanesulfonic acid (MSA), see Table 1-4 [25]. The addition of 1% MSA (1:1, MSA:Ru) resulted in full conversion of benzanilide with excellent selectivity for the secondary amine (entry 1) [25]. Interestingly, the product distribution also changed. In the absence of acid, aniline **49** was found to be the major side product, whereas the corresponding tertiary amine, *N*,*N*-dibenzylaniline **50**, was produced as the major side product in the presence of acid, which presumably arises from the transamidation of the starting benzanilide **46** with product *N*-benzylaniline **47**, followed by hydrogenation of the tertiary amide **51**.

The amount of acid added is also vital, as too much (10% MSA) leads to the formation of *N*-phenylpyrrolidine, derived from a reaction of the solvent THF with aniline, as well as tertiary amide **51** (entry 3). The optimum MSA for this reaction appears to be between 0.5 and 1%, although 1.5% is found to be optimal for the hydrogenation of acetanilide. Under 5 bar of H_2 pressure, the reaction proceeded with full conversion (entry 4) and a respectable selectivity of 76% after 62 h. Last but not the least, the reaction is sensitive to the type of autoclave used, and the best results are achieved in Hastelloy C-276,[1] as opposed to stainless steel, which was found to promote the formation of alcohol and hydrogenation of the phenyl ring (entry 6). Under optimized conditions, the best selectivity of 92% can be achieved in 8 h, with only 4% of the tertiary amine as side product (entry 8).

The optimized reaction conditions were subsequently applied to the hydrogenation of a series of *para*-substituted acetanilides (Table 1-5), where it was found that 98–100% conversion can be achieved with methoxy-, methyl-, and fluoro-substituted acetanilides (entries 1, 2, and 4), which was comparable to acetanilide itself (entry 3). The selectivity, however, was lower at 60–68%, compared with 85% for acetanilide. Conversely, the hydrogenation of more electron-withdrawing substituents did not proceed cleanly (entries 6 and 7); an insoluble precipitate was obtained from the CF_3-substituted acetanilide. While the nitro-substituted substrate showed some reactivity, there was evidence of alcohol hydrogenation and phenyl ring hydrogenation.

[1] Nickel-based alloy incorporating chromium and molybdenum.

TABLE 1-5. Study of *para*-Substituted Acetanilides, Incorporating Electron-Withdrawing and -Donating Groups[a]

Entry	*p*-Subst.	Amide (%)	2° Amine (%)	Imine (%)	Aniline (%)	3° Amine (%)	Pyrrolidine (%)	Conversion (%)	Selectivity (%)
1	OMe	3	66	1	6	24	0	98	66
2	Me	1	68	0	6	22	4	99	68
3	H	0	86	0	4	8	3	100	86
4	F	1	60	1	4	23	10	99	60
5	CF_3	Insoluble precipitate, reaction outcome unclear							
6	NO_2	Side reactions of ring hydrogenation and alcohol hydrogenation							

[a]Conditions: *p*-substituted acetanilide (5 mmol), $[Ru(acac)_3]$ (1 mol%), Triphos (2 mol%), MSA (1%), 220 °C, 16 h, THF (10 ml), Hastelloy autoclave, product distribution calculated based on GC-FID.

1.5.3 Tertiary Amides

The observation of tertiary amine as a side product in the above system indicates that the system should also be active for the hydrogenation of tertiary amides, such as PhC(O)NBzPh, formed from the transamidation of the starting benzanilide with the product *N*-benzylaniline. Thus, the hydrogenation of the tertiary amine, *N,N*-diphenylacetanilide, was carried out in the presence of MSA (Table 1-6) [25]. In this case, a much higher conversion was achieved at 40 bar, compared with 10 bar (85% vs. 47%). In both cases, the selectivity was around 50%; a large proportion of diphenylamine was also present.

1.5.4 Scope of Ru/Triphos Amide Hydrogenation

Accordingly, full conversion to hydrogenation products can be obtained with a primary amide substrate, such as butanamide, if MSA is added to the reaction mixture [25]. As was observed in the initial findings [23], secondary amines are the major products unless ammonia is added, in which case the selectivity for the primary amine can be as high as 80%.

The general scope of the hydrogenation of amines using Ru/Triphos in the presence of MSA is outlined in Figure 1-3.

1.5.5 Hydrogenation of Diacids in the Presence of Amines

Building on previous results (Section 1.5.1), the hydrogenation of dicarboxylic acids and esters was also attempted in the presence of amines. For long-chain diesters such as dimethyl 1,19-nonadecanedioate, reduction in the presence of aniline produces an oligoamide (nylon 19) [26]. Intermediate-chain length diacids such as 1,8-octanedioic acid

TABLE 1-6. Hydrogenation of *N,N*-Diphenylacetanilide[a]

Entry	pH_2 (bar)	Ph_2NH (%)	Amine (%)	Amide (%)	Conversion (%)	Selectivity (%)
1	10	24	23	53	47	49
2	40	46	38	15	85	45

[a]Conditions: *N,N*-diphenylacetanilide (5 mmol), $[Ru(acac)_3]$ (1 mol%), Triphos (2 mol%), MSA (1%), 220 °C, 16 h, THF (10 ml), Hastelloy autoclave, product distribution calculated based on GC-FID.

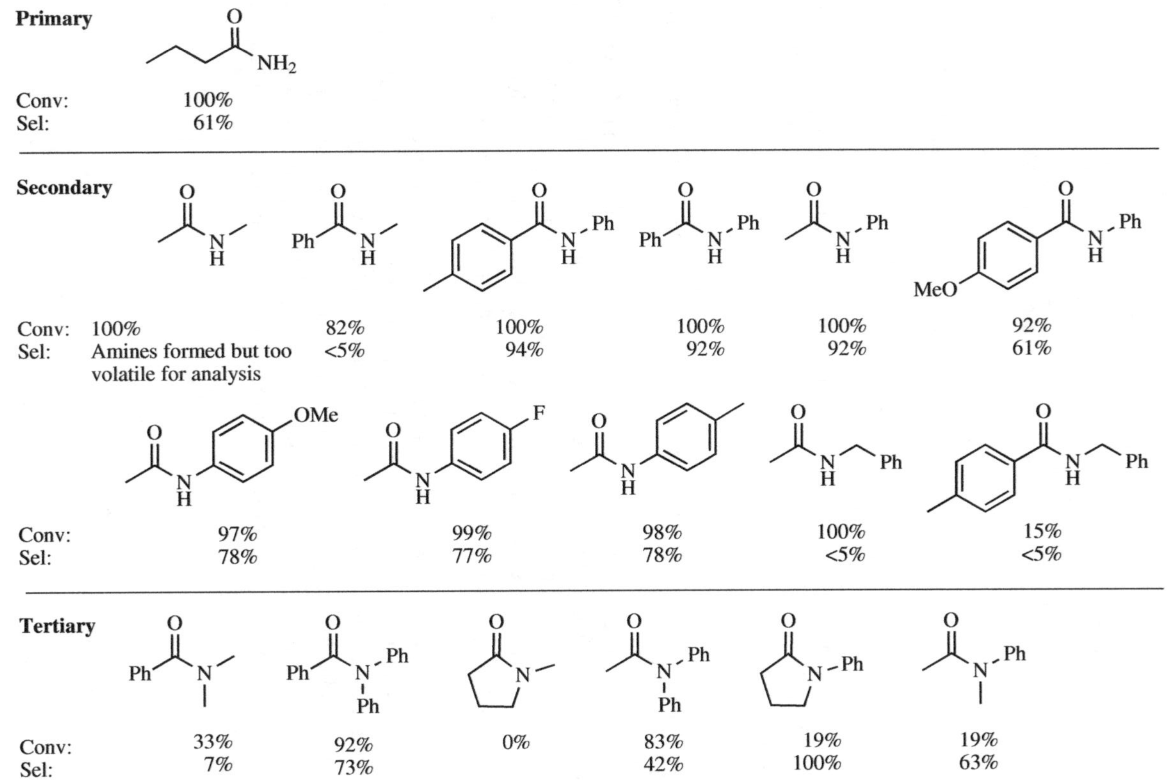

FIGURE 1-3. Substrate scope for the hydrogenation of amides catalyzed by $[Ru(acac)_3]$/Triphos/MSA. Substrate (5 mmol), $[Ru(acac)_3]$ (0.05 mmol, 1 mol%), Triphos (0.10 mmol), MSA (3 μl), 1.5 mol%), 220°C, H_2 (10–40 bar), THF (16 h) [25].

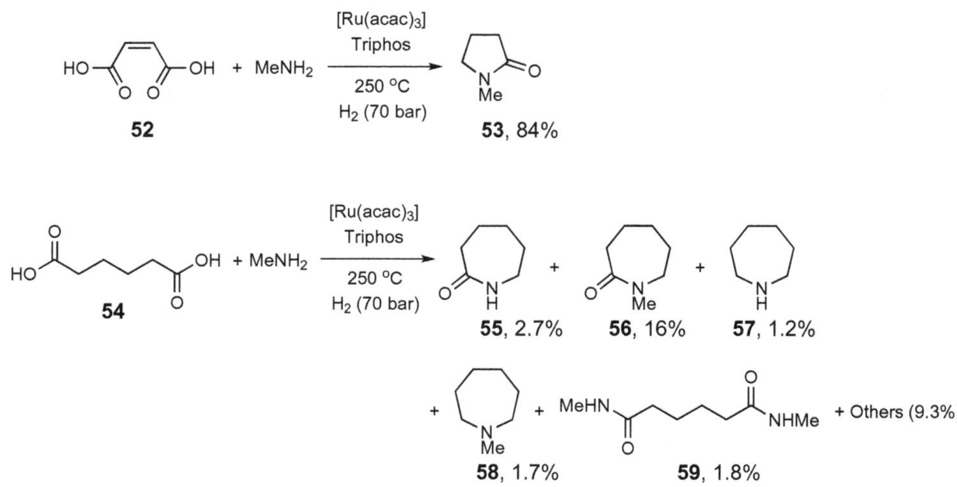

SCHEME 1-19. Catalytic formation of cyclic lactams from the hydrogenation of diacids. The product distribution obtained from the reaction with adipic acid **54** was derived from the integration of GC peak areas (TCD) [28].

give linear α,ω-amino alcohols and diamines [27], while shorter-chain diacids give heterocyclic molecules [27].

Earlier, Crabtree and coworkers had reported that the hydrogenation of maleic acid **52** or adipic acid **54** in the presence of water and methylamine (at 250 °C and 70 bar) gave the corresponding cyclic lactam (**53** and **56,** respectively) with selectivity up to 83.5% [28]. The fact that *N*-methylpyrrolidone is also produced from *N*-methylsuccinimide under similar conditions without any methylamine suggests that it might be an intermediate in the reaction (Scheme 1-19).

More recently, these reactions were reexamined [27] using aniline as the amine substrate, which were shown to give a variety of products (Scheme 1-20, Figure 1-4), which do not change with extended reaction time. However, if MSA is added, the major product is the heterocyclic *N*-phenyl hexahydroazepine **60**. Hence, five- to eight-membered heterocycles are obtained by varying the chain length, but longer chain diacids give amino alcohols. The reaction also works when ammonia or benzylamine was employed, but less selectively.

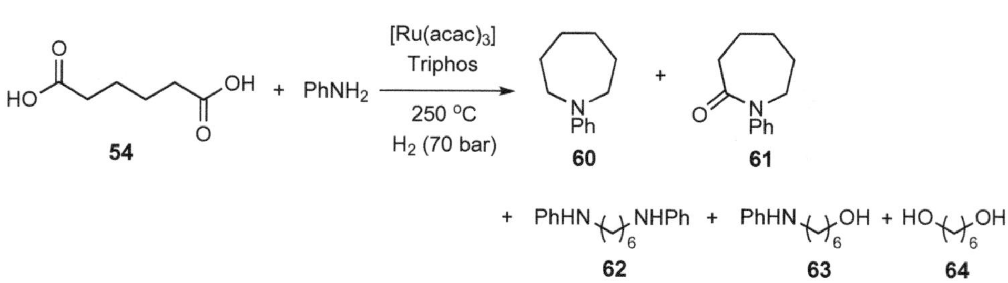

SCHEME 1-20. Products obtained from the hydrogenation of adipic acid **54** in the presence of aniline, $[Ru(acac)_3]$, Triphos, and optionally MSA [27].

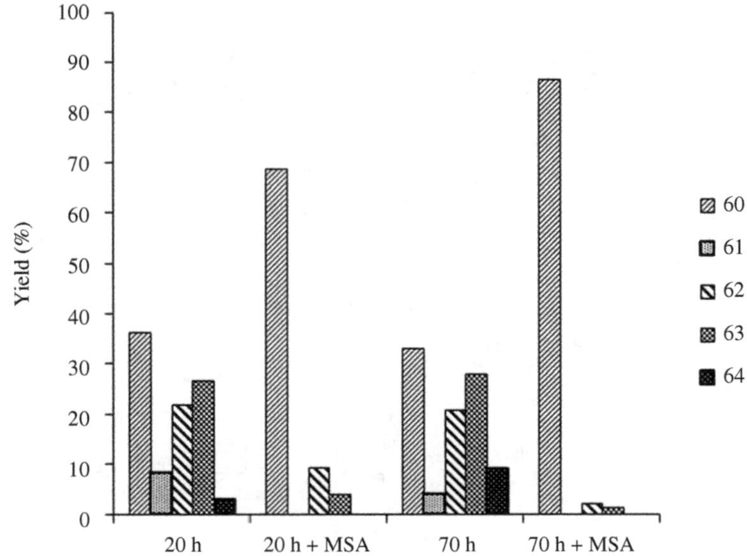

FIGURE 1-4. Product distribution obtained from the hydrogenation of adipic acid **54** in the presence of aniline, $[Ru(acac)_3]$, Triphos, and optionally MSA [27].

Since heterocyclic moieties are important in many pharmaceutical compounds, this low-waste route from readily available diacids may prove to be important for future applications.

In a related reaction, Bruneau and coworkers reported the hydrogenation of cyclic imides to lactams, employing $[RuCl_2(p\text{-cymene})]_2$ and $[Ru_4H_6(p\text{-cymene})_4]Cl_2$ as catalyst precursors at 60 bar H_2 and 90 °C [29]. The use of water as the solvent further improves the green credentials of this reaction. A series of substrates was tested, and excellent conversions (full conversion in many cases) were achieved. Succinimide and phthalimide (**65**, Scheme 1-21) were both successfully mono-reduced in yields of 87% and 78%, respectively. One drawback with this method, however, is that unsaturated moieties such as alkenes and aromatic rings were also hydrogenated under these conditions.

1.5.6 Homogeneous Amide Hydrogenation Mechanism

The proposed mechanism is based on one described by Milstein and coworkers for the hydrogenation of esters (Scheme 1-22) [30]. The two reduction steps are thought to

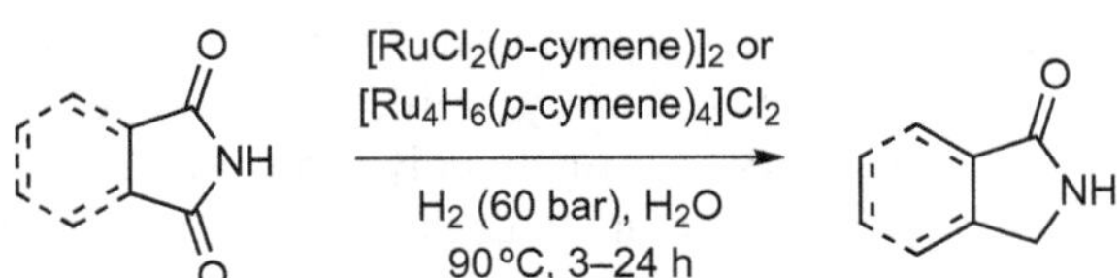

SCHEME 1-21. Monoreduction of succinimide and phthalimide (**65**, without and with fused benzo ring, respectively) [29].

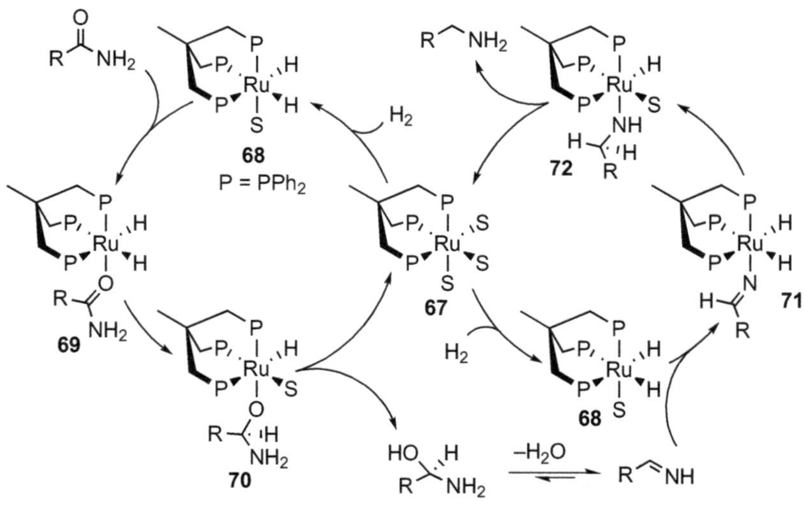

SCHEME 1-22. Tentative mechanism for the Ru-catalyzed hydrogenation of amides, based on a similar ester hydrogenation mechanism proposed by Milstein (S = coordinating solvent) [30].

proceed through a common catalytic precursor, **67**, which first undergoes oxidative addition with hydrogen to give **68**. Next, coordination of the amide (**69**) is followed by its hydrogenation. Finally, reductive elimination of the intermediate **70** generates the uncoordinated hemiaminal, which is highly unstable and rapidly eliminates water to give the imine. The imine can then enter the second catalytic cycle by coordination to the dihydride species **68**. Hydrogenation of **71** followed by reductive elimination of **72** gives the expected amine.

It is not clear yet what role the tridentate ligand plays, other than restricting the geometry in a *facial* fashion. It is also not clear what the nature of the moieties occupying the other coordination sites is. Additionally, tripodal ligands have been shown to exhibit hemilability [31]. Leitner and coworkers [32] have reported calculations relating to the mechanism of hydrogenation of itaconic acid and levulinic acid using the same catalyst system [33]. Proposing that protonated (cationic) complexes may be important in the mechanism of hydrogenation [32], they demonstrated that activation barriers obtained for the reaction are consistent with those obtained from the calculations using protonated intermediates, although a neutral cycle could not be ruled out. Adapting their mechanism to amide hydrogenation would give the mechanism shown in Scheme 1-23. An alternative pathway involving a loss of amine from the 1,1-aminoalcohol to give aldehyde, followed by Schiff base condensation, seems less likely since, as the reaction proceeds, Schiff base formation with the product amine would be favored. This would be expected to lead to the formation of tertiary amines in the case of secondary amide substrates. However, these products are usually only formed in low yield and may arise from transamidation of the product amine with the substrate amide. Furthermore, this mechanism could not operate for *N,N*-disubstituted (tertiary) amides, since it is not possible to eliminate water from 1,1-aminoalcohol. If they have enolizable α-hydrogens, they could proceed through the vinyl amine, but this is not possible for a substrate such as *N,N′*-diphenylbenzamide.

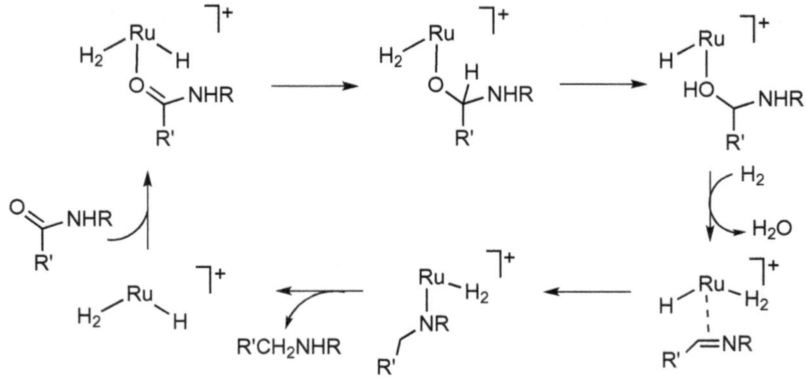

SCHEME 1-23. Possible mechanism for the hydrogenation of primary (R = H) and secondary amides based on that proposed by Leitner and coworkers [32] for the hydrogenation of carboxylic acids. Three coordination sites are always occupied by the Triphos ligand.

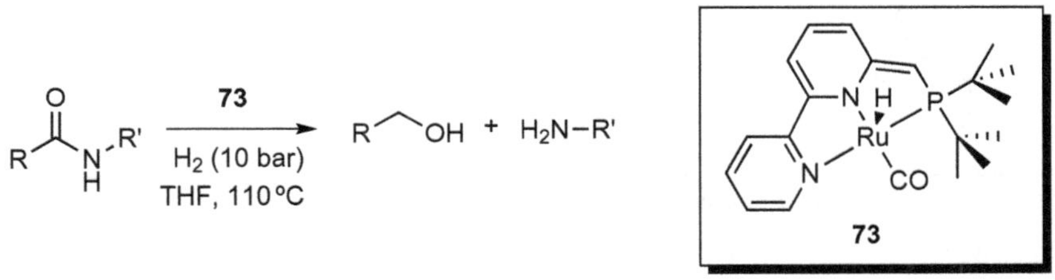

SCHEME 1-24. Ru-catalyzed C–N cleavage of amides employing pincer complex **73**.

1.5.7 Amide C–N Cleavage by Hydrogenation

Using a ruthenium complex **73** containing the 6-di-*tert*-butylphosphinomethyl-2,2′-bipyridine (BPy-*t*-PNN) ligand, secondary amides undergo hydrogenation to alcohols and primary amines at 10 bar and 110 °C (Scheme 1-24) [24]. This reaction also occurs under anhydrous conditions, so it is not thought to be a result of hydrolytic cleavage of the amide. It is not an atom efficient route to the desired amine as 1 equiv. of the corresponding alcohol is generated, which must be removed from the reaction mixture. In addition, the amine formed in this way is generally less valuable than the starting amide.

Similar findings were reported by Ikariya and coworkers where a series of lactams **74** underwent hydrogenation via C–N cleavage to give linear amino alcohols **75** (Scheme 1-25) [34].

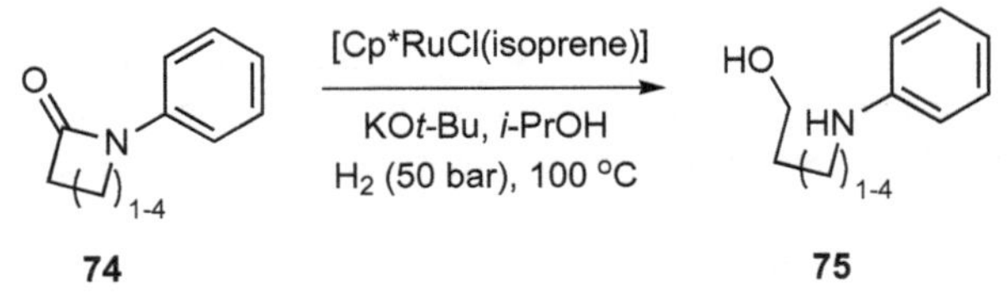

SCHEME 1-25. Hydrogenation of lactams **74** to linear amino alcohols **75**.

TABLE 1-7. C–N Cleavage of Lactams via Ru-Catalyzed Hydrogenation (Scheme 1-25) [34][a]

Entry	Substrate		*t* (h)	Conversion (%)	Yield (%)[b]
1	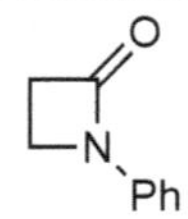		24	>99	83
2	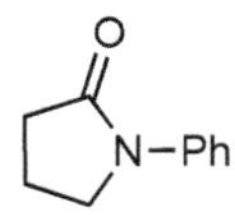		24	84	73
3	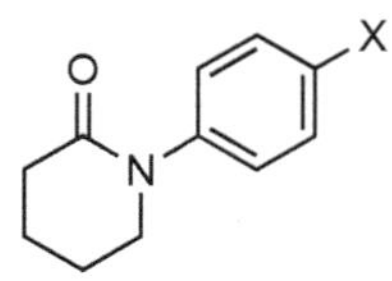	X = H	24	>99	96
4		X = CF_3	24	>99	90
5		X = OMe	24	88	78
6	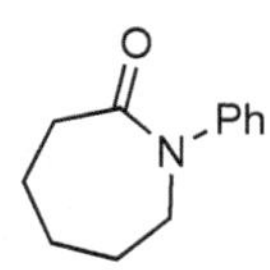		72	60	60[c]

[a]Conditions: lactam (1.0 M in 2-propanol), [Cp*RuCl(isoprene)] (10 mol%), KO*t*-Bu (25 mol%), 100°C, pH_2 (50 bar).
[b]Isolated yield.
[c]Product yield determined by ^1H NMR spectroscopy, using triphenylmethane as an internal standard.

Table 1-7 summarizes the hydrogenation of a series of heterocycles. Maximum conversion is obtained with heterocycles containing four- and six-membered rings (entries 1 and 3). A lower conversion of 84% was obtained in the case of the five-membered ring (entry 2), and only 60% for the seven-membered ring, even after an extended reaction period (entry 6). Electronic influences were also investigated (entries 3–5) and an electron-donating substituent (*p*-OMe) had a detrimental effect on the conversion. No such effect was observed with an electron-withdrawing substituent (*p*-CF_3). The reduction was investigated for a series of *N*-protected pyrrolidones, and the reactivity was found to increase in the following order: CBz < Boc ≈ Ts < CO_2CH_3 < Ms.

The procedure was applied to the asymmetric synthesis of paroxetine (Scheme 1-26) [35], whereby the imide **76** was desymmetrized to give optical active **77** as a result of C–N cleavage. Excellent enantioselectivity was achieved (>99% ee), which was maintained throughout subsequent steps toward paroxetine precursor **78**.

John and Bergens have reported similar findings employing a similar Ru complex, $[Ru(Ph_2P(CH_2)_2NH_2)_2(\eta^3\text{-}C_3H_5)]BF_4$ **79**, to hydrogenate lactams and tertiary amines to their C–N cleavage products [36]. Scheme 1-27 shows the hydrogenation of *N*-phenylpyrrolidin-2-one **80** to the linear amino alcohol **81**, which proceeds in 100% conversion and a TON of 1000.

The analogous six-membered lactam was also hydrogenated with full conversion as were a number of acyclic tertiary amides. The TON was found to increase to ca. 7000 when $[(TMS)_2]NK$ is replaced with NaOMe. The proposed mechanism of this system

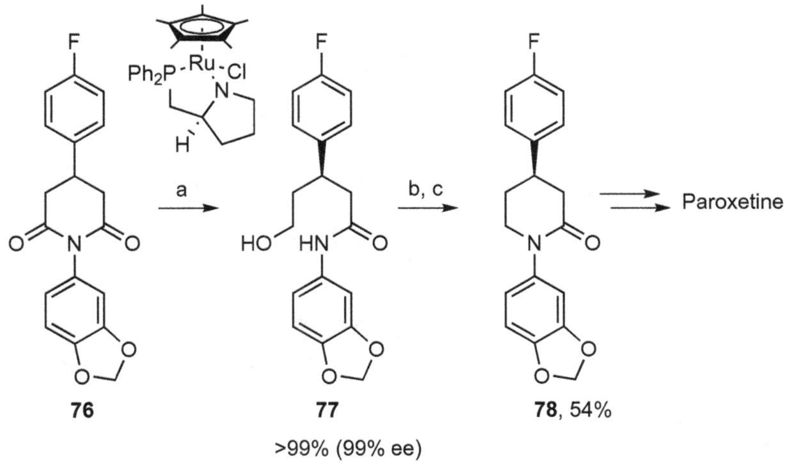

SCHEME 1-26. Asymmetric synthesis of paroxetine by amide reduction: (a) H_2 (30 bar), KO*t*-Bu, 60 °C, 18 h; (b) CBr_4, PPh_3; (c) NaH.

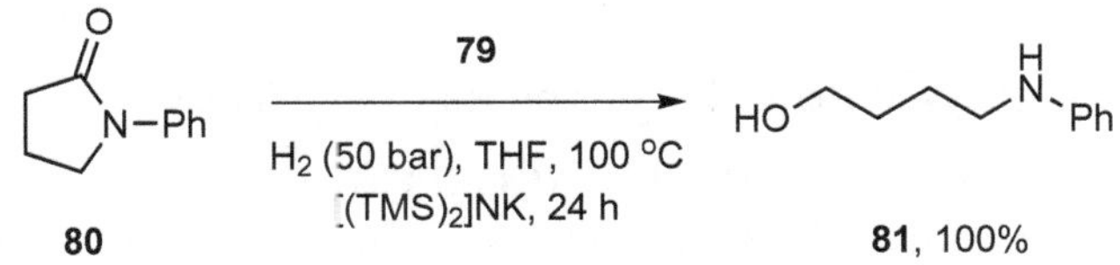

SCHEME 1-27. Hydrogenation of *N*-phenylpyrrolidin-2-one **80** to 4-(phenylamino)butan-1-ol **81**.

(Scheme 1-28) involves the bifunctional-type addition to the dihydride species **82**, resulting in ruthenium hemiaminal **83**. Base-assisted elimination generates the free hemiaminal **85** and "ruthenium amide" **84**. The hemiaminal eliminates aldehyde that undergoes hydrogenation to give the alcohol and the amine products. The dihydride **82** is regenerated following the addition of H_2 to **84**.

1.6 HYDROSILATION

The use of silanes as a hydride source in amide reductions has received a lot of interest in recent years. It has proven to be highly effective and tolerant to a variety of functional groups. It should be noted, however, that there are hazards associated with the use of triethoxysilane and Lewis acids, as there is the potential for highly pyrophoric and toxic silane gas (SiH_4) to form [37]. Methyldiethoxysilane [38] or polymethylhydrosiloxane (PMHS), which do not lead to the formation of silane, are safer alternatives, particularly for larger scale syntheses [39].

Igarashi and Fuchikami [40] reported a series of metal-catalyzed amide reductions by employing group 7–10 transition metals (Mn, Re, Ru, Os, Rh, Ir, Pd, Pt) and trisubstituted silanes. Other cocatalysts such as ethyl iodide and diethylamine were required in some cases to achieve good yields.

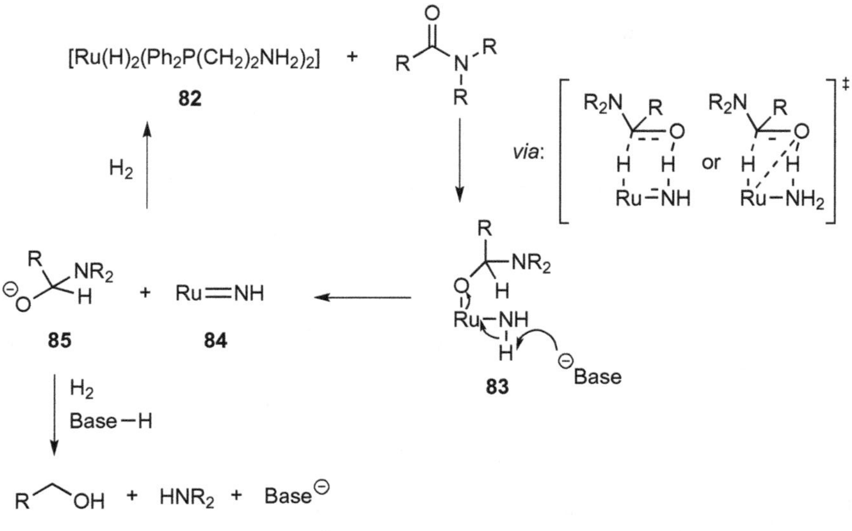

SCHEME 1-28. Proposed mechanism for amide hydrogenation [36].

1.6.1 Rhodium-Catalyzed Reduction of Amides Using Silanes

Used in conjunction with Ph_2SiH_2, $[RhH(CO)(PPh_3)_3]$ was found to be the most effective for the reduction of tertiary amides to amines, where yields of 65–98% were achieved at room temperature over periods of 0.5–48 h [41]. Notably, ester and epoxide functionalities remained inert under these reaction conditions. A catalytic cycle was proposed (Scheme 1-29), where the first step is thought to involve the oxidative addition of hydrosilane to the rhodium complex **86**, generating the activated hydrido silane complex **87**. The amide is then believed to insert into the Rh–Si bond, leading to

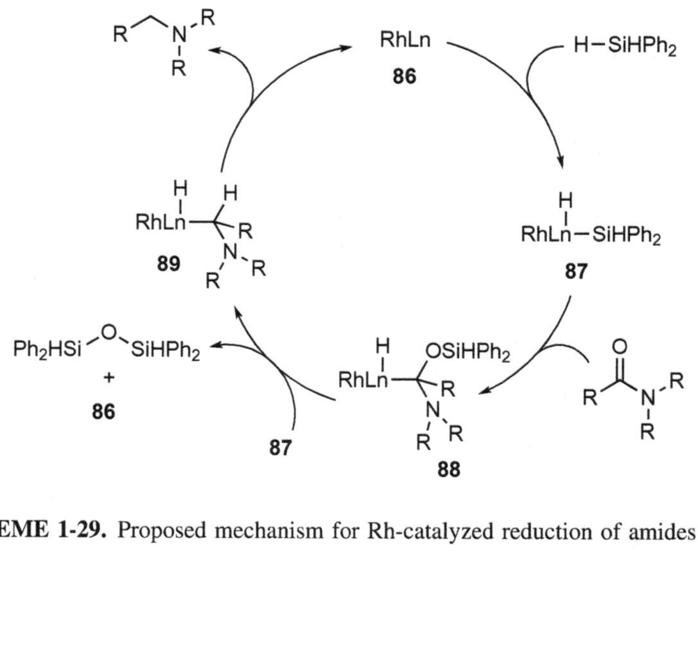

SCHEME 1-29. Proposed mechanism for Rh-catalyzed reduction of amides [41].

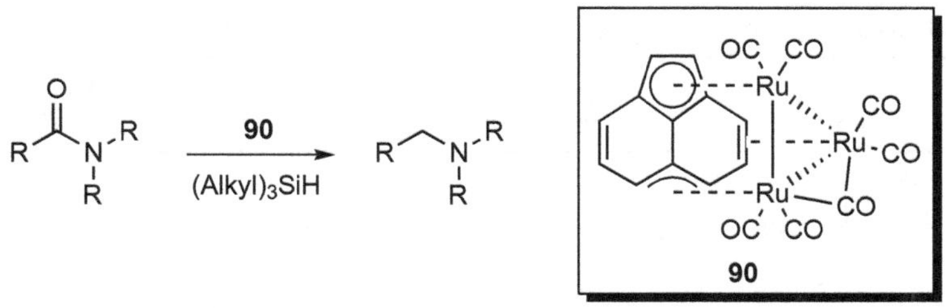

SCHEME 1-30. [(μ_3, η^2; η^3; η^5-acenaphthylene)$Ru_3(CO)_7$] (**90**)-catalyzed silane-mediated reduction of amides.

species **88**. Hydride transfer from activated silane complex **87** then allows selective C–O cleavage, siloxane formation, and regeneration of the catalyst **86**. The resultant amino complex **89** then reductively eliminates the product amine.

1.6.2 Ruthenium-Catalyzed Reduction of Amides Using Silanes

Nagashima and coworkers [42] first reported the use of a ruthenium cluster catalyst, [(μ_3, η^2; η^3; η^5-acenaphthylene)$Ru_3(CO)_7$] (**90**), in silane-mediated amide reduction in 2002 (Scheme 1-30). Three tertiary amides were reduced in 90–100% conversion, employing $EtMe_2SiH$ at 20 °C for 30 min.

They also addressed the problem of lengthy and wasteful workup procedures by encapsulating the catalyst within a polymer [43]. Using PMHS as the reducing agent, gel formation occurs as the reaction proceeds, as the oxygen molecules produced aid crosslinking of the polymer. At the end of the reaction, the products are easily separated from the catalyst, which is encapsulated within the gel. ICP-mass analysis indicated that <15 ppm of ruthenium was present in the amine (compared with ca. 10% when acid–base workup is used). Additionally, the encapsulated catalyst remains active and can be recycled. This is an excellent example of a green approach to waste disposal, and removing the need for acid–base workups.

The ruthenium cluster **90** is also able to reduce the more challenging secondary amides in >99% conversion under optimized reaction conditions by employing a disiloxane, $(H(Me)_2Si)_2O$ [44].

Using the same catalyst, selective reduction of amides can also be achieved in the presence of ketones and esters (Table 1-8) [45]. The presence of an amine, NEt_3, poisons the catalyst toward ketone and ester reduction, resulting in selective reduction of amides (entries 1 and 2). In many cases, it was found that the addition of an amine was not necessary as the amino groups produced during the reaction were sufficient to poison the catalytic activity (entries 3–6).

An interesting observation was made when the reduction of a primary amide **91** was attempted with this system, where dehydration to the nitrile **92** was observed with the evolution of hydrogen gas, rather than the expected reduction (Scheme 1-31) [46].

1.6.3 Platinum-Catalyzed Reduction of Amides Using Silanes

Nagashima and coworkers also applied their self-encapsulation strategy to platinum-catalyzed silane reduction of tertiary amides [47,48]. The most effective system involved

TABLE 1-8. Reduction of Tertiary Amides in the Presence of Ketones and Esters[a]

Entry	Substrate	Si–H (equiv.)	*t* (h)	Product	Yield (%)[b]	Selectivity (%)[c]
1	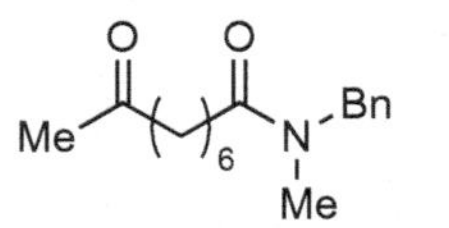	2.5	1	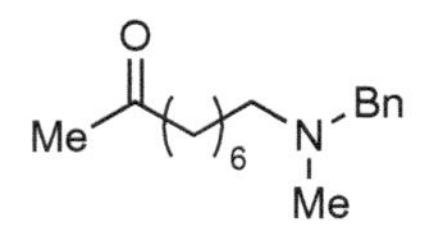	91	>99
2[d]	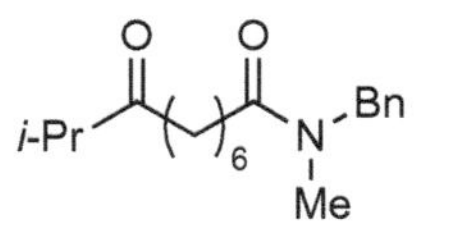	2.5	1.5	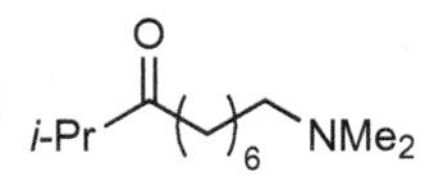	94	>99
3[e]	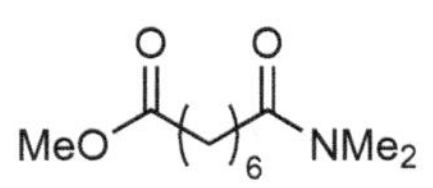	3.0	1.5	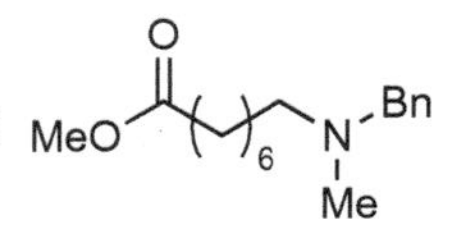	91	>99
4[e]	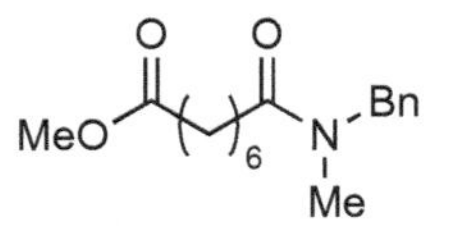	3.0	0.5	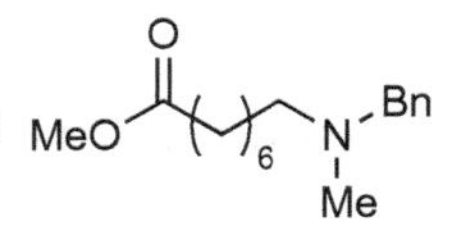	96	>99
5[e]	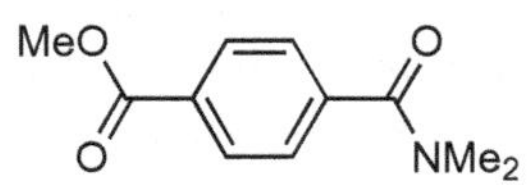	3.0	3.5	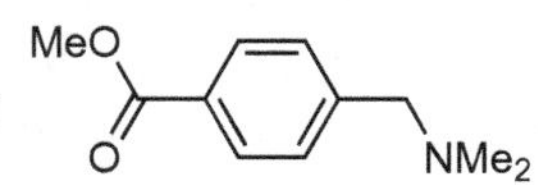	81	>99
6[e]	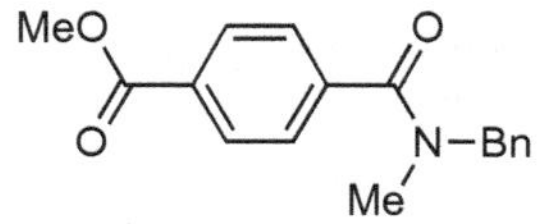	3.5	18	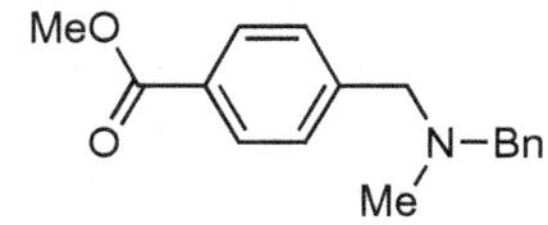	88	>99

[a]Conditions (unless otherwise indicated): Amide (1 mmol), **90** (0.01 mmol), $PhMe_2SiH$ (2.5–3.5 mmol), NEt_3 (1 mmol), rt.
[b]Isolated yield (determined by ^1H NMR).
[c]Determined by ^1H NMR spectroscopy of crude reaction mixtures.
[d]0.5 mmol substrate, 0.005 mmol catalyst.
[e]In the absence of NEt_3.

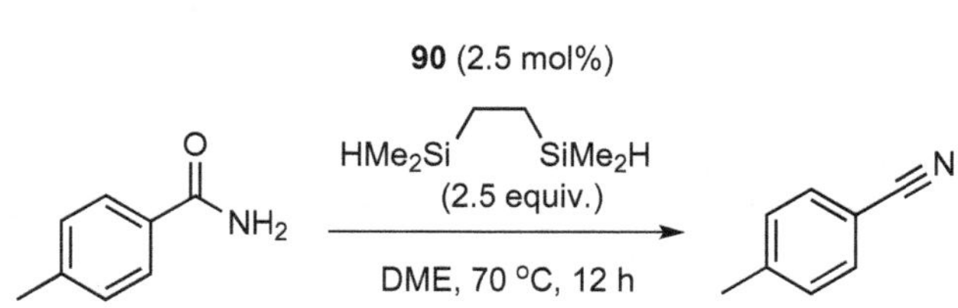

91 **92**, >95%

SCHEME 1-31. Dehydration of primary amide **91** to nitrile **92**.

$H_2PtCl_6{\cdot}H_2O$ and PMHS, which formed a gel to encapsulate the platinum catalyst. After purification on a short Al_2O_3 column, amines were obtained with <1 ppm of Pt residue. The Pt-containing resin was still active, and could be recycled, although the conversion was somewhat diminished (first: 100%, second: 84%, third: 76%).

TABLE 1-9. Structural Effect of Hydrosilanes on the Reduction of *N,N*-Dimethyl-3-Phenylpropanamide[a]

Entry	Hydrosilane	Yield (%)[b]	Entry	Hydrosilane	Yield (%)[b]
1	$PhMe_2SiH$	<1	6	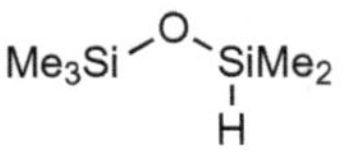	>98 (90)[c]
2	$EtMe_2SiH$	<1	7[d]	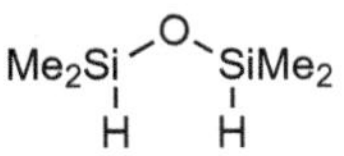	>98 (63)[c]
3	$(EtO)_3SiH$	<1	8[e,f]	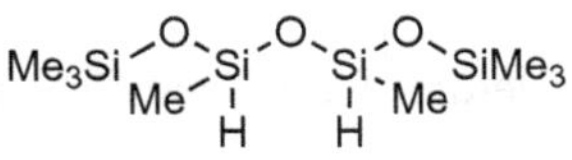	>98
4	Ph_2SiH_2	<1	9[f]	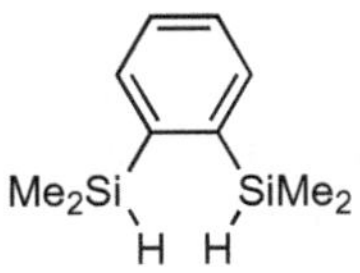	<1
5	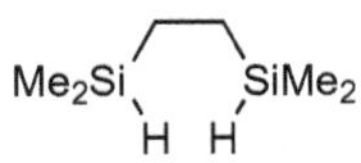	<1			

[a]Conditions (unless otherwise indicated): Amide (1 mmol), $H_2PtCl_6{\cdot}H_2O$ (0.01 mmol), silane (3 equiv.), 50°C, 3 h.
[b]Determined by ^1H NMR.
[c]Isolated yield after alumina column chromatography.
[d]10 h.
[e]8 h.
[f]4 equiv. silane.

More recently in 2009, the "dual Si–H effect" was described with respect to platinum-catalyzed hydrosilation [48]. A series of mono- and disilanes and siloxanes was tested (Table 1-9) with varying numbers of Si–H and proximate Si–H bonds in the hydrogenation of *N,N*-dimethyl-3-phenylpropanamide **93** to amine **94** (Scheme 1-32). Silanes and siloxanes containing one or two Si–H bonds gave <1% conversion (Table 1-9, entries 1–5). In comparison, siloxanes with two proximal Si–H units resulted in >98% yield (entries 6–8), whereas the related disilane (entry 9) showed no conversion.

From this study, 1,1,3,3-tetramethyldisiloxane (TMDS) was chosen as the optimal reductant due to its commercial availability and low cost. It was used in 5 equiv. to reduce a variety of tertiary amides in excellent yields of up to 96%, in the presence of sensitive functional groups including nitro, nitrile, ester, ether, and di- and trisubstituted alkenes. Selected secondary amides were also reduced, although higher catalyst and silane loadings were required (3 mol% and 10 equiv., respectively).

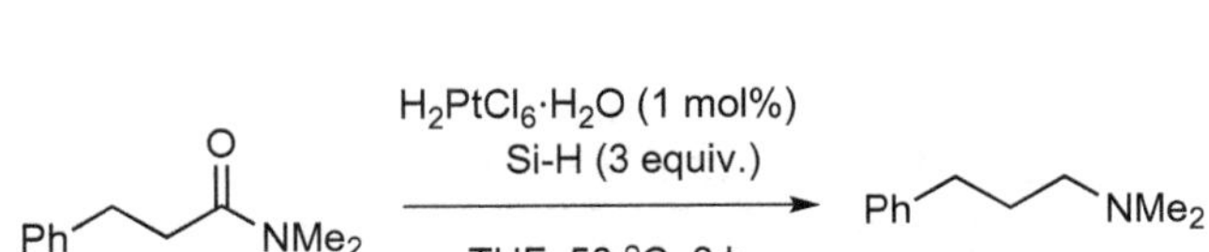

93 **94**

SCHEME 1-32. Pt-catalyzed reduction of *N,N*-dimethyl-3-phenylpropanamide **93** to amine **94**.

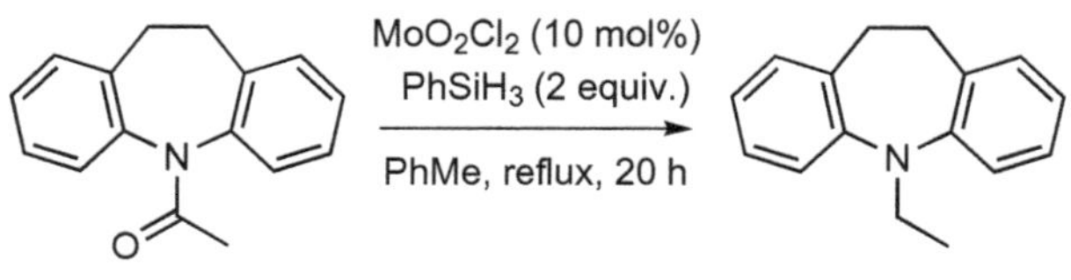

SCHEME 1-33. Mo-catalyzed reduction of 1-(10,11-dihydro-5*H*-dibenzo[*b,f*]azepin-5-yl)ethanone **95**.

1.6.4 Molybdenum-Catalyzed Reduction of Amides Using Silanes

Fernandes and Romão [49] reported MoO_2Cl_2-catalyzed reductions of secondary and tertiary amides in yields of 40–78% and 81–87%, respectively. The silane $PhSiH_3$ was chosen, in combination with 10 mol% of the catalyst, to reduce the amide **95** to **96** in very high yields (Scheme 1-33), for example. The mechanism is thought to proceed via the imine, as confirmed by deuterium studies.

1.6.5 Indium Bromide-Catalyzed Reduction of Amides Using Silanes

A system for amide reduction described by Sakai et al. [50] employs indium bromide in conjunction with triethylsilane (Scheme 1-34).

This system has been shown to reduce a series of tertiary amides with yields of up to 94% in the case of *N*-benzoyl morpholine. Aliphatic amides gave much lower yields, as did secondary amides, where acetanilide and benzanilide gave yields of 45% and 72%, respectively. Some functional group tolerance was also observed as the reaction proceeded in the presence of a nitrile group (85%).

1.6.6 Iron-Catalyzed Reduction of Amides Using Silanes

In 2009, two groups, Beller and coworkers [51] and Nagashima [52], simultaneously and independently reported the catalyzed hydrosilane reduction of tertiary amides catalyzed by iron. Beller and coworkers reported a system using $[Fe_3(CO)_{12}]$ with 4 equiv. of PMHS at 100 °C for 24 h (Table 1-10). Most substrates based on *N,N*-dibenzylbenzamide were reduced with excellent yields (>90%), which included a variety of functional groups such as halides, ethers, cyclopropane, and alkenes (entries 1–9, 11–17). The only exception was an ester-substituted amide (entry 10), which gave a rather low yield of 58%. The presence of sterically demanding groups requires further equivalents of silane to achieve good conversions, for example, substrates with a *tert*-butyl on the C=O or *i*-propyl groups on nitrogen required 8 equiv. of silane (entries 17 and 22). The conditions for this reaction are fairly mild, as they were conducted at ambient pressure at 100 °C.

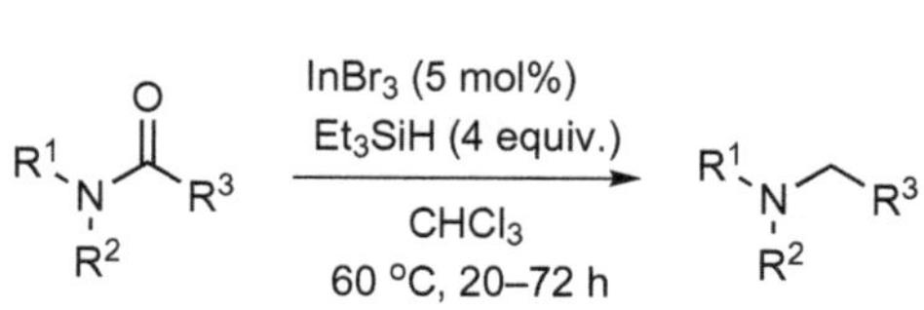

SCHEME 1-34. In-catalyzed reduction of tertiary amides.

TABLE 1-10. Iron-Catalyzed Reduction of Tertiary Amides in the Presence of Silanes[a]

Entry	Amide		Cat. (%)	Si–H (equiv.)	Yield (%)[b]
1	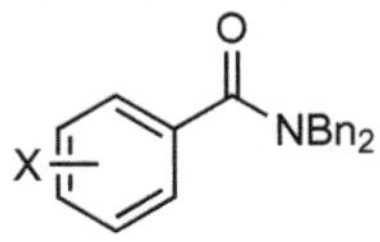	X = H	4	4	89
2		X = *p*-Me	4	4	93
3		X = *m*-Me	4	4	93
4		X = *o*-Me	8	4	94
5		X = *p*-OMe	4	4	93
6		X = *p*-CF_3	4	4	84
7		X = *p*-Cl	4	4	95
8		X = *p*-Br	4	4	92
9[c]		X = *p*-I	4	4	83
10[d]		X = *p*-CO_2Me	4	4	58
11[d]	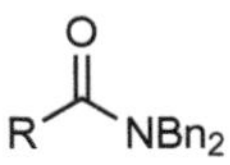	R = Me	4	4	89
12		R = 1-naphthalene	4	4	77
13		R = 2-naphthalene	4	4	84
14		R = 2-furan	10	8	82
15		R = 2-thiophene	10	8	84
16		R = Cyclopropane	6	6	82
17		R = *i*-Bu	8	8	98
18[d]	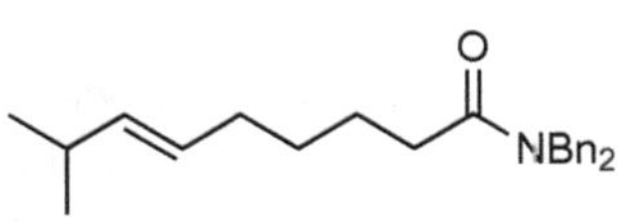		4	4	76
19[c]	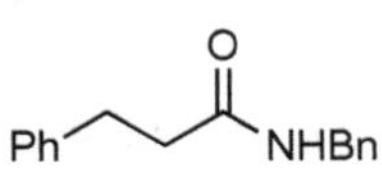		4	4	63
20	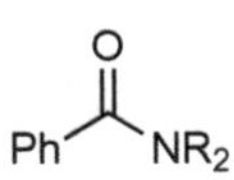	R = $-(CH_2)_5-$	3	4	94
21		R = $-(CH_2)_2O(CH_2)_2-$	3	4	80
22		R = *i*-Pr	10	8	59
23		R = Me, Ph	4	4	81
24		R = Ph	4	4	50
25[d]		R = Me, H	2.5[e]	99[f]	
26[d]	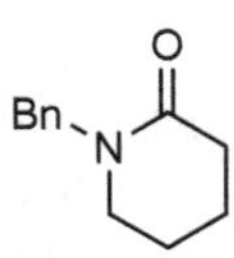		3	4	90
27[d,g]	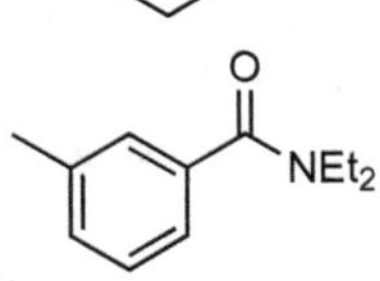		2	PMHS (4)	85

[a]Conditions (unless otherwise indicated): Amide (1 mmol), $[Fe_3(CO)_{12}]$ (2–10 mol%), PMHS (4–8 mmol), *n*-Bu_2O (5 ml), 100°C, 24 h.
[b]Isolated yield.
[c]Contained 10% tribenzylamine.
[d]Toluene.
[e]$PhSiH_3$.
[f]Determined by GC.
[g]100 mmol scale.

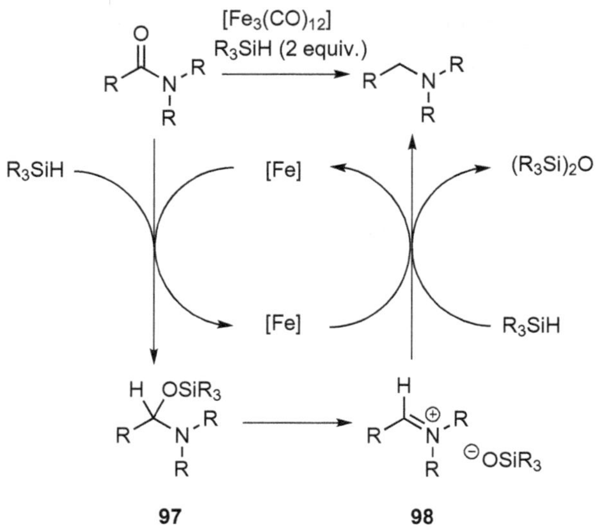

SCHEME 1-35. Proposed mechanism for Fe-catalyzed hydrosilation of tertiary amides.

Beller and coworkers proposed a mechanism based on that previously described by Nagashima for the ruthenium-catalyzed reaction (Scheme 1-35). This proceeds through the *O*-silylated intermediate **97**, a result of the reaction of the amide with an activated iron/silane species, which was transformed to the iminium species **98**. A second equivalent of the activated iron species is then required to regenerate the catalyst, release the newly formed amine, and generate 1 equiv. of siloxane. This mechanism is supported by deuteration studies employing Ph_2SiD_2, which showed the incorporation of two deuterium atoms in place of the carbonyl moiety.

On the other hand, Nagashima and coworkers found that $[Fe(CO)_5]$ and $[Fe_3(CO)_{12}]$ were both suitable catalyst precursors for this reaction under similar conditions to that adopted by Beller (100 °C, 24 h, toluene, TMDS). Furthermore, the reduction of a series of tertiary amides could also be achieved at room temperature under photoassisted conditions. The utility of the thermal and photochemical reactions was compared (Table 1-11). Overall, good-to-high yields were achieved with all the substrates tested, including another reducible carbonyl (entry 7), halogens (entries 5 and 6) and a cyclic amide (entry 3). In contrast, a benzyl chloride substrate resulted in a dehalogenated side product in the thermal reaction, but then proceeded to the desired amine product under photoassisted conditions (entry 8).

As described earlier in the Ru and Pt examples, it is also possible to conduct the reaction using PMHS, using the encapsulation technique to remove residual iron from the sample.

Nagashima also noted that the iron-catalyzed reactions behaved differently toward nitro groups. In contrast to ruthenium and platinum systems reported previously, nitroarenes were readily reduced instead of amides, which remained intact even with an excess of silane. It was thought that aniline products effectively inhibit the catalyst, preventing further reaction with the amide.

TABLE 1-11. Thermal and Photoassisted Iron-Catalyzed Silane Reductions of Tertiary Amides[a]

			Thermal Reduction[b]		Photoassisted Reduction[c]	
Entry	Amide		$[Fe(CO)_5]$ (%)	$[Fe_3(CO)_{12}]$ (%)	$[Fe(CO)_5]$ (%)	$[Fe_3(CO)_{12}]$ (%)
1	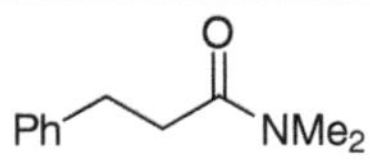		86	85	94	73
2	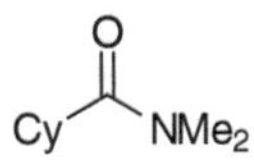		67	81	69	68
3	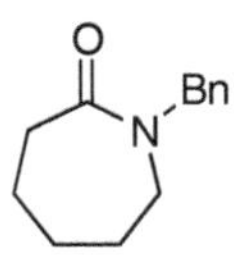		95	84	81	81
4	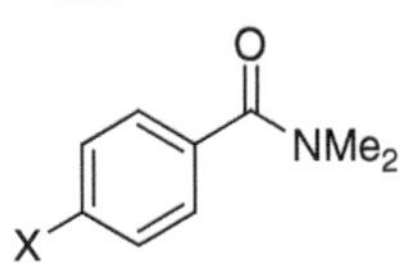	X = OMe	81	82	82	93
5		X = Cl	66	85	78	90
6		X = Br	98	96	77	95
7		X = CO_2Me	88	84	79	94
8	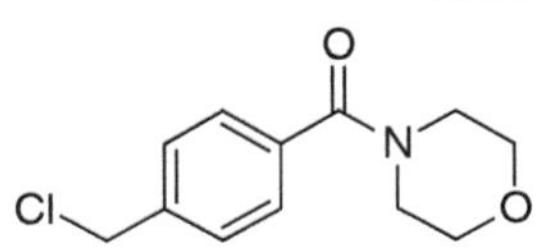		–	21[d]	–	84

[a]Conditions (unless otherwise indicated): Amide (1 mmol), iron catalyst (0.1 mmol), TMDS (2.2 mmol, Si–H = 4.4 mmol), toluene (0.5 ml).
[b]100°C, 24 h.
[c]rt, 9 h, 400 W high-pressure Hg lamp.
[d]Dehalogenated product formed in 68% yield.

1.6.7 Zinc-Catalyzed Reduction of Amides Using Silanes

Zinc-catalyzed reductions of tertiary amide in the presence of silanes were reported by Beller and coworkers in 2010 [38]. By employing zinc acetate in the presence of silanes such as $PhSiH_3$, Ph_2SiH_2, $(EtO)_2MeSiH$, and $(EtO)_3SiH$, amides could be reduced to amines, even at room temperature in the case of $(EtO)_3SiH$ (Scheme 1-36).

A variety of substituted tertiary amides were tested, and excellent yields (73–99%) were obtained (Table 1-12). Steric and electronic effects of piperidinyl-substituted amides were investigated, where it was found that aromatic, aliphatic, and heterocyclic amides can be

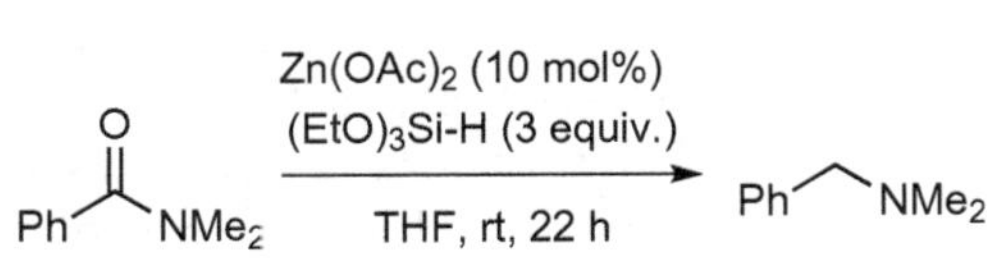

SCHEME 1-36. Zinc acetate-catalyzed reduction of *N,N*-dimethylbenzamide.

TABLE 1-12. Zinc Acetate-Catalyzed Reduction of Tertiary Amides[a]

Entry	Amide		Yield (%)
1	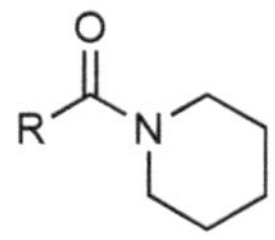	R = Ph	85
2		R = Cy	86
3		R = 2-naphthalene	99
4		R = 2-thiophene	82
5[b,c]		R = 2-furan	73
6[d]		R = CH_2CH_2Ph	73
7	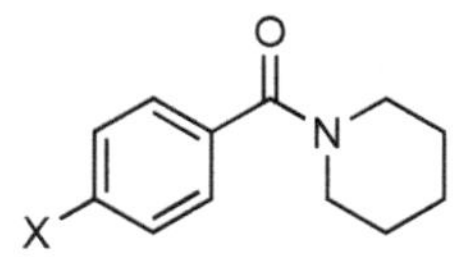	X = *p*-F	92
8		X = *p*-Br	97
9[d]		X = *p*-OMe	80
10[e]	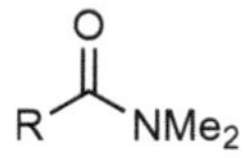	R = Me	96
11[e]		R = Ph	97
12		R = Ph-*o*-Cl	87
13	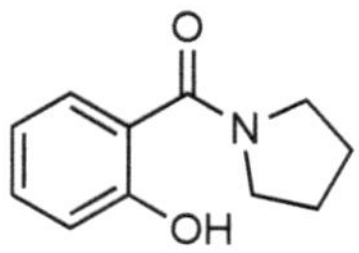		87
14[b,c]	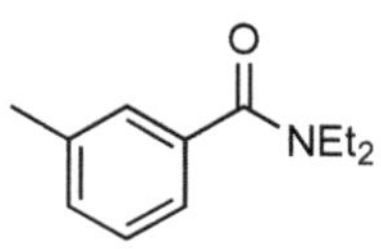		75

[a]Conditions (unless otherwise indicated): Amide (1 mmol), $Zn(OAc)_2$ (10 mol%), $(EtO)_3SiH$ (3 mmol), rt or 40 °C, 22 h, THF (3 ml), yields were isolated.
[b]3 mmol scale.
[c]Purified by distillation.
[d]Purified by column chromatography.
[e]Not isolated yield.

successfully reduced (entries 1–6), including the naphthyl-substituted amide in 99% yield (entry 3). Electron-donating aromatic groups were found to give poorer yields than electron-withdrawing groups (compare entries 7–9).

The reaction also showed excellent functional group tolerance (Table 1-13); amide reduction proceeded with >99% selectivity in the presence of azo, alkene, nitro, ketone, benzyl, nitrile, and ester groups.

Zinc acetate was found to activate the silane as opposed to the carbonyl of the amide. This was confirmed by IR studies of the carbonyl stretch (which showed no change) and ^{29}Si NMR spectroscopy (which indicated the formation of a new species). The proposed mechanism of the reduction is shown below (Scheme 1-37), beginning with reaction of the silane with the zinc to generate the activated species **99**. Next, coordination of the amide occurs at the metal center in **99**, which results in the hemiaminal species **101** (via **100**). An imine species, **102**, is then liberated from the release of the anionic zinc siloxide, **103**. A second equivalent of silane is then required to convert the imine to the amine, producing siloxane and regenerating the catalyst.

TABLE 1-13. Zinc Acetate-Catalyzed Reduction of Tertiary Amides to Demonstrate Functional Group Tolerance[a]

Entry	Amide	Yield (%)	Entry	Amide	Yield (%)
1	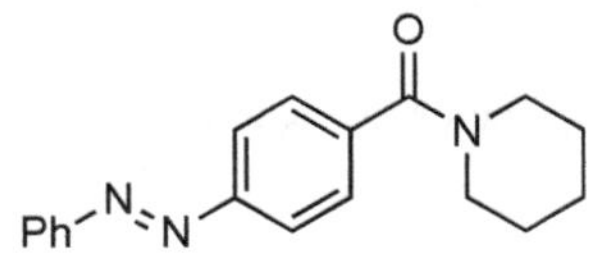	91	5	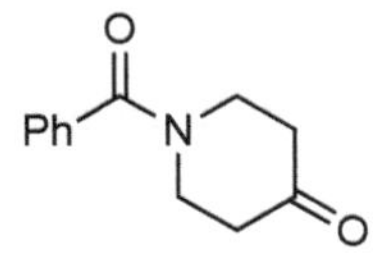	83
2	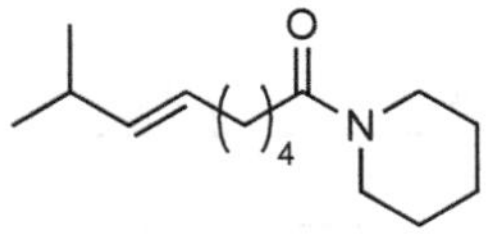	87	6[b]	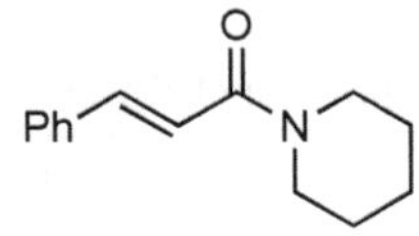	78
3	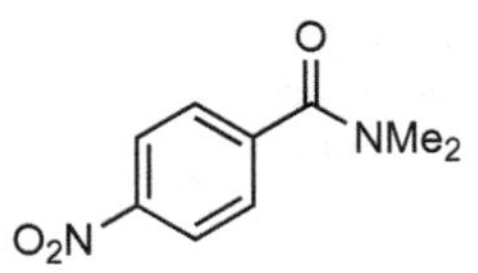	85	7[b]	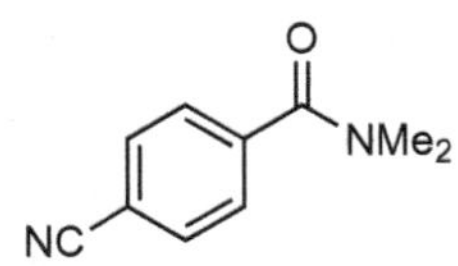	72
4	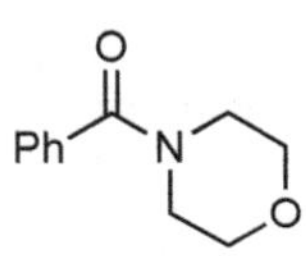	84	8[b]	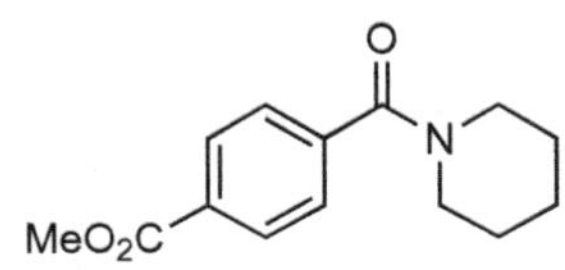	65

[a]Conditions: Amide (1.0 mmol), $Zn(OAc)_2$ (0.1 mmol), $(EtO)_3SiH$ (3.0 mmol), rt, 22 h, THF (3 ml).
[b]Purified by column chromatography.

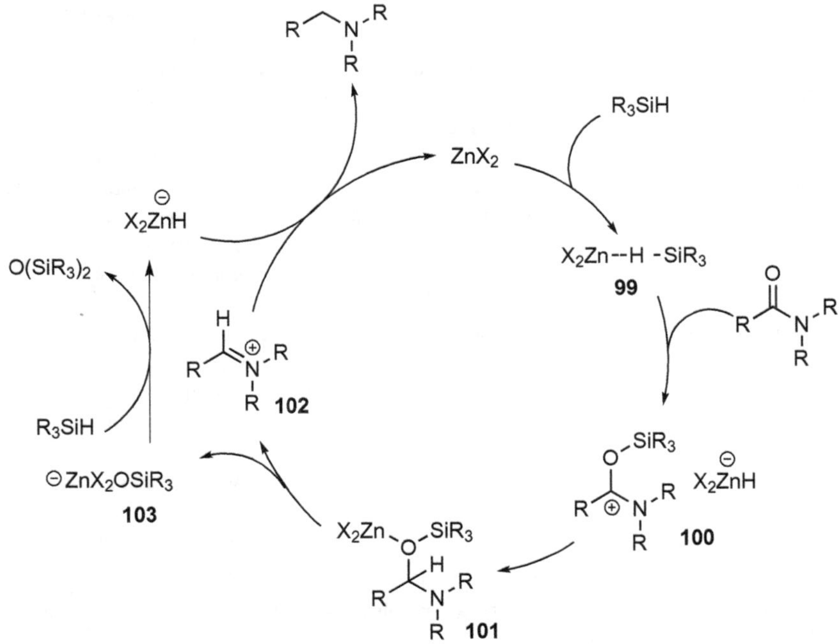

SCHEME 1-37. Proposed mechanism of Zn-catalyzed hydrosilation of amides [38].

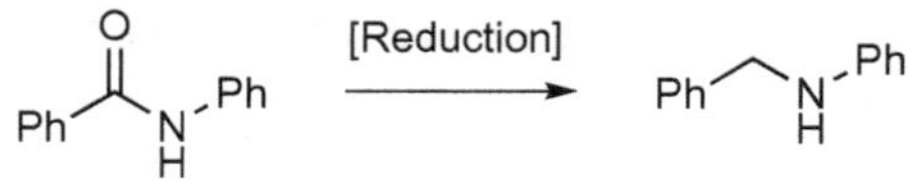

SCHEME 1-38. Reduction of benzanilide **104** to phenyl benzylamine **105**.

TABLE 1-14. Comparison of Different Amide Reduction Methods

Method	Advantages	Disadvantages	AE[a]	E[b]
Homogeneous hydrogenation (Ru/Triphos)	The only by-product is water 1°, 2° and 3° amides Low pressures Excellent selectivity of 1° and 2° amides	High temperatures Need specialist equipment	91	0
Heterogeneous hydrogenation (bimetallic)	Simple catalyst separation Excellent substrate scope (1°, 2°, 3° amides) Recyclability	High temperatures and pressures Incompatible with aromatics Need specialist equipment	91	0
Heterogeneous hydrogenation (copper chromite)	Simple catalyst separation	Harsh reaction conditions Poor selectivity Low functional group tolerance	91	0
Homogeneous hydrosilation (e.g., Zn or Fe)	Mild conditions Excellent functional group tolerance	Excess silane required Potential for SiH_4 formation Mainly 3° amides	35	1.9
B_2H_6 (stoichiometric)	Efficient Good functional group tolerance	Difficult to handle Low atom economy Waste disposal Complex workup procedure	68	0.5
$LiAlH_4$ (stoichiometric)	Efficient	Difficult to handle Low atom economy Waste disposal Complex workup procedure Low functional group tolerance Low reactivity toward 1° amides	46	1.2

[a]AE = atom economy = (mass product/mass starting materials), based on theoretical (100% yield) hydrogenation of benzanilide to phenylbenzylamine.
[b]E = environmental factor = mass of waste products/mass of desired product, based on 100% theoretical yield, does not take into account solvents or workup.

1.7 CONCLUSIONS AND FUTURE PERSPECTIVES

The homogeneous hydrogenation of amides is still in its infancy; however, great strides are being made toward milder reaction conditions, improved selectivity, and functional group tolerance. It is possible to hydrogenate primary, secondary, and tertiary amides, in the presence of sensitive functional groups, but there are still some trade-offs to be made. Currently, heterogeneous bimetallic catalysts afford excellent conversions of a variety of substrates with straightforward catalyst removal, but there are problems associated with the overreduction of aromatic groups and the relatively harsh operating conditions. In contrast, metal-catalyzed reductions with silanes offer relatively mild transformations of tertiary amides to tertiary amines with excellent functional group tolerance, but the additional waste disposal and safety implications associated with stoichiometric amounts of silanes are undesirable. Homogeneous hydrogenation with molecular hydrogen represents a truly green approach, as water is generated as the only by-product. However, current advances still require relatively harsh reaction conditions. Nevertheless, it is possible to hydrogenate primary and secondary amides to give primary and secondary amines with excellent conversions and selectivities. Selective reduction of tertiary amides remains a challenge in this area.

The green credentials of different methods for the reduction of amides to amines are compared in terms of AE and E (environmental) factor (Scheme 1-38 and Table 1-14) [53] based on the theoretical reduction of benzanilide **104** to the desired product **105** in 100% conversion (and does not take into account workup procedures and solvents).

Future challenges include hydrogenation reactions that can be conducted at much lower temperatures and pressures, and preferably without the need of specialist equipment. For widespread application, particularly in the pharmaceutical industry, the reaction must also have excellent selectivity and functional group tolerance. Although this has been achieved to some extent in the case of tertiary amides employing hydrosilation technology, an entirely green solution is some way off. With an ever-increasing interest in this area and further investments, more breakthroughs are sure to follow.

REFERENCES

1. Salvatore RN, Yoon CH, Jung KW (2001). Synthesis of secondary amines. *Tetrahedron* 57, 7785–7811.
2. Constable DJC, Dunn PJ, Hayler JD, Humphrey GR, Leazer, Jr., JL, Linderman RJ, Lorenz K, Manley J, Pearlman BA, Wells A, Zaks A, Zhang TY (2007). Key green chemistry research areas—a perspective from pharmaceutical manufacturers. *Green Chem.* 9, 411–420.
3. Carey JS, Laffan D, Thomson C, Williams MT (2006). Analysis of the reactions used for the preparation of drug candidate molecules. *Org. Biomol. Chem.* 4, 2337–2347.
4. Gill CD, Greenhalgh DA, Simpkins NS (2003). Application of the chiral base desymmetrisation of imides to the synthesis of the alkaloid jamtine and the antidepressant paroxetine. *Tetrahedron* 59, 9213–9230.
5. Andersson PG, Schink HE, Osterlund K (1998). Asymmetric total synthesis of (+)-tolterodine, a new muscarinic receptor antagonist, via copper-assisted asymmetric conjugate addition of aryl Grignard reagents to 3-phenyl-prop-2-enoyl-oxazolidinones. *J. Org. Chem.* 63, 8067–8070.
6. Gant TG, inventor; Auspex Pharmaceuticals Inc, assignee. Substituted benzamide modulators of dopamine receptor. US Patent US2010/0105755. April 29, 2010.

7. Giles ME, Thomson C, Eyley SC, Cole AJ, Goodwin CJ, Hurved PA, Morlin AJG, Tornos J, Atkinson S, Just C, Dean JC, Singleton JT, Longton AJ, Woodland I, Teasdale A, Gregertsen B, Else H, Athwal MS, Tatterton S, Knott JM, Thompson N, Smith SJ (2004). Development of a manufacturing process for sibenadet hydrochloride, the active ingredient of viozan. *Org. Process Res. Dev.* 8, 628–642.
8. Bannister RM, Brookes MH, Evans GR, Katz RB, Tyrrell ND (2000). A scaleable route to the pure enantiomers of verapamil. *Org. Process Res. Dev.* 4, 467–472.
9. Nakazato A, Ohta K, Sekiguchi Y, Okuyama S, Chaki S, Kawashima Y, Hatayama K (1999). Design, synthesis, structure-activity relationships, and biological characterization of novel arylalkoxyphenylalkylamine sigma ligands as potential antipsychotic drugs. *J. Med. Chem.* 42, 1076–1087.
10. Adkins H, Wojcik B (1934). Catalytic hydrogenation of amides to amines. *J. Am. Chem. Soc.* 56, 2419–2422.
11. King RM, inventor; The Proctor & Gamble Company, assignee. Catalytic hydrogenation of *N,N*-disubstituted amides to amines. US Patent 4,448,998. May 15, 1984.
12. Hirosawa C, Wakasa N, Fuchikami T (1996). Hydrogenation of amides by the use of bimetallic catalysts consisting of group 8 to 10, and group 6 or 7 metals. *Tetrahedron Lett.* 37, 6749–6752.
13. Smith AA, Dani P, Higginson PD, Pettman AJ, inventors; Avantium International B.V., assignee. Bimetallic and trimetallic non-alloy catalysts for the hydrogenation of amides into amines under mild reaction conditions. World Patent WO2005/066112. July 21, 2005.
14. Beamson G, Papworth AJ, Philipps C, Smith AM, Whyman R (2010). Selective hydrogenation of amides using Rh/Mo catalysts. *J. Catal.* 269, 93–102.
15. Beamson G, Papworth AJ, Philipps C, Smith AM, Whyman R (2010). Selective hydrogenation of amides using ruthenium/molybdenum catalysts. *Adv. Synth. Catal.* 352, 869–883.
16. Beamson G, Papworth AJ, Philipps C, Smith AM, Whyman R (2011). Selective hydrogenation of amides using bimetallic Ru/Re and Rh/Re catalysts. *J. Catal.* 278, 228–238.
17. Kilner M, Tyers DV, Crabtree SP, Wood MA, inventors; Davy Process Technology, assignee. Homogenous process for the hydrogenation of carboxylic acids and derivatives thereof. World Patent WO2003/093208. November 13, 2003.
18. Teunissen HT, Elsevier CJ (1997). Ruthenium catalyzed hydrogenation of dimethyl oxalate to ethylene glycol. *Chem. Commun.* 667–668.
19. Teunissen HT, Elsevier CJ (1998). Homogeneous ruthenium catalyzed hydrogenation of esters to alcohols. *Chem. Commun.* 1367–1368.
20. van Engelen MC, Teunissen HT, de Vries JG, Elsevier CJ (2003). Suitable ligands for homogeneous ruthenium-catalyzed hydrogenolysis of esters. *J. Mol. Catal. A: Chem.* 206, 185–192.
21. Sung KM, Huh S, Jun MJ (1999). Syntheses of ruthenium(II) complexes containing polyphosphine ligands and their applications in the homogeneous hydrogenation. *Polyhedron* 18, 469–479.
22. Bautista MT, Earl KA, Maltby PA, Morris RH, Schweitzer CT (1994). New dihydrogen complexes—the synthesis and spectroscopic properties of iron(II), ruthenium(II), and osmium(II) complexes containing the meso-tetraphos-1 ligand. *Can. J. Chem.* 72, 547–560.
23. Nùnez-Magro AA, Eastham GR, Cole-Hamilton DJ (2007). The synthesis of amines by the homogeneous hydrogenation of secondary and primary amides. *Chem. Commun.* 3154–3156.
24. Balaraman E, Gnanaprakasam B, Shimon LJW, Milstein D (2010). Direct hydrogenation of amides to alcohols and amines under mild conditions. *J. Am. Chem. Soc.* 132, 16756–16758.
25. (a) Dodds DL, Coetzee J, Brosinski S, Klankermeyer J, Leitner W, Cole-Hamilton DJ (2011). Published as an addition to reference 24: http://www.rsc.org/suppdata/cc/b7/b706635j/addition .htm. (b) Dodds DL, Coetzee J. Brosinski S, Klankermeyer J, Leitner W, Cole-Hamilton DJ. Upon the hydrogenation of amides to amines. *Chemistry. Eur. J.* Submitted.

26. Furst MRL, le Goff R, Quinzler D, Mecking S, Cole-Hamilton DJ (2012). Polymer precursors from catalytic reactions of natural oils. *Green Chem.* 14, 472–477.
27. Baxter EF, Lim K, Harvie M, Charkovskiy A, Pogorzelec PJ, Cole-Hamilton DJ. A new route to nitrogen heterocycles. Manuscript in preparation.
28. Crabtree SP, Tyers DV, Sharif, M, inventors; Davy process Technology, assignee. Process for the production of lactams. World Patent WO2005/051907. June 9, 2005.
29. Aoun R, Renaud JL, Dixneuf PH, Bruneau C (2005). Concomitant monoreduction and hydrogenation of unsaturated cyclic imides to lactams catalyzed by ruthenium compounds. *Angew. Chem., Int. Ed.* 44, 2021–2023.
30. Zhang J, Leitus G, Ben-David Y, Milstein D (2006). Efficient homogeneous catalytic hydrogenation of esters to alcohols. *Angew. Chem., Int. Ed.* 45, 1113–1115.
31. Landgrafe C, Sheldrick WS, Sudfeld H (1998). Fluxionality and phosphane arm-off reactions of octahedral ruthenium(II) complexes with the tripodal polyphosphane $MeC(CH_2PPh_2)_3$. *Eur. J. Inorg. Chem.* 407–414.
32. Geilen FMA, Engendahl B, Holscher M, Klankermayer J, Leitner W (2011). Selective homogeneous hydrogenation of biogenic carboxylic acids with $[Ru(TriPhos)H]^+$—a mechanistic study. *J. Am. Chem. Soc.* 133, 14349–14358.
33. Geilen FMA, Engendahl B, Harwardt A, Marquardt W, Klankermayer J, Leitner W (2010). Selective and flexible transformation of biomass-derived platform chemicals by a multifunctional catalytic system. *Angew. Chem., Int. Ed.* 49, 5510–5514.
34. (a) Ito M, Ootsuka T, Watari R, Shiibashi A, Himizu A, Ikariya T (2011). Catalytic hydrogenation of carboxamides and esters by well-defined Cp^*Ru complexes bearing a protic amine ligand. *J. Am. Chem. Soc.* 133, 4240–4242. (b) Ito M, Koo LW, Himizu A, Kobayashi C, Sakaguchi A, Ikariya T (2009). Hydrogenation of *N*-acylcarbamates and *N*-acylsulfonamides catalyzed by a bifunctional $[Cp^*Ru(PN)]$ complex. *Angew. Chem., Int. Ed.* 48, 1324–1327.
35. Ito M, Sakaguchi A, Kobayashi C, Ikariya T (2007). Chemoselective hydrogenation of imides catalyzed by $Cp^*Ru(PN)$ complexes and its application to the asymmetric synthesis of paroxetine. *J. Am. Chem. Soc.* 129, 290–291.
36. John JM, Bergens SH (2011). A highly active catalyst for the hydrogenation of amides to alcohols and amines. *Angew. Chem., Int. Ed.* 50, 10377–10380.
37. Wells AS (2010). On the perils of unexpected silane generation. *Org. Process Res. Dev.* 14, 484.
38. Das S, Addis D, Zhou S, Junge K, Beller M (2010). Zinc-catalyzed reduction of amides: unprecedented selectivity and functional group tolerance. *J. Am. Chem. Soc.* 132, 1770–1771.
39. Lawrence NJ, Drew MD, Bushell SM (1999). Polymethylhydrosiloxane: a versatile reducing agent for organic synthesis. *J. Chem. Soc., Perkin Trans. 1* 3381–3391.
40. Igarashi M, Fuchikami T (2001). Transition-metal complex-catalyzed reduction of amides with hydrosilanes: a facile transformation of amides to amines. *Tetrahedron Lett.* 42, 1945–1947.
41. Kuwano R, Takahashi M, Ito Y (1998). Reduction of amides to amines via catalytic hydrosilylation by a rhodium complex. *Tetrahedron Lett.* 39, 1017–1020.
42. Matsubara K, Iura T, Maki T, Nagashima H (2002). A triruthenium carbonyl cluster bearing a bridging acenaphthylene ligand: an efficient catalyst for reduction of esters, carboxylic acids, and amides by trialkylsilanes. *J. Org. Chem.* 67, 4985–4988.
43. Motoyama Y, Mitsui K, Ishida T, Nagashima H (2005). Self-encapsulation of homogeneous catalyst species into polymer gel leading to a facile and efficient separation system of amine products in the Ru-catalyzed reduction of carboxamides with polymethylhydrosiloxane (PMHS). *J. Am. Chem. Soc.* 127, 13150–13151.
44. Hanada S, Ishida T, Motoyama Y, Nagashima H (2007). The ruthenium-catalyzed reduction and reductive *n*-alkylation of secondary amides with hydrosilanes: practical synthesis

of secondary and tertiary amines by judicious choice of hydrosilanes. *J. Org. Chem.* 72, 7551–7599.

45. Sasakuma H, Motoyama Y, Nagashima H (2007). Functional group-selective poisoning of molecular catalysts: a ruthenium cluster-catalyzed highly amide-selective silane reduction that does not affect ketones or esters. *Chem. Commun.* 4916–4918.
46. Hanada S, Motoyama Y, Nagashima H (2008). Hydrosilanes are not always reducing agents for carbonyl compounds but can also induce dehydration: a ruthenium-catalyzed conversion of primary amides to nitriles. *Eur. J. Org. Chem.* 4097–4100.
47. Hanada S, Motoyama Y, Nagashima H (2006). Dual Si−H effects in platinum-catalyzed silane reduction of carboxamides leading to a practical synthetic process of tertiary-amines involving self-encapsulation of the catalyst species into the insoluble silicone resin formed. *Tetrahedron Lett.* 47, 6173–6177.
48. Hanada S, Tsutsumi E, Motoyama Y, Nagashima H (2009). Practical access to amines by platinum-catalyzed reduction of carboxamides with hydrosilanes: synergy of dual Si−H groups leads to high efficiency and selectivity. *J. Am. Chem. Soc.* 131, 15032–15040.
49. Fernandes AC, Romão CC (2007). Reduction of amides with silanes catalyzed by MoO_2Cl_2. *J. Mol. Catal. A: Chem.* 272, 60–63.
50. Sakai N, Fujii K, Konakahara T (2008). One-step conversion to tertiary amines: $InBr_3/Et_3SiH$-mediated reductive deoxygenation of tertiary amides. *Tetrahedron Lett.* 49, 6873–6875.
51. Zhou S, Junge K, Addis D, Das S, Beller M (2009). A convenient and general iron-catalyzed reduction of amides to amines. *Angew. Chem., Int. Ed.* 48, 9507–9510.
52. Sunada Y, Kawakami H, Imaoka T, Motoyama Y, Nagashima H (2009). Hydrosilane reduction of tertiary carboxamides by iron carbonyl catalysts. *Angew. Chem., Int. Ed.* 48, 9511–9514.
53. Constable DJC, Curzons AD, Cunningham VL (2002). Metrics to 'green' chemistry—which are the best? *Green Chem.* 4, 521–527.

2

HYDROGENATION OF ESTERS

Lionel A. Saudan
Corporate R&D Division, Firmenich SA, Geneva, Switzerland

2.1 INTRODUCTION

Alcohols are important ingredients or building blocks in the synthesis of perfumery ingredients. Some examples of alcohols used in perfumery are shown in Figure 2-1 with their olfactory properties [1].

To date, the reduction of carboxylic esters to alcohols by hydrogenation with heterogeneous catalysts [2] represents the only industrial alternative to the use of stoichiometric hazardous metal hydrides (e.g., $LiAlH_4$, Vitride^{TM}, and the silicon hydride polymethylhydrosiloxane, PMHS). Despite the harsh conditions (>100°C, >100 bar), this process is used for the large-scale synthesis of fatty alcohols [3], an important component of surfactants (C_{12}–C_{18}). Surfactants are important constituents of cleaning products, for example detergents (powder or liquids), and can represent 5–30% w/w of the final product [4]. There are two main classes of surfactants based on natural fatty alcohols that are used by the fragrance industry. Natural fatty alcohols obtained from palm oils and coco oils are becoming more important in recent years, due to the high cost of petroleum-derived precursors and increasing consumer demand for a "green" label [4b]. The first class of surfactants is nonionic, made up of either alkyl polyglucoside or alkylated polyethylene glycol, as shown in Figure 2-2. Some units of unsaturation are sometimes present in the linear alkane chain.

The second class of surfactant is made up of sulfate esters of fatty alcohols, alkyl sulfates (AS), as shown in Figure 2-3; these tend to replace the linear alkane benzene sulfonate (LABS) surfactants derived from petroleum sources.

Compared to the hydrogenation of olefins [5] and ketones [6], the hydrogenation of esters is very much underexplored. Driven by the need for more user- and environment-friendly processes, this particular field of research has witnessed a new resurgence of

Sustainable Catalysis: Challenges and Practices for the Pharmaceutical and Fine Chemical Industries, First Edition. Edited by Peter J. Dunn, K. K. (Mimi) Hii, Michael J. Krische, and Michael T. Williams.

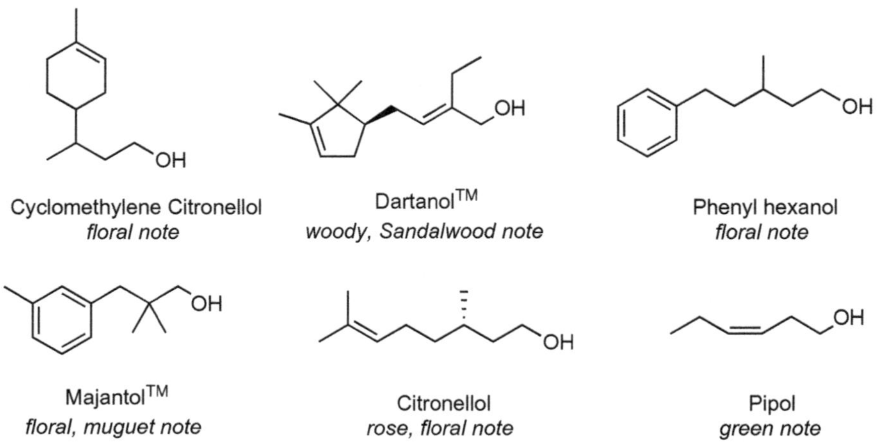

FIGURE 2-1. Examples of perfumery alcohols.

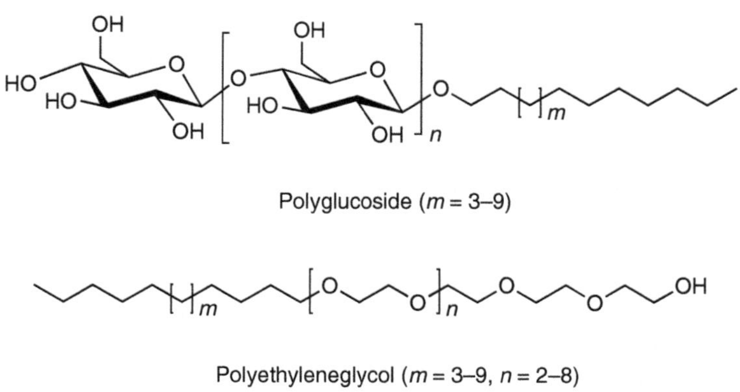

FIGURE 2-2. Examples of nonionic surfactants.

interest in recent years. In this account, a survey of newly developed homogeneous catalysts will be described [7]. This chapter is divided by substrate classes, with a discussion on the catalyst choice and efficiency for each class of substrate studied, and some examples of application of the target alcohols or diols are also given.

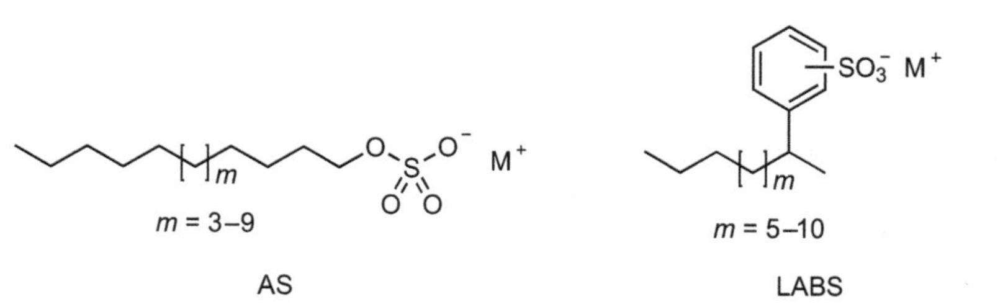

FIGURE 2-3. Examples of anionic surfactants.

2.2 HYDROGENATION OF ALIPHATIC ESTERS

In 2006, almost 10 years after Elsevier and Teunissen reported the first example of the hydrogenation of an aliphatic ester with $Ru(acac)_3$/Triphos/Et_3N [8,9], Milstein and his coworkers described a well-defined ruthenium catalyst **1** (Figure 2-4) for the hydrogenation of aliphatic esters under low H_2 pressure (5.4 bar) without any basic additives [10]. Although a high catalyst loading was needed (1 mol%), the desired alcohols were obtained in high yield (82–99%) in 4–12 h (Table 2-1, entries 1, 6, and 10), except *tert*-butylacetate, which was hydrogenated in low yield (11%) even after 24 h (entry 4), which is most certainly due to unfavorable steric interactions between the substrate and the bis-*t*-Bu-phosphine moiety. By replacing the bis-*t*-Bu-phosphine donor group with an *N*-heterocyclic carbene (NHC), the ruthenium complex **2** showed a much higher catalytic activity, although the presence of a base (KO*t*-Bu: 8 equiv./[Ru]) is required (Table 2-1, entries 2 and 3) [11]. However, sterically demanding esters, such as the *tert*-butyl acetate, can be hydrogenated with a higher yield (92%) and in a shorter period of time (entry 5), compared to the use of catalyst **1** (entry 4). Recently, Milstein described a highly active ruthenium catalyst containing a phosphine–bipyridine ligand for the direct hydrogenation of amides to amines and alcohols [12]. This was followed by the preparation of the analogous complex **3** for the hydrogenation of esters, where the phosphine moiety is replaced by an NHC group [13]. Complex **3** was highly active in the hydrogenation of aliphatic esters, in the presence of a catalytic amount of a strong base (KO*t*-Bu, 1 mol%) (entries 7 and 8). Using 1 mol% of the catalyst, the reaction proceeded smoothly at low hydrogen pressure (5.5 bar) at 135 °C in a short period of time (2 h). The catalyst loading could be decreased to 0.025 mol% at the expense of a higher hydrogen pressure (50 bar), affording a good conversion of 72% after 12 h [13].

Shortly after Milstein's first report, an industrial research team at Firmenich reported the use of ruthenium complexes containing aminophosphine (P^N) ligands **4** and **5** for the catalytic hydrogenation of esters [14]. These catalysts are conveniently obtained [15] from the commercially available $RuCl_2(PPh_3)_3$ and the corresponding ligand in one step and

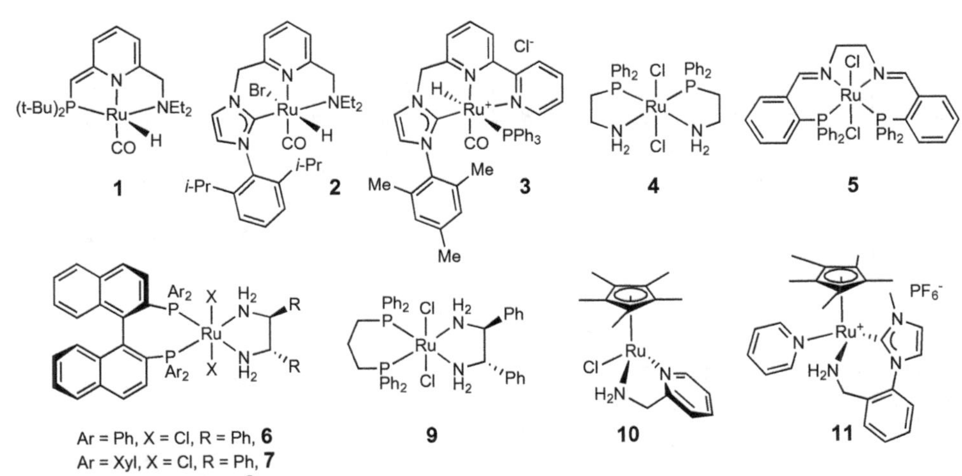

FIGURE 2-4. Examples of ruthenium complexes known to hydrogenate esters.

TABLE 2-1. Comparative Examples of Hydrogenation of Aliphatic Esters with Various Ruthenium Complexes

Entry	Starting Ester	[Ru]	Base (mol%)	pH_2 (bar)	T (°C)	t (h)	Conversion (%)	Alcohol	Yield (%)	Reference
1	CH_3CO_2Et	**1**[a]	–	5.4	115	12	86	EtOH	86	[10]
2	CH_3CO_2Et	**2**[b]	KO*t*-Bu (8)	5.3	105	2	99	EtOH	99	[11]
3	CH_3CO_2Bn	**2**[b]	KO*t*-Bu (8)	5.3	105	3	90	EtOH	87	[11]
4	CH_3CO_2*t*-Bu	**1**[a]	–	5.4	115	24	11	EtOH	11	[10]
5	CH_3CO_2*t*-Bu	**2**[b]	KO*t*-Bu (8)	5.3	105	2	93	EtOH	93	[11]
6	*n*-PrCO_2Et	**1**[a]	–	5.4	115	4	>99	*n*-BuOH	98	[10]
7	*n*-PrCO_2Et	**3**[c]	KO*t*-Bu (1)	5.5	135	2	98	*n*-BuOH	97	[13]
8	*n*-BuCO_2(*n*-Pent)	**3**[c]	KO*t*-Bu (1)	5.5	135	2	96	*n*-PentOH	96	[13]
9	*n*-BuCO_2Et	**8**[d]	KO*t*-Bu (9)	4.1	30	3	–	*n*-PentOH	61	[18]
10	*n*-PentCO_2(*n*-Hex)	**1**[a]	–	5.4	115	5	82	*n*-HexOH	82	[10]
11	*n*-HepCO_2Me	**4**[e]	NaOMe (10)	50	100	2.5	>99	*n*-OctOH	94	[14]
12	*n*-HepCO_2Me	**5**[e]	NaOMe (10)	50	100	2.5	>99	*n*-OctOH	87	[14]
13	*n*-HepCO_2Me	**7**[e]	NaOMe (5)	50	60	20	95	*n*-OctOH	78	(C.M. Saudan, L.A. Saudan, Firmenich, Geneva, unpublished results)
14	*n*-HepCO_2Me	**9**[f]	KO*t*-Bu (4)	50	100	16	98	*n*-OctOH	93	[17]
15	CyCO_2Me	**4**[e]	NaOMe (10)	50	100	2.5	98	CyOH	88	[14]
16	CyCO_2Me	**5**[e]	NaOMe (10)	50	100	2.5	98	CyOH	89	[14]
17	CyCO_2Me	**7**[e]	NaOMe (5)	50	60	20	73	CyOH	41	(C.M. Saudan, L.A. Saudan, Firmenich, Geneva, unpublished results)
18	CyCO_2Me	**9**[e]	NaOMe (5)	50	60	20	98	CyOH	84	(C.M. Saudan, L.A. Saudan, Firmenich, Geneva, unpublished results)
19	BnCO_2Et	**2**[b]	KO*t*-Bu (8)	5.3	105	2	>99	BnOH	>99	[11]
20	BnCO_2Me	**4**[e]	NaOMe (10)	50	100	2.5	99	BnOH	82	[14]
21	BnCO_2Me	**5**[e]	NaOMe (10)	50	100	2.5	99	BnOH	83	[14]

22	$BnCO_2Me$	**7**[e]	NaOMe (5)	50	60	20	>99	BnOH	96	(C.M. Saudan, L.A. Saudan, Firmenich, Geneva, unpublished results)
23	$BnCO_2Me$	**9**[e]	NaOMe (5)	50	60	20	>99	BnOH	91	(C.M. Saudan, L.A. Saudan, Firmenich, Geneva, unpublished results)
24	$BnCH_2CO_2Me$	**4**[e]	NaOMe (10)	50	100	2.5	>99	$BnCH_2OH$	96	[14]
25	$BnCH_2CO_2Me$	**5**[e]	NaOMe (10)	50	100	2.5	>99	$BnCH_2OH$	90	[14]
26	$BnCH_2CO_2Me$	**7**[e]	NaOMe (5)	50	60	20	>99	$BnCH_2OH$	94	(C.M. Saudan, L.A. Saudan, S.A. Firmenich, Geneva, unpublished results)

Conditions:

[a]Ester (2 mmol), [Ru] (1 mol%), dioxane (2 ml).

[b]Ester (1 mmol), [Ru] (1 mol%), KO*t*-Bu (8 mol%), toluene (2 ml).

[c]Ester (1 mmol), [Ru] (1 mol%), KO*t*-Bu (1 mol%), toluene (2 ml).

[d]Ester (1 mmol), [Ru] (2 mol%), THF (2 ml).

[e]Ester (20 mmol), [Ru] (0.05 mol%), NaOMe (5–10 mol%), THF (10–14 ml).

[f]Ester (4 mmol), [Ru] (0.2 mol%), KO*t*-Bu (4 mol%), THF (5.3 ml).

were highly active in the presence of a strong base (NaOMe: 20–100 equiv./[Ru]) under relatively mild conditions (50 bar H_2, 100 °C). The reduction of a range of aliphatic esters was completed (>99%) in a short period of time (1–2.5 h) using a low catalyst loading (0.05 mol%), to furnish the desired alcohol in high yield (82–96%) (Table 2-1, entries 11–12, 15–16, 20–21, and 24–25).

Realizing that the presence of NH_2 donor groups is responsible for the high catalytic activity of these complexes, the potential of Noyori's ruthenium complexes of the type $RuCl_2$(P^P)(N^N) was quickly explored as an active catalyst. Indeed, research groups at Firmenich [16], Takasago [17], and Bergens's group in Alberta [18] had all independently shown that ruthenium complexes **6–9** containing a diphosphine ligand, such as BINAP or DPPP, with a diamine such as 1,2-diphenylethylenediamine or ethylenediamine (Figure 2-4), were also active in the hydrogenation of esters. Judicious combination of the diphosphine and the diamine is crucial for the reactivity of such complexes [16,17]. For example, the use of DPPE or DPPB instead of DPPP gave catalysts with a much lower activity [17]. Furthermore, by using the dihydride complex **8**, Bergens showed that the hydrogenation of an aliphatic ester could be performed at room temperature under low hydrogen pressure (4 bar), giving the desired alcohol in moderate yield in 3 h (Table 2-1, entry 9), showing the high activity of such complexes [18]. On the other hand, linear aliphatic esters were also hydrogenated in high yields with complexes **7** and **9** under relatively mild conditions (60–100 °C, 50 bar H_2) with a low catalyst loading (0.05 mol%), in the presence of a strong base (NaOMe or KO*t*-Bu) (entries 13–14, 22–23, and 26). Moreover, for an α-branched ester (entries 17 and 18), complex **7** gave a much lower yield of the desired alcohol (41%) compared to complex (**9**) (84%), presumably due to the use of the less sterically demanding diphosphine DPPP, rather than Xyl-BINAP.

One of the challenges associated with the hydrogenation of ester into alcohol is the selectivity of the reaction, particularly the stability of other functional groups under the reaction conditions (60–135 °C and 4–50 bar H_2). For example, a C=C double bond could undergo competitive hydrogenation. Indeed, none of the complexes **1**, **2**, and **3** were tested for the hydrogenation of unsaturated esters. However, chemoselective hydrogenation of unsaturated esters is certainly feasible, where the desired unsaturated alcohol can be obtained with a high selectivity in high yield. This was demonstrated by complex **5**, bearing a tetradentate P_2N_2 ligand, which was shown to catalyze the selective hydrogenation of esters in the presence of remote C=C double bonds (Table 2-2) [14a]. The study showed that the number of substituents on the C=C double bond has a crucial role on the chemoselectivity; while esters containing cyclic or acyclic di- and trisubstituted C=C double bonds were reduced to their corresponding unsaturated alcohols in high yield and with excellent chemoselectivity (>98:2, entries 1–3), reactions with esters containing a terminal C=C double bond were much less selective (entry 4). α,β-Unsaturated esters are particularly challenging substrates, where a fast conjugated reduction was observed (entry 5).

With access to chiral ruthenium complexes **6–9** that can hydrogenate esters in high yield, the possibility of performing kinetic resolution of a chiral ester was envisaged. Racemic esters with an α-stereogenic center were first tested, taking advantage of the basic reaction conditions required for the hydrogenation. In the presence of a strong base, the ester bearing an α-CH bond ($pK_a \approx 24$–30 in DMSO) could racemize rapidly, thereby allowing dynamic kinetic resolution (DKR) to occur [19]. The hydrogenation of 2-methyl-phenylacetic esters was thus attempted with different chiral $RuCl_2$(*R*)-BINAP) (N^N) complexes under basic conditions, and the results are shown in Scheme 2-1 (C.M. Saudan, L.A. Saudan, Firmenich, Geneva, unpublished results).

TABLE 2-2. Hydrogenation of Unsaturated Esters Catalyzed by Complex 5[a]

Entry	Starting Ester	Conversion (%)	Unsaturated Alcohol	Saturated Alcohol	Ratio[b]	Yield (%)[c]
1	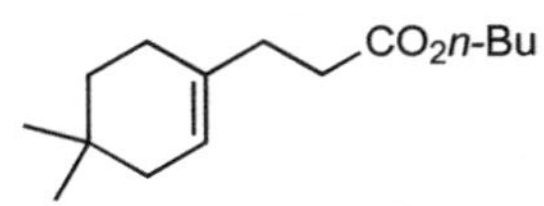	>99	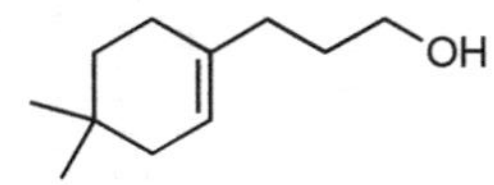	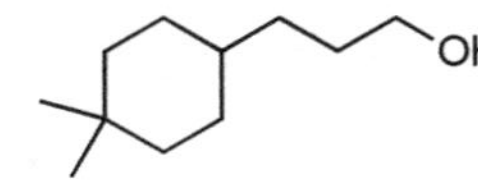	>98:2	95
2	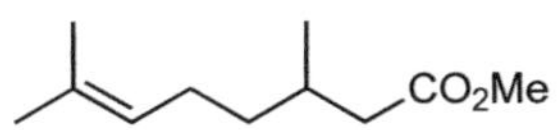	99	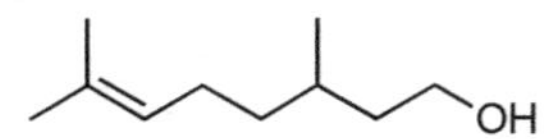	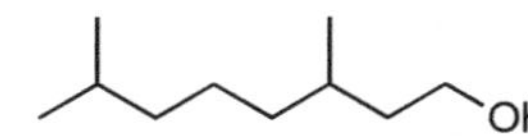	99:1	94
3	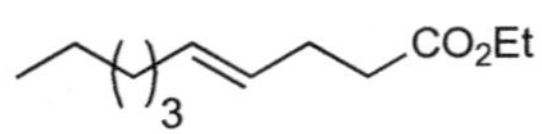	>99	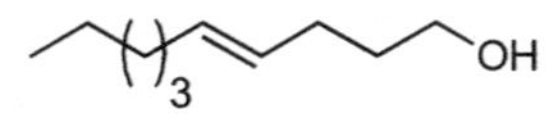	n-$C_{10}H_{21}OH$	99:1	93
4	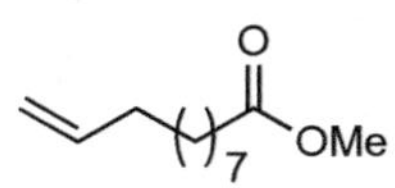	>99		n-$C_{11}H_{23}OH$	35:65[d]	94
5	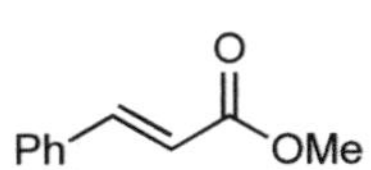	>99	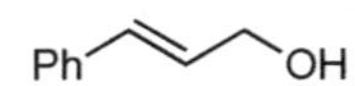	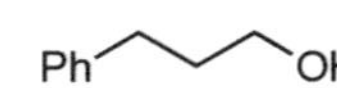	12:88	87

[a]Conditions: ester (20 mmol), (**5**) (0.05 mol%), NaOMe (5 mol%), THF (10 ml), H_2 (50 bar), 100 °C, 2.5 h, unless otherwise indicated.
[b]The product distribution (unsaturated alcohol/saturated alcohol) as determined by GC analysis of the crude reaction mixture.
[c]Yield of the mixture of unsaturated alcohol/saturated alcohol after isolation by chromatography or distillation.
[d]The product contained the isomerized double bond at position −9 (6%).

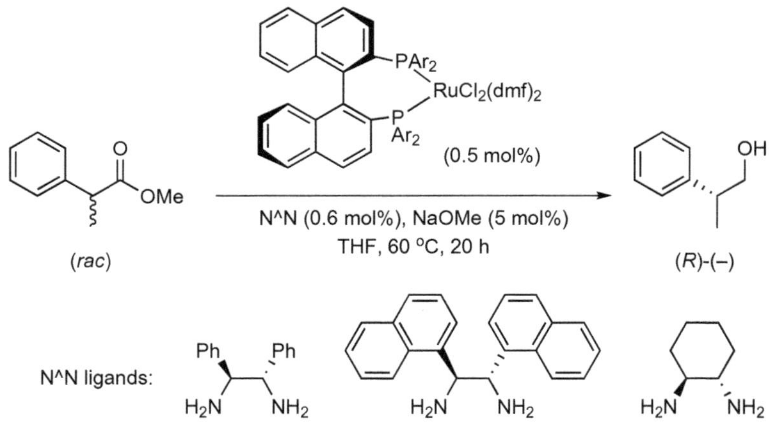

P^P ligands:			
Ar = Ph:	93% conv. 58% ee	98% conv. 58% ee	93% conv. 42% ee
Ar = *3,5*-Xyl:	99% conv. 45% ee	>99% conv. 39% ee	99% conv. 42% ee

SCHEME 2-1. Examples of kinetic resolution during hydrogenation of a chiral ester.

The desired enantiomerically enriched chiral alcohol was obtained with a high conversion, by using a combination of either (*R*)-BINAP or (*R*)-Xyl-BINAP with a disubstituted chiral 1,2-diamine in the catalyst. Interestingly, the ester substituent was found to influence the selectivity of the process; the enantiomeric excess increased from 46% to 60% by replacing the methyl ester by an isopropyl (Scheme 2-2). Although the enantioselectivity observed had so far remained modest, the feasibility of preparing

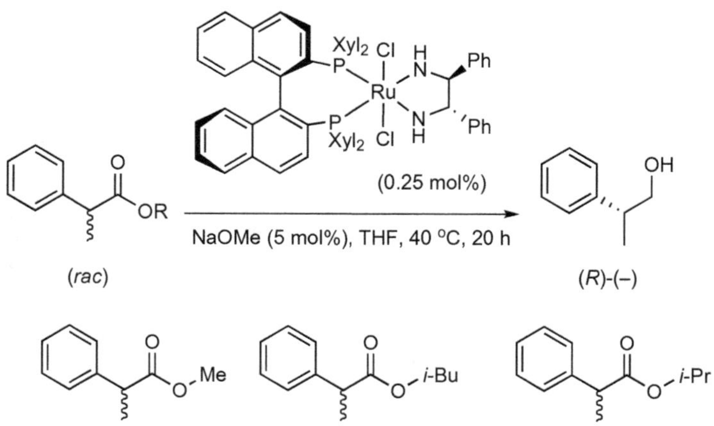

99% conv., 46% ee   95% conv., 52% ee   98% conv., 60% ee

SCHEME 2-2. Influence of the ester substituent on the DKR selectivity.

enantiomerically enriched chiral β-branched alcohols by hydrogenation of their corresponding esters was demonstrated.

2.3 HYDROGENATION OF LACTONES

Cyclic ethers are important perfumery ingredients, some of the most well-known and used being Ambrox®, Galaxolide®, Theaspirane, Rose Oxide, and Rhubafuran® (Figure 2-5) [20].[1] As shown from these examples, their olfactive descriptors are very diverse, ranging from ambergris to floral. The synthesis of these cyclic ethers relies, in most cases, on an acid-catalyzed cyclization from the corresponding diol, or the unsaturated alcohol [1b,d].

The acid-catalyzed cyclization of diols to form cyclic ethers [21] is an important step in the synthesis of THF from 1,4-butanediol (BDO) (Reppe process) (Scheme 2-3) [3]. THF is a commodity organic solvent produced at the multi-tonne scale, for the preparation of elastomers (e.g., LYCRA®)[2] from its polymerization to polytetramethylene glycol, and also for its use as a solvent in the polymer and pharmaceutical industry [22]. Recently, 2-MeTHF (2-methyl-THF) has emerged as an alternative organic solvent to THF [23]. It is also more eco-friendly, as it can be obtained from renewable sources (e.g., cellulose from agricultural plant wastes) by complete hydrogenation of 2-furaldehyde [24], or by successive hydrogenation/acid-catalyzed cyclization of levulinic acid using heterogeneous catalysts [25]. Similarly, 3-MeTHF is also emerging as an attractive substitute of THF, which can be obtained from renewable sources such as itaconic acid (IA) by using successive heterogeneous hydrogenation/acid-catalyzed cyclization [26], or from the hydrogenation of 3-furaldehyde [27]. 3-MeTHF can also be obtained by the direct hydrogenation of 2-methyl-γ-butyrolactone (MGBL) with an acid-supported catalyst [28,29].

A convenient method to prepare substituted 1,4-diols is by hydrogenation of the corresponding substituted γ-lactones generated in these processes (Scheme 2-3). For example, the group of Leitner [30] reinvestigated the use of homogeneous ruthenium–phosphine complexes [31] for the one-pot hydrogenation of levulinic acid directly to 2-MeTHF (Scheme 2-4). By modifying different components of the catalytic system (the phosphine ligands, the ionic and acidic additives), as well as the temperature and H_2 pressure, the authors were able to direct the selectivity from the possible intermediates,

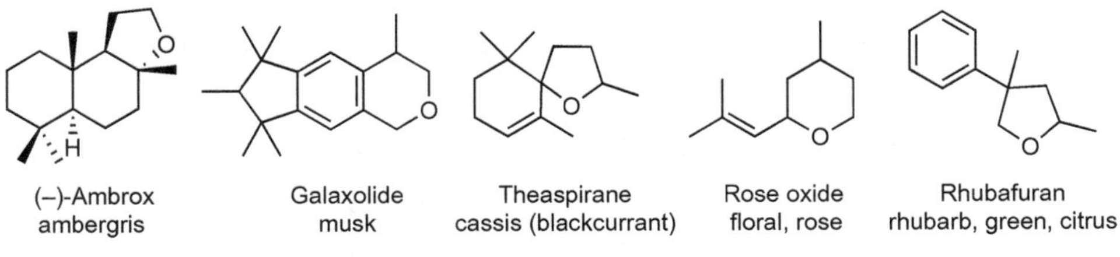

FIGURE 2-5. Examples of cyclic ethers used in perfumery.

[1] *Ambrox*® is a registered trademark of *Firmenich SA*; *Galaxolide*® is a registered trademark of *IFF*, and *Rhubafuran*® is a registered trademark of *Givaudan SA*.
[2] LYCRA® is a registered trade name of *Invista*.

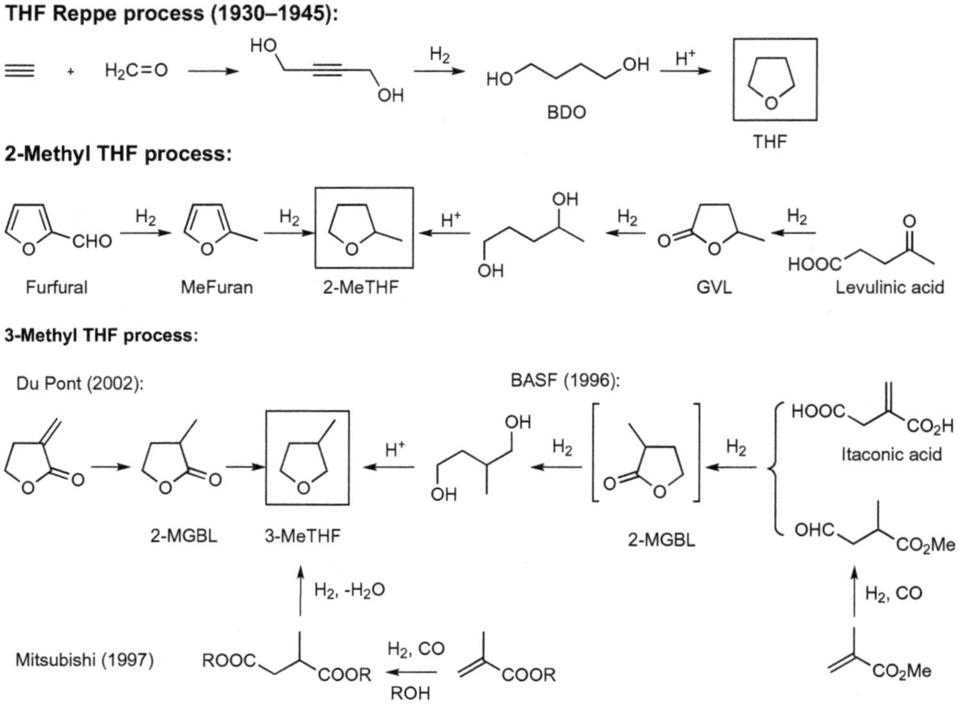

SCHEME 2-3. Processes to prepare THF and its 2-methyl and 3-methyl analogs.

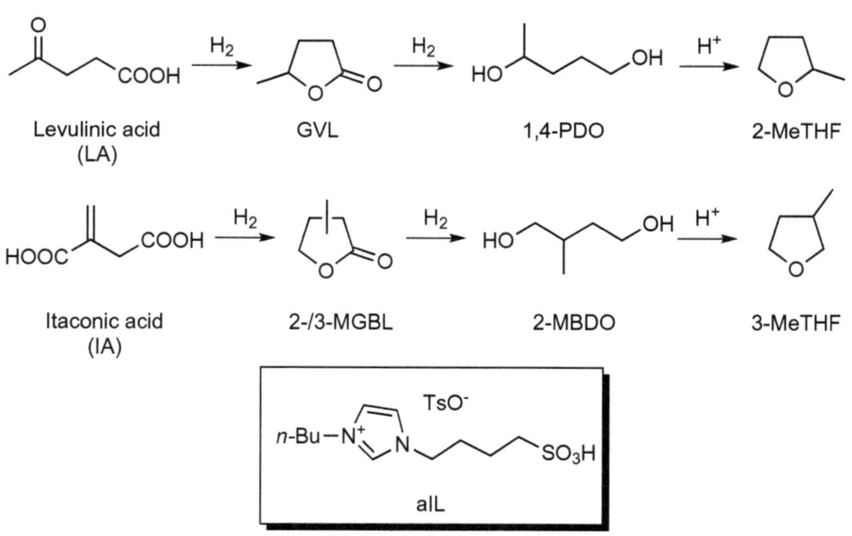

SCHEME 2-4. Hydrogenation processes to 2-methyl and 3-methyl THF.

TABLE 2-3. Effect of Changing Reaction Parameters on the Hydrogenation of Levulinic Acid to 2-MeTHF

Entry	Ligand	T (°C)	Additive	GVL (%)	1,4-PDO (%)	2-MeTHF (%)
1	$P(n\text{-}Oct)_3$	160	NH_4PF_6	>99	0	0
2	Triphos	160	–	3	95	0
3	Triphos	160	NH_4PF_6 + aIL	1	0	92

Conditions: levulinic acid (10 mmol), $[Ru(acac)_3]$ (0.1 mol%), $P(n\text{-}Oct)_3$ (1 mol%), Triphos (0.2 mol%), additive (1 mol%), H_2 (100 bar), 18 h.

TABLE 2-4. Influence of Changing Reaction Parameters on the Hydrogenation of Itaconic Acid to 3-MeTHF

Entry	Ligand	Solvent	T (°C)	Additive	MGBL (%)	2-MBDO (%)	3-MeTHF (%)
1	DPPB	THF	195	–	93	2	0
2	Triphos	Dioxane	195	–	6	93	0
3	Triphos	Dioxane	195	NH_4PF_6 + pTsOH	0	0	97

Conditions: IA (2.4 mmol), $[Ru(acac)_3]$ (0.5 mol%), Triphos (0.6 mol%), additive (2 mol%), solvent (2 ml), H_2 (100 bar), 18 h.

for example, γ-lactones and 1,4-diols, toward the formation of 2-MeTHF (Table 2-3). The system employed a ruthenium complex with the Triphos ligand [8] in the presence of an acidic additive (aIL), allowing for the rapid hydrogenation of levulinic acid into 2-MeTHF in high yield (92%).

Encouraged by these results, the authors tested their catalytic system for the more challenging hydrogenation of methyl succinic acid (MSA), generated by the facile C=C bond hydrogenation of IA, into 3-MeTHF (Table 2-4). Only a slightly higher reaction temperature was necessary to achieve a full conversion of IA into MGBL, using DPPB as a ligand (entry 1). Here too, the Triphos ligand was beneficial in generating a more active catalyst, and the reaction proceeds further to the diol stage, with the formation of 2-methyl butanediol (2-MBDO) in high yield (93%, entry 2). Finally, dehydrative cyclization was also possible with the addition of acidic additives (pTsOH + NH_4PF_6), producing 3-MeTHF in high yield (97%, entry 3).

Despite significant advances, the reaction conditions needed to hydrogenate substituted γ-lactones for the preparation of substituted THF are quite drastic ($H_2 > 50$ bar, $T > 100$ °C) and are not tolerant of other functional groups (C=O and C=C bonds are hydrogenated). Therefore, more active ruthenium complexes are required for the preparation of more elaborated products. To this end, the ruthenium complexes $RuCl_2(DPPEA)_2$ (**4**) (DPPEA = 2-(diphenylphosphino)ethylamine) and $RuCl_2(P_2N_2)$ (**5**) (Figure 2-4) were highly efficient in the hydrogenation of aliphatic and aromatic lactones (Table 2-5) [14]. In these cases, the desired diols were obtained in high yields with a high S/C ratio of 2000 in a short reaction time (3–4 h). Moreover, relatively mild reaction conditions (50 bar H_2, 100 °C) can be employed. For the reduction of lactones, the replacement of THF by toluene and NaOMe by KOMe was found to be beneficial, allowing the formation of the corresponding diol in high yield.

Other $RuCl_2$(P^P)(N^N) complexes were also shown to catalyze the hydrogenation of lactones to diols. Bergens showed that complex **8** catalyzed the hydrogenation of five- and seven-membered lactones in high yields (Table 2-6) [18]. The reactions could be run at

TABLE 2-5. Hydrogenation of Lactones with $[RuCl_2(dppea)_2]$ (4) and $[RuCl_2(P2N2)]$ (5) Complexes[a]

Entry	Starting Ester	[Ru]	*t* (h)	Conversion (%)	Diol	Yield (%)[b]
1	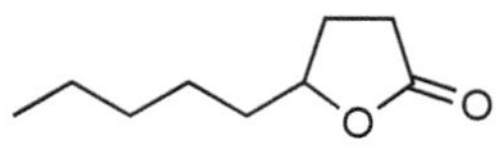	**4** **5**	4 4	98 98	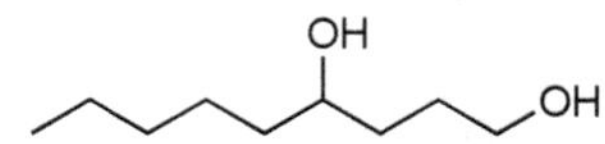	87 91
2	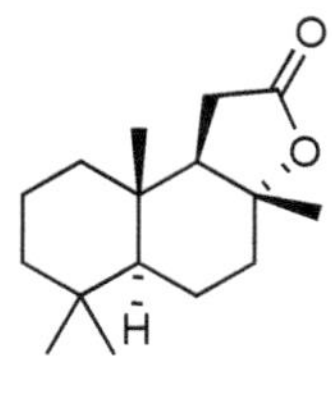	**4**	6	>99	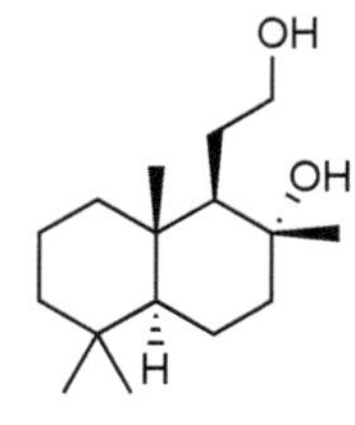	91[c]
3	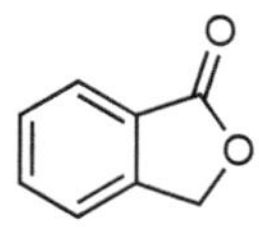	**4** **5**	4 4	96 98	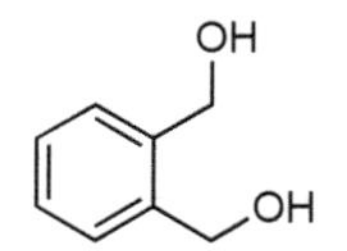	80 87
4	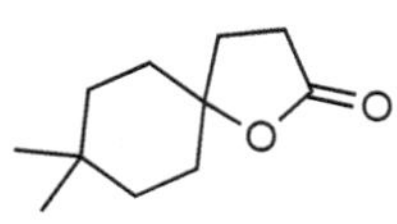	**4** **5**	4 4	96 99	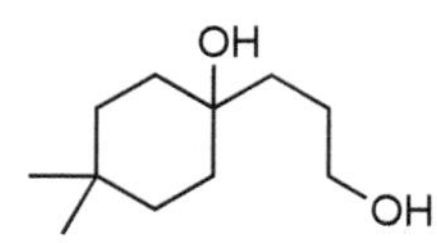	80 85
5	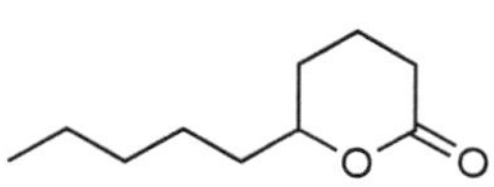	**4** **5**	4 4	99 97	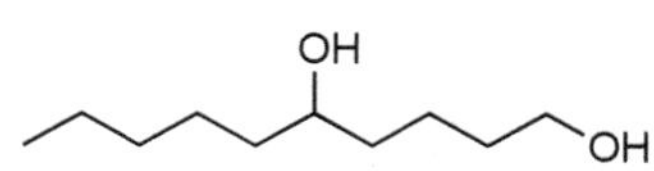	93 86
6[d]	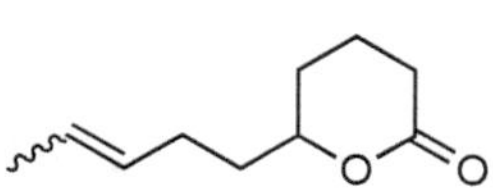	**5**[e]	3	>99	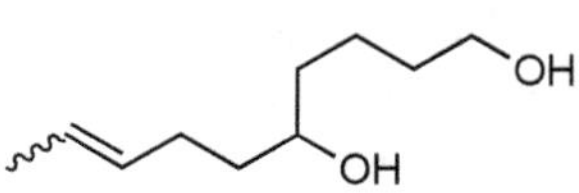	85[f]

[a]Conditions (unless otherwise indicated): ester (20 mmol), [Ru] (0.05 mol%), KOMe (10 mol%), THF (14 ml), H_2 (50 bar), 100 °C.

[b]Diol isolated by chromatography or distillation, unless otherwise indicated.

[c]H_2 (30 bar).

[d]Starting material is an *E*/*Z* mixture (85/15).

[e]KOMe (5 mol%) and THF (10 ml) were used.

[f]Yield of mixture of unsaturated alcohol/saturated alcohol after isolation by chromatography. The product distribution (unsaturated alcohol/saturated alcohol) of 98.5/1.5 was determined by GC analysis of the crude reaction mixture.

near-ambient temperature (30 °C) with 1 mol% of catalyst under low H_2 pressure (4 bar), which was attributed to slower product inhibition of the catalyst by lactones, compared to acyclic esters.

Ruthenium complexes of ligands containing the ancillary Cp* (pentamethylcyclopentadienyl) and a protic amine, that is, $Cp^*Ru(L\text{-}NH_2)Cl$, have been shown to catalyze the hydrogenation of aromatic esters [32] and lactones in the presence of an excess inorganic base, KO*t*-Bu (base/catalyst = 25) [33]. This allowed the authors to expand the substrate scope to include polar functional groups [34]. The screening of several amino ligands ($L\text{-}NH_2$) revealed that the most active catalyst was derived from the 2-picolylamine (PICA). Using [Cp*Ru(PICA)Cl] **10** (1 mol%) and KO*t*-Bu (25 mol%), several five- and six-membered lactones were hydrogenated to their corresponding diols

TABLE 2-6. Hydrogenation of Lactones with Catalyst 8[a]

Entry	Starting Ester	t (h)	Diol	Yield (%)[b]
1	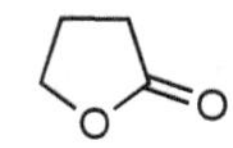	3	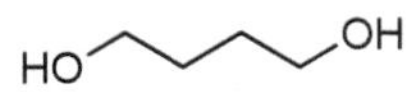	>99
2	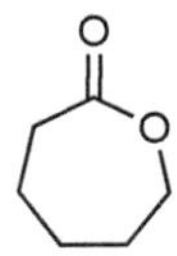	3	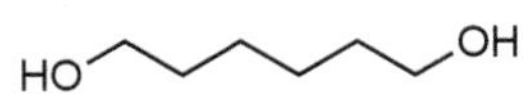	>99
3	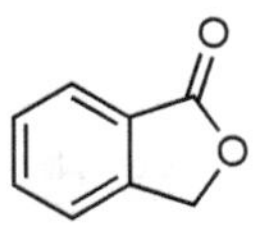	4	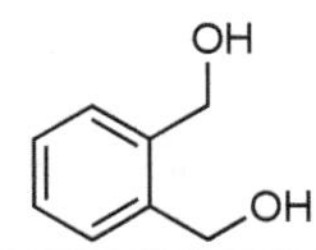	92

[a]Conditions: ester (1 mmol), [Ru] (1 mol%), KOt-Bu (9 mol%), THF (2 ml), H_2 (4 bar), 30 °C.
[b]Yield determined by ^1H NMR spectroscopy.

in high yields in 2-propanol (Table 2-7). However, it is noted that under the same conditions (100 °C, 50 bar H_2), catalyst **10** is less efficient (≈100×) than the ruthenium complexes **4** and **5** bearing the aminophosphine ligands, for example, 5-pentyl-γ-butyrolactone was reduced by **4** and **5** in 4 h instead of 14 h, employing a higher S/C ratio of 2000 instead of 100 (see Tables 2-5 and 2-7).

TABLE 2-7. Hydrogenation of Lactones with [Cp*Ru(PICA)Cl], 10[a]

Entry	Starting Ester	t (h)	Conversion (%)	Diol	Yield (%)[b]
1	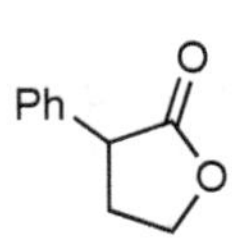	6	>99	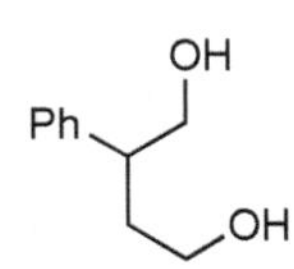	82
2	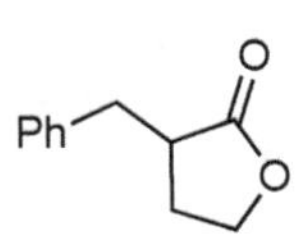	15	>99	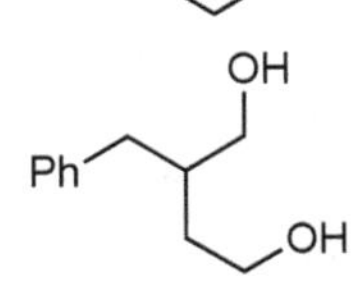	74
3	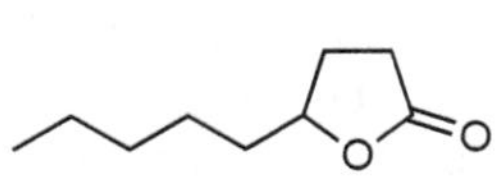	14	>99	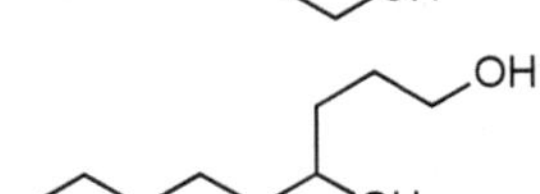	73
4	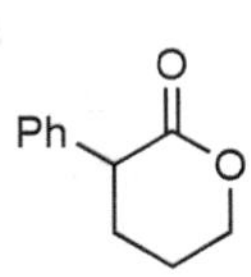	12	>99	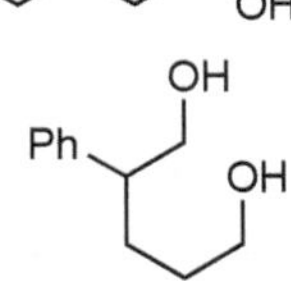	87
5	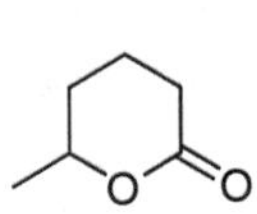	15	>99	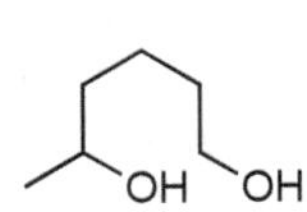	81

[a]Conditions: [Ru] (1 mol%), KOt-Bu (25 mol%), i-PrOH, H_2 (50 bar), 100 °C.
[b]Isolated yield.

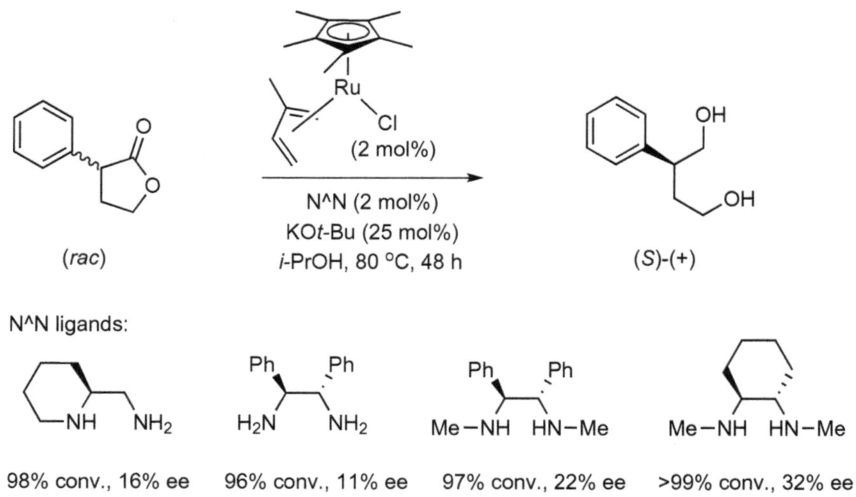

SCHEME 2-5. Examples of kinetic resolution in the hydrogenation of a chiral lactone.

Under basic conditions, an α-acidic C–H bond of the lactone substrate can be deprotonated. Recently, Ikariya and coworkers [33] exploited this feature to perform DKR of a racemic lactone using a chiral ruthenium complex. Despite the low enantioselectivity observed (Scheme 2-5), the reaction provides enantiomerically enriched chiral diols from racemic lactones.

2.4 HYDROGENATION OF AROMATIC ESTERS

Complexes **1–11** are able to hydrogenate aromatic esters under milder conditions (lower reaction temperature or lower H_2 pressure) or more efficiently (higher S/C ratio, shorter reaction time), compared to the Ru(acac)$_3$/Triphos catalytic system described by Elsevier and Teunissen [8]. The hydrogenation reactions of methyl and ethyl benzoate esters catalyzed by complexes **1–11** are summarized in Table 2-8, which allowed the reactivity of these complexes to be evaluated under various reaction conditions.

Generally, low hydrogen pressure (4–5 bar) required a high catalyst loading (S/C = 100) in order to achieve conversion in a reasonable period of time (1–4 h). Under these conditions, complex **8** catalyzed the hydrogenation of methyl benzoate at lower temperature (50 °C vs. 115 °C) than complex **1**, but only in the presence of a strong base (9 mol% of KO*t*-Bu) (Table 2-8, entries 1 and 9). Ethyl benzoate was also hydrogenated at low H_2 pressure (≈ 5 bar) in high yields with complexes **1–3** (entries 12–14). An average order of activity can be obtained from an average calculated TOF and is found to be: **2** ≈ **3** > **1**,[3] showing the beneficial replacement of the bis-*tert*-butylphosphine of **1** by the NHC ligand in **2**. Nevertheless, the experiments were not performed in the same solvent (dioxane vs. toluene), nor at the same temperature with **1**, **2**, and **3** (115, 105, and 135 °C, respectively).

[3] Calculated average TOF in h^{-1} at H_2 (≈5 bar) and 100–135 °C: 24 (**1** at 115 °C), 49 (**2** at 105 °C), 49 (**3** at 135 °C).

TABLE 2-8. Hydrogenation of Methyl and Ethyl Benzoate Esters

Entry	Starting Ester	[Ru] (mol%)	Additive (mol%)	pH_2 (bar)	T (°C)	t (h)	Conversion (%)[a]	Yield (%)[a]	Reference
1	$PhCO_2Me^b$	**1** (1)	–	5.4	115	4	>99	97	[10]
2	$PhCO_2Me^c$	**4** (0.05)	NaOMe (5)	50	100	1	>99	99	[14]
3	$PhCO_2Me^c$	**4** (0.05)	NaOMe (5)	30	100	1	97	96	[14]
4	$PhCO_2Me^c$	**4** (0.05)	NaOMe (5)	50	60	2	92	90	[14]
5	$PhCO_2Me^c$	**5** (0.05)	NaOMe (5)	50	100	1	>99	99	[14]
6	$PhCO_2Me^c$	**6** (0.05)	NaOMe (5)	50	100	1	77	70	[16]
7	$PhCO_2Me^c$	**6** (0.05)	NaOMe (5)	50	60	8	97	94	[16]
8	$PhCO_2Me^c$	**7** (0.05)	NaOMe (5)	50	60	20	78	73	[16]
9	$PhCO_2Me^d$	**8** (1)	KO*t*-Bu (9)	4	50	3	>99	>99	[18]
10	$PhCO_2Me^e$	**9** (0.2)	NaOMe (10)	50	100	3	–	97	[17]
11	$PhCO_2Me^f$	**11** (0.066)	KO*t*-Bu (0.53)	25	50	2	78	–	[32]
12	$PhCO_2Et^b$	**1** (1)	–	5.4	115	4	99	96	[10]
13	$PhCO_2Et^g$	**2** (1)	KO*t*-Bu (8)	5.3	105	2	>99	98	[11]
14	$PhCO_2Et^g$	**3** (1)	KO*t*-Bu (1)	5.5	135	2	97	97	[13]
15	$PhCO_2Et^g$	**3** (0.025)	KO*t*-Bu (0.025)	51	135	12	72	71	[13]
16	$PhCO_2Et^c$	**4** (0.05)	NaOMe (5)	50	100	1	>99	99	[14]

Conditions:
[a]Conversion and yield were determined by GC.
[b]Ester (2 mmol), dioxane (2 ml).
[c]Ester (20 mmol), THF (10 ml).
[d]Ester (1 mmol), THF (2 ml).
[e]Ester (4 mmol), THF (3 ml).
[f]Ester (6.9 mmol), THF (6 ml).
[g]Ester (1 mmol), toluene (2 ml).

At higher H_2 pressure (50 bar) and at 100 °C, the order of activity for the hydrogenation of methyl benzoate follows the trend: **4** ≈ **5** > **6** > **9** (entries 2, 5–6, and 10),[4] showing that the PN-ligated ruthenium complexes **4** and **5** are much more robust at high temperature than the $RuCl_2$(P^P)(N^N) complexes **6** and **9**. At lower temperature (60 °C), a high yield of benzyl alcohol was still observed with complexes **4** and **6** (entries 4 and 7), but a longer reaction time was necessary with complex **7** to achieve an acceptable yield (entry 8), probably due to the presence of the sterically more demanding diphosphine in complex **7** (Xyl-BINAP vs. BINAP). Based on average TOF, the following order was observed at 60 °C for the complexes: **4** > **6** > **7**.[5] With complex **11**, a lower yield (78%) of benzyl alcohol was obtained at 25 bar of H_2 (entry 11) [32], while on the other hand, complex **4** gave a much higher yield (96%) under a similar H_2 pressure (30 bar) but a higher temperature (100 °C vs. 50 °C) was needed (entry 3). Similarly, ethyl benzoate was readily hydrogenated (99%) with complex **4** at 100 °C

[4] Calculated average TOF in h^{-1} at H_2 (50 bar) and 100 °C: 1980 (**4**), 1980 (**5**), 1400 (**6**), 162 (**9**).
[5] Calculated average TOF in h^{-1} at H_2 (50 bar) and 60 °C: 900 (**4**), 235 (**6**), 73 (**7**).

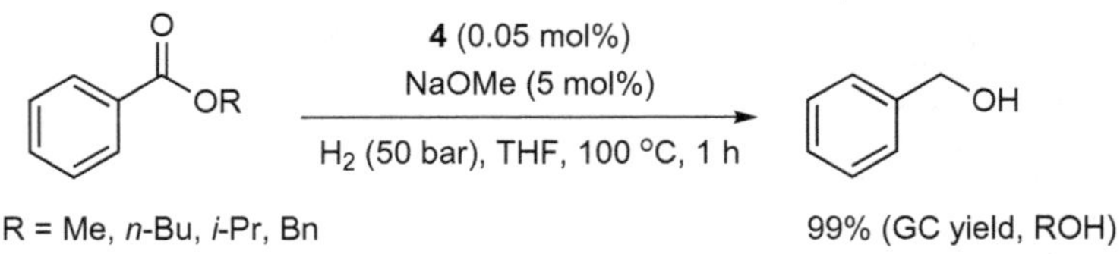

SCHEME 2-6. Effect of the ester substituent on the yield of the hydrogenation catalyzed by **4**.

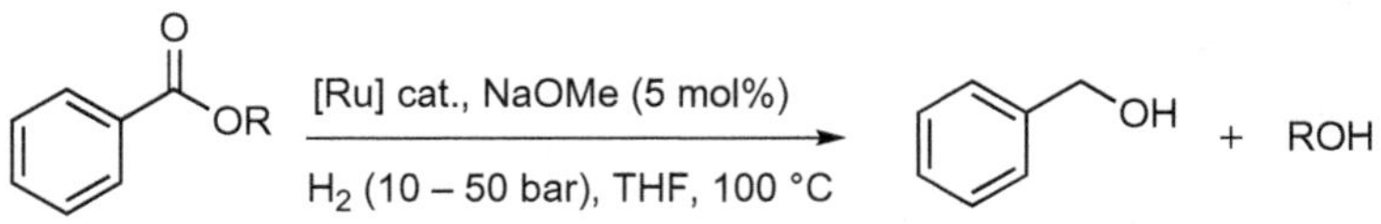

SCHEME 2-7. Effect of the ester substituent and H_2 pressure on the yield of the hydrogenation catalyzed by **4**.

under 50 bar of H_2 under similar conditions (entry 16), while complex **3** gave a lower yield of the alcohol (72%), but with a much lower amount of base (0.025 mol% vs. 5 mol%) (entry 15).

Using complex **4** as catalyst (0.05 mol%), the nature of the alkyl substituent on the benzoate esters had no effect on the product yield when the hydrogenation was conducted at 100 °C under 50 bar of H_2 (Scheme 2-6); in all cases, 99% yield was obtained.

However, for the hydrogenation of methyl benzoate, lower yields of the alcohol were observed when the reaction was performed with a lower amount of complex **4** (0.01 mol%), or at the lower H_2 pressure of 10 bar (Scheme 2-7, Table 2-9, entries 1–3) [14a]. This could be attributed to product [18] and/or to the amount of MeOH liberated during the reaction that could deactivate the catalyst, either by the formation of nonproductive carbonyl complexes or by an increase of the relative acidity of the reaction medium. With this in mind, the hydrogenation of the corresponding isopropyl benzoate was investigated, as the isopropyl alcohol formed should not deactivate the catalyst (no CO formation) and also has a higher pK_a than MeOH (27.9 and 29.3 for MeOH and *i*-PrOH, respectively) [35]. Indeed, full conversion was observed at either low catalyst loading (0.01 mol%) or at low H_2 (10 bar) for the hydrogenation (entries 4–6) [14].

TABLE 2-9. Comparative Examples of Hydrogenations Catalyzed by Complex 4 (Scheme 2-7)[a]

Entry	Starting Ester	**4** (mol%)	pH_2 (bar)	*t* (h)	Conversion (%)[b]	Yield (%)[b]
1	$PhCO_2Me$	0.05	50	1	>99	99
2	$PhCO_2Me$	0.01	50	4	92	89
3	$PhCO_2Me$	0.05	10	4	58	47
4	$PhCO_2i$-Pr	0.05	50	1	>99	99
5	$PhCO_2i$-Pr	0.01	50	4	>99	99
6	$PhCO_2i$-Pr	0.05	10	4	>99	99

[a]Conditions: esters (20 mmol), THF (10 ml).
[b]Conversion and yield determined by GC, using *n*-tridecane as an internal standard.

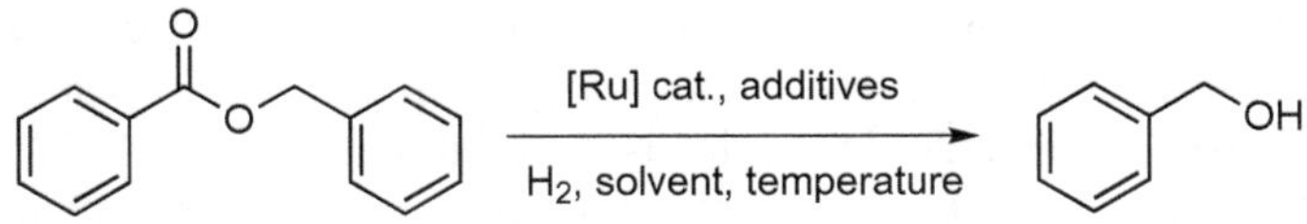

SCHEME 2-8. Hydrogenation of benzyl benzoate with various ruthenium complexes.

TABLE 2-10. Comparison of Catalytic Activities of Ru(acac)$_3$/Triphos, 1 and 5, in the Hydrogenation of Benzyl Benzoate (Scheme 2-8)

Entry	[Ru] (mol%)	Additives (mol%)	pH_2 (bar)	T (°C)	t (h)	Conversion (%)[a]	Yield (%)[a]	TOF (h^{-1})[b]
1	Ru(acac)$_3$[c] (0.05)	Triphos (0.06)	85	120	16	97	95	119
2	**5**[d] (0.05)	NaOMe (5)	85	120	1	>99	>99	2000
3	**1**[e] (1)	–	5.4	115	7	98.5	98	14
4	**5**[d] (0.05)	NaOMe (5)	10	115	7	>99	>99	286

[a]Conversion and yield (based on benzyl alcohol) were determined by GC.
[b]Average TOF = TON/time.
[c]Conditions: benzyl benzoate (28.4 mmol), Et_3N (9 mol%), $(CF_3)_2CHOH$ (15 ml).
[d]Conditions: benzyl benzoate (20 mmol), THF (10 ml).
[e]Conditions: benzyl benzoate (2 mmol), dioxane (2 ml).

In order to assess the efficiency of the P^N–ruthenium complexes, the hydrogenation of benzyl benzoate was performed with the tetradentate bis-imino–ruthenium complex **5**, under similar reaction conditions using Ru(acac)$_3$/Triphos [8] and complex **1** [10] (Scheme 2-8 and Table 2-10). The ruthenium complex **5** was shown to be the most effectiveness catalyst in this study, as it catalyzed the hydrogenation of benzyl benzoate to benzyl alcohol in a much shorter period of time under the conditions previously optimized for Ru(acac)$_3$/Triphos (1 h instead of 16 h, entries 1 vs. 2), even at a lower H_2 pressure (entries 1 vs. 4). Furthermore, a higher S/C ratio (2000 instead of 14) can be employed compared to complex **1** (entries 2 vs. 3).

2.5 HYDROGENATION OF FURANOIC ESTERS

Substituted furans [36] are particularly interesting, not only because they are found in several natural products, such as (−)-furodysin [37], dendrolasin [38], pallescensin-A [39], and one of the odoriferous constituent of rose oil, rosefuran [40] (Figure 2-6), but

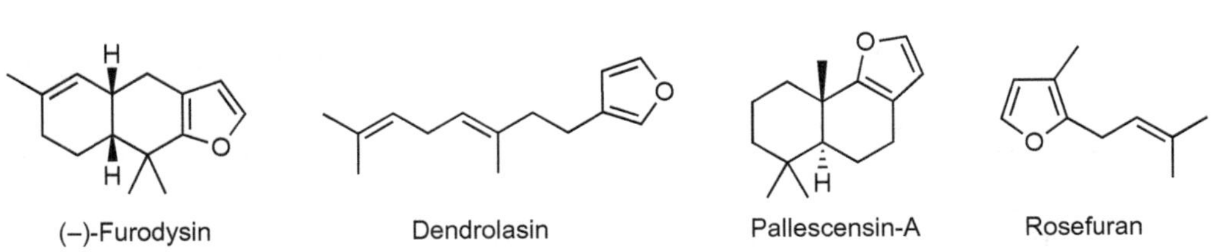

FIGURE 2-6. Examples of furan-containing natural products.

they can also serve as building blocks for the synthesis of more elaborated molecules by exploiting the heterocycle as a latent functional group equivalent [41]. For example, furan derivatives can undergo facile inter- and intramolecular [4+2] Diels–Alder reactions [41], or [4+3] cycloaddition reactions with rhodium carbenoids [42]. They can also be oxidized to 1,4-diketones with bromine [41], or upon oxidation of silylated furans, to give γ-hydroxybutenolides [43] or Δ^2-/Δ^3-butenolides [44], depending on the oxidant used. They can also serve as masked pyranones upon oxidation of the substituted furfurol derivatives [45].

Furthermore, as the furan ring can be easily hydrogenated with heterogeneous catalysts (Pd/C [46], PdO [47], Rh/C [48], Cu–Cr [24a]), it would be interesting to see if the ruthenium complexes known were able to selectively reduce furanoic esters [49]. In Table 2-11, the results obtained with complex **5** under the conditions described above for aromatic esters (S/C = 2000, S/B = 20, THF, 100 °C, 50 bar H_2) are shown (C.M. Saudan, L.A. Saudan, S.A. Firmenich, Geneva, unpublished results). The reactions proceeded smoothly, giving the desired furfuryl alcohols in excellent isolated yields (86–96%) without compromising the furan ring. The presence of a methyl group in the 3-position did not influence the rate of the reaction, and the ester was reduced with the same high yield (96%) (entries 1 and 2). On the other hand, 2-methyl-3-furanoic acid methyl ester was reduced at a slower reaction rate; nevertheless, the desired substituted 3-furfuryl alcohol was obtained in excellent yield (86%) after 6 h (entry 5). In contrast to

TABLE 2-11. Hydrogenation of Furan Esters with Complex 5[a]

Entry	Starting Ester	*t* (h)	Conversion (%)	Alcohol	Yield (%)[b]
1	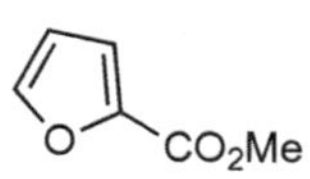	2	>99	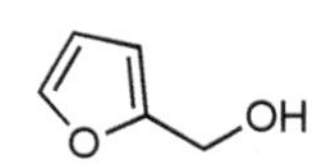	96
2	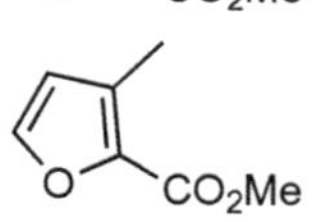	2	>99	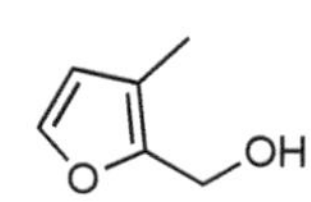	96
3[c]	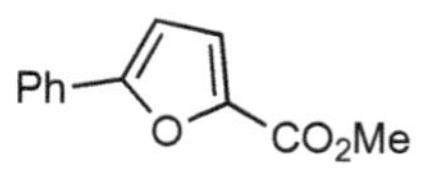	2	>99	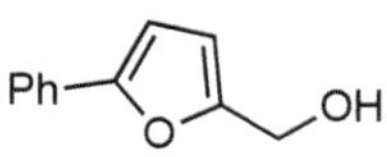	93
4[c]	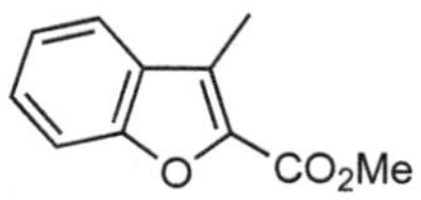	2	>99	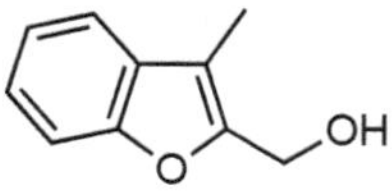	93
5	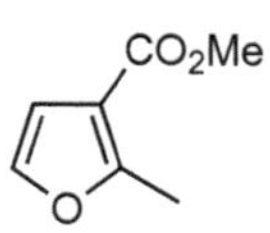	2 6	82 97	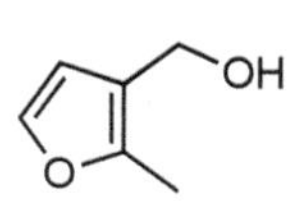	61 86
6	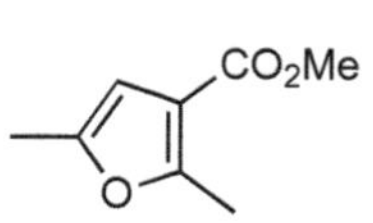	6	98[d]	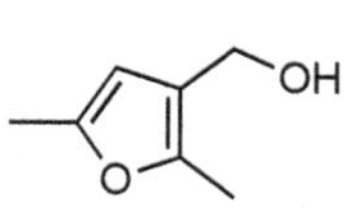	86

[a]Conditions (unless otherwise indicated): ester (20 mmol), [Ru] cat. (0.05 mol%), NaOMe (5 mol%), THF (10 ml), H_2 (50 bar), 100 °C.
[b]Yield of the isolated alcohol after chromatography or distillation.
[c]Reaction performed with ester (10 mmol) and THF (5 ml).
[d]2-Dihydrofuran alcohol was observed by GC (1%) to be present in the crude reaction mixture.

the 2-furanoic ester derivatives (entries 1–4), the 3-isomer showed some hydrogenation of the furan ring with the formation of 2-dihydrofuran alcohol (1%) (entry 6).

2.6 HYDROGENATION OF CHIRAL ESTERS (BASE-FREE CONDITIONS)

Following the work of Noyori, where the hydrogenation of base-sensitive ketones can be achieved using ruthenium hydrido-BH_4 complexes [50], the Takasago group reported the use of several hydrido-BH_4 complexes (Figure 2-7) that were able to catalyze the hydrogenation of esters without the addition of base [17,51]. This is particularly attractive for the hydrogenation of α-substituted chiral esters, which could racemize in the presence of base. Accordingly, chiral α-alkyl, α-*N*-protected amine (Boc), and α-*O*-protected hydroxyl (benzyl) substituted esters were hydrogenated in high yields, without any erosion of their optical purity (Table 2-12). Although the use of amine-

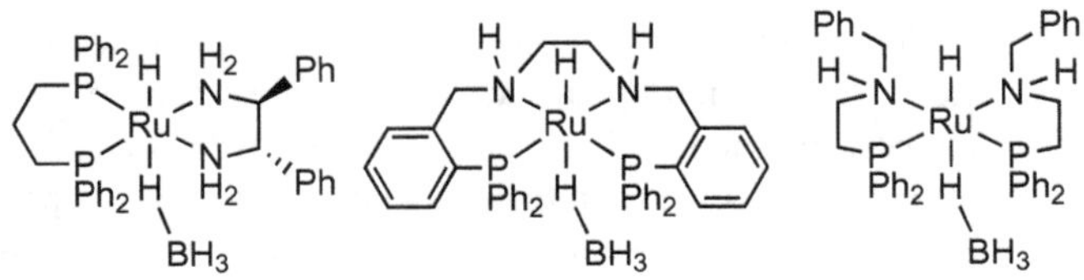

12 **13** **14**

FIGURE 2-7. Examples of ruthenium complexes reported for base-free hydrogenation of esters.

TABLE 2-12. Hydrogenation of Base-Sensitive Esters with $RuH(BH_4)(L)_n$ Complexes[a]

Starting Ester	[Ru] (mol%)	Conversion (%)	Alcohol	Yield[b] (%)	Δee	Reference
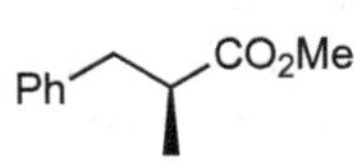	**12** (1)	–		95	1	[17]
	13 (0.2)	99		99[c]	<1	[51]
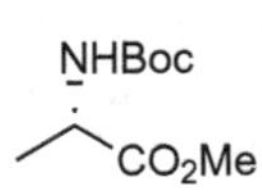	**12** (1)	–	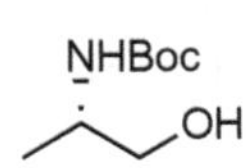	96	<1	[17]
	13 (0.6)	68		97[c]	–	[51]
	14 (0.2)	78		80[c]	<1	[51]
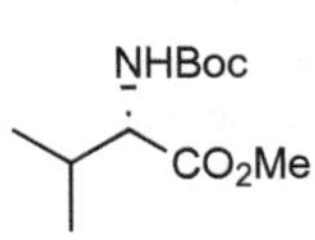	**12** (1)	–	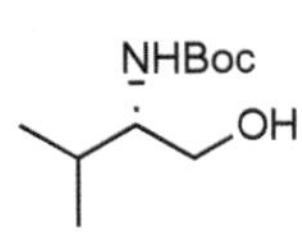	94	<1	[17]
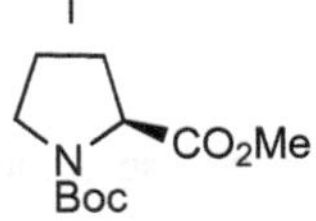	**12** (1)	–	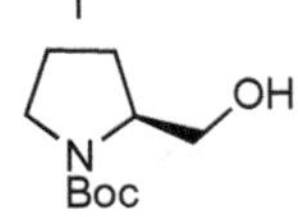	89	<1	[17]
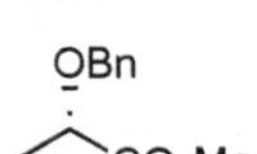	**12** (1)	–	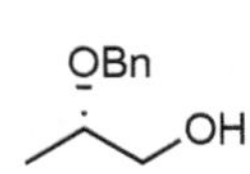	90	<1	[17]

[a]Conditions: ester (5 mmol), THF (2 ml), H_2 (50 bar), 80 °C, 16 h.
[b]Isolated yield, unless otherwise indicated.
[c]Selectivity defined as ratio (%) of GC area (%) of product versus all other products, except the starting material.

protecting groups (Boc) cannot be avoided, this method allows chiral amino alcohols to be obtained in high yield and in high enantiomeric purity from chiral amino acids without the use of stoichiometric hydride reagents and associated tedious workup.

2.7 CONCLUSIONS

Catalytic hydrogenation of esters is a very attractive methodology to replace the use of hazardous hydride reagents based on aluminum, silicon, and boron, for the reduction of esters on a large scale. By using hydrogen as a gaseous reagent, the manipulation of stoichiometric reagents, and the subsequent need for a tedious workup procedure to recover the desired product from the generated inorganic salts, can be avoided. The field has shown tremendous success in the last three decades, starting from the early examples of Grey and Pez [52], to the new catalysts based on Cp* [32,33] and NHC ligands [11,13]. However, despite the increasing number of catalysts reported for the hydrogenation of esters to alcohols, industrial implementation of such processes remains to be accomplished, which will depend on several factors, including catalyst cost, catalyst activity, productivity, and selectivity. To date, the best reported TON for such catalysts is between 2000 and 10,000, which is not sufficient to justify their use in the production of low-value products on large scale where a TON >50,000 is usually needed, but could be enough for high-value products where a TON > 1000 may suffice [53]. The measure of the catalyst productivity, given by the TOF (h^{-1}), is also an important factor for their use on large scale, as the occupation time of the reactor (cost/h) could dictate the choice of the process. The best catalytic processes for the hydrogenation of esters have TOF around 2000 h^{-1}, which are again insufficient for large-scale production (TOF > 10,000 h^{-1}), but certainly enough for small scale, high-value products [53]. The TOF is sometimes also related to the hydrogen pressure at which the reaction is performed (higher rate with higher H_2 pressure), so a compromise between the cost of the equipment (higher cost with higher pressure equipment) and the rate has to be found. The selectivity of the catalyst, more specifically its ability to tolerate other functional groups, is also of prime importance; even more so if the catalyst is to be used to reduce esters in high-value chemicals that incorporate several functional groups. So far, the substrate scope for the hydrogenation of esters with ruthenium catalysts has been focused merely on simple aliphatic and aromatic esters; for a future industrial implementation of these catalysts, the scope of esters has to be widened to more complex substrates used in the pharmaceutical and agrochemical industries.

Another important aspect for the successful implementation of a hydrogenation process on the industrial scale is the robustness of the process toward the purity of the substrate, the quality of the reagents (solvent, hydrogen, cocatalysts), and also the autoclave [54]. Usually, trace impurities, especially acidic ones, are deleterious for catalytic hydrogenation where a basic cocatalyst is used, and prior distillation of the feedstocks over base is usually required [54].

As cost and environmental constraints are increasing on industrial processes, the development of catalytic processes for the industrial hydrogenations of esters would become more important. Challenges still remain to increase the reactivity and productivity of the catalysts in order to make them cost-effective and also to improve their tolerance toward other functional groups, despite encouraging results that have already been achieved with unsaturated esters [14a].

REFERENCES

1. For reviews on the chemistry of odorants, see (a) Mimoun H (1996). Catalytic opportunities in the flavor and fragrance industry. *Chimia* 50, 620–625. (b) Fráter G, Bajgrowicz JA, Kraft P (1998). Fragrance chemistry. *Tetrahedron* 54, 7633–7703. (c) Yamamoto T. Advances in the transition metal (Rh, Ru)-BINAP-catalyzed asymmetric synthesis of chiral compounds for flavours and fragrances and their associated sensory properties. In: Swift KAD, editor. *Current Topics in Flavour and Fragrances*. Takasago Int. Co.; 1999, pp 33–58. (d) Kraft P, Bajgrowicz JA, Denis C, Fráter G (2000). Odds and trends: recent developments in the chemistry of odorants. *Angew. Chem., Int. Ed.* 39, 2980–3010. (e) Chapuis C, Jacoby D (2001). Catalysis in the preparation of fragrances and flavours. *Appl. Catal. A: Gen.* 221, 93–117. (f) Swift KAD (2004). Catalytic transformations of the major terpene feedstocks. *Top. Catal.* 27, 143–155. (g) Saudan LA (2007). Hydrogenation processes in the synthesis of perfumery ingredients. *Acc. Chem. Res.* 40, 1309–1319. (h) Birkbeck AA. Challenges in the synthesis of natural and non-natural volatiles. In: Herrmann A, editor. *The Chemistry and Biology of Volatiles*. Wiley, Chichester; 2010, pp 173–202.
2. (a) Adkins H, Folkers K (1931). The catalytic hydrogenation of esters to alcohols. *J. Am. Chem. Soc.* 53, 1095–1097. (b) Pouilloux Y, Autin F, Barrault J (2000). Selective hydrogenation of methyl oleate into unsaturated alcohols. Relationships between catalytic properties and composition of cobalt-tin catalysts. *Catal. Today* 63, 87–100 and references therein.
3. Weissermel K, Arpe HJ. *Industrial Organic Chemistry*. VCH Publishers, New York; 1993.
4. (a) Ho LTT. *Formulating Detergents and Personal Care Products*. AOCS Press, New York; 2000. (b) Hayes DG. Biobased surfactants: Overview and industrial state of the art. In: Hayes DG, Kitamoto D, Solaiman DKY, Ashby RD, editors. *Biobased Surfactants and Detergents*. AOCS Press, Urbana; 2009, pp 3–25.
5. (a) Minnaard AJ, Feringa BL, Lefort L, De Vries JG (2007). Asymmetric hydrogenation using monodentate phosphoramidite ligands. *Acc. Chem. Res.* 40, 1267–1277. (b) Roseblade SJ, Pfaltz A (2007). Iridium-catalyzed asymmetric hydrogenation of olefins. *Acc. Chem. Res.* 40, 1402–1411. (c) Oro LA, Carmona D. Rhodium. In: de Vries JG, Elsevier CJ, editors. *The Handbook of Homogeneous Hydrogenation*, Wiley-VCH, Weinheim; 2007, pp 3–30.
6. (a) Noyori R, Ohkuma T (2001). Asymmetric catalysis by architectural and functional molecular engineering: practical chemo- and stereoselective hydrogenation of ketones. *Angew. Chem., Int. Ed.* 40, 40–73. (b) Ohkuma T, Noyori R. Enantioselective ketone and β-keto ester hydrogenations (including mechanisms). In: de Vries JG, Elsevier CJ, editors. *The Handbook of Homogeneous Hydrogenation*, Wiley-VCH, Weinheim; 2007, pp 1105–1163. (c) Blaser HU, Spindler F, Thommen M. Industrial applications. In: de Vries, JG, Elsevier CJ, editors. *The Handbook of Homogeneous Hydrogenation*, Wiley-VCH, Weinheim; 2007, pp 1279–1324. (d) Blaser HU, Malan C, Pugin B, Spindler F, Steiner H, Studer M (2003). Selective hydrogenation for fine chemicals: recent trends and new developments. *Adv. Synth. Catal.* 345, 103–151. (e) de Koning PD, Jackson M, Lennon IC (2006). Use of achiral (diphosphine) $RuCl_2$(diamine) precatalysts as a practical alternative to sodium borohydride for ketone reduction. *Org. Process Res. Dev.* 10, 1054–1058. (f) Tanielyan SK, Marin N, Alvez G, Augustine RL (2006). Two efficient enantioselective syntheses of 2-amino-1-phenylethanol. *Org. Process Res. Dev.* 10, 893–898. (g) Johnson NB, Lennon IC, Moran PH (2007). Industrial-scale synthesis and applications of asymmetric hydrogenation catalysts. *Acc. Chem. Res.* 40, 1291–1299. (h) Ito M, Ikariya T (2007). Catalytic hydrogenation of polar organic functionalities based on Ru-mediated heterolytic dihydrogen cleavage. *Chem. Commun.* 5134–5142. (i) Tsutsumi K, Katayama K, Utsumi N, Murata K, Arai N, Kurono N, Ohkuma T (2009). Practical asymmetric hydrogenation of 3-quinuclidinone catalyzed by the XylSkewphos/PICA-ruthenium(II) complex. *Org. Process Res. Dev.* 13, 625–628.

7. For a review on previous ruthenium catalyzed hydrogenation of esters, see Clarke ML, Roff GJ. Homogeneous hydrogenation of aldehydes, ketones, imines and carboxylic acid derivatives: chemoselectivity and catalytic activity. In: de Vries JG, Elsevier CJ, editors. *The Handbook of Homogeneous Hydrogenation*, Wiley-VCH, Weinheim; 2007, pp 413–454.
8. (a) Teunissen HT, Elsevier CJ (1997). Ruthenium catalyzed hydrogenation of dimethyl oxalate to ethylene glycol. *Chem. Commun.* 667–668. (b) Teunissen HT, Elsevier CJ. (1998). Homogeneous ruthenium catalyzed hydrogenation of esters to alcohols. *Chem. Commun.* 1367–1368. (c) van Engelen MC, Teunissen HT, de Vries JG, Elsevier CJ (2003). Suitable ligands for homogeneous ruthenium-catalyzed hydrogenolysis of esters. *J. Mol. Catal. A: Chem.* 206, 185–192.
9. For the use of other tripodal ligands in ester hydrogenation, see (a) Boardman B, Hanton MJ, van Rensburg H, Tooze RP (2006). A tripodal sulfur ligand for the selective ruthenium-catalyzed hydrogenation of dimethyl oxalate. *Chem. Commun.* 2289–2291. (b) Hanton MJ, Tin S, Boardman BJ, Miller P (2011). Ruthenium-catalyzed hydrogenation of esters using tripodal phosphine ligands. *J. Mol. Catal. A: Chem.* 346, 70–78.
10. (a) Zhang J, Leitus G, Ben-David Y, Milstein D (2006). Efficient homogeneous catalytic hydrogenation of esters to alcohols. *Angew. Chem., Int. Ed.* 45, 1113–1115. (b) Milstein D (2010). Discovery of environmentally benign catalytic reactions of alcohols catalyzed by pyridine-based pincer Ru complexes, based on metal cooperation. *Top. Catal.* 53, 915–923.
11. Sun YS, Koehler C, Tan RY, Annibale VT, Song DT (2011). Ester hydrogenation catalyzed by Ru-CNN pincer complexes. *Chem. Commun.* 47, 8349–8351.
12. Balaraman E, Gnanaprakasam B, Shimon LJW, Milstein D (2010). Direct hydrogenation of amides to alcohols and amines under mild conditions. *J. Am. Chem. Soc.* 132, 16756–16758.
13. Fogler E, Balaraman E, Ben-David Y, Leitus G, Shimon LJW, Milstein D (2011). New CNN-type ruthenium pincer NHC complexes. Mild, efficient catalytic hydrogenation of esters. *Organometallics* 30, 3826–3833.
14. (a) Saudan LA, Saudan CM, Debieux C, Wyss P (2007). Dihydrogen reduction of carboxylic esters to alcohols under the catalysis of homogeneous ruthenium complexes: high efficiency and unprecedented chemoselectivity. *Angew. Chem., Int. Ed.* 46, 7473. (b) Saudan L, Dupau P, Riedhauser JJ, Wyss P, inventors; Firmenich SA, assignee. Hydrogenation of esters with Ru/bidentate ligands complexes. World Patent WO2006/106483. April 5, 2005. (c) Saudan L, Dupau P, Riedhauser JJ, Wyss P, inventors; Firmenich SA, assignee. Hydrogenation of esters with Ru/tetradentate ligands complexes. World Patent WO2006/106484. April 5, 2005.
15. (a) For the preparation of complex **4**, see Abdur-Rashid K, Guo RW, Lough AJ, Morris RH, Song DT (2005). Synthesis of ruthenium hydride complexes containing beta-aminophosphine ligands derived from amino acids and their use in the H_2-hydrogenation of ketones and imines. *Adv. Synth. Catal.* 347, 571–579. (b) For the preparation of complex **5**, see Gao JX, Wan HL, Wong WK, Tse MC, Wong WT (1996). Synthesis and characterization of iron(2+) and ruthenium(2+) diimino-, diamino, and diamido-diphosphine complexes. X-ray crystal structure of *trans*-$RuCl_2(P_2N_2C_2H_4)\cdot CHCl_3$. *Polyhedron* 15, 1241–1251.
16. Saudan L, Saudan C, inventors; Firmenich SA, assignee. Hydrogenation of esters with Ru/bidentate ligands complexes. World Patent WO2008/065588. November 27, 2007.
17. (a) Ino Y, Yoshida A, Kuriyama W, inventors; Takasago Int. Corp., assignee. Method for producing alcohols. European Patent EP1970360. March 10, 2008. (b) Kuriyama W, Ino Y, Ogata O, Sayo N, Saito T (2010). A homogeneous catalyst for reduction of optically active esters to the corresponding chiral alcohols without loss of optical purities. *Adv. Synth. Catal.* 352, 92–96. (c) Ino Y, Kuriyama W, Ogata O, Matsumoto T (2010). An efficient synthesis of chiral alcohols via catalytic hydrogenation of esters. *Top. Catal.* 53, 1019–1024.
18. Takebayashi S, Bergens SH (2009). Facile bifunctional addition of lactones and esters at low temperatures. The first intermediates in lactone/ester hydrogenations. *Organometallics* 28, 2349–2351.

19. For the DKR of *alpha*-branched aldehydes during catalytic hydrogenation catalyzed by ruthenium complexes, see (a) Li XG, List B (2007). Catalytic asymmetric hydrogenation of aldehydes. *Chem. Commun.* 1739–1741. (b) Xie JH, Zhou ZT, Kong WL, Zhou QL (2007). Ru-catalyzed asymmetric hydrogenation of racemic aldehydes via dynamic kinetic resolution: efficient synthesis of optically active primary alcohols. *J. Am. Chem. Soc.* 129, 1868–1869. (c) Zhou ZT, Xie JH, Zhou QL (2009). Enantioselective synthesis of chiral β-aryloxy alcohols by asymmetric hydrogenation of α-aryloxy aldehydes via dynamic kinetic resolution. *Adv. Synth. Catal.* 351, 363–366.

20. (a) Narula AP (2003). The search for new fragrance ingredients. *Perfumer Flavorist* 62–73. (b) Mouhib H, Stahl W, Lüthy M, Büchel M, Kraft P (2011). Cassis odor through microwave eyes: olfactory properties and gas-phase structures of all the cassyrane stereoisomers and its dihydro derivatives. *Angew. Chem., Int. Ed.* 50, 5576–5580.

21. (a) Olah GA, Fung AP, Malhotra R (1981). Synthetic methods and reactions; 99. Preparation of cyclic ethers over superacidic perfluorinated resinsulfonic acid (Nafion-H) catalyst. *Synthesis* 474–476. (b) Hunter SE, Ehrenberger CE, Savage PE (2006). Kinetics and mechanism of tetrahydrofuran synthesis via 1, 4-butanediol dehydration in high-temperature water. *J. Org. Chem.* 71, 6229–6239.

22. ChemSystems PERP Report Abstract PERP06/07-4 (2009).

23. Aycock DF (2007). Solvent applications of 2-methyltetrahydrofuran in organometallic and biphasic reactions. *Org. Process. Res. Dev.* 11, 156–159.

24. (a) Schniepp LE, Geller HH, Von Korff RW (1947). The preparation of acetopropyl alcohol and 1, 4-pentanediol from methylfuran. *J. Am. Chem. Soc.* 69, 672–674. (b) Ahmed I, inventor; Pure Energy Corp., assignee. Processes for the preparation of 2-methylfuran and 2-methyltetrahydrofuran. US Patent 2003/0018205. August 21, 2002. (c) Wabnitz T, Breuninger D, Heimann J, Backes R, Pinkos R, inventors; BASF SE, assignee. Process for one-stage preparation of 2-methyltetrahydrofuran from furfural over a catalyst. US Patent 2010/0099895. June 24, 2008.

25. (a) Elliot D, Frye JG, inventors; Battelle Memorial Inst., assignee. Hydrogenated 5-carbon compound and method of making. US Patent 1999/5883266. January 16, 1998. (b) Ayoub PM, Lange JP, inventors; Shell Int. Res. Maatschappij B.V., assignee. World Patent WO2008/14127. May 21, 2008.

26. (a) Weyer HJ, Fischer R, Merger F, Frank J, Henkelmann J, Siegel H, Ruehl T, inventors; BASF Aktiengesellschaft, assignee. Preparation of 2-methyl-1,4-butanediol and 3-methyltetrahydrofuran. US Patent 5,536,854. January 26, 1995. (b) Masaki T, Yoshikazu S, Takafumi A, inventors; Mitsubishi Gas Chem. Company Inc, assignee. Process for preparing 3-methyltetrahydrofuran. European Patent 0,773,215. November 7, 1995.

27. Ignatchenko AV, Beavers WA, Liu Z, inventors; Eastman Chem. Comp., assignee. Process for the preparation of 3-methyltetrahydrofuran. US Patent 6,521,765. April 18, 2002.

28. Manzer LE, inventor; E I Du Pont de Nemours and Co., assignee. US Patent US2003/0109724. July 9, 2002.

29. The synthesis of 2- and 3-Me-THF has also been reported by rearrangement of the corresponding vinyl oxiranes derived from isoprene and piperylene followed by hydrogenation of the dihydrofuran formed, see (a) Batory LA, McInnis CE, Njardarson JT (2006). Copper-catalyzed rearrangement of vinyl oxiranes. *J. Am. Chem. Soc.* 128, 16054–16055. (b) Njardarson J, Batory L, Brichacheck M, Bauer R, Rogers E, inventors; Cornell Research Foundation Inc., assignee. Production of 2,5-dihydrofurans and analogous compounds. World Patent WO2007/108999. March 17, 2006.

30. Geilen FMA, Engendahl B, Harwardt A, Marquardt W, Klankermayer J, Leitner W (2010). Selective and flexible transformation of biomass-derived platform chemicals by a multifunctional catalytic system. *Angew. Chem., Int. Ed.* 49, 5510–5514.

31. (a) Hara Y, Wada K (1991). Hydrogenation reaction of carboxylic anhydrides catalyzed by a new and highly active cationic ruthenium complex. *Chem. Lett.* 553–554. (b) Hara Y, Inagaki H, Nishimura S, Wada K (1992). Selective hydrogenation of cyclic ester to α,ω-diols catalyzed by cationic ruthenium complexes with trialkylphosphine ligands. *Chem. Lett.* 1983–1986.
32. Wylie WNO, Lough AJ, Morris RH (2010). The hydrogenation of molecules with polar bonds catalyzed by a ruthenium(II) complex bearing a chelating *N*-heterocyclic carbene with a primary amine donor. *Chem. Commun.* 8240–8242.
33. (a) Ikariya T, Ito M, Shibashi A, Ootsuka T, inventors; Tokyo Institute of Technology, assignee. European Patent 2,298,723. August 7, 2008. (b) Ito M, Ootsuka T, Watari R, Shiibashi A, Himizu A, Ikariya T (2011). Catalytic hydrogenation of carboxamides and esters by well-defined Cp*Ru complexes bearing a protic amine ligand. *J. Am. Chem. Soc.* 133, 4240–4242.
34. For the hydrogenation of other carboxylic acid derivatives catalyzed by Cp*Ru(L-NH_2)Cl: imides: (a) Ito M, Sakaguchi A, Kobayashi C, Ikariya T (2007). Chemoselective hydrogenation of imides catalyzed by Cp*Ru(PN) complexes and its application to the asymmetric synthesis of paroxetine. *J. Am. Chem. Soc.* 129, 290–291. (b) Ito M, Kobayashi C, Himizu A, Ikariya T (2010). Highly enantioselective hydrogenative desymmetrization of bicyclic imides leading to multiply functionalized chiral cyclic compounds. *J. Am. Chem. Soc.* 132, 11414–11415. *N*-Acylcarbamates and *N*-acylsulfonamides: (c) Ito M, Koo LW, Himizu A, Kobayashi C, Sakaguchi A, Ikariya T (2009). Hydrogenation of *N*-acylcarbamates and *N*-acylsulfonamides catalyzed by a bifunctional [Cp*Ru(PN)] complex. *Angew. Chem., Int. Ed.* 48, 1324–1327.
35. For a comprehensive compilation of Bordwell pK_a data, see http://www.chem.wisc.edu/areas/reich/pkatable/index.htm (last accessed 5 January 2012).
36. For the synthesis of substituted furans, see (a) Hou XL, Cheung HY, Hon TY, Kwan PL, Lo TH, Tong SY, Wong HNC (1998). Regioselective syntheses of substituted furans. *Tetrahedron* 54, 1955–2020. (b) Keay BA (1999). Synthesis of multi-substituted furan rings: the role of silicon. *Chem. Soc. Rev.* 28, 209–215.
37. Ho TL, Lee KY (1995). A new synthesis of (−)-furodysin. *Tetrahedron Lett.* 36, 947–948.
38. Tanis SP (1982). A simple synthesis of 3-substituted furans. The preparations of dendrolasin, perillene and congeners. *Tetrahedron Lett.* 23, 3115–3118.
39. Tanis SP, Herrinton PM (1983). Furans in synthesis. 3. Furans as terminators in cationic cyclization. *J. Org. Chem.* 48, 4572–4580.
40. Marshall JA, DuBay WJ (1993). Synthesis of rosefuran. A new route to furans through base-catalyzed cyclization of hydroxy alkynoates and alkynones. *J. Org. Chem.* 58, 3602–3603.
41. Lipshutz BH (1986). Five-membered heteroaromatic rings as intermediates in organic synthesis. *Chem. Rev.* 86, 795–819.
42. Davies HML, Calvo RL, Townsend RJ, Ren PD, Churchill RM (2000). An exploratory study of type II [3+4] cycloadditions between vinylcarbenoids and dienes. *J. Org. Chem.* 65, 4261–4268.
43. Tanaka K, Hata T, Hara H, Katsumura S (2003). Development of new Wittig reagent, silylfuranmethylid, and its reactivity. *Tetrahedron* 59, 4945–4952.
44. (a) Kuwajima I, Urabe H (1981). Regioselective synthesis of Δ^3-butenolides via oxidation of 2-trimethylsilylfurans. *Tetrahedron Lett.* 22, 5191–5194. (b) Goldsmith D, Liotta D, Saindane M, Waykole L, Bowen P (1983). 3-Alkylfurans as useful synthetic equivalents for substituted Δ^2-butenolides. *Tetrahedron Lett.* 24, 5835–5838.
45. (a) DeShong P, Waltermire RE, Ammon HL (1988). A general approach to the stereoselective synthesis of spiroketals. A total synthesis of the pheromones of the olive fruit fly and related compounds. *J. Am. Chem. Soc.* 110, 1901–1910. (b) Ireland RE, Wipf P, Miltz W, Vanasse B (1990). Methodology for the enantioselective synthesis of aldols and other 1,3-dioxygenated

systems. *J. Org. Chem.* 55, 1423–1424. (c) Cheng K, Kelly AR, Kohn RA, Dweck JF, Walsh PJ (2009). One-pot catalytic asymmetric synthesis of pyranones. *Org. Lett.* 11, 2703–2706.

46. Stamm S, Linden A, Heimgartner H (2003). Chiral heterospirocyclic 2*H*-azirin-3-amines as synthons for 3-amino-2,3,4, 5-tetrahydrofuran-3-carboxylic acid and their use in peptide synthesis. *Helv. Chim. Acta* 86, 1371–1396.

47. Starr D, Hixon RM (1943). Tetrahydrofuran. *Org. Synth. Coll.* 2, 566–568.

48. Maegawa T, Akashi A, Sajiki H (2006). A mild and facile method for complete hydrogenation of aromatic nuclei in water. *Synlett* 1140–1442.

49. The selective hydrogenation of heteroaromtic ketones with ruthenium complexes has been reported, see (a) Ohkuma T, Koizumi M, Yoshida M, Noyori R (2000). General asymmetric hydrogenation of hetero-aromatic ketones. *Org. Lett.* 2, 1749–1751. (b) Carpenter I, Clarke ML (2011). Enantioselective reduction of benzofuranyl aryl ketones. *Synlett* 65–68.

50. Ohkuma T, Koizumi M, Muñiz K, Hilt G, Kabuto C, Noyori R (2002). *trans*-RuH(η^1-BH_4)(binap)(1, 2-diamine): a catalyst for asymmetric hydrogenation of simple ketones under base-free conditions. *J. Am. Chem. Soc.* 124, 6508–6509.

51. Kuriyama W, Ino, Y, Ogata O, inventors; Takasago Int. Corp., assignee. Alcohol production method by reducing ester or lactone with hydrogen. European Patent EP 2,161,251. September 9, 2008.

52. (a) Grey RA, Pez GP, Wallo A, Corsi J (1980). Homogeneous catalytic hydrogenation of carboxylic acid esters to alcohols. *Chem. Commun.* 783–784. (b) Grey RA, Pez GP, Wallo A (1981). Anionic metal hydride catalysts. 2. Application to the hydrogenation of ketones, aldehydes, carboxylic acid esters and nitriles. *J. Am. Chem. Soc.* 103, 7536–7542. (c) Grey RA, Pez GP, Wallo A, Corsi J (1983). The hydrogenation of carbonyl compounds catalyzed by anionic ruthenium hydride complexes and alkali-doped supported group VIII metals. *Ann. N. Y. Acad. Sci.* 415, 235–243.

53. Blaser HU (2003). Enantioselective catalysis in fine chemicals production. *Chem. Commun.* 293–296.

54. Naud F, Spindler F, Rueggeberg CJ, Schmidt AT, Blaser HU (2007). Enantioselective ketone hydrogenation: from R&D to pilot scale with industrially viable Ru/phosphine-oxazoline complexes. *Org. Process Res. Dev.* 11, 519–523.

3

SYNTHESIS OF CHIRAL AMINES USING TRANSAMINASES

NICHOLAS J. TURNER[1] AND MATTHEW D. TRUPPO[2]

[1]*Manchester Institute for Biotechnology, School of Chemistry, University of Manchester, Manchester, UK*

[2]*Merck Research Laboratories, Rahway, NJ, USA*

3.1 IMPORTANCE OF CHIRAL AMINES

Chiral amines, in their enantiomerically pure forms, have proven to be useful building blocks in stereoselective synthesis [1], and are among the most important and commonly found subunits in chiral drugs [2–5]. Indeed, 75% of drugs and drug candidates across a wide range of therapeutic areas contain chiral amines (including antihypertensives, thyroid hormone replacements, antibiotics, antivirals, antidepressants, antidiabetics, treatments for Parkinson's disease, antihistamines, and treatments for urinary incontinence) [6]. These amines can be 1°, 2°, or 3° in structure and, hence, ideally new synthetic methods for their preparation should be versatile enough to encompass all structural subtypes. In practice, however, this is rarely the case and, hence, specific methods tend to be best suited to one group of chiral amine target. As with all types of chiral functional groups, there is an increased tendency to avoid stoichiometric use of reagents, but instead develop catalytic (both chemo- and bio-) methods for their preparation, particularly on scale. This chapter reviews the use of transaminases, an emerging class of biocatalyst that are finding increased application in the direct conversion of ketones to (chiral) amines.

3.1.1 Challenges with Chemocatalytic Synthesis of Chiral Amines

Highly enantioselective reductive amination starting from prochiral ketones has been a challenge for chemocatalysis [7]. Normally, this transformation is achieved by the catalytic asymmetric hydrogenation of imines (Scheme 3-1) or enamine derivatives using late transition metal complexes based on chiral phosphine ligands [8,9]. Note also that in a

Sustainable Catalysis: Challenges and Practices for the Pharmaceutical and Fine Chemical Industries, First Edition. Edited by Peter J. Dunn, K. K. (Mimi) Hii, Michael J. Krische, and Michael T. Williams.

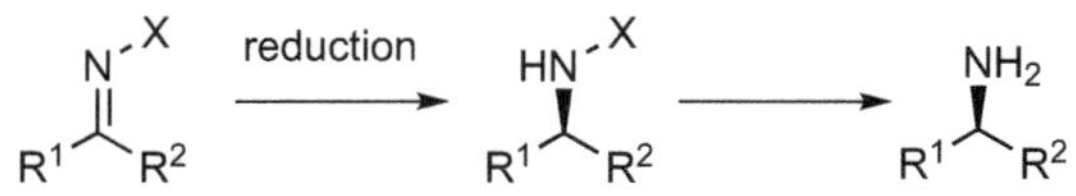

SCHEME 3-1. Asymmetric hydrogenation of imines to form chiral amines.

related sense, the asymmetric reduction of enantiopure imines using zinc borohydride has also been demonstrated [10]. The imine or enamine precursors can be made by condensation of ketones with appropriate amines. This multistep approach also requires isolation of the substrates for the hydrogenation reaction [11]. The difficulty in this approach lies with the synthesis of the prochiral imine precursor, and the cleavage of the auxiliary group X to give the free amine [12].

3.2 TRANSAMINASES

Transaminases belong to one of four families of B_6 (pyridoxal phosphate (PLP))-dependent enzymes that participate in the metabolism of amino acids [13]. The four families are able to catalyze decarboxylations, replacements and eliminations, transaminations, and racemizations. PLP-dependent enzymes serve to labilize one of the three bonds (hydrogen[a], CO_2[b], and R-group[c]) of the α-carbon atom of an amino acid substrate (Figure 3-1) [14].

3.2.1 Transaminase Mechanism

The reaction mechanism and structural properties of transaminases have been studied in great detail [15–18]. The transamination reaction proceeds by a ping-pong bi-bi mechanism and consists of two half-reactions (Scheme 3-2). First, an internal Schiff base between an active site lysine residue (via its ε-amino group) and the aldehyde group of a PLP bound to the enzyme (internal aldimine) is replaced by a Schiff base between the amino donor and the PLP, forming an external aldimine. The lysine residue that is released by this process serves as the base for the next step and catalyzes a 1,3-hydrogen shift involving the abstraction of the α-proton from the external aldimine and reprotonation to the imine of the coenzyme. Internal rearrangement proceeds through a quinoid intermediate, yielding a

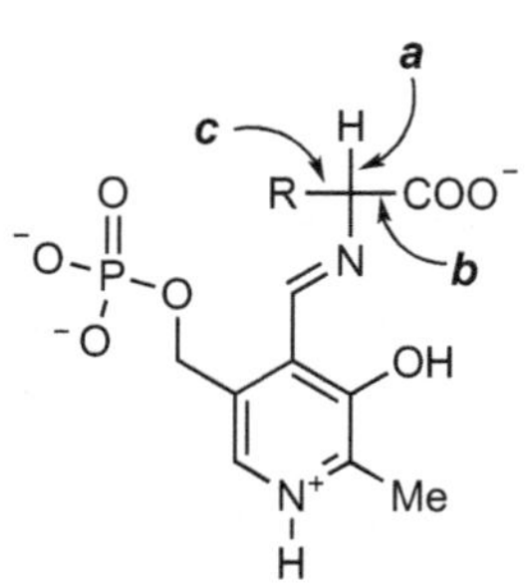

FIGURE 3-1. Transformations catalyzed by PLP-dependent enzymes.

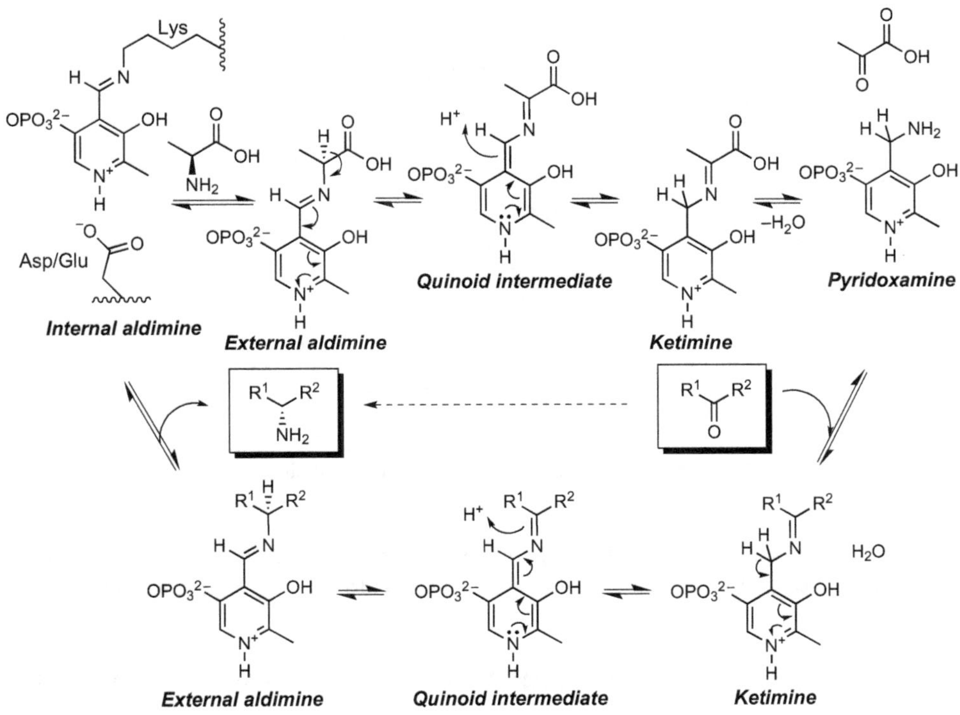

SCHEME 3-2. Transaminase (bi-bi) mechanism.

ketimine stereoselectively. This intermediate is stabilized by a tyrosine, which hydrogen bonds with the phenolic oxygen of the PLP and the Schiff base imine group of the internal/external aldimine. It is also stabilized by a carboxylate side chain from aspartate or glutamate, which forms an ion pair with the pyridine nitrogen of the cofactor. This causes a stabilization of the positive charge in the aromatic ring (forming an electron sink). Hydrolysis of the ketimine intermediate yields pyridoxamine 5′-phosphate (PMP) and releases the product ketone. Formation of a Michaelis complex between the PMP bound to the enzyme and a different amino acceptor initiates the reverse second half of the reaction and regenerates the enzyme-bound PLP form.

3.2.2 Transaminase Selectivity

Homology active site models [19] and 3D crystal structure determination [20,21] have been used to elucidate how the enzyme controls substrate specificity and stereoselectivity [22,23]. Shin and Kim [24] proposed a two-binding site model in which a large (L) pocket serves as a carboxylate trap and a small (S) pocket accommodates the substrate side chain (Figure 3-2). Strong repulsion for a carboxylate in the S pocket is a key determinant in their model for substrate specificity and stereoselectivity. Based on substrate–structure mechanistic analysis, there have been several studies investigating the production of artificial transaminases, ranging from converting the catalytic activity of an alanine racemase to a transaminase [25,26], to genetic modification of a protein cavity and insertion of a chemically generated catalytically active group [27,28], to using entirely synthetic chiral

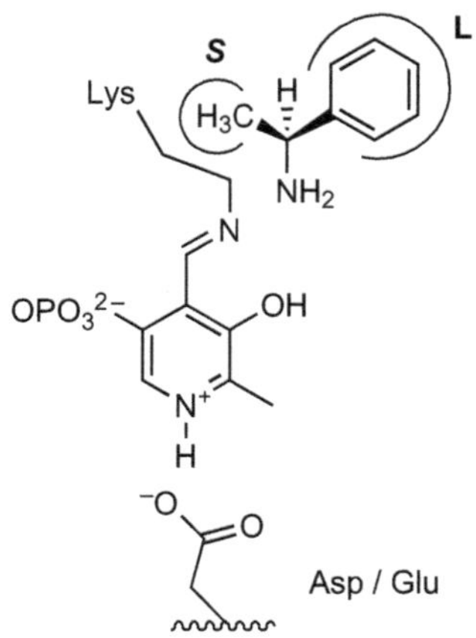

FIGURE 3-2. Transaminase selectivity active site model.

Lewis acid catalysts [29]. All of these studies demonstrate that the enzyme structure is necessary for orienting the substrate in such a way as to give high stereoselectivity.

3.3 TRANSAMINASE-CATALYZED RESOLUTION OF RACEMIC AMINES

The late 1990s saw the development of transaminase-mediated enzymatic resolution of racemic amines. Celgene developed transaminase technology for the enantioselective conversion of chiral amines to ketones [30]. Low-molecular weight aldehydes, such as propionaldehyde, were used as the amine-group acceptor (Scheme 3-3). This process has been used on a 2.5 m^3 scale and has been demonstrated on several different aliphatic and aromatic amines. A disadvantage of this approach was the relatively low product concentrations attainable in an all aqueous reaction system with hydrophobic compounds. Another disadvantage was the significant inhibition of the transaminases by both the amine substrates and ketone products.

Shin and Kim [31,32] developed various methods aimed at increasing the product concentrations of transaminase-catalyzed amine resolutions, through the continuous removal of product ketone from the reactions. The application of an aqueous/organic two-phase system to the ω-transaminase-catalyzed resolution of racemic α-methylbenzyl-amine (MBA) was found to be superior to an aqueous-only system in product concentration obtained (Scheme 3-4). A drawback of the biphasic system was an increased enzyme deactivation rate compared with the aqueous-only system, due to the formation of an aqueous/organic emulsion. Another disadvantage was the significant inhibition of the transaminase by the amine-acceptor substrate (pyruvate) used.

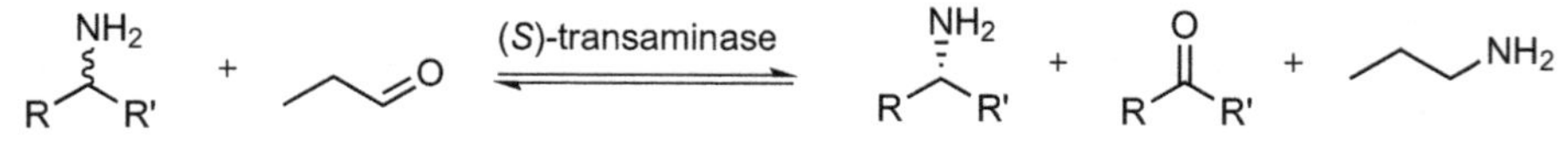

SCHEME 3-3. Transaminase-catalyzed resolution of amines using propionaldehyde as the amine-acceptor group.

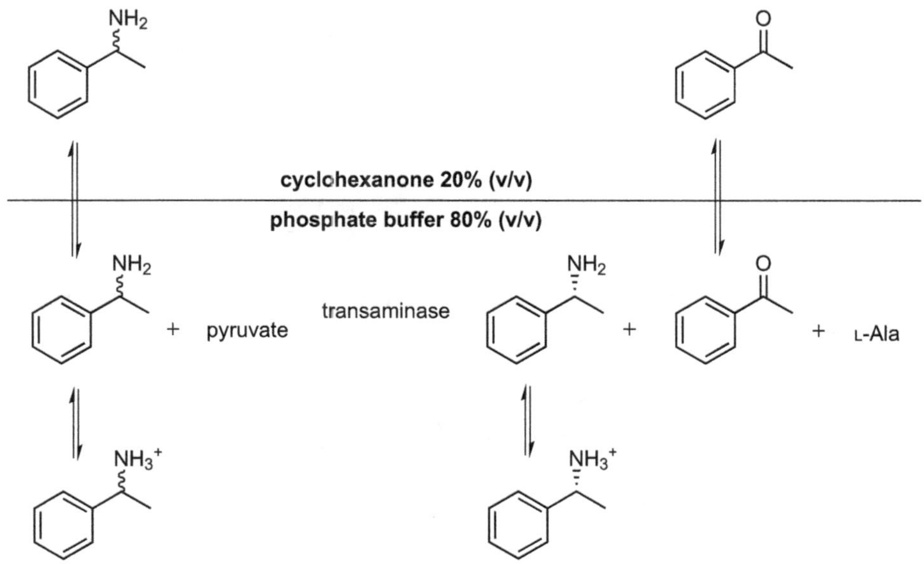

SCHEME 3-4. Aqueous/organic biphasic reaction system for the transaminase-catalyzed resolution of MBA.

In a further development, the kinetic resolution of α-MBA was demonstrated using an enzyme-membrane reactor and a hollow-fiber membrane contactor for the continuous removal of the inhibitory product acetophenone from the reaction system without incurring the enzyme deactivation penalty of the biphasic emulsion system [33]. An alternative system utilizing a packed-bed reactor with a membrane contactor also showed promise [34]. Entrapped whole cells of *Vibrio fluvialis* in Ca-alginate gel beads were packed in a plug-flow column reactor. Fresh substrate solution was passed through the column reactor, and product was removed by passing the column effluent through the membrane contactor. The substrate scope was enhanced via the redesign of the substrate specificity of ω-transaminase to include the kinetic resolution of aliphatic chiral amines [35].

In 2008, Hanson et al. [36] identified an (*S*)-amine transaminase from *Bacillus megaterium* that was used for the preparation of (*R*)-1-cyclopropylethylamine and (*R*)-*sec*-butylamine by resolution of the racemic amines (Scheme 3-5). Pyruvate was used as the amine acceptor in this study.

Hoehne et al. [37] have demonstrated a substrate-protection strategy that enhances both the rate and the enantioselectivity of transaminase-catalyzed kinetic resolution reactions. The ω-transaminase-catalyzed resolution of the pharmaceutically important synthons

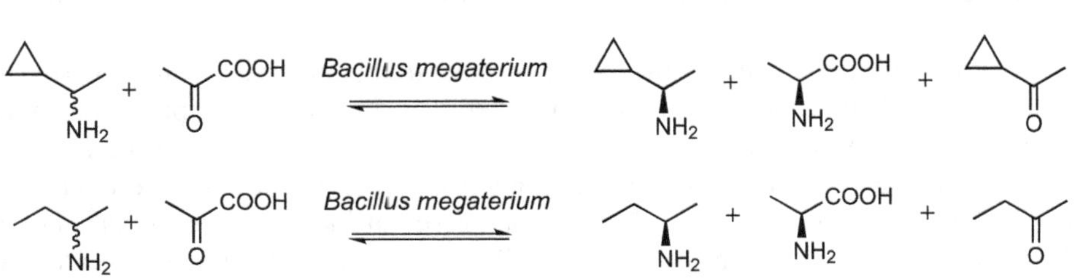

SCHEME 3-5. Transaminase-catalyzed resolution of 1-cyclopropylethylamine and *sec*-butylamine.

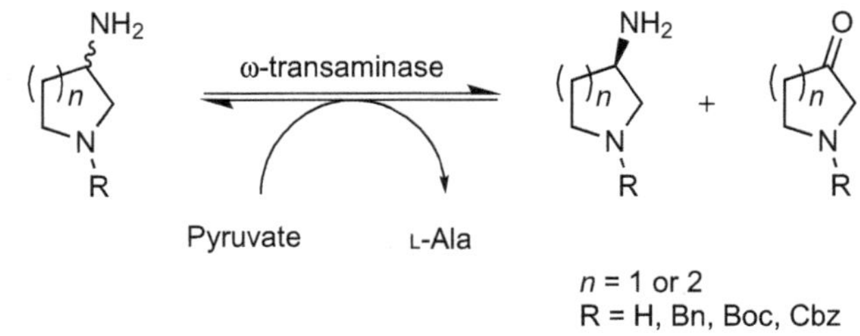

SCHEME 3-6. Protection strategy for increasing the rate and enantioselectivity of transaminase-catalyzed amine resolutions.

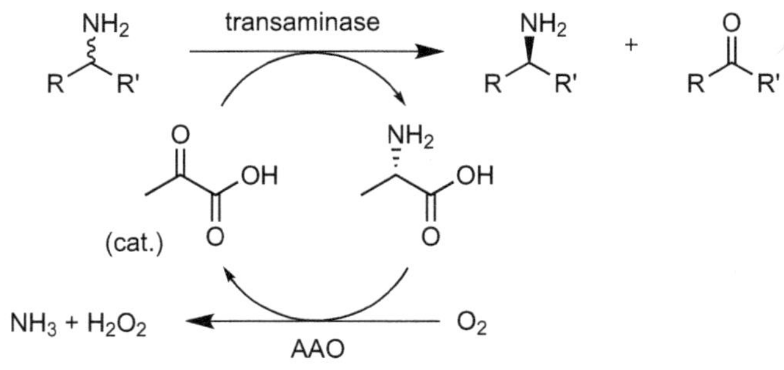

SCHEME 3-7. Use of catalytic pyruvate and AAO for the regeneration of PLP.

3-aminopyrrolidine and 3-aminopiperidine was improved by employing *N*-protecting groups (Scheme 3-6). Reaction rates were improved by up to 50-fold, and product ee was improved from 86 to 99%.

Recently, an alternative method for the resolution of racemic amines that uses amino acid oxidase (AAO) with catalytic quantities of pyruvate to regenerate the PLP cofactor has been developed [38]. AAO catalyzes the *in situ* oxidation of alanine back to pyruvate in the presence of molecular oxygen (Scheme 3-7). As a result, the AAO approach eliminates the use of stoichiometric amounts of pyruvate for the regeneration of PLP. This serves to reduce the cost of the process (making O_2 the only reagent consumed in the cofactor regeneration), and also eliminates transaminase inhibition due to high concentrations of pyruvate.

3.4 TRANSAMINASE-CATALYZED ASYMMETRIC SYNTHESIS OF AMINES

The generation of chiral amines from prochiral ketones using transaminases has received lots of attention, and is now the preferred way of generating chiral amines at preparative scale. Transaminases have long been used for the production of amino acids starting from keto acids, and a primary goal has been to extend their use to the production of simple amines from prochiral ketones [1,39]. The first reported transaminase capable of transferring an amine to a substrate without a carboxylic acid group was a diamine-ketoglutaric transaminase identified in 1964 [40]. However, it was not until the end of the 1980s that transaminase technology became an industrially viable option for the synthesis of chiral amines.

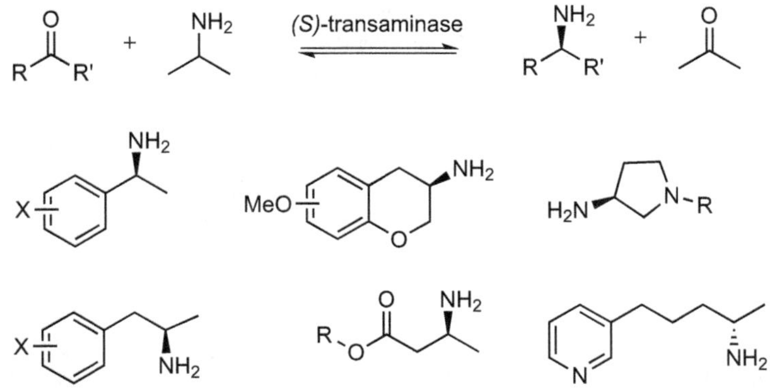

SCHEME 3-8. Celgene transamination of ketones using isopropylamine as the amine-donor change.

Researchers at Celgene developed both (*R*)- and (*S*)-selective transaminases that were active on a range of aliphatic and aromatic ketones and amines [30,41–43]. Two methodologies were employed: a synthesis approach and a kinetic resolution approach. The kinetic resolution approach has been discussed previously (Section 3.1.3). The synthesis approach started with a prochiral ketone substrate that served as an amine acceptor, and isopropylamine as the amine-group donor (Scheme 3-8). The product of the reaction was the desired chiral amine and acetone. The Celgene processes have been run successfully at 2.5 m^3 scale; however, poor equilibrium necessitates the use of excess reagents to drive the reactions to completion, and low product concentrations were obtained with hydrophobic substrates.

Several investigations into the challenges facing transaminase technology were conducted by Shin and Kim in the late 1990s. Using an (*S*)-selective ω-transaminase isolated from *V. fluvialis*, Shin and Kim demonstrated the synthesis of aliphatic and aryl chiral amines using L-alanine as the amine donor (Scheme 3-9) [44,45].

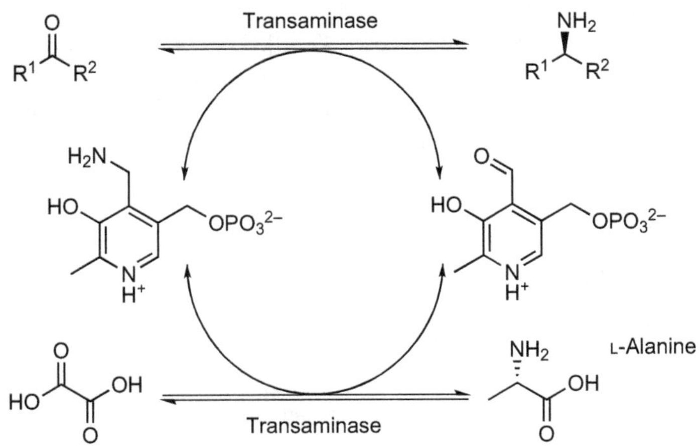

SCHEME 3-9. Enantioselective transamination of ketones to chiral amines using alanine amine donor to regenerate PLP.

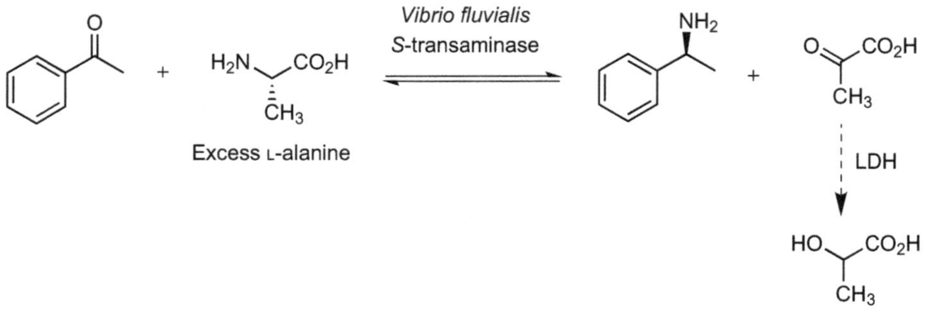

SCHEME 3-10. Transamination using L-alanine as the amine donor with removal of the inhibitory pyruvate product via reduction using lactate dehydrogenase.

Severe inhibition of the transaminase by both the amine and the keto-acid reaction products, as well as extremely poor equilibrium, were observed. Even with a large excess of amine donor, reactions proceeded in $<2\%$ conversion. The continuous *in situ* removal of the inhibitory pyruvate product (Scheme 3-10) via reduction using lactate dehydrogenase (LDH) boosted reaction yields to $>90\%$, but inhibition by the desired amine product still resulted in poor productivity. Studies comparing ω-transaminases from different microorganisms further demonstrated the extreme inhibition of transaminases by reaction products [46] and also identified the rarity of (*R*)-selective transaminases [47].

In situ removal of pyruvate has also been accomplished using pyruvate decarboxylase (Scheme 3-11) [48]. A potential advantage of employing pyruvate decarboxylase over LDH is that it requires no cofactor recycling system (LDH is an NAD(P)H-dependent enzyme). However, a significant disadvantage is the competitive inhibition of the transaminase by the aldehyde by-product, which can also serve as a substrate for transamination.

An alternative method for alleviating inhibition by the coproduct pyruvate, developed by Truppo and Turner, uses catalytic alanine to regenerate the PLP cofactor coupled with an amino acid dehydrogenase (AADH) and ammonia for alanine recycling (Scheme 3-12). AADH is a nicotinamide adenine dinucleotide (NADH)-dependent enzyme and requires NADH regeneration using GDH and glucose [49,50]. The constant removal and recycling of pyruvate back to alanine eliminates transamination inhibition by pyruvate, and allows for only catalytic amounts of alanine to be used. Therefore, the inexpensive cosubstrate ammonia becomes the substrate consumed in stoichiometric quantities for PLP cofactor regeneration.

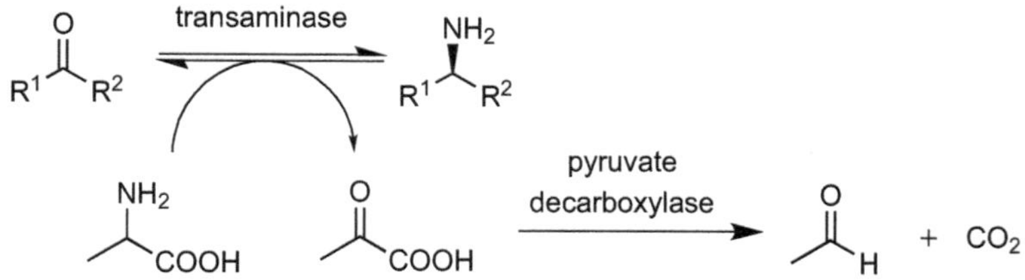

SCHEME 3-11. Removal of pyruvate with pyruvate decarboxylase.

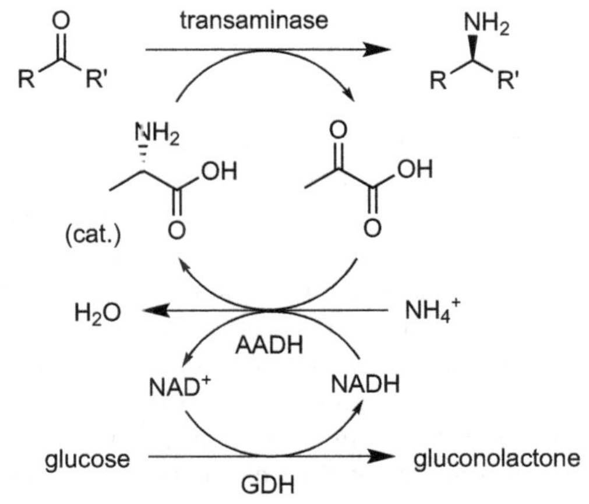

SCHEME 3-12. Use of catalytic alanine and AADH for the regeneration of PLP.

Thus far, the most economical approach for cofactor regeneration in transaminase-mediated amine synthesis has been the directed evolution of transaminases to accept and be active in high concentrations of isopropylamine. Isopropylamine serves as the amine-group donor to regenerate the cofactor, and the acetone by-product can be removed via vacuum distillation, driving the reaction equilibrium toward the desired chiral amine product. This single-enzyme system is proving itself to be the most cost effective way to generate chiral amines, with volumetric productivities routinely exceeding $50\,g\,l^{-1}$ [51]. Most notably, Merck and Codexis have demonstrated the large-scale production of the antidiabetic compound sitagliptin (Januvia) using isopropylamine for the regeneration of the PLP cofactor (Scheme 3-13) [52]. Productivities $>200\,g\,l^{-1}\,day^{-1}$ have been reported (see Chapter 4).

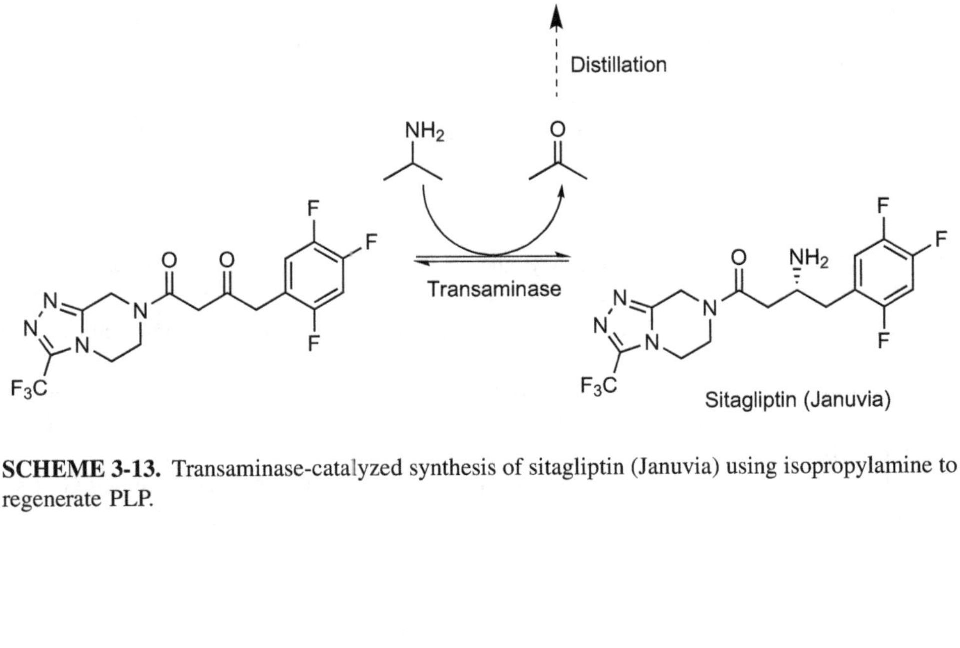

SCHEME 3-13. Transaminase-catalyzed synthesis of sitagliptin (Januvia) using isopropylamine to regenerate PLP.

3.5 CONCLUSIONS

Despite being initially investigated in the late 1980s, transaminases represent a relatively recent addition to the biocatalytic toolbox of most synthetic organic chemists. In the last 5 years, there has been a dramatic rise in the number of papers and patents concerning transaminases, covering various aspects including enzyme discovery, protein engineering, high-throughput screening, process development, and, ultimately, application in the manufacture of amines on scale. Directed evolution has proven a powerful strategy for broadening the substrate specificity of transaminases, such that they are now able to catalyze amination of a structurally broad range of ketones. Ongoing developments, particularly with respect to engineering transaminases that are suitably robust to be used with high-substrate concentration, and in the presence of organic solvents, suggest that these enzymes may soon become the method of choice for converting ketones to chiral primary amines.

REFERENCES

1. Stewart JD (2001). Dehydrogenases and transaminases in asymmetric synthesis. *Curr. Opin. Chem. Biol.* 5, 120–129.
2. Enders D, Reinhold U (1997). Asymmetric synthesis of amines by nucleophilic 1,2-addition of organometallic reagents to the CN-double bond. *Tetrahedron Asymmetry* 8, 1895–1946.
3. Bloch R (1998). Additions of organometallic reagents to C=N bonds: reactivity and selectivity. *Chem. Rev.* 98, 1407–1438.
4. Kobayashi S, Ishitani H (1999). Catalytic enantioselective addition to imines. *Chem. Rev.* 99, 1069–1094.
5. Devalia JL, DeVos C, Hanotte F, Baltes E (2001). A randomized, double-blind, crossover comparison among cetirizine, levocetirizine, and ucb 28557 on histamine-induced cutaneous responses in healthy adult volunteers. *Allergy* 56, 50–57.
6. Fernandez I, Valdivia V, Khiar N (2008). *N*-Isopropylsulfinylimines as useful intermediates in the synthesis of chiral amines: expeditive asymmetric synthesis of the calcimimetic (+)-NPS R-568. *J. Org. Chem.* 73, 745–748.
7. Gomez S, Peters JA, Maschmeyer T (2002). The reductive amination of aldehydes and ketones and the hydrogenation of nitriles: mechanistic aspects and selectivity control. *Adv. Synth. Catal.* 344, 1037–1057.
8. Boezio AA, Pytkowicz J, Cote A, Charette AB (2003). Asymmetric, catalytic synthesis of alpha-chiral amines using a novel *bis*(phosphine) monoxide chiral ligand. *J. Am. Chem. Soc.* 125, 14260–14261.
9. James BR (1997). Synthesis of chiral amines catalyzed homogeneously by metal complexes. *Catal. Today* 37, 209–221.
10. Cimarelli C, Palmieri G (2000). Asymmetric reduction of enantiopure imines with zinc borohydride: stereoselective synthesis of chiral amines. *Tetrahedron Asymmetry* 11, 2555–2563.
11. Tararov VI, Borner A (2005). Approaching highly enantioselective reductive amination. *Synlett* 203–211.
12. Breuer M, Ditrich K, Habicher T, Hauer B, Kesseler M, Sturmer R, Zelinski T (2004). Industrial methods for the production of optically active intermediates. *Angew. Chem., Int. Ed.* 43, 788–824.
13. Salzmann D, Christen P, Mehta PK, Sanndmeier E (2000). Rates of evolution of pyridoxal-5′-phosphate-dependent enzymes. *Biochem. Biophys. Res. Commun.* 270, 576–580.

14. Bachmann S, Knudsen KR, Jorgensen KA (2004). Mimicking enzymatic transaminations: attempts to understand and develop a catalytic asymmetric approach to chiral alpha-amino acids. *Org. Biomol. Chem.* 2, 2044–2049.
15. Metzler DE, Jansonius JN, Arnone A, Martinez-Carrion M, Manning JM (1982). Transaminases. *Fed. Proc.* 41, 2432–2436.
16. Yonaha K, Toyama S, Kagamiyama H (1983). Properties of the bound co-enzyme and subunit structure of omega-amino acid:pyruvate aminotransferase. *J. Biol. Chem.* 258, 2260–2265.
17. Yonaha K, Nishie M, Aibara S (1992). The primary structure of omega-amino acid:pyruvate aminotransferase. *J. Biol. Chem.* 267, 12506–12510.
18. Rozzell JD, Bommarus AS. Transaminations. In: Drauz K, Waldmann H, editors. *Enzyme Catalysis in Organic Synthesis*. Wiley-VCH, Weinheim; 2002, pp 873–892.
19. Toney MD, Pascarella S, Debiase D (1995). Active-site model for gamma-aminobutyrate aminotransferase explains substrate-specificity and inhibitor reactivities. *Protein Sci.* 4, 2366–2374.
20. Sugio S, Petsko GA, Manning JM, Soda K, Ringe D (1995). Crystal-structure of a D-amino-acid aminotransferase—how the protein controls stereoselectivity. *Biochemistry* 34, 9661–9669.
21. Peisach D, Chipman DM, Van Ophem PW, Manning JM, Ringe D (1998). D-Cycloserine inactivation of D-amino acid aminotransferase leads to a stable noncovalent protein complex with an aromatic cycloserine-PLP derivative. *J. Am. Chem. Soc.* 120, 2268–2274.
22. Yoshimura T, Nishimura K, Ito J, Esaki N, Kagamiyami H, Manning JM, Soda K (1993). Unique stereospecificity of D-amino-acid aminotransferase and branched-chain L-amino-acid aminotransferase for C-4′ hydrogen-transfer of the coenzyme. *J. Am. Chem. Soc.* 115, 3897–3900.
23. Jhee KH, Yoshimura T, Esaki N, Soda K (1996). Stereospecificity of thermostable ornithine 5-aminotransferase for the hydrogen transfer in the L- and D-ornithine transamination. *Biochemistry* 35, 9792–9796.
24. Shin JS, Kim BG (2002). Exploring the active site of amine:pyruvate aminotransferase on the basis of the substrate structure–reactivity relationship: how the enzyme controls substrate specificity and stereo selectivity. *J. Org. Chem.* 67, 2848–2853.
25. Yow GY, Watanabe A, Yoshimura T, Esaki N (2003). Conversion of the catalytic specificity of alanine racemase to a D-amino acid aminotransferase activity by a double active-site mutation. *J. Mol. Catal. B: Enzym.* 23, 311–319.
26. Kurokawa Y, Watanabe A, Yoshimura T, Esaki N, Soda K (1998). Transamination as a side-reaction catalyzed by alanine racemase of *Bacillus stearothermophilus*. *J. Biochem.* 124, 1163–1169.
27. Haring D, Kuang H, Qi D, Distefano MD (2001). Converting a fatty acid binding protein to an artificial transaminase: novel catalysts by chemical and genetic modification of a protein cavity. *J. Mol. Catal. B: Enzym.* 11, 967–970.
28. Haring D, Distefano MD (2002). Exploring routes to stabilize a cationic pyridoxamine in an artificial transaminase: site-directed mutagenesis versus synthetic cofactors. *Protein Eng.* 15, 603–610.
29. Knudsen KR, Bachmann S, Jorgensen KA (2003). Catalytic enantioselective transamination of alpha-keto esters: an organic approach to enzymatic reactions. *Chem. Commun.* 2602–2603.
30. Matcham GW, Bowen ARS (1996). Biocatalysis for chiral intermediates: meeting commercial and technical challenges. *Chim. Oggi* 14, 20–24.
31. Shin JS, Kim BG (1997). Kinetic resolution of alpha-methylbenzylamine with omega-transaminase screened from soil microorganisms: application of a biphasic system to overcame product inhibition. *Biotechnol. Bioeng.* 55, 348–358.
32. Shin JS, Kim BG (1998). Kinetic modeling of omega-transamination for enzymatic kinetic resolution of alpha-methylbenzylamine. *Biotechnol. Bioeng.* 60, 534–540.

33. Shin JS, Kim BG, Liese A, Wandrey C (2001). Kinetic resolution of chiral amines with omega-transaminase using an enzyme–membrane reactor. *Biotechnol. Bioeng.* 73, 179–187.
34. Shin JS, Kim BG, Shin DH (2001). Kinetic resolution of chiral amines using packed-bed reactor. *Enzyme Microb. Technol.* 29, 232–239.
35. Cho BK, Park HY, Seo JH, Kim J, Kang TJ, Lee BS, Kim BG, (2008). Redesigning the substrate specificity of omega-aminotransferase for the kinetic resolution of aliphatic chiral amines. *Biotechnol. Bioeng.* 99, 275–284.
36. Hanson RL, Davis BL, Chen Y, Goldberg SL, Parker WL, Tully TP, Montana MA, Patel RN (2008). Preparation of (*R*)-amines from racemic amines with an (*S*)-amine transaminase from *Bacillus megaterium*. *Adv. Synth. Catal.* 350, 1367–1375.
37. Hoehne M, Robins K, Bornscheuer UT (2008). A protection strategy substantially enhances rate and enantioselectivity in omega-transaminase-catalyzed kinetic resolutions. *Adv. Synth. Catal.* 350, 807–812.
38. Truppo MD, Turner NJ, Rozzell JD (2009). Efficient kinetic resolution of racemic amines using a transaminase in combination with an amino acid oxidase. *Chem. Commun.* 2127–2129.
39. Turner NJ, Fotheringham IG, Speight R (2004). Novel biocatalyst technology for the preparation of chiral amines. *Innovations Pharm. Technol.* 4, 114–116.
40. Kim KH (1964). Purification and properties of a diamine alpha-ketoglutarate transaminase from *Escherichia coli*. *J Biol. Chem.* 239, 783–786.
41. Sterling DI, Zeitlin AL, Matcham GW, inventors; Celgene Corp, assignee. Enantiomeric enrichment and stereoselective synthesis of chiral amines. US Patent 4,950,606. August 21, 1990.
42. Matcham GW, Rozzell JD, Sterling DI, Zeitlin AL, inventors; Celgene Corp, assignee. Enantiomeric enrichment and stereoselective synthesis of chiral amines. US Patent 5,169,780. December 8, 1992.
43. Matcham GW, Sterling DI, Zeitlin AL, inventors; Celgene Corp, assignee. Enantiomeric enrichment and stereoselective synthesis of chiral amines. US Patent 5,300,437. April 5, 1994.
44. Shin JS, Kim BG (1999). Asymmetric synthesis of chiral amines with omega-transaminase. *Biotechnol. Bioeng.* 65, 206–211.
45. Shin JS, Yun H, Jang JW, Park I, Kim BG (2003). Purification, characterization, and molecular cloning of a novel amine: pyruvate transaminase from *Vibrio fluvialis* JS17. *Appl. Microbiol. Biotechnol.* 61, 463–471.
46. Shin JS, Kim BG (2002). Substrate inhibition mode of omega-transaminase from *Vibrio fluvialis* JS17 is dependent on the chirality of substrate. *Biotechnol. Bioeng.* 77, 832–837.
47. Shin JS, Kim BG (2001). Comparison of the omega-transaminases from different microorganisms and application to production of chiral amines. *Biosci. Biotechnol. Biochem.* 65, 1782–1788.
48. Hohne M, Kuhl S, Robins K, Bornscheuer UT (2008). Efficient asymmetric synthesis of chiral amines by combining transaminase and pyruvate decarboxylase. *Chem. Bio. Chem.* 9, 363–365.
49. Truppo MD, Rozzell JD, Moore JC, Turner NJ (2009). Rapid screening and scale-up of transaminase catalyzed reactions. *Org. Biomol. Chem.* 7, 395–398.
50. Truppo MD, Turner NJ (2010). Micro-scale process development of transaminase catalyzed reactions. *Org. Biomol. Chem.* 8, 1280–1283.
51. Truppo MD, Rozzell JD, Turner NJ (2010). Efficient production of enantiomerically pure chiral amines at concentrations of 50 g/L using transaminases. *Org. Process Res. Dev.* 14, 234–237.
52. Savile CK, Janey JM, Mundorff EC, Moore JC, Tam S, Jarvis WR, Colbeck JC, Kreber A, Fleitz FJ, Brands J, Devine PN, Huisman GW, Hughes GJ (2010). Biocatalytic asymmetric synthesis of chiral amines from ketones applied to sitagliptin manufacture. *Science* 329, 305–309.

4

DEVELOPMENT OF A SITAGLIPTIN TRANSAMINASE

JACOB M. JANEY

Department of Process Research, Merck Research Laboratories, Merck & Co Inc., Rahway; Chemical Development, Bristol-Myers Squibb, New Brunswick, NJ, USA

4.1 INTRODUCTION

Sitagliptin constitutes the active pharmaceutical ingredient (API) in *Januvia*® and *Janumet*® [1,2], an oral DPP-IV inhibitor for the treatment of type II diabetes produced by Merck and Co., Inc. Type II diabetes is estimated to affect over 220 million people worldwide [1]; hence this therapeutic class of drugs represents an important treatment option to control blood glucose without the risk of hypoglycemia associated with conventional insulin injections.

The increasing prevalence of this disease in addition to once-daily dosage of 100 mg of sitagliptin, increasing demand for API, and general cost pressures led process chemists at Merck to reevaluate the current manufacturing route used to produce sitagliptin phosphate monohydrate (Scheme 4-1).

The current manufacturing route is centered upon a fairly unprecedented asymmetric hydrogenation of an unprotected enamine amide **2** using a Rh-Josiphos catalyst under 250 psi of hydrogen [3] (Josiphos = 1-[2-(diphenylphosphino)ferrocenyl]ethyldicyclohexylphosphine). While highly efficient in terms of assay yield (95–98%) and catalyst loading (0.15 mol%), the enantioselectivity typically is 97% ee and the toxic (and precious) rhodium metal needs to be removed and recovered with a carbon treatment on the crude reaction stream. The low enantioselectivity thus requires a crystallization of sitagliptin free base **3** from *i*-PrOH/heptanes to reject the minor enantiomer and reach an

[1] http://www.who.int/mediacentre/factsheets/fs312/en/ (last accessed May 3, 2012).

Sustainable Catalysis: Challenges and Practices for the Pharmaceutical and Fine Chemical Industries, First Edition. Edited by Peter J. Dunn, K. K. (Mimi) Hii, Michael J. Krische, and Michael T. Williams.

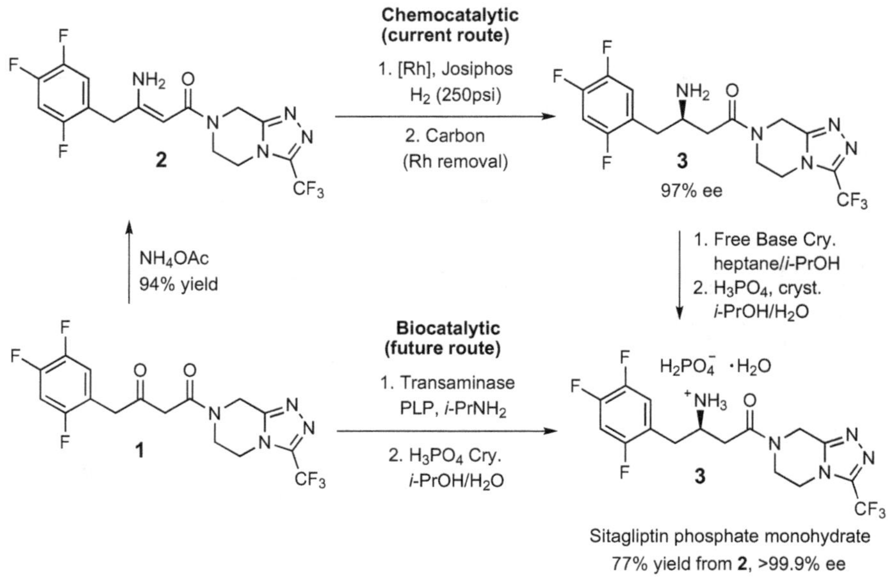

SCHEME 4-1. Current versus future sitagliptin manufacture.

ee of >99.9%. This crystalline free base is then converted to the active sitagliptin phosphate monohydrate through a second crystallization process. Overall, this manufacturing route involves two isolations and four operations, giving the desired API **3** in 77% yield.

In order to improve the overall efficiency of this manufacturing route, it was clear that elimination of the carbon treatment and ee upgrade crystallization operations would be highly advantageous in terms of yield and overall time. One technology that had the potential to address these issues was a direct transamination of keto amide **1**, which would deliver sitagliptin free base **3** in >99.9% ee. In addition, a biocatalytic route would eliminate the use of expensive rhodium metal and hazardous high-pressure hydrogen. Unfortunately, at the time that this was considered, all known transaminases were strictly limited to either cyclic, unsubstituted ketones or methyl ketones. Hence, a collaboration between Merck and Codexis was initiated to evolve a transaminase which not only met this initial goal, but the series of enzymes that were the result of this collaboration proved to be broadly useful solution for the conversion of most ketones to (*R*)-amines [4,5]. This highly efficient manufacturing process was recognized with the 2010 Environmental Protection Agency (EPA) Presidential Green Chemistry Award.

4.2 CREATING ACTIVITY

As stated previously, known transaminases were expected to be unreactive to keto amide **1** [6–10]. This fact was borne out during laboratory screens of **1** with all commercial and academic transaminases available where sitagliptin **3** was never detected by careful analytical analysis. An approach was thus formulated based on the concept of "substrate walking." The idea is to evolve activity toward an analog that is expected to be an

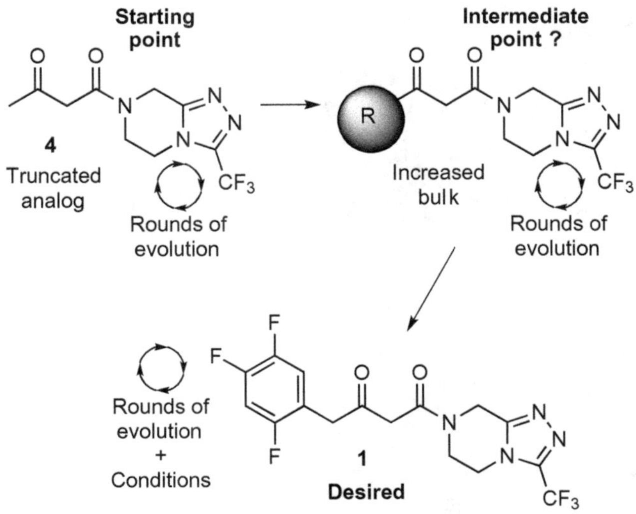

SCHEME 4-2. "Substrate walking" with directed evolution.

acceptable substrate for wild-type enzymes; in this case the methyl ketone analog **4** would be evolved for. After rounds of evolution, a more sterically challenging substrate would be introduced, perhaps the ethyl or cyclopropyl analog. The evolutionary process would be repeated and then applied to the actual desired ketone **1** (Scheme 4-2).

To carry out this strategy, an appropriate enzyme starting point was required and an (*R*)-selective transaminase (ATA-117, a homolog of an enzyme from *Arthrobacter* sp.) [11], was selected, as this particular transaminase had proven broad utility for cyclic and methyl ketones [8,10]. Based on an *in silico* analysis of a structural homology model of ATA-117 docked with the prositagliptin ketoamide **1**, it was clear that there were severe nonbonding interactions in both the small and the large pockets of the active site (Figure 4-1a). For the "substrate walking" strategy, the methyl ketone analog **4** was optimized in the active site by first engineering the large pocket to accommodate the large heterocyclic amide portion [4].

Empirical validation of the calculated homology model active site occurred upon evaluation of a site saturation library of the residues lining the proposed active site large pocket. While the parent ATA-117 gave only ~4% conversion of **4** ($2\,\mathrm{g\,l^{-1}}$) to amine product, a 10-fold improvement in activity was observed in variants containing an S223P mutation. To further validate this model, improve activity toward **4**, and perhaps create activity toward the desired ketoamide **1**, a focused library of variations (V69, F122, T283, and A284) at residues predicted to interact with the trifluorophenyl moiety in the small pocket was generated and tested. A major breakthrough was observed in this small pocket mutation library when a variant containing four mutations was identified that gave the first detectable transaminase ($10\,\mathrm{g\,l^{-1}}$) activity toward **1** ($2\,\mathrm{g\,l^{-1}}$) giving 0.7% conversion to sitagliptin **3**. The homology model and sequence analysis revealed that this initial activity was due to an F122I, a V, or an L mutation in combination with V69G or A284G that presumably reduced steric interactions and strain in the small pocket (Figure 4-1b). Having generated activity where once there was none, the process of evolving the enzyme to meet manufacturing targets and capabilities was initiated.

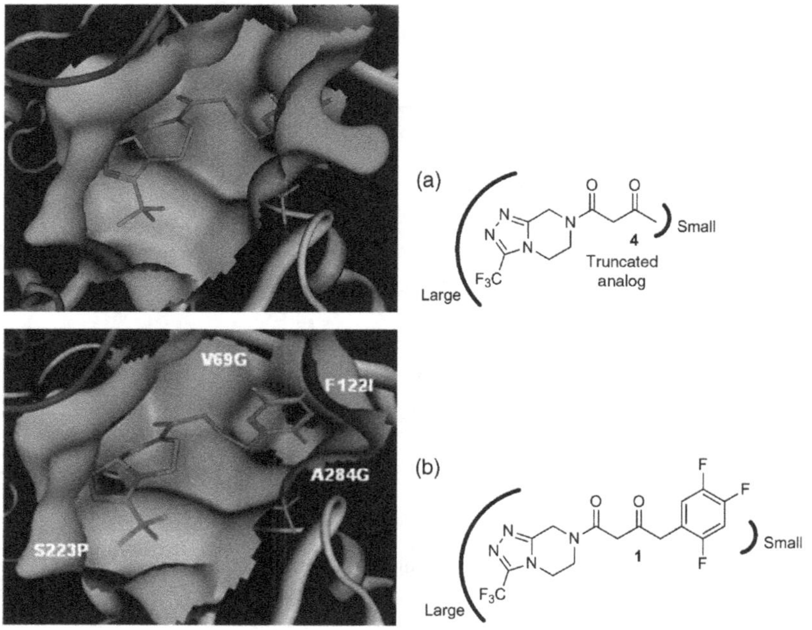

FIGURE 4-1. (a) Wild-type (ATA-117) homology model. (b) Model with V69G, F122I, A284G, and S223P. Reprinted with permission from Ref. 4. (*See the color version of this figure in Color Plate section.*)

4.3 TRANSAMINASE EVOLUTION

At the outset of the directed evolution process, a set of initial targets was created defining a minimally acceptable level of reaction efficiency, which would compete with the current manufacturing process, namely $100\,\mathrm{g}\,\mathrm{l}^{-1}$ substrate **1** and $<5\,\mathrm{g}\,\mathrm{l}^{-1}$ transaminase using isopropylamine as the donor. The initial stages of evolution were focused on increasing innate activity toward **1**, whereas the later stages were conducted in parallel with reaction conditions development work focused on cosolvent stability, temperature, pH, and loadings.

For this first-stage optimizing activity, the variant with the highest activity from the round 1b library (the previously described focused library based on hits from the first library created) was selected as the parent for the second round. Furthermore, all of the identified beneficial mutations from the first rounds were combined into this round 2. The resulting combinatorial round 2 library resulted in a transaminase with a 75-fold improvement over the round 2 parent. This second round winner contained 12 mutations, of which 10 aligned with the active site in the homology model.

The next stage of evolution began to focus on substrate concentration and overall enzyme stability to increasingly challenging reaction conditions. It's important to note that the reaction conditions and processing were developed in parallel to the enzyme evolution. After each round, the top variants underwent a series of larger scale validations, testing, and reaction optimization. The information generated from this was then fed into the next library in order to identify the library parents and best mutations and to modify the reaction screening conditions to closer reflect the desired reaction conditions.

TABLE 4-1. Screening Conditions

Round	1 and 2	3	4	5	6	7–9	10 and 11
Screen [**1**] (gl^{-1})	2	5	10	40	100	100	50
Scale-up [**1**] (gl^{-1})	N/A	N/A	N/A	50	100–200	200	200–300
[*i*-PrNH$_2$] (M)	0.5	0.5	1.0	1.0	1.0	1.0	1.0
Cosolvent	5% DMSO	5% MeOH	5% MeOH	10% MeOH	20% MeOH	25% DMSO	50% DMSO
pH	7.5	7.5	8.5	8.5	8.5	8.5	8.5
T (°C)	22	30	30	45	45	45	45

Using this approach, iterative rounds of directed evolution were conducted by generating mutants, sorting (ProSAR) [12], and recombining across the entire protein sequence (Table 4-1). It is important to note that the minor enantiomer of sitagliptin **3** was *never observed* at detectable levels in select variants from each library; hence the ee was always >99.95% for all top variants in every round.

Some important modifications to the reaction conditions were made along the way. In round 4, the pH was increased from 7.5 to 8.5, resulting in a 2.5-fold rate improvement. The temperature was also increased from room temperature to 45 °C. Interestingly, while the later rounds generated variants capable of withstanding temperature of >60 °C and pH >10.5, ketoamide **1** readily decomposes at a competitive rate under these conditions. The primary challenge to increasing the substrate **1** concentration is due to ketoamide **1** aqueous solubility (<1 $g\,l^{-1}$). Hence, increasing amounts of organic cosolvents were needed. Solvent screening revealed both methanol and DMSO to be acceptable solvents to increase **1** solubility while minimizing enzyme denaturation. However, scale-up studies of the round 6 variants showed that MeOH was easily lost to evaporation under the reaction conditions and caused oiling out of ketoamide **1**, creating a two-phase solution. MeOH also increased the rate and amount of ketoamide **1** degradation. Thus, DMSO was selected as the preferred cosolvent to increase substrate concentration. In addition, scale-up studies on 30–50 ml reaction volume revealed that screening could be conducted at higher substrate concentrations (100 gl^{-1} **1** in rounds 6–9), although this was decreased to 50 gl^{-1} **1** in rounds 10–11 as it ensured a more robust screen with no chance of a second-phase separating.

The concentration of the isopropylamine donor was also increased throughout the evolution cycle. The early rounds (1–2) utilized 1 M *i*-PrNH$_2$, but this was found to decrease the enzyme's stability; hence the screening was conducted with 0.5 M *i*-PrNH$_2$. This was later increased back to 1 M after sufficient stability was engineered in order to push the reaction equilibrium forward. It was also noted that the accumulation of acetone over the course of the reaction pushes the equilibrium backward; thus running small-scale reactions loosely capped was advantageous to remove acetone.

In addition to the more stringent screening conditions with each subsequent round, the later rounds were also optimized for mutations that would maximize expression levels in the *E. coli* manufacturing host. Typically, the lyophilized enzyme powder is targeted at 60% w/w active transaminase (the remainder being fermentation debris); hence mutations that ensure this robust manufacturing were selected for as well.

A head-to-head comparison of the top variants from each round of evolution under the same manufacturing like conditions is shown in Figure 4-2. In the early rounds (1–5), there is a clear trend where increased initial reaction rates were observed in the first 5 h, but extended reaction times show that conversions plateau at <50%. Later rounds (6–9) do not show a real increase in

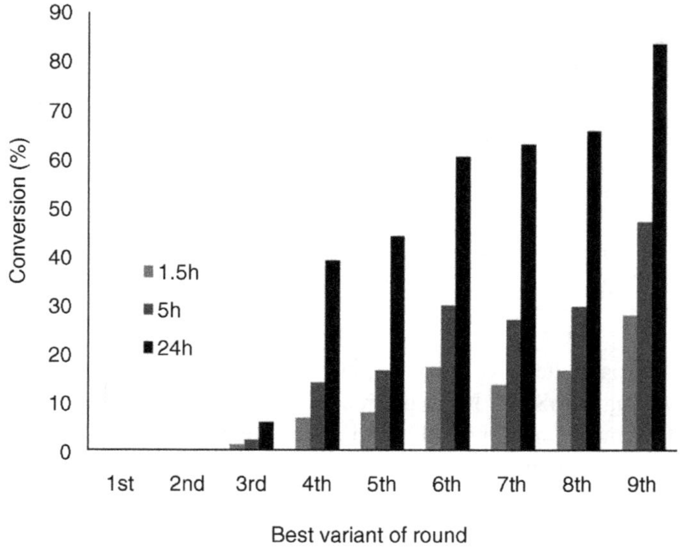

FIGURE 4-2. Head-to-head comparison of top variants from each evolution round at 5 gl^{-1} enzyme, 50 gl^{-1} ketoamide **1**, 1 M *i*-$PrNH_2$, 1 mM PLP, 50% DMSO, 50% 0.1 M triethanolamine buffered to pH 8.5, 45 °C, no pH control.

rates during the first 5 h of reaction, but allowing the reaction to age 24 h gave much better final conversions (>60%). This trend clearly demonstrates how the early rounds of evolution were focused on increased activity, whereas later rounds were screened under increasingly challenging process conditions that selected for more robust, stable enzyme variants.

Overall, the final round 11 variant contained a total of 27 mutations. The homology model predicted that 17 noncatalytically active residues in the active site were interacting with the substrate and 10 of these were mutated in the final catalyst. These 27 key mutations were identified through an arsenal of techniques at Codexis, including homology modeling, random mutagenesis, and site saturation. Mapping of these mutations readily revealed that the early round mutations were centered upon the active site, whereas the later rounds contained mutations along the active catalyst dimer interface region and solvent-exposed areas.

The primary lesson learned from this directed evolution-enabled screening was that multiple factors could be optimized in parallel: activity, cosolvent, temperature, isopropylamine, substrate **1** concentration, inhibition (acetone, substrate, and product), and expression in *E. coli*. All this can be accomplished by validating the winners from each screen in scale-up confirmation studies and modifying the reaction conditions, and hence the next round screening conditions, to reflect the desired process goals. A key component to this continued challenge to the screening conditions is to have an understanding of what the final, optimized process should be and what desired traits the enzyme must possess.

4.4 PROCESS OPTIMIZATION

A number of process chemistry challenges were identified and addressed during the evolution of the transaminase and during subsequent scale-up studies. The primary challenge for this particular enzymatic reaction was the poor mass transfer of ketoamide

1. This substrate **1** has only trace solubility in aqueous systems (<1 gl^{-1}), and even the use of 50 vol% DMSO as a cosolvent gives only 9–10 gl^{-1}**1**. This necessitated the use of a slow feed (2–4 h addition time) of ketoamide **1** dissolved in DMSO to a solution of enzyme, pyridoxal-5′-phosphate (PLP 1 gl^{-1}), and *i*-PrNH_2 HCl dissolved in 37 vol% DMSO in water. At the end of this charge of **1**, the final DMSO content is ~50 vol%. If the addition of **1** is too fast (or if the final DMSO content is <40 vol%), a large amount of "oiling out" was observed. This second phasing consists of a mixture of sitagliptin **3**, ketoamide **1**, and a readily reversible imine dimer between **1** and **3**. This protocol ensures a fairly homogeneous reaction course and minimizes exposure of the enzyme to high DMSO content. Also, running the reaction at temperatures between 45 and 55°C increased solubility, and hence the overall reaction rate. Using this slow addition, up to 275 gl^{-1} of ketoamide **1** can be converted to sitagliptin **3** in 24 h using 3% w/w of the final enzyme variant.

A second challenge for this enzymatic process is the need for pH control during the reaction course. The pK_a of sitagliptin **3** is ~9.5, whereas the isopropylamine that is being consumed and converted to acetone has a pK_a of ~11.5 in this DMSO/water solvent system. This means that as the reaction proceeds, the pH continually drops until it reaches a steady equilibrium value of ~8. This necessitated the use of a pH-controlled feedback wherein a 4 M solution of isopropylamine in water was added to the batch to keep the pH at 8.5 ± 0.5. The reaction and enzyme itself tolerates a fairly broad pH range from 8.0 to 11.0, although ketoamide **1** was found to be less stable at elevated pH and higher temperatures (>60 °C). In addition, isopropylamine is quite volatile (b.p. 32 °C) and is easily lost to the vapor phase throughout the reaction. It was also found that the optional use of a 0.1–0.2 M triethanolamine buffer allowed for less rapid pH changes. A typical reaction consumes 1–2 molar equiv. of the 4 M isopropylamine throughout its course in addition to the isopropylamine in the initial buffer make-up.

Transaminations are equilibrium-based processes, hence the need to shift equilibria to the right to reach high levels of conversion (Figure 4-3). For this process, the inherent equilibrium using 4–10 equiv. of isopropylamine with 200 gl^{-1}**1** at pH 8.5 gives an equilibrium ratio of ~87:13 between sitagliptin 3 and ketoamide **1** in a sealed reactor (1.8 vol% acetone generated). This implied that in order to reach conversion levels in excess of 95%, acetone had to be removed to <0.8 vol%. To meet this requirement, *in situ* IR was used to observe both acetone build-up during the reaction along with sitagliptin **3** formation. Both of these responses were calibrated by GC and HPLC samples. Using this tool to monitor the reaction progress allowed us to determine that a combination of rapid stirring, mild vacuum (100–350 torr), and a nitrogen sweep of the reactor headspace (2–6 fps) allowed for >70% acetone removal. This had the desired effect of driving conversions above 95% and also increased the reaction rate. The efficiency of acetone removal was found to be highly scale dependent and had to be adjusted accordingly.

The final process has a wide range of operating conditions:

- Ketoamide loading: Up to 275 gl^{-1} in <24 h
- Cosolvent: 0–60 vol% DMSO (or MeOH)
- Temperature: 30–60 °C
- Enzyme loading: 3–6 wt.% (~0.01 mol%)
- pH range: 9–11
- 15.6 $gl^{-1}h^{-1}$ productivity
- Ketone ~9 gl^{-1} solubility (mass transfer limited kinetics)

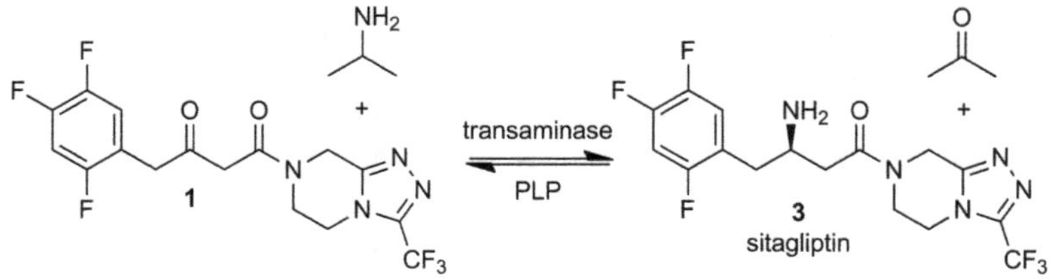

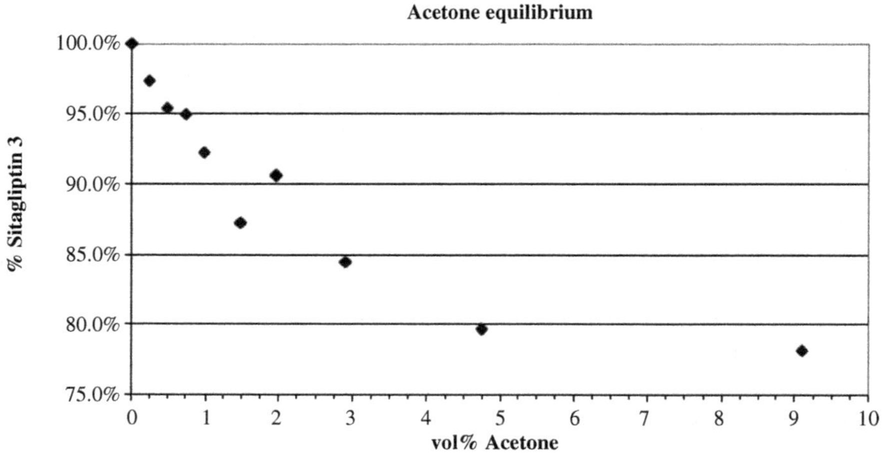

FIGURE 4-3. Acetone equilibrium study at pH 8.5.

Importantly, the transaminase is highly robust and very resistant to denaturation under forcing conditions of up to 50 vol% DMSO, pH ~11, and 60 °C. Under the typical operating conditions, the kinetics appears to be dominated by mass transfer rates of the ketoamide **1**, due to its limited solubility. This was also reflected during the evolution process (Figure 4-2), where productivity gains in the later rounds were more due to optimization of enzyme stability rather than increasing the rate of catalyst turnover.

One of the major challenges for scaling of any enzymatic transformation is the workup due to the propensity for emulsification during extractions. In a GMP environment, the removal of enzyme becomes a critical operation and requirement. On a smaller scale, a simple extraction using centrifugation easily handles even fairly intractable emulsions. On a larger scale, this becomes increasingly impractical and time consuming. Thus, two generalized enzyme removal procedures were developed.

The first involves a filtration with the transaminase precipitated out of the reaction prior to organic extraction. For this reaction, a flocculant (Solka-Floc® or celite) is first charged to the vessel followed by a pH adjustment to ~2 with 12 M HCl. This acidification not only causes the enzyme to readily precipitate onto the flocculant, but it breaks up any imine dimers between **1** and **3** that may have accumulated. This slurry is then sent through a cloth filter with aqueous rinsing of the filter cake. To the filtrate is then added an extraction solvent, isopropyl acetate (IPAc), followed by a pH adjustment to ~11 using 10 N NaOH. While this filtration protocol works well, the filtration can be slow on scale and the filter medium can often blind; hence a simpler extraction procedure was developed.

The direct extraction is advantageous in that it is fast and operationally simple. The key to its development was the recognition that the extracting solvent density plays a key part in enzyme partitioning and emulsification. Simply adjusting the pH to ∼11 and then using a 1/1 solvent ratio of isopropyl alcohol and IPAc gives a very clean phase cut with all of the enzyme suspended in the alcohol-rich aqueous layer and sitagliptin **3** free base in the IPAc-rich organic layer. This is followed by brine washes of the organic extracts that serve to remove residual DMSO and water.

The overall process represents a marked improvement over the asymmetric hydrogenation process (Scheme 4-1). A comparison of the two processes reveals that the new biocatalytic route is a through process from ketoamide **1** to sitagliptin phosphate monohydrate **3** in >99.95% ee and 88% overall yield. This is a 7% increase in overall yield that translates to a ∼30% increase in productivity (based on yield, volumetric productivity, and time), along with a decrease in product mass intensity (kg waste/kg **3**) from 38 down to 26. Due to these improvements, and the technical achievement they represent, this process was recognized by the 2010 EPA Presidential Green Chemistry Award[2]. It also validates the investment in new technologies to drive cost savings.

4.5 A GENERAL AMINATION METHODOLOGY

The general, direct conversion of a ketone to an enantiopure (*R*)-amine is a highly desirable transformation. Approximately two-thirds of the top 200 generic and top 200 branded pharmaceuticals that contain a stereogenic amine center are of *R*-configuration[3]. The previously described sitagliptin transamination conditions ($1\ gl^{-1}$ PLP, 1–10 wt.% transaminase, 1 M *i*-$PrNH_2$ in 1/1 water/DMSO at 45–60 °C and pH 8.5–11) are indeed a fairly general amination methodology.

Typically, engineered enzymes are viewed as highly specific catalysts tied to a particular substrate. This may hold true over extensive rounds of mutation and selection, but in the case of the *R*-transaminases produced for manufacturing sitagliptin, the top variants from the middle to late rounds proved to be highly general amination catalysts that can serve as evolutionary starting points for other transformations as well. As shown in the unoptimized results in Figure 4-4, the sitagliptin variants show a marked increase in activity for ketones that were previously unreactive with the wild-type-derived ATA-117 (three-enzyme system using alanine as the amine donor). Of particular note is the transamination of trifluoromethyl acetophenone to give access to enantiopure *(S)*-trifluoromethyl amines that are a structural motif of interest for medicinal chemists [13]. Also, these conditions are compatible with leaving groups, such as chloride, to give 2-aryl pyrroles via an *in situ* cyclization of the chiral amine that is generated during transamination [14]. Given these results and the ability to tailor enzymes with directed evolution from these highly evolved starting points, transamination is the preferred industrial methodology for the conversion of ketones to *(R)*-amines.

4.6 CONCLUSION AND OUTLOOK

Nature is capable of amazing and subtle chemical transformations under conditions that are generally benign. Historically, harnessing these catalysts for industrial applications, such as Baker's yeast reductions and hydrolase resolutions, has primarily relied upon naturally

[2] http://www.epa.gov/gcc/pubs/pgcc/winners/grca10.html (last accessed May 3, 2012).

[3] http://cbc.arizona.edu/njardarson/group/top-pharmaceuticals-poster (last accessed May 3, 2012).

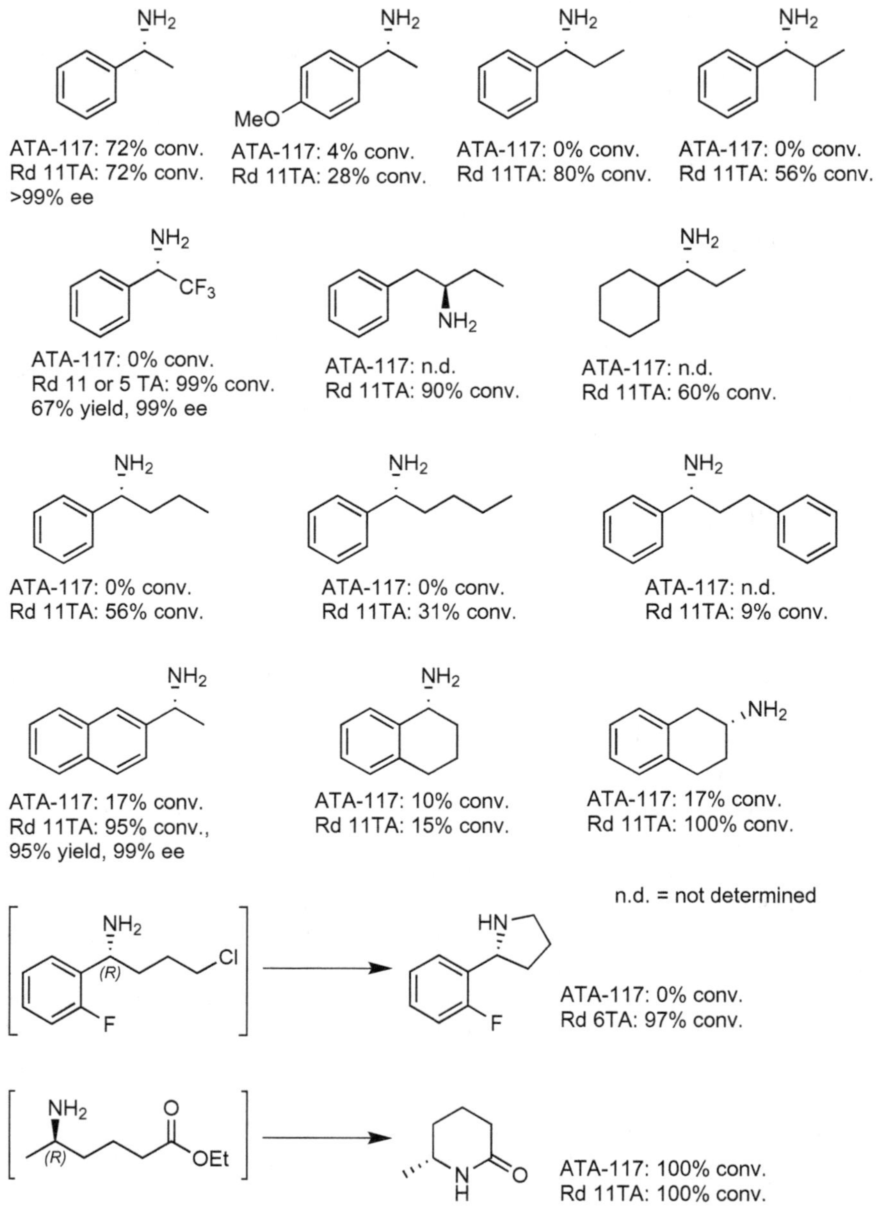

FIGURE 4-4. Substrate survey comparing ATA-117 (wild type using 3-enzyme recycling conditions) versus top variants from sitagliptin evolution (single enzyme *i*-PrNH_2 recycle system in DMSO/water). Reprinted with permission from Ref. 4.

occurring diversity in order to identify catalysts that accept the desired substrate and operate under conditions applicable to manufacture. Unfortunately, nature rarely tailors enzymes to operate on compounds and intermediates that are of broad interest to synthetic organic chemists, and these enzymes rarely tolerate solvents, temperatures, and pHs that are often required to process these compounds. The intersection of a number of key, breakthrough technologies has changed this paradigm, namely computational power, rapid sequencing, bioinformatics, and the techniques of directed evolution. No longer tied to natural diversity, a chemist has free rein to rapidly generate and evolve reactivity. From a desktop computer, a chemist can identify a promising sequence from a database, perhaps based on homology, order the DNA, have it expressed and fermented in a production host, and then test the resulting enzyme variant all in a matter of 4–6 weeks. Coupling this with high-throughput experimentation makes the creation of thousands of related mutants and their screening an industrial process.

The present case study in enzyme evolution represents a step change in industrial biocatalyst development. For the first time, enzymatic activity was created where none had previously existed, and this was taken on to a manufacturing level. Notably, this new biocatalytic process was more efficient than the previous asymmetric hydrogenation manufacturing route. The implication of this is that any given enzymatic transformation performed by nature can, in theory, be evolved to operate on any substrate of interest to a synthetic organic chemist, given enough effort and resource commitment. Currently, hydrolytic resolutions, ketone reductions, and now ketone aminations are well served by biocatalysts. Immobilization of these catalysts will further extend their scope and generality. Additional challenges are now being addressed to evolve enzymes for selective oxidations [15], carbon–carbon bond formations [16], and halogenations [17,18]. The next generation of enzyme engineering is the creation of reactivity not currently found in nature, such as Diels–Alderases [19]. This expansion of nature's toolbox will require greater understanding of enzymatic catalysis and more sophisticated computation, but the foundations are already being established.

4.7 PROCEDURES

4.7.1 Transaminase Reaction at 1 kg Reaction Scale

The reaction was run in a vessel fitted with a mechanical stirrer, temperature probe, pH probe, and base addition line. The base addition line was used to control the pH between 8.4 and 8.6 using a feed of 4 M isopropylamine free base in water. To the vessel was added 1.92 l water, followed by 109 ml (0.82 mol, 0.33 equiv.) triethanolamine and 1.64 l (6.56 mol, 2.67 equiv.) of 4 M isopropylamine solution. The pH was then adjusted to 8.5 using 12 N HCl (424 ml). The reactor was then charged with 6.7 g (0.027 mol, 0.011 equiv.) of PLP followed by 40 g of CDX-017, and the mixture was carefully dissolved with gentle agitation. The vessel was placed on the reactor block with the temperature probe, base addition line, pH probe, and stirrer set to 400 rpm. (Note: pH control loop is off at this point.) Next, 2.22 l of DMSO was added into the stirring solution and the reactor was heated to 45 °C. When the temperature stabilized, the pH control loop was turned on and the pH was adjusted to 8.5 (pH controlled with 4 M i-$PrNH_2$ in water). At this point, stirring was increased to 600 rpm, but tip speed was kept below 2 m s^{-1} to avoid vortexing. Then, 1.0 kg (corrected weight is 1 kg, as received ketoamide is typically 96–98 wt.% as a hemihydrate; 2.46 mol, 1.00 equiv.) of ketoamide was dissolved into 1.11 l of DMSO, and this solution was added to the reactor over 2–3 h. The reactor was then stirred at 45 °C with the pH maintained between 8.6 and 8.4 for another ~13 h, with acetone removal being accomplished with 300 torr vacuum and 2 fps nitrogen sweep. After ~15 h total reaction time (1.3–2.0 equiv. isopropylamine uptake), the reaction was at 90–95% conv. as judged by reverse-phase HPLC analysis.

4.7.2 Filtration Workup

The pH control loop was turned off, and 13 g of Solka-Floc was added to the vessel followed by 12 M HCl until pH 2–3. The reaction was aged 1–2 h at 45 °C and 1000 rpm. The slurry was then passed through a filter (e.g., fritted plastic Buchner with filter paper on 1 kg scale, or sparkle filter with no recycle loop on pilot plant scale). The vessel and filter were rinsed with 1 l of 0.01 N HCl. To this aqueous acidic filtrate was then added 3 l of IPAc, and the pH of the aqueous phase was then adjusted to 11 with 19 N NaOH. The layers were agitated with stirring and then allowed to settle and separated (mild heat or vacuum accelerates phase separation). This was repeated twice more with 3 l of IPAc, and the combined organics were then washed with 3 l of brine (at pH 11). The resulting IPAc solution of the sitagliptin free base was then assayed for yield (typically 88–92% assay yield; 882–922 g) and solvent switched to isopropanol (IPA) for downstream processing to sitagliptin phosphate monohydrate.

4.7.3 Direct Extraction Workup

The pH control loop was turned off, and 12 M HCl was added until pH 2–3. The reaction was aged 1–2 h at 45 °C and 1000 rpm. The batch was cooled to room temperature, and then 3 l of IPA was added, followed by 3 l of IPAc. The pH of the aqueous layer was then adjusted to 11 with 19 N NaOH. The mixture was agitated at 20–45 °C (heat may be used to break the emulsion) and then allowed to settle and separate. The IPAc/IPA layer was set aside, and the aqueous layer was extracted with 3 l of 80/20 (v/v) IPAc/IPA. The combined IPAc/IPA extracts were then washed with 3 l of brine. The resulting IPAc/IPA solution of the sitagliptin free base was then assayed for yield (typically 87–90% assay yield, 872–902 g) and solvent switched to IPA for downstream processing to sitagliptin phosphate monohydrate.

REFERENCES

1. Thornberry NA, Weber AE (2007). Discovery of JANUVIA® (sitagliptin), a selective dipeptidyl peptidase IV inhibitor for the treatment of type 2 diabetes. *Curr. Top. Med. Chem.* 7, 557–568.
2. Engel SS, Williams-Herman DE, Golm GT, Clay RJ, Machotka SV, Kaufman KD, Goldstein BJ (2010). Sitagliptin: review of preclinical and clinical data regarding incidence of pancreatitis. *Int. J. Clin. Pract.* 64, 984–990.
3. Hansen KB, Hsiao Y, Xu F, Rivera N, Clausen A, Kubryk M, Krska S, Rosner T, Simmons B, Balsells J, Ikemoto N, Sun Y, Spindler F, Malan C, Grabowski EJJ, Armstrong JD (2009). Highly efficient asymmetric synthesis of sitagliptin. *J. Am. Chem. Soc.* 131, 8798–8804.
4. Savile CK, Janey JM, Mundorff EC, Moore JC, Tam S, Jarvis WR, Colbeck JC, Krebber A, Fleitz FJ, Brands J, Devine PN, Huisman GW, Hughes GJ (2010). Biocatalytic asymmetric synthesis of chiral amines from ketones applied to sitagliptin manufacture. *Science* 329, 305–309.
5. Savile CK, Mundorff EC, Moore JC, Devine PN, Janey JM, inventors; Codexis, inc., assignee. Transaminase biocatalyst. World Patent WO/2010/099501. February 26, 2010.

6. Hohne M, Kuhl S, Robins K, Bornscheuer UT (2008). Efficient asymmetric synthesis of chiral amines by combining transaminase and pyruvate decarboxylase. *ChemBioChem* 9, 363–365.
7. Koszelewski D, Lavandera I, Clay D, Rozzell D, Kroutil W (2008). Asymmetric synthesis of optically pure pharmacologically relevant amines employing omega-transaminases. *Adv. Synth. Catal.* 350, 2761–2766.
8. Koszelewski D, Clay D, Rozzell D, Kroutil W (2009). Deracemisation of alpha-chiral primary amines by a one-pot, two-step cascade reaction catalyzed by omega-transaminases. *Eur. J. Org. Chem.* 14, 2289–2292.
9. Truppo MD, Rozzell DJ, Moore JC, Turner NJ (2009). Rapid screening and scale-up of transaminase catalyzed reactions. *Org. Biomol. Chem.* 7, 395–398.
10. Truppo MD, Turner NJ, Rozzell JD (2009). Efficient kinetic resolution of racemic amines using a transaminase in combination with an amino acid oxidase. *Chem. Commun.* 2127–2129.
11. Iwasaki A, Yamada N, Kizaki N, Ikenaka Y, Hasegawa J (2005). Microbial synthesis of chiral amines by (*R*)-specific transamination with Arthrobacter sp KNK168. *Appl. Microbiol. Biotechnol.* 69, 499–505.
12. Fox RJ, Davis SC, Mundorff EC, Newman LM, Gavrilovic V, Ma SK, Chung LM, Ching C, Tam S, Muley S, Grate J, Gruber J, Whitman JC, Sheldon RA, Huisman GW (2007). Improving catalytic function by ProSAR-driven enzyme evolution. *Nat. Biotechnol.* 25, 338–344.
13. Gosselin F, O'Shea PD, Roy S, Reamer RA, Chen C, Volante RP (2005). Unprecedented catalytic asymmetric reduction of N-H imines. *Org. Lett.* 7, 355–358.
14. Campos KR, Klapers A, Waldman JH, Dormer PG, Chen C (2006). Enantioselective, palladium-catalyzed alpha-arylation of *N*-Boc-pyrrolidine. *J. Am. Chem. Soc.* 128, 3538–3539.
15. Jung ST, Lauchli R, Arnold FH (2011). Cytochrome P450: taming a wild type enzyme. *Curr. Opin. Biotechnol.* 22, 809–817.
16. Clapes P, Fessner WD, Sprenger GA, Samland AK (2010). Recent progress in stereoselective synthesis with aldolases. *Curr. Opin. Chem. Biol.* 14, 154–167.
17. Lang A, Polnick S, Nicke T, William P, Patallo EP, Naismith JH, Van Pee KH (2011). Changing the regioselectivity of the tryptophan 7-halogenase PrnA by site-directed mutagenesis. *Angew. Chem., Int. Ed.* 50, 2951–2953.
18. Wong C, Fujimori DG, Walsh CT, Drennan CL (2009). Structural analysis of an open active site conformation of nonheme iron halogenase CytC3. *J. Am. Chem. Soc.* 131, 4872–4879.
19. Siegel JB, Zanghellini A, Lovick HM, Kiss G, Lambert AR, St. Clair JL, Gallaher JL, Hilvert D, Gelb MH, Stoddard BL, Houk KN, Michael FE, Baker D (2010). Computational design of an enzyme catalyst for a stereoselective bimolecular Diels–Alder reaction. *Science* 329, 309–313.

5

DIRECT AMIDE FORMATION AVOIDING POOR ATOM ECONOMY REAGENTS

Benjamin M. Monks and Andrew Whiting
Department of Chemistry, Durham University, Durham, UK

5.1 INTRODUCTION

The carboxylic acid amide is not only a common functionality in organic compounds, it is also ubiquitous in the natural chemical and biochemical world. Indeed, analysis of the Comprehensive Medicinal Chemistry database reveals that the amide motif arises in around 25% of all synthetic pharmaceuticals [1], and as a result, amides are produced on a wide range of scales from milligrams to tonnes. The processes by which amides are made are therefore important, allowing both the rapid and high yielding synthesis of amides and providing routes that are amenable to scale-up and optimization.

In 2005, the American Chemical Society, Green Chemistry Institute, Pharmaceutical Roundtable, which consisted of several prominent pharmaceutical companies, was formed in order to encourage and implement green chemical principles into the global pharmaceutical and chemical industry [2]. The Roundtable highlighted the amide formation reaction as a key process that invariably involved poor atom economy reagents. It was, therefore, highlighted as a key challenge to synthetic chemists to develop greener direct amide formation reactions, starting from carboxylic acids and amines, without the use of stoichiometric and high molecular weight reagents. In this review, recent developments that address these issues are surveyed.

The most attractive method of accessing amides without attendant waste would be to directly react a carboxylic acid **1** and an amine **2**, and simply eliminate water via a seemingly straightforward condensation. Such a reaction would clearly be environmentally benign and would eliminate all waste. However, we are generally led to believe that this process is "seldom of preparative value" according to one leading advanced organic chemistry textbook [3]. In fact, the formation of an amide bond is thermodynamically

Sustainable Catalysis: Challenges and Practices for the Pharmaceutical and Fine Chemical Industries, First Edition. Edited by Peter J. Dunn, K. K. (Mimi) Hii, Michael J. Krische, and Michael T. Williams.

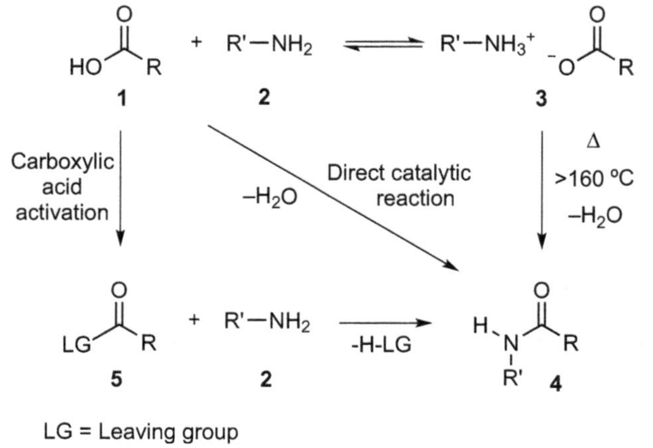

SCHEME 5-1. Different direct amide formation pathways from a carboxylic acid and an amine.

favorable and is, therefore, expected to be facile. The main problems with the condensation reaction between a carboxylic acid and an amine are: (1) that it is impeded by a high energy of activation barrier [4,5], which results in a reaction that is not widely understood from a mechanistic point of view [6,7] and is certainly underdeveloped; and (2) since the pK_a of most carboxylic acids is lower than that of many protonated amines, an ammonium carboxylate salt **3** forms preferentially, which is more resistant to the desired dehydration reaction [6,7]. Indeed, the ammonium carboxylate salt to amide conversion requires temperatures in excess of 160 °C in order to force the direct thermal reaction [4,6]. In consequence, there are no general methods for producing amides **4** directly at ambient temperatures, without resorting to the use of coupling or activating agents (Scheme 5-1) that are not only employed in stoichiometric amounts but also result in the formation of considerable amounts of waste by-products [4,6,8,9]. Hence, the common condensation strategies have poor atom economy, generally high E-factors [10] (mass of waste/mass of product) and poor effective mass yields [11].

Despite the lack of generally applicable reaction conditions for direct amide formation without the use of condensing agents, in recent years, promising catalytic systems have been developed, which allow the direct transformation of a carboxylic acid **1** and an amine **2** to provide the amide **3** with water as the only by-product (Scheme 5-1). Although truly generally applicable, ambient temperature, highly active, catalytic solutions to direct amide formation are still a little way-off, it is clear that chemists are moving in the right direction, and in certain cases, especially with more reactive carboxylic acid–amine combinations, this reaction can be synthetically viable. In this chapter, the current state-of-the-art in direct amide formation is examined, mainly involving boronic and boric acid-based systems, being perhaps most developed, though also examining other catalytic solutions.

5.2 MECHANISM FOR BORONIC AND BORIC ACID CATALYSIS

The exact mechanism by which either boronic or boric acids catalysis operates is far from clear, and hence, there is no agreement on the rate-determining step (RDS) or the exact nature of the acylating agent involved. The involvement of either mono-, di-, or in the case

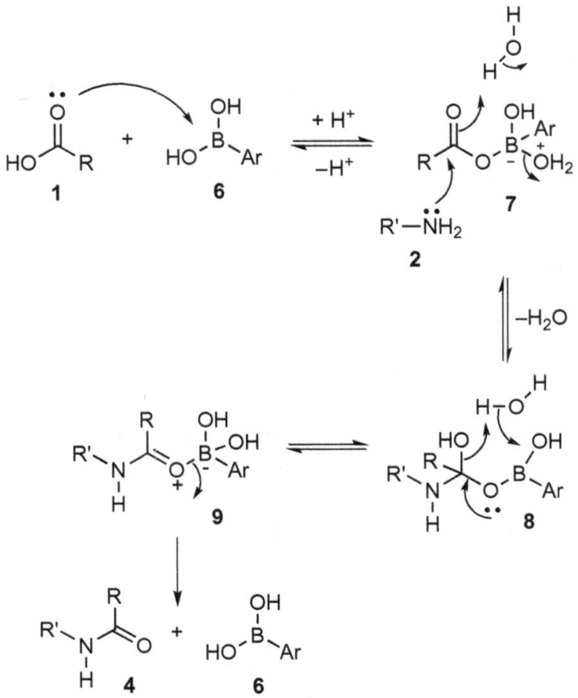

SCHEME 5-2. Proposed mechanism of boronic acid-catalyzed direct amide formation [12].

of boric acid, even triacyloxyboron species are theoretically possible [4,8,12,13]. Perhaps the most convincing hypothesis to date based on density functional theory (DFT) calculations using a large basis set is summarized in Scheme 5-2 [12].

The catalyst **6** reacts initially with a carboxylic acid **1** to form an acyloxyboronate species **7**, which remains closely associated with water and is accompanied by a concerted proton transfer that creates a good leaving group (water) attached to the boron. The acyloxyboronate species **7** is, therefore, the activated form of the carboxylic acid, enhanced through boron conjugation and intramolecular hydrogen bonding. Subsequent nucleophilic attack by the amine **2** creates a tetrahedral intermediate **8** from which the boron-complexed amide derivative **9** is generated following a water-assisted dehydration step [12,4]. Finally, dissociation of the amide product **4** regenerates the catalyst **6**. Suggestions for which step is rate determining include the formation of the catalyst-bound amide by the assisted dehydration step (i.e., **8** to **9**, Scheme 5-2) [12] and the generation of a monoacyloxyboronate complex species such as **7** [13], though the involvement of polyacyloxy boronate [6,8], and borate species, has yet to be ruled out. It may even prove to be the case that the RDS is the dissociation of the ammonium salt **3** back to the free acid **1** and amine **2**. Evidence of this stems from the fact that electron-rich carboxylic acids (higher pK_a) undergo direct amidation more readily than their electron-deficient (lower pK_a) analogs [7], and hence, electron-deficient carboxylic acids are more likely to react to form less reversible ammonium salts. However, more electron-rich carboxylic acids are presumably also more prone to form acyloxyboronate species being more nucleophilic. In addition, more electron-deficient boronic acid catalysts are generally more reactive [13,14]; they have higher electrophilicity (Lewis acidity) and are presumably more prone to acylation. The electrophilicity of boron is clearly not the only factor involved [4], and mechanism studies are further complicated by the fact that noncatalyzed, direct reaction between amines and carboxylic acids can proceed [7] (especially at elevated temperatures) to a much greater level than is typically appreciated and is highly dependent upon the carboxylic acid and amine concerned [7,8].

5.3 BORIC ACID-BASED CATALYSIS

In 2005, Tang [15] documented the application of inexpensive boric acid as an effective catalyst for direct amide formation. The reaction is operationally simple involving azeotropic water removal to accomplish condensation between simple amines (such as benzylamine and cyclic aliphatic amines) and simple carboxylic acids using 5 mol% catalyst loading. Even aniline derivatives with electron-withdrawing groups attached were reactive, though catalyst loading needed to be increased to 25 mol%. For example, reacting 4-phenylbutyric acid **10** with benzylamine **11** in refluxing heptane (b.p. ~98 °C) gave a 99% yield of the amide **12** in 15 h (Scheme 5-3), though it should be noted that for these high reactivity reactants, the direct, noncatalyzed reaction contributes to a considerable extent [8]. This yield of amide only decreases marginally to 95% when using the less nucleophilic 3,5-dimethylpiperidine in refluxing toluene (b.p. ~110 °C) [14]. Hence, at these temperatures, the reactivity of boric acid is excellent; however, it drops considerably at lower reaction temperatures [8,15].

The application of more complex boric acid-level oxidation-state catalysts has a longer history than the work of Tang [15]. In 1978, Ganem et al. [16] reported the condensation of carboxylic acids and amines in THF between −78 °C and room temperature using stoichiometric amounts of catecholborane **13**. The reaction was proposed to proceed via an intermediate acyloxyborate **14** that reacts with an amine **2** to provide the amide **4**, as shown in Scheme 5-4 [16]. Interestingly, a second equivalent of base (i.e., amine **2**) is required to drive the reaction to completion presumably due to competing reaction of the amine with the Lewis acidic catechol by-product resulting in an amino complex of type **15**.

Of the different acyloxyborates tested by Ganem, only catecholborane exhibited real potential as an amide condensation agent; other systems resulted in fragmentation to hydroxyl containing species that attacked the active intermediate **14**. The success of catecholborane **13** as an amide formation reagent is that the leaving group resulting from intermediate **14**, that is, benzo[*d*][1,3,2]dioxaborol-2-ol, shows a low susceptibility to disproportionation, forming nonnucleophilic products. Another advantage of **14** is that it is derived from a relatively electron-deficient borate intermediate with enhanced reactivity as an active ester intermediate or mixed anhydride-like intermediate. This reduces the

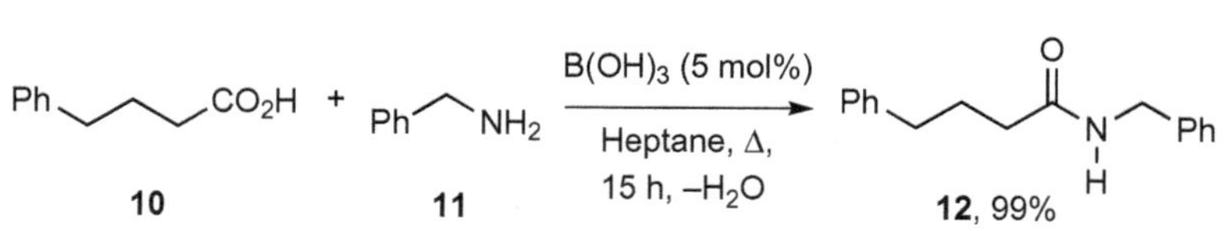

SCHEME 5-3. Tang's coupling of 4-phenylbutyric acid **10** with benzylamine **11** catalyzed by boric acid.

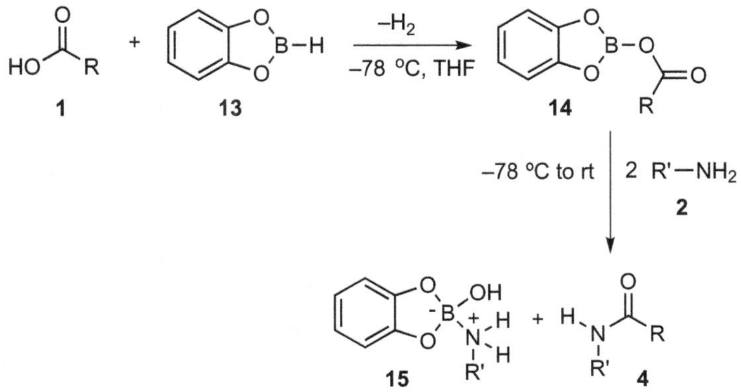

SCHEME 5-4. Ganem's amide condensation using catecholborane **13** [16].

occurrence of unwanted side reactions, and the addition of appropriate substituents to the catechol ring provides an opportunity for developing a catalytic application and increases versatility (see below). However, due to the stoichiometric requirements of catecholborane **13** and the inability of the unstable benzo[*d*][1,3,2]-dioxaborol-2-ol to undergo dehydration to reform acyloxycatecholate esters, the catecholborane-derived species cannot be used catalytically. It can though be an effective amidation agent, for example condensing chiral amino acid derivatives such as *N*–benzoyl-L-leucine with amines with no measurable loss (<2%) of enantiomeric purity during the amidation reaction [16]. In addition, a resin-bound version of **13** catecholborane has been reported [17] and is capable of forming the same type of active ester intermediate of type **14**. The solid-supported catecholborane is shaken with the carboxylic acid, followed by adding the amine and leaving it to react, resulting in modest yields of amides compared to the Ganem procedure.

Despite the poor atom economy procedure of the Ganem process [16], this report is significant because in 2006, the Ishihara group [18] was able to devise a catalytic system based on the original stoichiometric reaction. This was thought to be possible by devising a catechol borate ester **16** that could react further with carboxylic acid **1** through loss of water, rather than reacting with amine **2**, resulting in reformation of the active acylating agent **14** and therefore, resulting in a catalytic process, as outlined in Scheme 5-5.

In order to accomplish the process outlined in Scheme 5-5, it was necessary to increase the Lewis acidity at boron which was achieved through the introduction of chlorine substituents on the catechol ring, that is, by using 4,5,6,7-tetrachlorobenzo-[*d*][1,2,3] dioxaborol-2-ol, **17** [18,19] (Figure 5-1).

Catechol borate ester **17** is an effective catalyst for condensing bulky carboxylic acids and amines in nonpolar solvents [18,19], and as with all catalytic direct amide formation reactions, azeotropic or activated molecular sieve (MS)-mediated removal of water is required to drive the reaction to completion. Hence, for the reaction between non-bulky 4-phenylbutyric acid and benzylamine in refluxing toluene, with 5 mol% catalyst loading, benzo[*d*][1,3,2]dioxaborol-2-ol (**16**) gave an incomplete yield of 61% in 1 h and produced an unreactive dimer following amidation [19]. Boric acid was even less active, producing only 31% of the amide over the same period. In contrast, chlorocatechol derivative **17** gave 93% yield -which is comparable to that 96% afforded by 3,5-bis (trifluoromethyl)phenylboronic acid **18** (see below), as shown in Scheme 5-6. The main

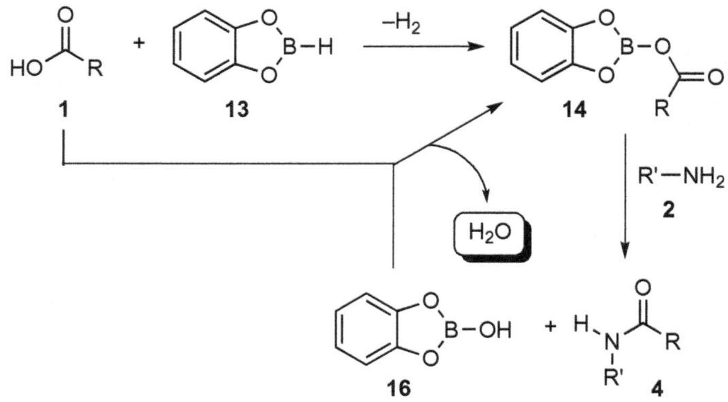

SCHEME 5-5. Catalytic process developed by Ishihara *et al.* [18] based on the Ganem direct amidation using catecholborane **13** [16].

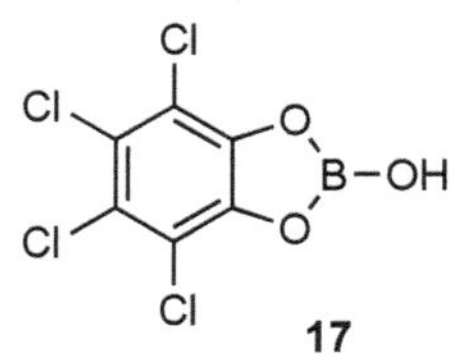

FIGURE 5-1. Structure of compound **17**.

advantage of the chlorocatechol borate catalyst **17** over the electron-deficient boronic acid **18** is that it is sterically less hindered and has increased Lewis acidity, which is claimed to aid the regeneration of the catalyst through reformation of the acyloxyboron species of type **14** from the hydrated intermediate.

For the more sterically demanding combinations of acid and amine, such as cyclohexanecarboxylic acid and benzylamine, catalyst **17** showed considerable improvement over **18** in terms of yield, giving a conversion of 52% relative to 32% in the presence of **18**. 4,5,6,

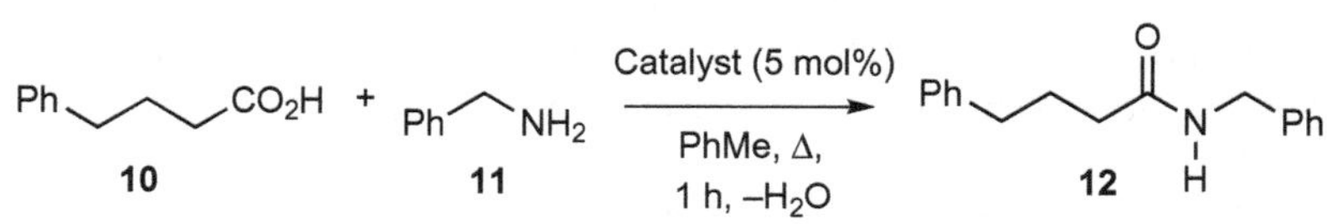

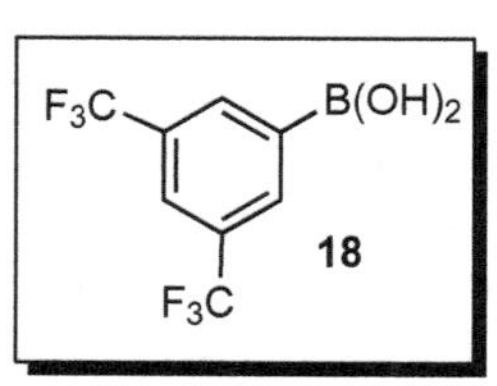

Catalyst	Yield (%)
16	61
$B(OH)_3$	31
17	93
18	96

SCHEME 5-6. Comparisons of catalytic activities of different boric acids for the coupling between **10** and **11**.

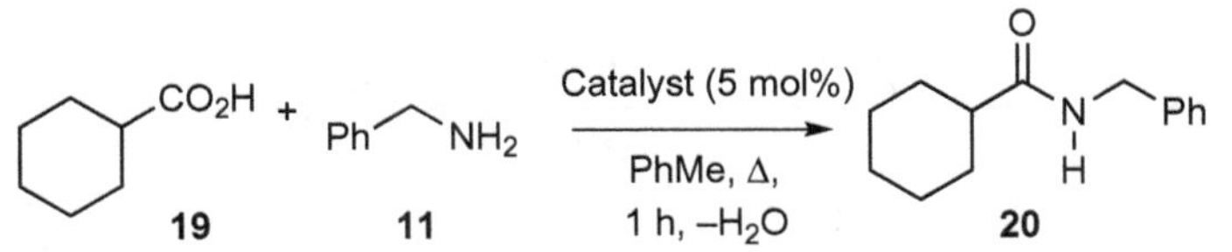

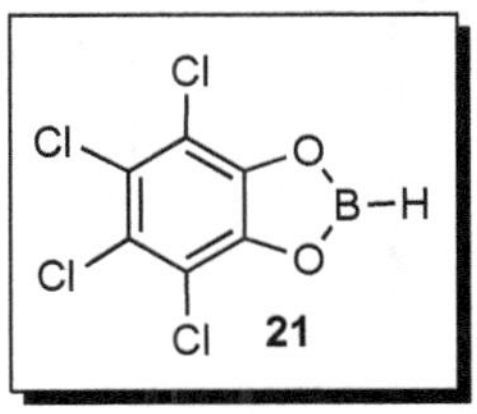

Catalyst	Yield (%)
$B(OH)_3$	2
17	52
18	32
21	62

SCHEME 5-7. Comparisons of catalytic activities of different boric acids for the coupling between **11** and **19**.

7-Tetrachlorobenzo[*d*][1,3,2]dioxaborole **21** induced a conversion of 62%, higher than that of **17** and far in excess of boric acid (2%) (Scheme 5-7). The catecholborane **21** was also shown to condense small aliphatic carboxylic acids in reasonable yields, albeit slightly less effectively than **18**; however, it was considerably more effective with the larger and more functionalized substrates such as *N*-protected alanine, for which conversions in the region of 90% were obtained using 5 mol% catalyst over 24 h in either refluxing toluene or *o*-xylene. Interestingly, direct amidation was not observed when tetrachlorocatechol, boric acid, cyclohexanecarboxylic acid, and benzylamine were simultaneously reacted together, which implies that *in situ* catalyst formation is required before substrate addition [19].

5.4 BORONIC ACID-BASED CATALYSIS

In 1996, Ishihara et al. [13] reported that *meta*- and *para*-substituted benzeneboronic acids bearing electron-withdrawing groups were effective catalysts for direct amide formation in nonpolar solvents. In addition, they claimed that the advantage of these types of commercially available catalysts is that they are thermally stable and also water-, acid-, and base-tolerant. Hence, the condensation of 4-phenylbutyric acid **10** and 3,5-dimethylpiperidine **22** in refluxing toluene (using 4 Å MS to remove water) with 5 mol% of catalyst generated yields of 74% with 3,4,5-trifluorobenzeneboronic acid **24** and 56% with 3,5-bis(trifluoromethyl)benzeneboronic acid **18** within an hour; this was a substantial enhancement compared to using unsubstituted benzeneboronic acids, which gave only 23% conversion and the uncatalyzed reaction gave only <2% yield (Scheme 5-8). Under similar conditions, 1 mol% of catalyst **24** also condensed 4-phenylbutyric acid **10** with benzylamine **11** to give the amide **12** in 96% yield over 18 h and 4-phenylbutyric acid **10** with 3,5-dimethylpiperidine **22** in >99% yield over 16 h.

Secondary amines and aromatic substrates could also be efficiently reacted in refluxing toluene in the presence of **24**; however, some of the more demanding substrates, such as aniline, required higher reaction temperatures to be used, that is, refluxing mesitylene (b.p. ~165 °C) to give the corresponding amides in high yields and

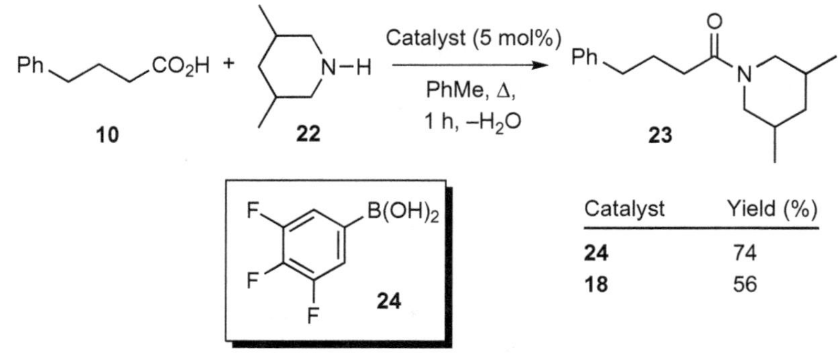

Catalyst	Yield (%)
24	74
18	56

SCHEME 5-8. Comparisons of catalysts **18** and **24** for the coupling between **10** and **22**.

with good purity directly after aqueous workup. In addition, there was no measurable loss of enantiomeric purity when carrying out direct amide formation using α-hydroxycarboxylic acid substrates [13].

Ishihara et al. went on to show [20] that trifluorophenylboronic acid **24** was also an effective catalyst for the direct polycondensation of carboxylic acids and amines. This was demonstrated by the reaction of hexamethylenediamine and adipic acid in refluxing *o*-xylene (4 Å MS) to provide nylon-6,6 in 89% yield over 20 h and using a 10 mol% catalyst loading. The number and average molecular weights of the polymer were fairly low; however, they could be increased substantially upon changing the solvent from *o*-xylene to a 1:3 mixture of *m*-cresol and *o*-xylene while maintaining an 85% yield [20].

More recently, Ishihara et al. [21] identified that arylboronic acid catalysts **18** and **24** could also achieve condensation between carboxylic acids with relatively nonnucleophilic ureas (Scheme 5-9). Under azeotropic water removal conditions in refluxing toluene and 5 mol% catalyst loading, *N*-monoacylurea products were formed, for example an equimolar mixture of 4-phenylbutyric acid **10** and urea over 26 h resulted in acyl urea **25** in yields of 92 and 88% using **18** and **24**, respectively [21].

In 2001, Ishihara et al. [22] synthesized 3,5-bis(perfluorodecyl)phenylboronic acid **26** and 4-(perfluorodecyl)phenylboronic acid **27**, which were found to be less active than catalysts **18** and **24**, but they had the advantage of being recoverable by extraction with a fluorous solvent such as perfluoromethylcyclohexane. These catalysts were compared using the reaction of 4-phenylbutyric acid **10** with 3,5-dimethylpiperidine **22** in refluxing

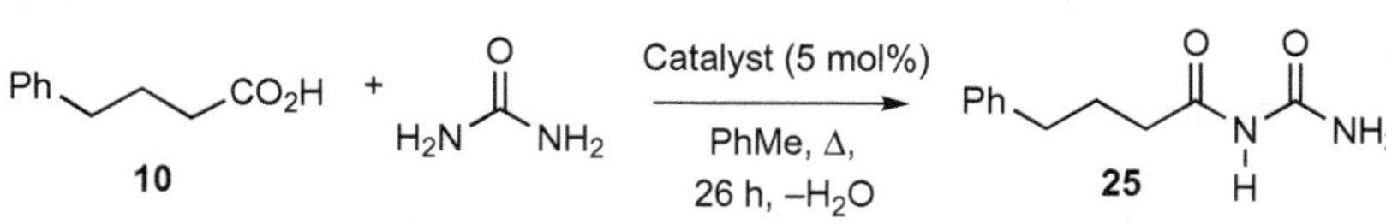

Catalyst	Yield (%)
18	92
24	88

SCHEME 5-9. Coupling between **10** and urea catalyzed by **18** and **24**.

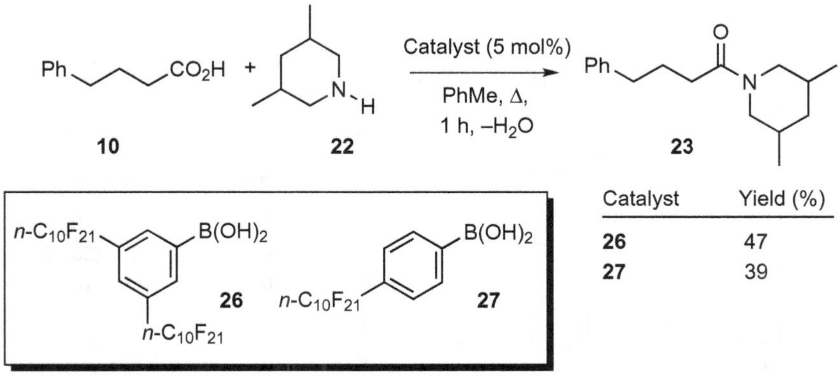

Catalyst	Yield (%)
26	47
27	39

SCHEME 5-10. Amide coupling between **10** and **22** catalyzed by **26** and **27**.

toluene (4 Å MS) with 5 mol% catalyst loading (Scheme 5-10). Within an hour, **26** had achieved 47% conversion to amide **23**, rising to 95% in 15 h. Catalyst **27** performed slightly less effectively, giving 39% conversion to **23** in 1 h, which is still considerably superior to the uncatalyzed reaction that proceeded to <2% yield. In comparison, catalysts **18** and **24** achieved conversions of 60 and 59%, respectively for the same reaction and over the same time period. Use of the more reactive benzylamine rather than 3,5-dimethylpiperidine **22** required only 2 mol% catalyst **26** loading to react with 4-phenylbutyric acid **10** in a 99% yield over 4 h.

The activity and recyclability of the fluorinated catalyst **26** was tested under azeotropic reflux with water removal in a 1:1:1 mixture of *o*-xylene, xylene, and perfluorodecalin over 12 h. Under these conditions, 3 mol% of catalyst **26** was sufficient to react cyclohexanecarboxylic acid with benzylamine in >99% yield. The catalyst was recovered from the fluorous phase in 98% yield, and even following a fifth cycle, showed no loss of activity [22]. While recyclability is a useful method for catalyst recovery, it necessarily requires the use of fluorous solvents, and hence, requires efficient solvent recycling.

The development of electron-deficient arylboronic acids as direct amide formation catalysts, activated by various fluorine functions, was clearly a considerable step forward in the search for more efficient direct amide formation processes. However, the drive to continue to reduce reaction temperatures and increase catalyst reactivity led to the development of boronopyridinium salts as potential catalysts [23]. In particular, their use in direct amide formation reactions carried out in polar solvents such as anisole, acetonitrile, and *N*-methylpyrrolidinone was envisaged. Wang et al. [24] reported the testing of compounds such as pyridine-3-boronic acid **28** and 3-borono-*N*-methylpyridinium iodide **29** (Scheme 5-11). Hence, the condensation of benzoic acid **31** with benzylamine **11** in refluxing toluene (4 Å MS) with 1 mol% catalyst loading gave yields of 92% of amide **32** in 19 h with **28** and 92% in 25 h with **29**. Catalyst **30**, a polystyrene-supported analog of **29**, gave yields of amide **32** of 65% in 16 h at 1 mol% loading and 96% in 8 h at 5 mol% loading, and the catalyst also showing no loss of activity after repeated use. This experiment highlighted the fact that improved yields and shorter reaction times were possible by simply increasing catalyst loading levels to around 5 mol% [24].

In 2005 Ishihara et al. [23] took this research further by examining the corresponding 4-pyridine and pyridinium analogs of catalysts **28–30**, that is, **33–35**. The suggestion was that 4-borono-*N*-methylpyridinium iodide **34** was also an effective catalyst for

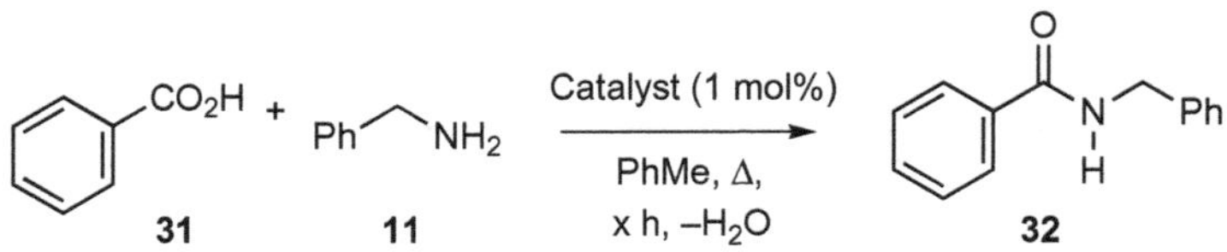

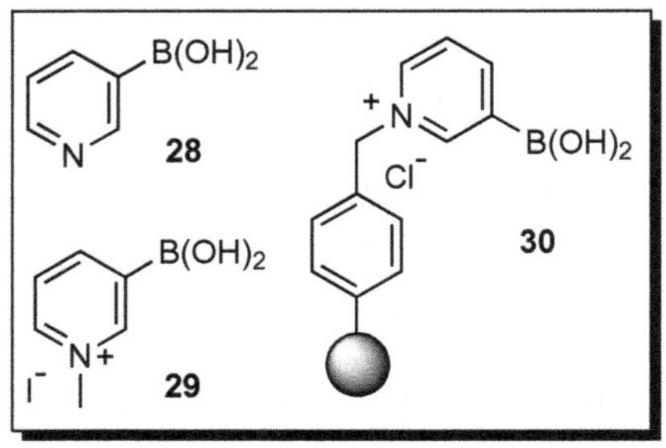

Catalyst	Yield (%)
28	92 (19 h)
29	92 (25 h)
30	65 (16 h)

SCHEME 5-11. Comparisons of catalysts **28–30** for the amide coupling between **11** and **31**.

direct amidation in polar solvents, and furthermore, it should possess greater thermal stability than the corresponding 3-boronopyridine derivatives [19,23]. In addition, pyridinium salt **34** can be reused if ionic liquid-toluene biphasic solvents are used, or a resin-bound variant **35** could also be employed as a heterogeneous alternative that could be readily recovered by filtration [20]. Ishihara et al. also proposed that charged boronopyridinium iodides **29** and **34** were more active than their uncharged pyridine-boronic acid counterparts **28** and **33**; an observation demonstrated by their activities in a 5:1 biphasic solvent of toluene and 1-ethyl-3-methylimidazolium trifluoromethanesulfonate, [emim][OTf], using azeotropic water removal (Scheme 5-12). In this solvent system, cyclohexanecarboxylic acid **19** was condensed with benzylamine **11** most

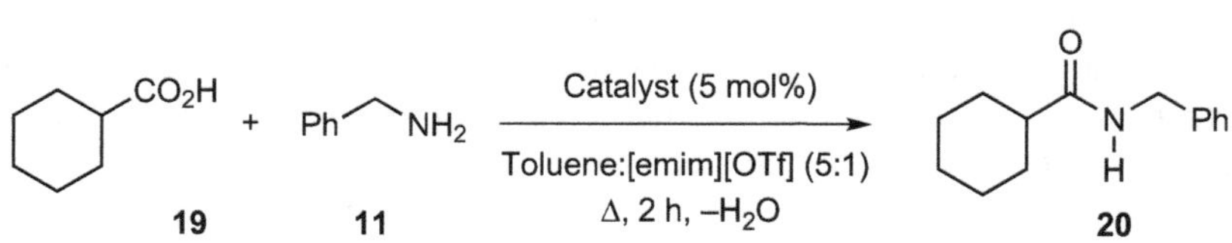

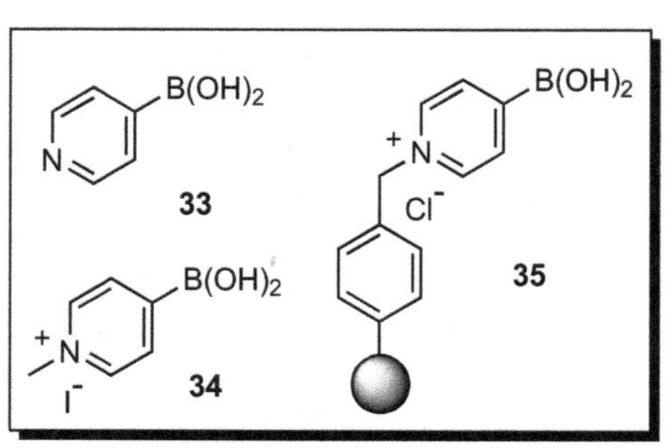

Catalyst	Yield (%)
28	43
29	54
33	8
34	59
35	59

SCHEME 5-12. Catalytic activities of pyridine-derived boric acids for the coupling between **11** and **19**.

efficiently with catalyst **34**, giving a yield of 59% of the amide **20** in 2 h, compared to 54% conversion for catalyst **29**, 43% using catalyst **28**, and only 8% using catalyst **33**. One contributor to the poor activity of catalyst **33** was identified as the poor dissolution in the solvent system employed.

In order to test their thermal stability, the catalysts **33–35** were heated for several hours at 120 °C in DMF. Over the course of an hour, catalyst **34** was transformed into a yellow precipitate that proved (by X-ray diffraction studies) to be a dodecameric derivative of **34**, denoted as $\mathbf{34_d}$. The activities of **34** and $\mathbf{34_d}$ were compared through the condensation of 4-phenylbutyric acid **10** with benzylamine **11** in refluxing toluene to give amide **12**. Over the course of an hour, 5 mol% of catalyst **34** generated the amide **12** in 41% yield compared to 15% conversion with $\mathbf{34_d}$. The activities dramatically improved upon changing to the biphasic solvent system (5:1, toluene:[emim][OTf]), showing conversions of 74 and 75%, respectively in 1 h which increased to >99% in 5 h with **34**. This improvement is likely to be due to the fact that **34** is more stable in [emim][OTf] and can be regenerated by the hydrolysis of $\mathbf{34_d}$ in this medium, which also suppresses the dodecamerization. The suppression of this dodecamerization is suggested as being a desirable quality of the [emim][OTf] solvent system, despite the impressive conversion levels achieved by $\mathbf{34_d}$ under these conditions. It seems more likely that the change in solvent polarity, and hence ease of water removal, in fact plays a larger role in yield determination. It is also useful to note that the amide **20** was recovered from the biphasic solvent system by extraction with ether, leaving catalyst **34** in the [emim][OTf] layer from which it could be reused with no loss of catalytic activity. Interestingly, 3,5-bis(perfluorodecyl)phenylboronic acid **26** also gave an impressive 88% conversion of **19** and **11** to amide **20** within an hour in [emim][OTf]; however, it remained in the toluene layer and was not reusable without further extraction.

Catalyst **34** also worked well with more functionalized molecules such as conjugated carboxylic acids and aromatic substrates, which was facilitated by changing the toluene component of the biphasic solvent to *o*-xylene in order to enable these more difficult amidation reactions. Hence, 5 mol% of catalyst **34** caused 4-phenylbutyric acid and *N*-benzylmethylamine **36** to condense in a 95% yield within 18 h in toluene, while the condensation of mandelic acid with benzylamine in *o*-xylene gave a 91% yield in the same time. Catalyst **35**, the polymer bound analog of **34**, showed no loss of activity through repeated use and maintained a high level of activity even in the absence of [emim][OTf] as the support prevented its dodecamerization. Condensing 4-phenylbutyric acid **10** with *N*-benzylmethylamine **36** in toluene with 5 mol% of **35** gave a 92% yield of amide **37** in 20 h (Scheme 5-13) [23].

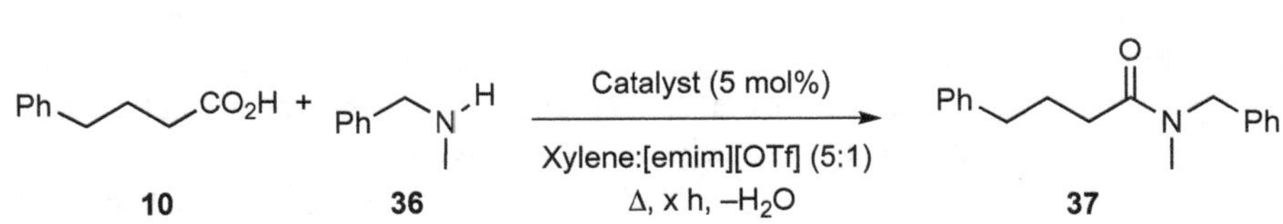

Catalyst	Yield (%)
34	95 (18 h)
35	92 (20 h)

SCHEME 5-13. Comparison of catalysts **34** and **35** for the coupling between **10** and **36**.

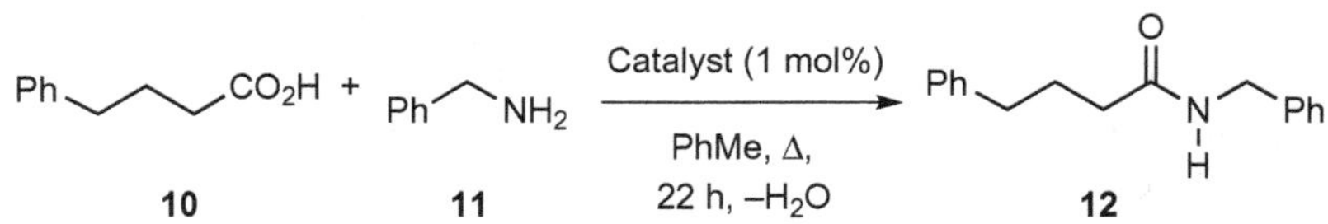

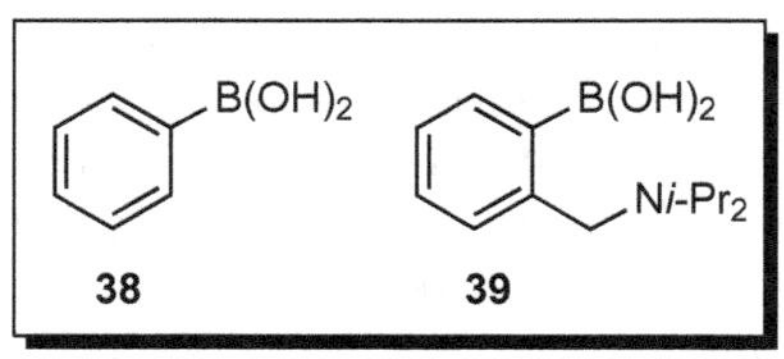

Catalyst	Yield (%)
No catalyst	60
$B(OH)_3$	73
24	80
38	78
39	65

SCHEME 5-14. Amide coupling between **10** and **11** in the absence/presence of catalysts.

In 2006, Whiting et al. [8] published kinetic studies, which showed that direct amidation could occur at modest temperatures, even in the absence of a catalyst, however, this was highly substrate dependent. They also showed that for less reactive substrate pairings, the addition of a boronic acid catalyst would, in many cases, vastly improve both reaction rate and yield. The condensation of 4-phenylbutyric acid **10** with benzylamine **11** in refluxing toluene (3 Å MS) showed a yield of ca. 60% within 22 h for the uncatalyzed (thermal) reaction, while boric acid, 3,4,5-trifluoro-benzeneboronic acid **24**, benzeneboronic acid **38** and the newly developed bifunctional catalyst *ortho-N,N*-di-isopropylbenzylaminoboronic acid **39** showed only marginal improvement upon this (65–80% yield) at 1 mol% loading (Scheme 5-14). Similarly, the reaction between 4-phenylbutyric acid **10** and 4-phenylbutylamine also showed a considerable thermal reaction, ca. 70% yield in 22 h, with only incremental improvements upon addition of the same catalysts (i.e., 85–100% yield). This led to the use of reaction conditions in which the thermal reaction was considerably less competitive. When benzoic acid and benzylamine were reacted under similar reaction conditions, it was found that this condensation was poorly catalyzed by most catalysts with the exception of boric acid and the new bifunctional catalyst **39**, which gave yields of ca. 70 and 40%, respectively over 22 h using 1 mol% catalyst loading.

In order to separate the catalytic effects from those of the uncatalyzed (thermal) reaction, it was necessary to retest each condensation at a lower reaction temperature to minimize thermal contributions. Hence, replacing refluxing toluene with fluorobenzene (b.p. ~85 °C) allowed a better catalyst evaluation to be carried out. The catalysts had a particularly notable effect upon yield, although an increased (10 mol%) catalyst loading was required in order to obtain reasonable reaction rates. Despite the reduced reaction temperature, the thermal reaction between highly reactive partners 4-phenylbutyric acid **10** and benzylamine **11** still managed to achieve 60% conversion over 24 h; a respectable yield, albeit inferior to the catalyzed yields of 80–85%. For example, 3,4,5-trifluorobenzeneboronic acid **24** was particularly reactive under these conditions, giving a rapid reaction and 80% yield of amide **12**, whereas boric acid was found to be generally less effective at the lower temperature (75%). Hence, for certain reactive reagent combinations, there was very little difference in catalyst activity; however, when the reagents became inherently less reactive, the bifunctional catalyst **39** displayed greater effectiveness. For example, for the condensation of benzoic acid **31** with benzylamine **11**, **39** showed the

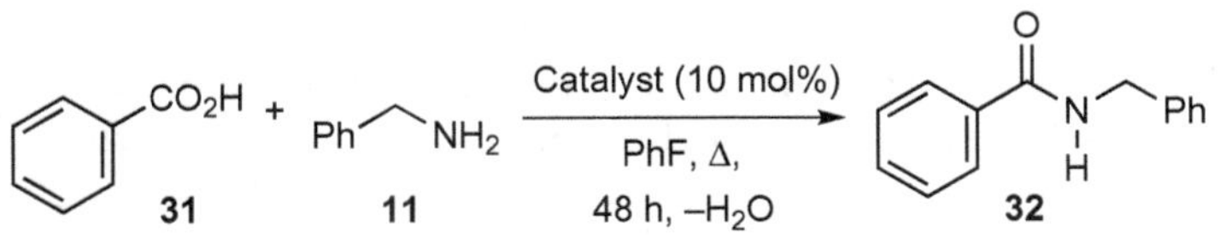

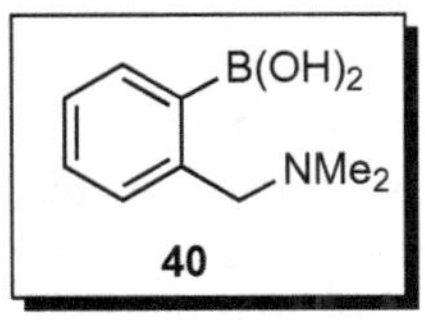

Catalyst	Yield (%)
No catalyst	0
$B(OH)_3$	40
24	65
39	80
40	58

SCHEME 5-15. Comparisons of catalysts for the amide coupling between **11** and **31** in fluorobenzene.

highest conversion of ca. 80% over 48 h, a huge improvement upon the other catalysts (Scheme 5-15) and the nonexistent thermal reaction.

Interestingly, the less basic and less hindered **40** showed similar activity to catalyst **24**, an effect likely due to the chelation between boron and nitrogen, which is not observed in the bulkier **39**. Catalyst **39** was also shown to induce the best conversion for the condensation of benzoic acid **31** with 4-phenylbutylamine, forming the corresponding amide in ca. 60% yield in 48 h, which highlights that catalyst **39** is generally more effective with less reactive carboxylic acids, such as benzoic acid, at low temperature, while the more Lewis acidic **24** is superior with more reactive reagents such as 4-phenylbutyric acid [8]. The effects of aryl substitution upon catalyst **39** were further exemplified [14] by studying the effects of tuning catalyst Lewis acidity through the introduction of electron-withdrawing and -donating groups on the aryl ring. This led to the synthesis of catalysts **41** and **42** that were compared under the conditions outlined in Scheme 5-16.

The catalysts were compared via the condensation of benzoic acid **31** and benzylamine **11** in refluxing fluorobenzene (dried using 3 Å MS) since there was no uncatalyzed

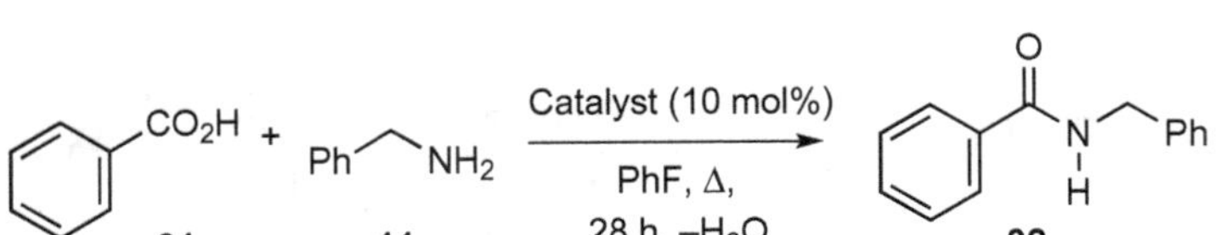

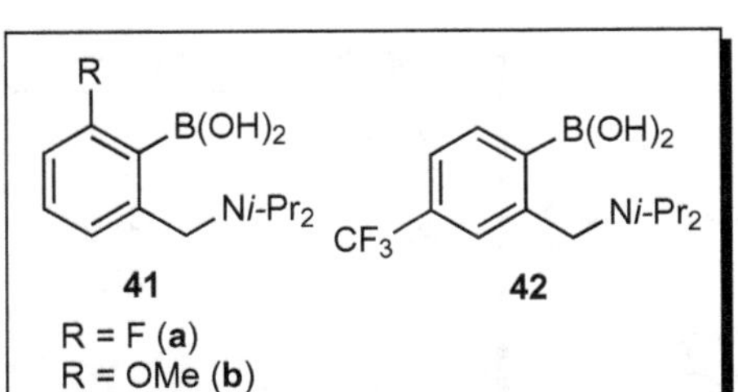

Catalyst	Yield (%)
No catalyst	0
41a	55
41b	28
42	63
39	58

SCHEME 5-16. Effect of aryl substitution on the catalytic activity of **39**.

amidation to give **32** under these conditions. The inclusion of a *para*-trifluoromethyl substituent, that is **42**, showed a small increase in reaction rate relative to the unsubstituted catalyst **39**, whereas adding an *ortho*-fluorine (**41a**) caused a slight reduction in rate. The inclusion of an electron-rich methoxy group (**41b**) caused a large reduction in rate, reinforcing the need to maintain a more Lewis acidic boronic acid to achieve higher rates of reaction and higher conversions [14].

Despite its slightly reduced activity compared to trifluoromethyl derivative **42**, catalyst **39** was commercially available and was therefore tested on a range of reagent combinations to evaluate its applicability [14]. For the majority of amidations, good-to-high isolated yields were obtained within 48 h in refluxing fluorobenzene using 10 mol% catalyst loading. Toluene was required for some of the less reactive pairings. For example, the condensation of benzoic acid **31** and α-methylbenzylamine showed a significant improvement when catalyzed by 10 mol% of **39**, with the yield rising from 52% over 48 h in refluxing fluorobenzene to 71% over 24 h in refluxing toluene [14].

Since direct amidation has generally been an elevated temperature process, it seemed unlikely that asymmetric induction would be possible using such a process; the problem of small energy differences between diastereomeric transition states would be amplified, and there were concerns about reagent degradation [6]. Despite this, the first report of a nonenzyme-catalyzed asymmetric direct amide formation appeared in 2008 [25] using a bifunctional ferrocenylboronic acid derivative **45**. Compound **45** was shown to give kinetic resolution of racemic α-substituted benzylamines through direct amidation with achiral carboxylic acids in refluxing fluorobenzene. Hence, benzoic acid **31** and α-methylbenzylamine **43** were condensed over a 48 h period with 10 mol% of catalyst **45** to give yields of 52, 21, and 38% for catalysts **24**, **45**, and **46**, respectively (Scheme 5-17). Although the conversion catalyzed by **45** was low, it was the only catalyst to show asymmetric induction, giving **44** in 41% ee Surprisingly, the asymmetric analog of the successful bifunctional catalyst **24**, that is, **46**, was ineffective at accomplishing kinetic resolution; the exact mechanism by which **45** is effective is yet to be elucidated. An increase in conversion to 73% was possible when the more reactive 4-phenylbutyric acid **10** was condensed with α-methylbenzylamine in the presence of **45**, albeit with a reduced 29% ee, due to competing protodeboronation and competitive boric acid-catalyzed direct amidation [25].

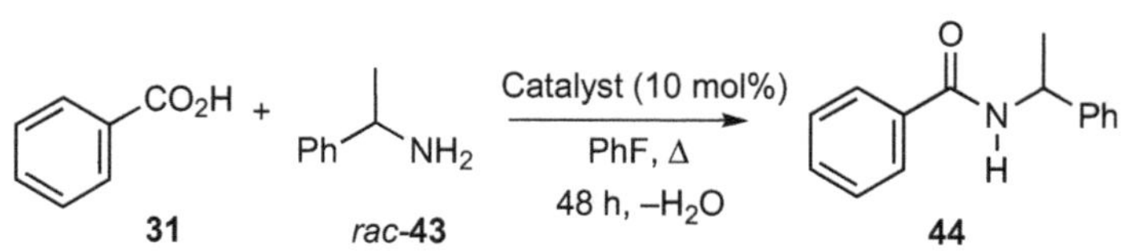

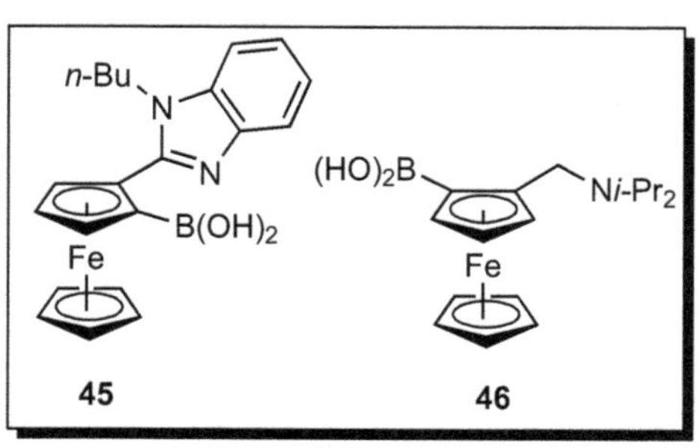

Catalyst	Yield (%)
No catalyst	0
24	52
45	21 (41% ee)
46	38 (0% ee)

SCHEME 5-17. Ferrocenylboronic acids as asymmetric catalysts for amide coupling reactions.

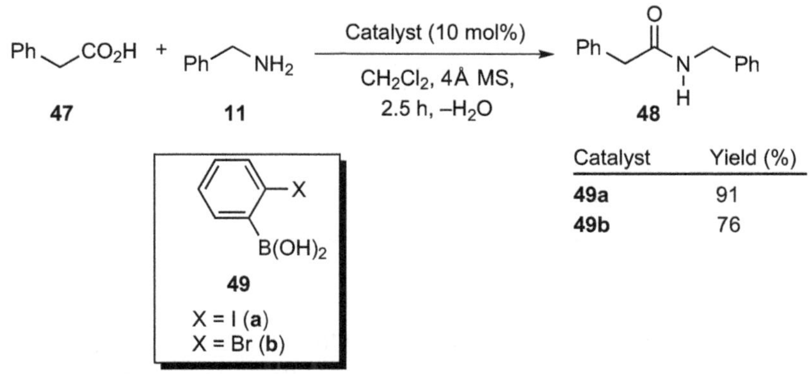

Catalyst	Yield (%)
49a	91
49b	76

SCHEME 5-18. Amide coupling between **11** and **48** catalyzed by **49**.

Also in 2008, Hall et al. [4] reported direct amide formation at room temperature using *ortho*-halogenated benzeneboronic acid catalysts **49**. These catalysts are more active than 3,4,5-trifluorophenylboronic acid **24**, which normally requires temperatures in excess of 110 °C to react [4,6,26]. Hence, a heterogeneous reaction of phenylacetic acid **47** and benzylamine **11** in dichloromethane with 4 Å MS at 25 °C yields amide **48** in 91 and 76% within 2.5 h using 10 mol% of **49a** and **49b**, respectively (Scheme 5-18); amidation proceeds to 99% in 48 h using catalyst **49b**. Generally, if the reactants involved are reactive carboxylic acids and primary amines, catalysts **49** achieve moderate-to-high yields of the corresponding amides within 48 h. This was also true of some secondary amines, provided they were cyclic. Aromatic carboxylic acids, on the other hand, were shown to condense with amines in the presence of catalysts **49** under slightly higher temperature conditions (50 °C) albeit in lower yield. Catalyst **49a** tends to be more effective in these more difficult applications. It is also notable that these catalysts can be recovered in high yield from the aqueous phase for reuse [4].

Hall et al. showed [4] that *para*- and *meta*-substituted benzeneboronic acids showed reduced activity as direct amide formation catalysts, which implies that the halide groups play an active role in the reaction mechanism. Hall speculated that the effectiveness of the iodo-substituent in **49a** could not be rationalized by inductive effects alone and could be due to subtle electronic or structural effects [4]. However, recently Marcelli [12] carried out DFT calculations in order to shed light upon the catalytic processes that might be operating these types of reactions and suggested that the *ortho*-substituent is involved in OH–halogen hydrogen bonding, that is, these types of catalysts behave as bifunctional systems and act through a more complex mechanism than simple acyloxyboronate intermediates. The insights revealed by this type of approach bode well not only for a more fundamental understanding of the direct amide formation process, but also for the design of improved, more generally effective boronic acid-based catalysts in future.

5.5 TRIAZINE-BASED REAGENTS

In 2005, Kunishima et al. [27] proposed the possibility of catalytic amidation in a micellar reaction media using amphiphilic 1,3,5-triazine-based dehydro-condensation agents, such

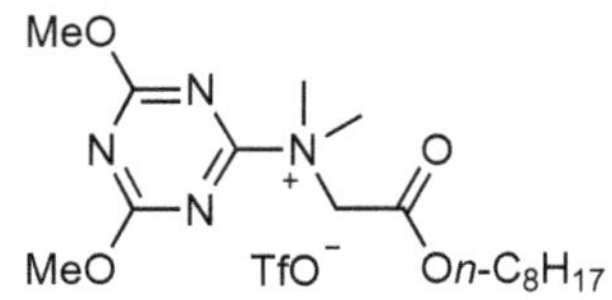

50

FIGURE 5-2. Structure of compound **50**.

as **50** (Figure 5-2), which were synthesized *in situ* from the corresponding chlorotriazine and examined in aqueous solution with a range of substrates. Dehydro-condensation is in theory ideally suited for rate enhancement in a colloidal environment as any reagent possessing a long hydrophobic alkyl chain can form a micelle or be incorporated into a surfactant-derived micellar system. In addition, because the triazine catalysts and carboxylate salts are both polar moieties, they can align at the micellar interface in a proximal manner, giving rise to high local concentrations and advantageous preorientational effects. The activated intermediates generated can then be attacked by an amine nucleophile also contained within the micellar core, giving rise to enhanced reaction rates.

The theory was tested through the simultaneous condensation of two carboxylic acids with excess *n*-butylamine **52** in the presence of 10 mol% of different triazine-based catalysts in a phosphate buffer at 25 °C. The competitive amidation of 1-octanoic acid and 1-butanoic acid showed 24% conversion within 4 h and gave a mixed product composition, which varied with catalyst. However, the mixed product did not occur to any appreciable extent when one of the acids had a longer alkyl "tail" and was hence able to form micelles. Kinetic studies showed that 1-dodecanoic acid **51** would react with *n*-butylamine **52** in the presence of **50** over 1700 times faster than with 1-butanoic acid, which is explained as being due to micellar effects. This was further corroborated by the an 88% yield of butyl dodecanamide in just over an hour at room temperature (Scheme 5-19) and the inclusion of 1-butanoic acid in the reaction give a 99.6% chemoselectivity for the longer chain carboxylic acid. Hence, triazine **50** is a highly effective catalyst under these conditions when applied to longer alkyl chain carboxylic acids [27] and is claimed to react via carboxylate displacement of the ammonium function to derive an acyloxy triazine derivative. This activated carboxylate reacts like a mixed anhydride with the incoming amine to give the amide product.

Despite the impressive yields, there has not been widespread application of catalyst **50** to date, due in large part to the substrate requirements of a large hydrophobic segment. Kunishima et al. have published the results for several other triazine-based reagents,

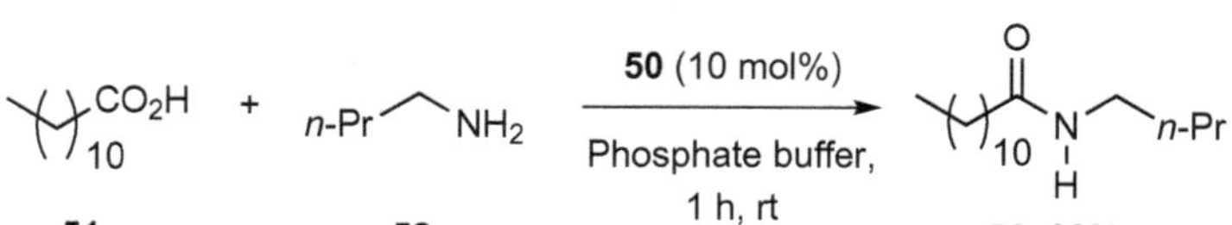

SCHEME 5-19. Amide coupling between an aliphatic carboxylic acid with aliphatic amine, catalyzed by **50**.

including 4-(4,6-dimethoxy-1,3,5-triazin-2-yl)-4-methylmorpholinium chloride, which operates in a more orthodox carboxylic acid-activating fashion and can effect amide formation in high yield [27,28,29]; however, these methods are stoichiometric and hence, not atom economic.

5.6 TITANIUM(IV)-BASED REAGENTS

In 1969, Weingarten et al. [30] showed that direct amidation could be induced by stoichiometric amounts of $TiCl_4$ in THF at –70 °C using a ratio of catalyst:acid:amine of 1:2:6, respectively. After warming to room temperature and leaving the reaction for several hours, good yields of amides can be directly isolated. For example, formic acid and ethylamine condense to give ethyl formamide in 82% yield after 8 h. This technique is applicable to the preparation of *N*-alkyl and *N*,*N*-dialkyl carboxamides, although the methods are less effective for weakly basic amines and sterically hindered acids [30]. More recently, Carlson et al. [31] suggested that $TiCl_4$ acted catalytically (despite stoichiometric loadings) in direct amide formation and that it was considerably more effective at condensing secondary amines than primary. They also showed that the addition of a drying agent had a negligible effect on yield (in stark contrast to the boronic acid-catalyzed reactions, see above) showing the dehydrating role of this system. The conditions used involved a 1:1:3 stoichiometry of catalyst:acid:amine, respectively in refluxing toluene and, for example, managed to condense diisopropylamine with benzoic acid in 66% yield and morpholine with benzoic acid in 88% yield in 10 h [31].

This work was followed up in 1988 by Mader et al. [32] who reported that $Ti(O\textit{i}\text{-}Pr)_4$ could also act as an effective catalyst for the formation of 5- and 6-membered ring lactams from simple primary and secondary amino acids. The $Ti(O\textit{i}\text{-}Pr)_4$ reagent produced high yields of amidation products; however, the reaction was still accomplished with high titanium loadings, that is, 50 mol%. This 2:1 amino acid to catalyst stoichiometry is consistent with titanium dioxide being the by-product of the reaction, which limits the reaction to essentially a stoichiometric process. It is not surprising, therefore, that ratios of 4:1 and 10:1 of amino acid to $Ti(O\textit{i}\text{-}Pr)_4$ were also effective, but gave poorer conversion [32].

The fact that the activity of titanium(IV) compounds is dependent upon stoichiometric loading and that they cannot be recycled limits their large-scale application in industry, despite their relatively low cost.

5.7 ANTIMONY-BASED REAGENTS

In 1986, Nomura et al. [33] carried out a series of experiments, which showed that triphenylstibine oxide could catalyze direct amide formation at temperatures in excess of 100 °C. This was followed in 1991 [34] by results detailing how a combination of triphenylstibine oxide and tetraphosphorus decasulfide could also catalyze direct amidation but at more ambient temperatures. Since triphenylstibine oxide had been found to catalyze direct amide formation by reaction of amines with thiocarboxylic acids and to accelerate the thiolation of carboxylic acids in the presence of tetraphosphorous decasulfide (P_4S_{10}), the development of the combined reagent system was an obvious development.

The reaction is carried out using equimolar quantities of the carboxylic acid (e.g., acetic acid) and amine (e.g., hexylamine) at 40 °C in the presence of 5 mol% of triphenylstibine oxide and 15 mol% of P_4S_{10}, which gives hexyl acetamide in 90% yield in 5 h. It is worth noting that there was no amidation in the absence of either catalyst, even at higher temperatures and over prolonged reaction times, and triphenylstibine oxide and tetraphosphorus decasulfide catalyze direct amide formation of a range of many simple aliphatic and aromatic amines in good yield. However, more forcing reaction conditions were required with sterically hindered substrates, for example, the condensation of pivalic acid with *tert*-butylamine required the reaction to be conducted at 80 °C with a large excess of amine over 12 h to 88% amide formation. This catalytic system was also effective at coupling *N*-protected amino acids with amino acid esters, and for the amidation of hydroxylated dipeptides [34].

This type of direct amide formation process was developed further in 1995 by Ishihara et al. [35] who reported a series of experiments using antimony(III) ethoxide as catalyst for converting spermine-derived tetraamino esters into 17-membered ring lactams [35,36]. In the presence of a high loading (120 mol%) of antimony(III) ethoxide, substrate **54a** (Figure 5-3) underwent intramolecular ester aminolysis in dry benzene to yield the corresponding lactam in 76% yield within 14 h, and under the same conditions, the phenyl-substituted substrate **54b** cyclized in a 90% yield over 9 h. This process is limited therefore by both catalyst loading and lack of wider applicability; for example while the yield of lactams produced from tetraamino esters was high, triamino esters only reacted in poor yields indicating that the antimony(III) ion prefers to template 17-membered lactam formation.

More importantly, Ishihara also studied the catalytic activity of antimony(III) ethoxide for application as an intermolecular amidation catalyst for use with both carboxylic acids and esters with amines. Primary amines showed high chemoselectivity and efficient conversions, especially with less hindered substrates. For example, 1-dodecanoic acid condensed with *n*-hexylamine in refluxing toluene to give 91% yield after several hours in the presence of 10 mol% of antimony(III) ethoxide (Scheme 5-20) [35,36].

5.8 HETEROGENEOUS CATALYSTS AND MICROWAVE-ASSISTED AMIDE SYNTHESIS

Various catalysts, especially heterogeneous systems (including K-10 montmorillonite [37], zeolite-HY [38], $TaCl_5$-silica [39], KF-alumina [40], and silica alone), have been applied to

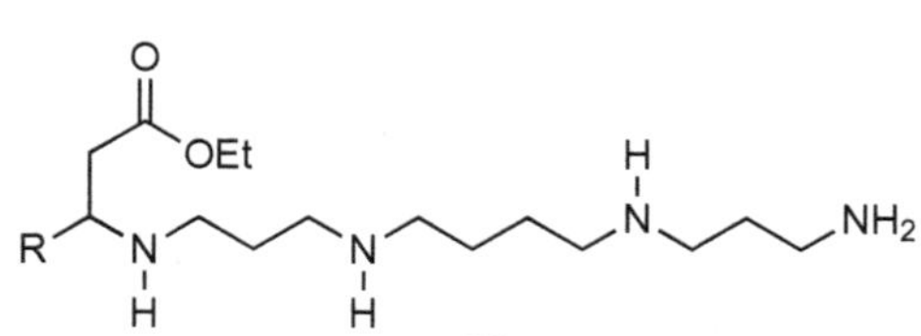

54

R = 4-MeOC$_6$H$_4$ (**a**)
R = Ph (**b**)

FIGURE 5-3. Structure of compound **54**.

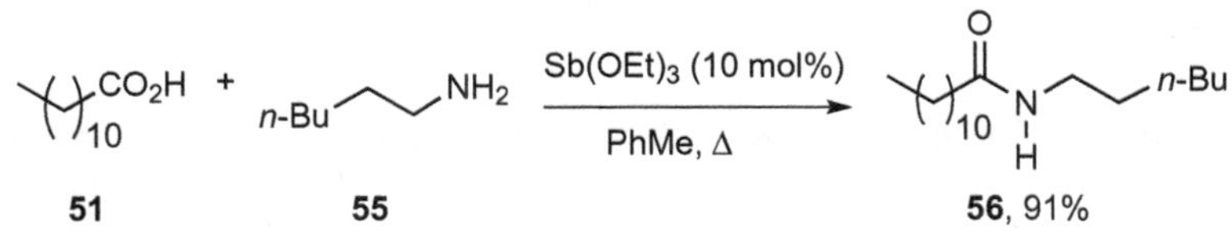

SCHEME 5-20. Amide coupling between linear acid and amine catalyzed by $Sb(OEt)_3$.

direct amide formation reactions carried out in domestic microwave ovens with no temperature or power control. This makes comparisons between the different reactions and further application difficult. However, more recently Loupy et al. [41] reported a systematic comparison between thermal and microwave-assisted reactions under solvent-free, heterogeneous conditions, that is, essentially pyrolysis reactions of carboxylate ammonium salts. This study highlighted observations that are general to both thermal and microwave-assisted reactions, that is, that (1) the reactivity order for carboxylic acids was n-$C_9H_{19}CO_2H > BnCO_2H >> PhCO_2H$; (2) the reactivity order for amines was $PhCH_2NH_2 > n$-$C_8H_{17}NH_2 > p$-$MeOPhNH_2 > PhNH_2$; (3) and the reactivity was not simply acid or amine $pK_{a/b}$ related, which presumably has important mechanistic implications that are not clear as yet [6]. However, their systematic work [41] showed that direct amide formation using microwave conditions was a practical method for certain amine-carboxylic acid combinations, giving yields up to 93% in 10–30 min, whereas standard thermal reaction yields were generally >20% less.

Silica gel (Kieselgel 60), preactivated at high temperature, has recently been reported to be a cost-effective and efficient heterogeneous catalyst for direct amide formation by Clark et al. [42], which is effective at catalyst loadings from 10–50 weight % depending upon the substrates. For example, direct amide formation between benzoic acid and aniline was be accomplished using 50 weight% catalyst in refluxing toluene to give phenyl benzamide in 47% yield. This catalyst system is not only suitable for standard reaction protocols but is also readily recycled. In addition, a mesoporous solid acid suitable for use under microwave-assisted conditions has also been developed by the same group [43]. This type of catalyst system, a Starbon® acid, is effective for the facile formation of acetamides at 130 °C, providing essentially quantitative yields of the amides derived from a wide range of amines.

5.9 SUMMARY AND FUTURE DIRECTIONS

In conclusion, a number of different catalytic systems have been developed, which can more or less efficiently catalyze direct amide formation between carboxylic acids and amines. However, it is also clear that all current methods suffer from drawbacks, which include high reaction temperature conditions, insufficient general substrate applicability, limited solvent compatibility, and insufficient catalyst activity. There is, therefore, major scope for the development of improved, higher reactivity, widely applicable direct amide formation catalysts, the development of which is only likely to occur with a deeper understanding of how direct amide formation works, and the mechanistic processes by which the different catalytic systems work; the first steps in this direction are being made [6,7,12].

REFERENCES

1. Ghosez AK, Viswanadhan VN, Wendoloski JJ (1999). A knowledge-based approach in designing combinatorial or medicinal chemistry libraries for drug discovery. 1. A qualitative and quantitative characterization of known drug databases. *J. Comb. Chem.* 1, 55–68.
2. Constable DJC, Dunn PJ, Hayler JD, Humphrey GR, Leazer, Jr., JL, Linderman RJ, Lorenz K, Manley J, Pearlman BA, Wells A, Zaks A, Zhang TY (2007). Key green chemistry research areas – a perspective from pharmaceutical manufacturers. *Green Chem.* 9, 411–420.
3. Smith MB, March J. *March's Advanced Organic Chemistry*. 6th edition, Wiley–Interscience, Hoboken, NJ, 2007.
4. Al-Zoubi RM, Marion O, Hall DG (2008). Direct and waste-free amidations and cycloadditions by organocatalytic activation of carboxylic acids at room temperature. *Angew. Chem., Int. Ed.* 47, 2876–2879.
5. Montalbetti CAGN, Falque V (2005). Amide bond formation and peptide coupling. *Tetrahedron* 61, 10827–10852.
6. Charville H, Jackson DA, Hodges G, Whiting A (2010). The thermal and boron-catalyzed direct amide formation reactions: mechanistically understudied yet important processes. *Chem. Commun.* 1813–1823.
7. Charville H, Jackson DA, Hodges G, Whiting A, Wilson MR (2011). The direct amide formation reaction: mechanism studies and the key role of carboxylic acid H-bonding. *Eur. J. Org. Chem.* 5981–5990.
8. Arnold K, Davies B, Giles R, Grosjean C, Smith G, Whiting A (2006). To catalyze or not to catalyze? An insight into direct amide bond formation from amines and carboxylic acids under both thermal and catalyzed conditions. *Adv. Synth. Catal.* 348, 813–820.
9. Carey JS, Laffan D, Thomson C, Williams MT (2006). Analysis of the reactions used for the preparation of drug candidate molecules. *Org. Biomol. Chem.* 4, 2337–2347.
10. Sheldon RA (2008). E factors, green chemistry and catalysis: an odyssey. *Chem. Commun.* 29, 3352–3365, and references therein.
11. Hudlicky T, Frey DA, Koroniak L, Claeboe CD, Brammer LE (1999). Toward a 'reagent-free' synthesis – tandem enzymatic and electrochemical methods for increased effective mass yield (EMY). *Green Chem.* 1, 57–59.
12. Marcelli T (2010). Mechanistic insights into the boronic acid-catalyzed direct amide formation: halogens as Lewis bases. *Angew. Chem., Int. Ed.* 49, 6840–6843.
13. Ishihara K, Ohara S, Yamamoto H (1996). 3,4,5-Trifluorobenzeneboronic acid as an extremely active amidation catalyst. *J. Org. Chem.* 61, 4196–4197.
14. Arnold K, Batsanov AS, Davies B, Whiting A (2008). Application of novel bifunctional *N,N*-diisopropylbenzylamineboronic acids catalysts for direct amide formation between carboxylic acids and amines. *Green Chem.* 10, 124–134.
15. Tang P (2005). Boric acid catalyzed amide formation from carboxylic acids and amines: *N*-benzyl-4-phenylbutyramide. *Org. Synth.* 81, 262–272.
16. Collum DB, Ganem B, Chen SC (1978). New synthesis of amides and macrocyclic lactams. *J. Org. Chem.* 43, 5291–5300.
17. Yang WQ, Gao XM, Springsteen G, Wang BH (2002). A detailed examination of boronic acid-diol complexation. *Tetrahedron Lett.* 43, 6339–6342.
18. Maki T, Ishihara K, Yamamoto H (2006). 4, 5,6,7-Tetrachlorobenzo-[*d*][1,3,2]dioxaborol-2-ol as an effective catalyst for the amide condensation of sterically demanding carboxylic acids. *Org. Lett.* 8, 1431–1434.
19. Ishihara K, Maki T, Yamamoto H (2007). New boron(III)-catalyzed amide and ester condensation reactions. *Tetrahedron* 63, 8645–8657.

20. Ishihara K, Ohara S, Yamamoto H (2000). Direct polycondensation of carboxylic acids with amines catalyzed by 3,4,5-trifluorophenylboronic acid. *Macromolecules* 33, 3511–3513.

21. Maki T, Ishihara K, Yamamoto H (2004). Arylboronic acid-catalyzed direct condensation of carboxylic acids with ureas. *Synlett* 1355–1358.

22. Ishihara K, Kondo S, Yamamoto H (2001). 3,5-Bis(perfluorodecyl)-phenylboronic acid as an easily recyclable direct amide condensation catalyst. *Synlett* 1371–1374.

23. Maki T, Ishihara K, Yamamoto H (2005). *N*-Alkyl-4-boronopyridinium salts as thermally stable and reusable amide condensation catalysts. *Org. Lett.* 7, 5043–5046.

24. Latta R, Springsteen G, Wang BH (2001). Development and synthesis of an arylboronic acid-based solid-phase amidation catalyst. *Synthesis* 1611–1613.

25. Arnold K, Davies B, Hérault D, Whiting A (2008). Asymmetric direct amide synthesis *via* kinetic amine resolution: a chiral bifunctional amino-boronic acid catalyzed reaction between a racemic amine and achiral carboxylic acid. *Angew. Chem., Int. Ed.* 47, 2673–2676.

26. Al-Zoubi RM, Hall DG (2010). Mild silver(I)-mediated regioselective iodination and bromination of arylboronic acids. *Org. Lett.* 12, 2480–2483.

27. Kunishima M, Imada H, Kikuchi K, Hioki K, Nishida J, Tani S (2005). Unusual rate enhancement of bimolecular dehydrocondensation to form amides at the interface of micelles of fatty acid salts. *Angew. Chem., Int. Ed.* 44, 7254–7257.

28. Kunishima M, Kawachi C, Morita J, Terao K, Iwasaki F, Tani S (1999). 4-(4,6-Dimethoxy-1,3,5-triazin-2-yl)-4-methyl-morpholinium chloride: an efficient condensing agent leading to the formation of amides and esters. *Tetrahedron* 55, 13159–13170.

29. Kunishima M, Kawachi C, Hioki K, Terao K, Tani S (2001). Formation of carboxamides by direct condensation of carboxylic acids and amines in alcohols using a new alcohol- and water-soluble condensing agent: DMT-MM. *Tetrahedron* 57, 1551–1558.

30. Wilson JD, Weingarten H (1970). Titanium tetrachloride promoted conversion of carboxylic acids to carboxamides. *Can. J. Chem.* 48, 983–986.

31. Carlson R, Nordahl A (1988). Carboxamides from carboxylic-acids by Lewis acid catalysis – use of principal properties for exploring different reaction conditions. *Acta Chem. Scand. Ser. B* 42, 28–34.

32. Mader M, Helquist P (1988). $Ti(O\text{-}iso\text{-}Pr)_4$ mediated lactamizations. *Tetrahedron Lett.* 29, 3049–3052.

33. Nomura R, Wada T, Yamada Y, Matsuda H (1986). Aminolysis of triphenylantimony dicarboxylates and its application to catalytic amidation. *Chem. Lett.* 15, 1901–1904.

34. Nomura R, Nakano T, Yamada Y, Matsuda H (1991). Facile one-pot amidation of carboxylic acids by amines catalyzed by triphenylstibene oxide tetraphosphorus decasulfide (Ph_3SbO/P_4S_{10}). *J. Org. Chem.* 56, 4076–4078.

35. Ishihara K, Kuroki Y, Hanaki N, Ohara S, Yamamoto H (1996). Antimony-templated macrolactamization of tetraamino esters. Facile synthesis of macrocyclic spermine alkaloids, (+/−)-buchnerine, (+/−)-verbacine, (+/−)-verbaskine, and (+/−)-verbascenine. *J. Am. Chem. Soc.* 118, 1569–1570.

36. Ishihara K, Kuroki Y, Hanaki N, Ohara S, Yamamoto H (1998). Metal-templated macrolactamization of triamino and tetramino esters. Facile synthesis of macrocyclic spermidine and spermine alkaloids, (*S*)-(+)-dihydroperiphylline, (+/−)-buchnerine, (+/−)-verbacine, (+/−)-verbaskine, and (+/−)-verbascenine. *Bull. Chem. Soc. Jpn.* 71, 1221–1230.

37. Ruault P, Pilard JF, Touaux B, Texier-Boullet F, Hamelin J (1994). Rapid generation of amines by microwave irradiation of ureas dispersed on clay. *Synlett* 935–936.

38. Gadhwal S, Dutta MP, Boruah A, Prajapati D, Sandhu JS (1998). Zeolite-HY: a selective and efficient catalyst for the synthesis of amides under microwave irradiations. *Indian J. Chem.* 37B, 725–727.

39. Chandrasekhar S, Takhi M, Uma G (1997). Solvent free *N*-alkyl and *N*-arylimides preparation from anhydrides catalyzed by $TaCl_5$ silica gel. *Tetrahedron Lett.* 38, 8089–8092.
40. Marquez H, Plutin A, Rodriguez Y, Perez E, Loupy A (2000). Efficient synthesis of 1-(4′-methylbenzoyl)-3,3-diethylthiourea under microwave irradiation using potassium fluoride on alumina. *Synth. Commun.* 30, 1067–1073.
41. Perreux L, Loupy A, Volatron F (2002). Solvent-free preparation of amides from acids and primary amines under microwave irradiation. *Tetrahedron* 58, 2155–2162.
42. Comerford JW, Clark JH, Macquarrie DJ, Breeden SW (2009). Clean, reusable and low cost heterogeneous catalyst for amide synthesis. *Chem. Commun.* 2562–2564.
43. Luque R, Budarin V, Clark JH, Macquarrie DJ (2009). Microwave-assisted preparation of amides using a stable and reusable mesoporous carbonaceous solid acid. *Green Chem.* 11, 459–461.

6

INDUSTRIAL APPLICATIONS OF BORIC ACID AND BORONIC ACID-CATALYZED DIRECT AMIDATION REACTIONS

JOANNE E. ANDERSON,[1] JANNINE COBB,[1] ROMAN DAVIS,[1] PETER J. DUNN,[2] RUSS N. FITZGERALD,[1] AND ALAN J. PETTMAN[3]

[1]*GlaxoSmithKline Inc., Research Triangle Park, NC, USA*
[2]*Pfizer Global Supply, Pfizer Ltd, Sandwich, Kent, UK*
[3]*Chemical Research and Development, Pfizer Ltd, Sandwich, Kent, UK*

6.1 INTRODUCTION

Amides are pervasive in nature (peptides, antibiotics) and technology as structural materials (nylon). Secondary structure, due in part to hydrogen bonding, provides structural rigidity, and the linkage is robust toward hydrolysis. The amide functional group appears in many pharmaceuticals and fine chemicals, frequently with other functionalities that are not compatible with simple stoichiometric activators. Unlike acid-catalyzed esterification, there have been few catalytic processes for converting carboxylic acids to amides. Stoichiometric activation of the carboxylic acid has almost always been employed, with metals, gases, or bases as typical by-products. Direct amide formation methodology has advantages of atom economy and avoids protection/deprotection steps. Boric acid-catalyzed direct amide formation exploits all these advantages in a potentially green and sustainable setting. Chapter 5 provided an excellent review of the academic work in this area, but it can be seen that most of the academic work uses simple aliphatic or aromatic acids or amines as substrates, such as 4-phenylbutyric acid, benzoic acid, or benzylamine. From an industrial point of view, it was of interest to know if the methodology also works for typical pharmaceutical intermediates, which are often rich in heteroatoms, and the examples given in this chapter (particularly in Sections 6.3 and 6.4) explore that question; examples of direct amidation from three companies are presented, a more

Sustainable Catalysis: Challenges and Practices for the Pharmaceutical and Fine Chemical Industries, First Edition. Edited by Peter J. Dunn, K. K. (Mimi) Hii, Michael J. Krische, and Michael T. Williams.

detailed case history from GlaxoSmithKline (GSK) covering the synthesis of an efaproxiral intermediate and several examples from Dr. Reddy's and Pfizer.

6.2 THE SYNTHESIS OF EFAPROXIRAL UTILIZING A DIRECT AMIDATION REACTION

In 2006, an opportunity arose to further showcase the scope and utility of GSK chemistry described in 2004 [1], by preparing efaproxiral [2]. Simultaneously, a (then) newly described amide formation, catalyzed by boric acid [3], looked especially attractive from both a waste management and throughput perspective. This catalytic amide formation was used effectively to prepare an efaproxiral intermediate while showcasing the selective chemistry. Boric acid is inexpensive, easy to recover and reuse, and environmentally neutral [4]. Boric acid is nontoxic to fish (>1000), aquatic invertebrates (133–226), birds (>2500), and bees (>36,000), reported as LC_{50} ($mg\,l^{-1}$ or $mg\,kg^{-1}$).

The synthesis of efaproxiral is shown in Scheme 6-1. Efaproxiral, which has been available since 2009 for metastatic cancer therapy, is an allosteric modifier of hemoglobulin [5]. Thus, it is designed to make tumors more responsive to radiation by increasing the amount of oxygen in tumor cells. Further studies of efaproxiral demonstrated improvement in exercise capacity [6]. It is an interesting comment on our social fears that efaproxiral was banned from athletic competition before it was available to patients, but that would be a subject for another book.

Azeotropic removal of water during the amide formation was essential for reaction completion, and rapid mixing or vigorous reflux shortened the reaction time. In addition to the low cost and the relatively low toxicity of toluene, its ability to form an azeotrope with water made it a suitable solvent for the amidation reaction. After screening (see Figures 6-1 and 6-2) for catalyst loading and reaction time, 7.7 mol% (4 wt%) and 18 h were selected. This catalyst loading provided a reasonable reaction time, since 24 h at reflux (two production shifts) was considered by GSK to be maximum. Even with a 9 wt% catalyst loading, the reaction time could not be shortened below 12 h (one shift). Interestingly, uncatalyzed water elimination required 3 days in refluxing xylene at 140 °C. As in all such cases, it is important to obtain early engineering involvement to optimize water elimination through efficient mass and heat transfer. Through the use of fixed-jacketed laboratory reactors designed to mimic plant equipment, the reaction was made ready for scale-up to pilot plant.

In a typical reaction, a mixture of 4-hydroxyphenylacetic acid (1.0 equiv.), 3,5-dimethylaniline (1.2 equiv.), boric acid (0.077 equiv.), and toluene (12.5 l for each 1 kg of 4-hydroxyphenylacetic acid) was heated at reflux with Dean–Stark (DS) water elimination for 18 h. The resulting mixture was cooled, the solid collected by filtration, washed with toluene and water, and dried to give 86–95% yield of *N*-(3,5-dimethylphenyl)-2-(4-hydroxyphenyl)acetamide.

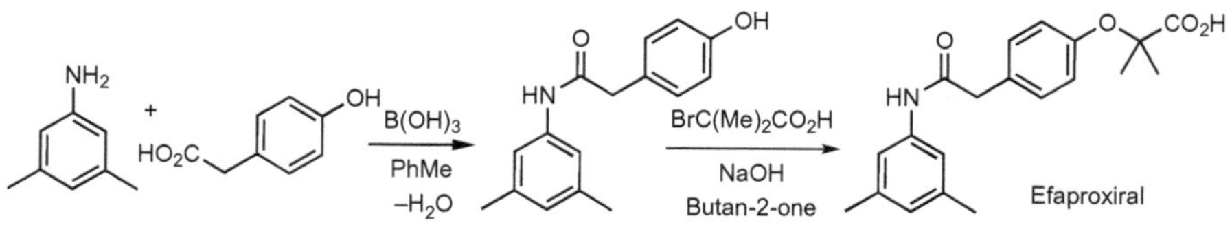

SCHEME 6-1. The synthesis of efaproxiral.

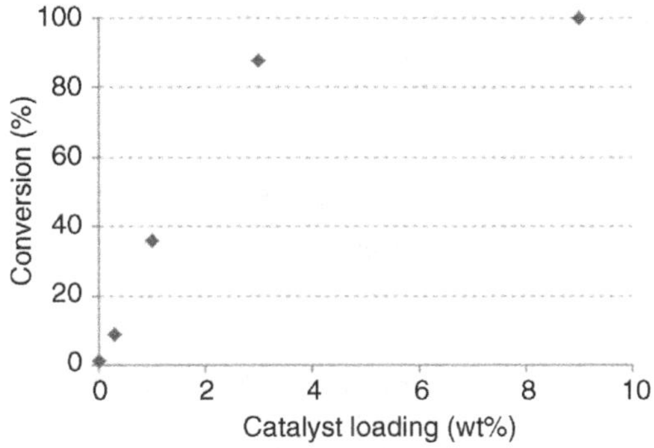

Load (wt%)	Conv.(%)
0	1
0.3	9
1	36
3	88
9	100

FIGURE 6-1. Reaction conversion as a function of catalyst loading at the 16 h time point.

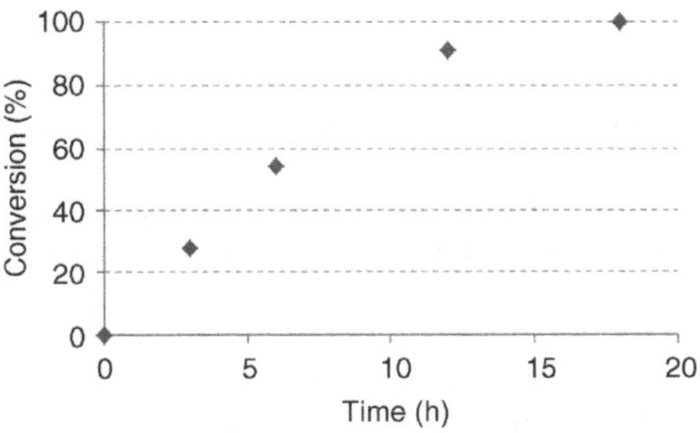

t (h)	Conv. (%)
0	0
3	28
6	54
12	91
18	100

FIGURE 6-2. Reaction profile at 4 wt% catalyst loading.

6.3 DIRECT AMIDATION EXAMPLES FROM DR. REDDY'S LABORATORIES

Perhaps the most detailed study of the synthesis of pharmaceutical intermediates using direct amidation methodology that is available in the literature has been reported by Dr. Reddy's Laboratories coworkers Bandichhor et al. [7]. The successful preparation of seven active pharmaceutical ingredients (APIs) or API intermediates was reported on a large laboratory scale using either boric acid or phenylboronic acid as catalyst. One of the examples has been scaled up to commercial scale, but Dr. Reddy's laboratories have not disclosed which example this is.

Galanthamine is marketed for the treatment of Alzheimer's disease, and the synthesis of the galanthamine intermediate **1** shown in Scheme 6-2 is particularly of note as the aryl

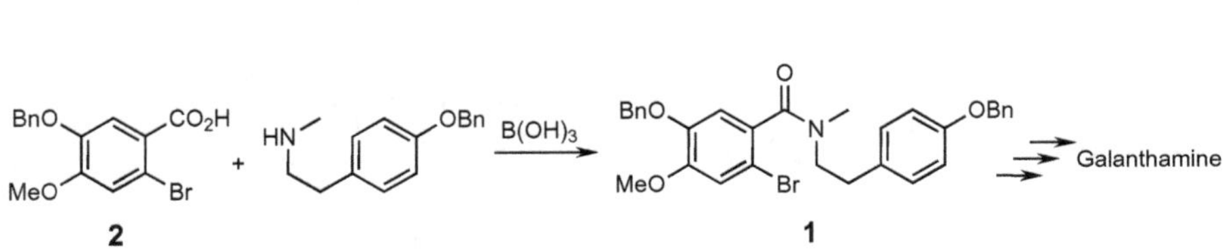

SCHEME 6-2. Synthesis of the galanthamine intermediate **1**.

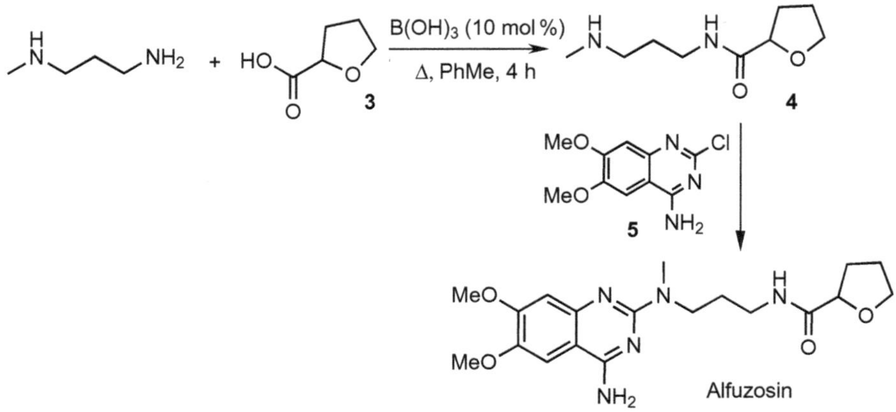

SCHEME 6-3. The synthesis of alfuzosin.

carboxyl group has a sterically demanding bromine atom in the *ortho*-position in a very hindered, tetrasubstituted aryl carboxylic acid **2** [7,8]. Unfortunately it is hard to know exactly what the best results are with this substrate as the results and discussion part of the paper describes a 93% yield in boiling toluene, whereas the experimental section details an 86% yield in boiling xylene [8].

Another impressive application of the use of direct amidation is in the synthesis of the alfuzosin intermediate **4** (Scheme 6-3). Alfuzosin has been marketed in the United States as Uroxatral as a treatment for benign prostatic hyperplasia since 2003 [9,10]. Tetrahydro-2-furanoic acid **3** is converted through to the alfuzosin intermediate **4** by a direct amidation reaction catalyzed by 10 mol% of boric acid in a reaction that is complete in just 4 h in boiling toluene with DS water removal. After distillation, a 95% yield of 94% pure material, by GC analysis, was obtained. The reaction shows excellent regioselectivity, with the primary amine reacting in strong preference to the secondary amine. The reaction of the amine **4** with the chloroquinazoline **5** gave alfuzosin [11], which is marketed as the hydrochloride salt.

The reaction of the chiral amine **6** with the arylacetic acid derivative **7** gave approximately 10% of the antidiabetic agent repaglinide in an uncatalyzed reaction in boiling toluene, but if the reaction was repeated with the addition of 10 mol % of boric acid for 18 h in boiling toluene, a more complete reaction was observed [7]. After cooling and removal of the boric acid by filtration, the filtrate was concentrated and triturated to give a 72% yield of repaglinide (Scheme 6-4). Chiral analysis indicated that no racemization had taken place.

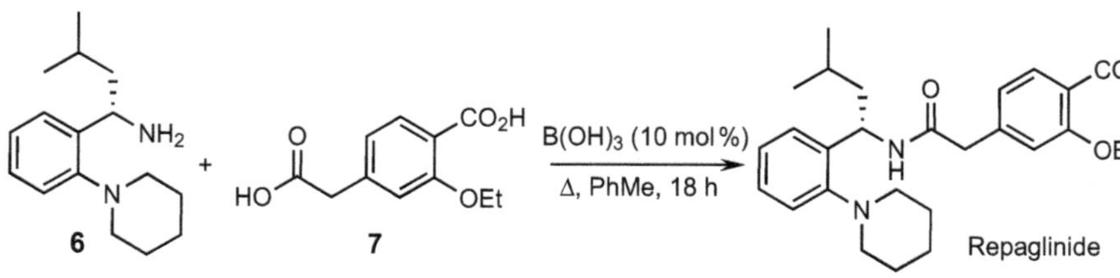

SCHEME 6-4. The synthesis of repaglinide by direct amidation.

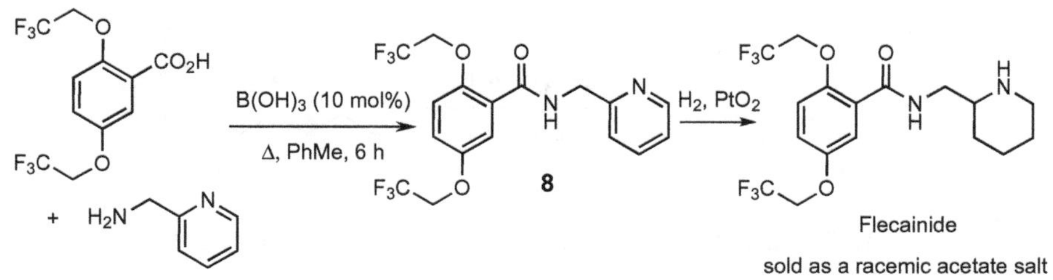

SCHEME 6-5. The synthesis of flecainide.

Flecainide acetate is marketed as an antiarrhythmic agent used to suppress ventricular arrhythmias [12,13]. It can be prepared via a direct amidation to give the amide **8** [7], followed by hydrogenation over Adams' catalyst (PtO_2) [14] to give the racemic piperidine derivative (Scheme 6-5). The direct amidation reaction proceeded in 6 h in boiling toluene. After cooling, the reaction mixture was poured directly into *n*-hexane to precipitate the product, which was washed with water to remove the boric acid and more *n*-hexane, giving the amide **8** in 87% yield and high purity.

6.4 DIRECT AMIDATION EXAMPLES FROM PFIZER

Pfizer has also been experimenting with direct amidation reactions, and one of the most promising substrates is the synthesis of the zamifenacin intermediate **9**. Zamifenacin (or UK-76,654) is an M_3 selective muscarinic antagonist that has been clinically evaluated for the treatment of overactive bladder and irritable bowel syndrome [15–17]. It was a particularly attractive candidate for direct amidation as it involves the reaction of an aryl acetic acid (normally a good substrate for direct amidation) with a nucleophilic piperidine **10**. The reaction proceeds quite well in boiling toluene without any catalyst but slows after 75% conversion as the concentrations of the two starting materials fall. For an effective industrial reaction, some catalysis is required and if the reaction was carried out in the presence of 2 mol% of boric acid in boiling toluene with DS water removal, the reaction was complete after 17 h to give a 96% yield of amide **9** of very high chemical purity. The amide **9** was reduced using diborane to give zamifenacin (Scheme 6-6) [18]. For this

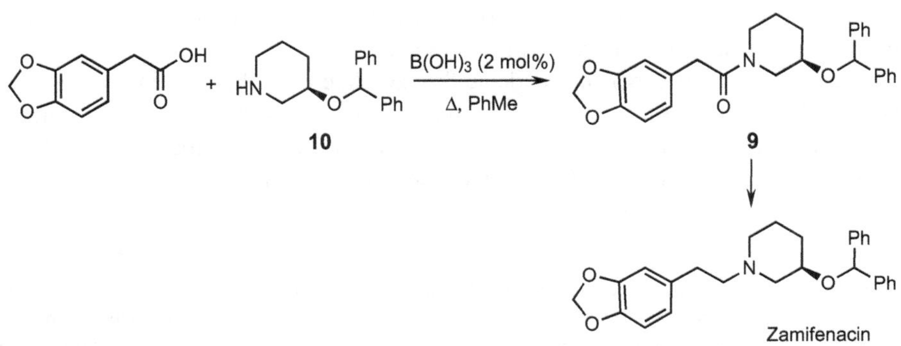

SCHEME 6-6. The synthesis of zamifenacin.

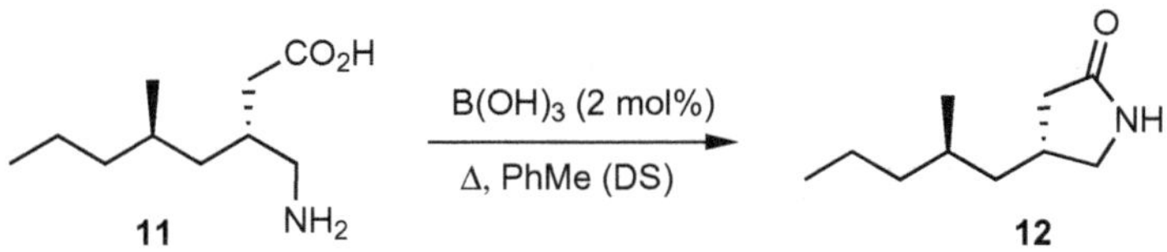

SCHEME 6-7. Intramolecular direct amidation to give lactam **12**.

substrate, boric acid was a more effective catalyst than either 3,4,5-trifluorophenylboronic acid or phenylboronic acid. It is noteworthy that the acid-sensitive benzhydryl group survived the reaction conditions completely intact.

Unfortunately in the 1990s, the zamifenacin program was put on hold and replaced by darifenacin, which went on to be a commercial product [19,20] for the treatment of overactive bladder; so the reaction depicted in Scheme 6-6 was never scaled up beyond laboratory scale.

Direct amidation reactions can also take place in an intramolecular fashion as shown by the example in Scheme 6-7. Treatment of the amino acid **11** with 2 mol% of boric acid in boiling toluene, with DS water removal, in an overnight reaction, cleanly gave the lactam **12** (I. McCarthy and M. Panesar, unpublished results). If the same reaction was performed in boiling toluene with DS water removal, in the absence of boric acid, some product was formed, but the reaction was much less clean and much slower.

Water removal using a DS trap, molecular sieves (MS), or an equivalent method, is critical for direct amidation with boric acid or boronic acid catalysis [21,22]. The use of MS in industry causes some problems, especially if the plant equipment cannot be dedicated to a particular reaction. In the pharmaceutical industry, most products are made in multi-purpose plants and equipment, which has to be rigorously cleaned between products. The presence of MS significantly complicates this cleaning process. If DS water removal is used, then toluene and xylene are excellent solvents for water removal due to the high percentage of water in their azeotropes (18% [23] and 40% [24], respectively). However, more polar compounds have limited solubility in toluene and xylene, and hence this is a limitation. To try to reduce this limitation, Pfizer has been experimenting with using chlorobenzene as solvent. Chlorobenzene has an excellent azeotrope with water (28% water) [25], but has the complication that it is heavier than water, and hence a reverse DS arm is required. In order to make laboratory screening experiments simpler to carry out, Pfizer used 4 Å MS[1] in its screening experiments, then moved to reverse DS water removal to explore the most promising results in more detail.

The reaction of 4-fluorobenzoic acid and the azetidine **13** (Scheme 6-8) was performed in boiling chlorobenzene for 16 h in the presence of 4 Å MS. Boric acid and nine boronic acids were screened,[2] and the most promising four results are summarized in Table 6-1 (N. Castro, D.A. Entwistle, and C. Slattery, unpublished results).

[1] Using molecular sieves at such a high temperature obviously reduces their ability to absorb water. Nevertheless these screening experiments were much easier to carry out than using a reverse DS water removal arm.

[2] The other boric acid derivatives that were screened but were less effective than the four examples given in Table 6.1 were: boric acid, phenylboronic acid, 3,4,5-(trifluoromethyl)phenylboronic acid, trimethyl borate, 2-furanylboronic acid, *N*-Boc-2-indanylboronic acid.

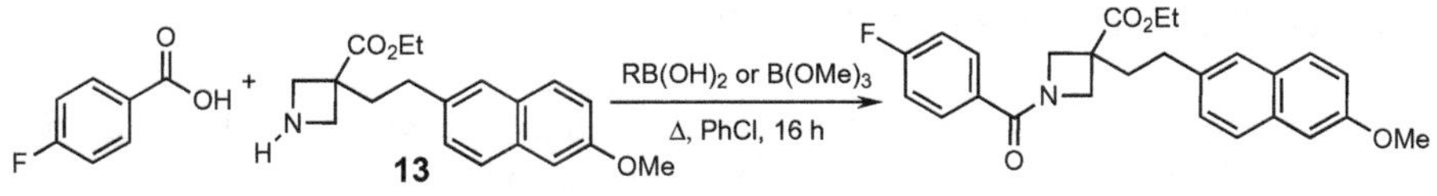

SCHEME 6-8. The acylation of azetidine **13**.

TABLE 6-1. Comparison of Catalysts for the Acylation of Azetidine 13

Catalyst	Conversion (%)
None	16
4-(Trifluoromethyl)phenylboronic acid	52
3,5-Bis(trifluoromethyl)phenylboronic acid	59
3,4,5-Trifluorophenylboronic acid	89
2-(*N*,*N*-Dimethylaminomethyl)phenylboronic acid	79

Azetidines react more readily with aliphatic acids as shown by the example in Scheme 6-9. The direct amidation of the acid **14** with the azetidine **15** proceeded in boiling toluene with just 2 mol% of boric acid to give the amide **16** in 87% isolated yield (D. Laity and B.R. Dillon, unpublished results).

The acid **17** and piperidine **18** (shown as partial structures) were almost completely insoluble in toluene and xylene, and attempts to perform direct amidations in those solvents failed completely due to poor solubility. However, it was possible to get the reaction to proceed in chlorobenzene, using a reverse DS arm in around 70% yield in 2 days using 100 mol% of boric acid (Scheme 6-10) (C. Grosjean and C.E. Thirsk, unpublished results).

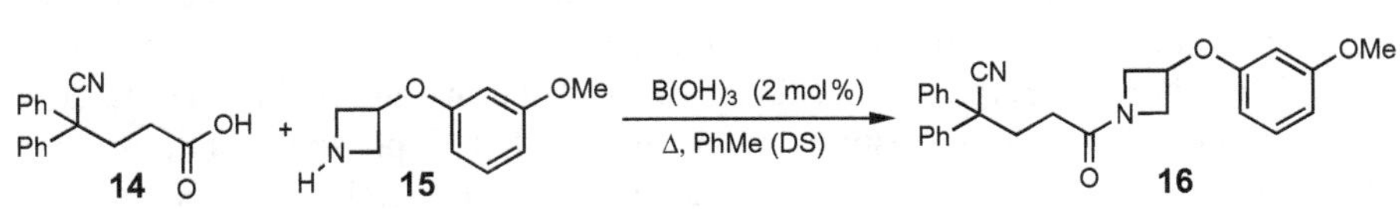

SCHEME 6-9. The synthesis of azetidine **16**.

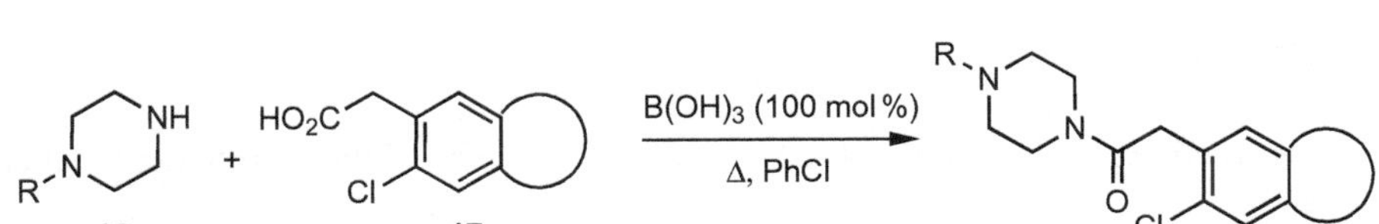

SCHEME 6-10. Direct amidation reaction in chlorobenzene as solvent.

6.5 POTENTIAL TOXICITY OF BORIC ACID

As previously mentioned, earlier reports [4] described the low toxicity of boric acid in the environment. However, more recently, there is growing evidence that boric acid should be considered a reproductive toxicant at high doses [26–28]. Researchers fed rats boric acid for three generations, and at the highest dose the rats did not reproduce. Pregnant rats fed with boric acid had offspring with lower body weights at all doses tested, and at the three highest doses, investigators detected more rib and brain changes [27]. Dosing of boric acid to pregnant rabbits led to more offspring, and heart defects at the highest doses [28].

In view of these and other toxicity findings, chemists should consider boric acid a reproductive toxicant and design their processes bearing this in mind. In addition, there is a move by the European Chemical Agency to place boric acid on the substances of very high concern list, which would significantly complicate its use in industrial processes in Europe.[3] In view of this, other direct amidation catalysts such as those developed by Clarke (activated silica) [29] and Akamanchi (sulfated tungstate) [30] are also worth exploring.

6.6 CONCLUSIONS

Industrial application of direct amidation is still in its infancy, but the strong interest of companies, such as GSK, Dr. Reddy's Laboratories, and Pfizer, shows that the methodology has very good potential. That said, there remain many areas for improvement. Water elimination, whether via MS or azeotropic reflux and separation in a DS trap, is not fully satisfactory. Molecular sieves are not amenable to batch reaction, requiring charging (and cleaning) of the reactor or pumping hot solutions through the sieves. Long azeotropic reflux offsets the "greenness" of the process, requires solubility of substrates in typically low polarity solvents, and stability of materials over long reflux times. While more powerful catalysts are needed to shorten reaction times, only very efficient and rapid water separation and elimination will drive efficient industrial processes. New visions are needed. As one example, a flow reactor packed with embedded catalyst, and built of materials that either absorb or shed water, should allow more varied solvents (both water soluble and insoluble), while eliminating the need to heat mixtures for long times. A recent publication stresses the importance of understanding the heat transfer in this type of process, in particular using intensified azeotropic distillation, and also suggests different reactor types that would maximize that heat transfer [31]. In summary, the current methodology, while interesting, is only applicable to a portion of industrial amide bond forming reactions. Further technical breakthroughs in academic, industrial, and mixed settings will be required to broaden the applicability of direct amidation. In spite of all the challenges outlined, we believe that there will be further industrial examples of the scale-up of direct amidation reactions over the next few years.

ACKNOWLEDGMENT

We thank Kevin R. Bartnik (Pfizer Inc.) for his expert review of the reprotoxicity data for boric acid.

[3] Boric Acid, Substance of Very High Concern (SVHC) Support Document http://echa.europa.eu/doc/candidate_list/svhc_supdoc_boric_acid_publication.pdf (last accessed 18th October, 2011).

REFERENCES

1. Davis R, Fitzgerald RN, Guo J (2004). Improved method for the synthesis of 2-methyl-2-aryloxypropanoic derivatives. *Synthesis* 1959–1962.
2. Anderson JE, Davis R, Fitzgerald RN, Haberman JM (2006). Selective phenol alkylation: an improved preparation of efaproxiral. *Synth. Commun.* 36, 2129–2133.
3. Tang P (2005). Boric acid catalyzed amide formation from carboxylic acids and amines: *N*-benzyl-4-phenylbutyramide. *Org. Synth.* 81, 262–268.
4. (a) Budavari S, editor. *The Merck Index—An Encyclopedia of Chemicals, Drugs, and Biologicals.* Merck and Co., Whitehouse Station, NJ; 1996. (b) National Pesticide Information Center (1998).
5. Kleinberg L, Grossman SA, Carson K, Lesser G, O'Neill A, Pearlman J, Phillips P, Herman T, Gerber M (2002). Survival of patients with newly diagnosed glioblastoma multiforme treated with RSR13 and radiotherapy: results of a phase II new approaches to brain tumor study therapy CNS consortium safety and efficacy study. *J. Clin. Oncol.* 20, 3149–3155.
6. Watabnabe T, Takeda T, Omiya S (2008). Reduction in haemoglobin-oxygen affinity results in the improvement of exercise capacity in mice with chronic heart failure. *J. Am. Coll. Cardiol.* 48, 779–786.
7. Mylavarapu RK, Kondaiah GCM, Kolla N, Veeramalla R, Koilkonda P, Bhattacharya A, Bandichhor R (2007). Boric acid catalyzed amidation in the synthesis of active pharmaceutical ingredients. *Org. Process Res. Dev.* 11, 1065–1068.
8. Reddy JM, Kumar KV, Raju V, Bhaskar BV, Himabindu V, Bhattacharya A, Sundaram V, Banerjee R, Reddy GM, Bandichhor R (2008). Alternative total synthesis of (−)-galanthamine hydrobromide. *Synth. Commun.* 38, 2138–2149.
9. Doggrell SA (2002). Present and future pharmacology for benign prostatic hyperplasia. *Drugs Future* 27, 973.
10. Lacerda AE, Kuryshev YA, Chen Y, Renganathan M, Eng H, Danthi SJ, Kramer JW, Yang T, Brown AM (2008). Alfuzosin delays cardiac repolarization by a novel mechanism. *J. Pharmacol. Exp. Ther.* 324, 427–433.
11. Manoury PM, Binet JL, Dumas AP, Lefevre-Borg F, Cavero I (1986). Synthesis and anti-hypertensive activity of a series of 4-amino-6,7-dimethoxyquinazoline derivatives. *J. Med. Chem.* 29, 19–25.
12. Rognerund CL, Ou CN (1989). Concurrent measurement of flecainide acetate and propranolol by normal-phase high-performance liquid chromatography. *Clin. Chem.* 35, 857–890.
13. Hudak JM, Banitt EH, Schmid JR (1984). Discovery and development of flecainide. *Am. J. Cardiol.* 53, 17B–20B.
14. Banitt EH, Bronn WR, Coyne WE (1977). Antiarrhymics. 2. Synthesis and antiarrhythmic acitivity of *N*-(piperidinylalkyl)trifluoroethoxybenzamides. *J. Med. Chem.* 20, 821–826.
15. Loughton LA, Rogers J, Whorwell PJ, Campbell FC, Williams NS, Goka J (1997). Zamifenacin (UK-76,654), a potent gut M_3 selective muscarinic antagonist, reduces colonic motor activity in patients with irritable bowel syndrome. *Aliment. Pharmacol. Ther.* 11, 561–568.
16. Wallis RM (1995). Pre-clinical and clinical pharmacology of selective muscarinic M_3 receptor antagonists. *Life Sci.* 56, 861–868.
17. Cosey J, Dumas C, Pardo DG (1997). A short efficient synthesis of zamifenacin a muscarinic M_3 receptor antagonist. *Bioorg. Med. Chem. Lett.* 7, 1343–1344.
18. Pettman AJ (2007). Application of biocatalysis and emerging synthetic methodologies in the synthesis of pharmaceuticals, presentation given at the Art of Catalysis in Process Chemistry meeting, Barcelona, Spain, April 12–13, 2007.
19. Chapple CR (2004). Darifenacin: a novel M_3 muscarinic selective receptor antagonist for the treatment of over active bladder. *Expert Opin. Investig. Drugs* 13, 1494–1500.

20. Dunn PJ, Matthews JG, Newbury TJ, O'Connor G, inventors; Novartis Int. Pharmaceutical Ltd, assignee. A stable hydrate of a muscarinic receptor antagonist. World Patent WO/2003/080599. October 2, 2003.
21. Al-Zoubi RM, Olivier M, Hall DG (2008). Direct and waste-free amidations and cycloadditions by organocatalytic activation of carboxylic acids at room temperature. *Angew. Chem., Int. Ed.* 47, 2876–2879.
22. Arnold K, Davies B, Giles RL, Grosjean C, Smith GE, Whiting A (2006). To catalyze or not to catalyze? Insight into direct amide bond formation from amines and carboxylic acids under thermal and catalyzed conditions. *Adv. Synth. Catal.* 348, 813–820.
23. Susarev MP, Gorbunov AN (1963). The liquid-vapor equilibrium in the system benzyl alcohol-toluene-water. *Zh. Prikl. Khim.* 36, 459–461.
24. Horsley LH. *Azeotropic Data-III*, 1973. *Advances in Chemistry Series*, 116, American Chemical Society, Washington, DC; 1973.
25. Prahl W, Mathes W (1934). Uber einige azetrope gemische im system chlorwasserstoff-wasser-chorbenzol-phenol. *Angew. Chem.* 47, 11–13.
26. National Pesticide Information Centre (2001). http://npic.orst.edu/factsheets/boricgen.pdf (last accessed 13 October, 2011).
27. Heindel JJ, Price CJ, Field EA, Marr MC, Myers CB, Morrissey RE, Schwetz BA (1992). Developmental toxicity of boric acid in mice and rats. *Fundam. Appl. Toxicol.* 18, 266–277.
28. Price CJ, Marr MC, Myers CB, Seely JC, Heindel JJ, Schwetz BA (1996). The developmental toxicity of boric acid in rabbits. *Fundam. Appl. Toxicol.* 32, 176–187.
29. Comerford JW, Clarke JH, Macquarrie DJ, Breeden SW (2009). Clean, reusable and low cost heterogeneous catalyst for amide synthesis. *Chem. Commun.* 2562–2564.
30. Chaudhari PS, Salim SD, Sawant RV, Akamanchi KG (2010). Sulfated tungstate: a new solid heterogeneous catalyst for amide synthesis. *Green Chem.* 12, 1707–1710.
31. Grosjean C, Parker J, Thirsk CE, Wright AR (2012). Intensified azeotropic distillation: a strategy for optimising direct amidation. *Org. Process Res. Dev.* 16, 781–787.

7

OH ACTIVATION FOR NUCLEOPHILIC SUBSTITUTION

JONATHAN M.J. WILLIAMS
Department of Chemistry, University of Bath, Claverton Down, Bath, UK

7.1 INTRODUCTION

The wide availability of alcohols makes them useful building blocks in synthesis. However, it is generally necessary to activate the alcohol in order to facilitate nucleophilic substitution reactions for the construction of new C–C or C–N bonds. The simplest method for activation is to enhance the leaving group ability of the hydroxyl group by protonation, although this strategy is often hampered by the fact that the nucleophile is deactivated under the acidic conditions. Conversion of the alcohol into an alkyl halide, alkyl tosylate, or related species is a commonly used strategy for activation of alcohols, although such alkylating agents are often highly toxic, which is a significant disadvantage. Activation of alcohols *in situ*, as seen in Mitsunobu coupling reactions [1], can also be used, but these methods require a stoichiometric amount of activating agent, which leads to poor atom economy and the need to separate the product from waste material.

Recently, an alternative approach to the activation of alcohols for nucleophilic substitution has been receiving considerable attention. This approach involves the temporary catalytic removal of hydrogen from the alcohol to form an intermediate carbonyl compound. Carbonyl compounds are better electrophiles than alcohols and can readily react with a range of nucleophiles. For suitable substrates, the hydrogen that has been removed can be returned to give a product with no net oxidation or reduction. This strategy has been termed "Borrowing Hydrogen Methodology" [2] and "The Hydrogen Autotransfer Process" [3]. The chemistry has been widely applied to both C–C and C–N bond formation from alcohols according to Schemes 7-1 and 7-2. For C–C bond formation, temporary removal of hydrogen from alcohol **1** generates an intermediate aldehyde **2**, which undergoes an alkene forming reaction, typically by an aldol reaction, to give the

Sustainable Catalysis: Challenges and Practices for the Pharmaceutical and Fine Chemical Industries, First Edition. Edited by Peter J. Dunn, K. K. (Mimi) Hii, Michael J. Krische, and Michael T. Williams.

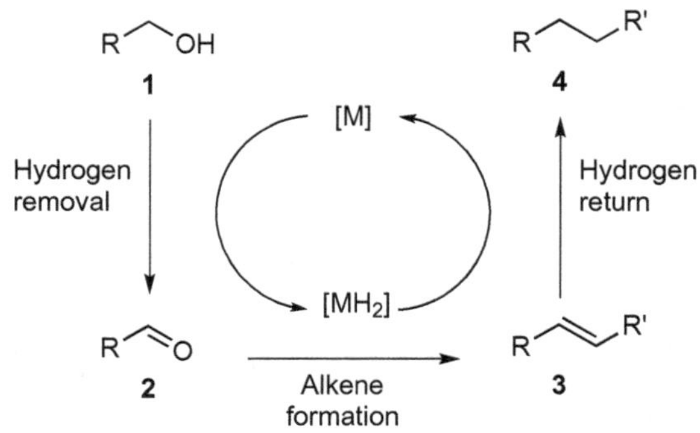

SCHEME 7-1. Metal-catalyzed C–C bond formation by temporary removal of hydrogen.

alkene **3**. Return of hydrogen gives the reduced product **4**, which is effectively the product of substitution of the original alcohol, formed indirectly (Scheme 7-1). Although represented as the temporary formation of a metal dihydride, alternative intermediates are possible, but in each case, it is the reversible removal of hydrogen that drives the reaction.

Similarly, for C–N bond formation, the pathway for reaction of the alcohol **1** proceeds through the aldehyde **2** (Scheme 7-2). The formation of imine **5** usually occurs by a condensation reaction with an amine **6**, where water is eliminated as the only by-product. Again, hydrogen is returned, and in this case a secondary amine **7** is formed, although the strategy has also been applied to the formation of tertiary amines (Section 7.3).

Currently, the main limitations of this methodology are that relatively high temperatures and high catalyst loadings are required, although progress is starting to be made to improve both of these issues. In addition, the majority of reports require the use of expensive platinum group metals, but there has been some recent attention given to cheaper alternative catalysts that are proving to be effective.

Several reviews have appeared which detail various aspects of metal-catalyzed activation of alcohols by temporary removal of hydrogen [4–8].

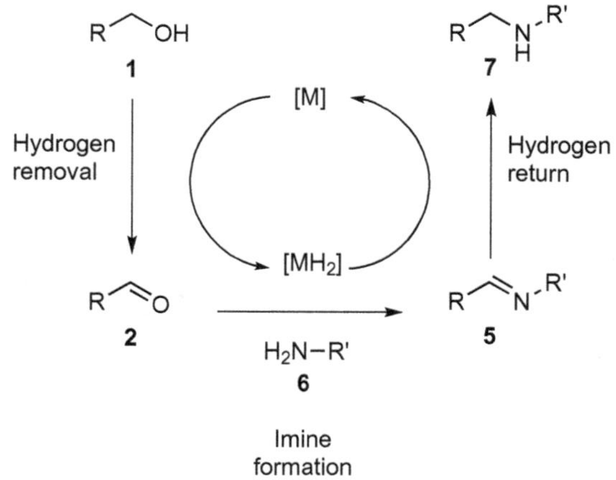

SCHEME 7-2. Metal-catalyzed C–N bond formation via temporary removal of hydrogen.

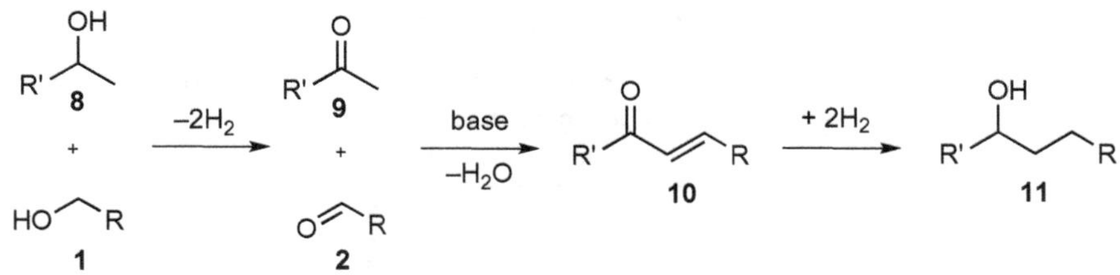

SCHEME 7-3. Dehydrative coupling of a primary alcohol with a secondary alcohol.

7.2 FORMATION OF C–C BONDS FROM ALCOHOLS

The construction of C–C bonds is fundamental to organic synthesis, and alcohols have been used in a range of C–C bond forming reactions using the metal-catalyzed temporary reversible removal of hydrogen. Primary and secondary alcohols can be coupled with loss of water according to the pathway shown in Scheme 7-3 . The primary alcohol **1** is oxidized to an aldehyde **2**, and typically this will be a nonenolizable aldehyde. The secondary alcohol **8** is oxidized to a ketone **9**, and then an aldol condensation reaction occurs to give an unsaturated ketone **10**. Return of 2 equiv. of H_2 completes the catalytic cycle to afford alcohol **11**, with water as the only reaction by-product.

In these reactions, the removal of hydrogen activates the alcohols in two ways; the aldehyde formation renders the compound more susceptible to nucleophilic addition, whereas ketone formation, and subsequent enolate formation, allows the original substrate to function as a nucleophile. Perhaps surprisingly, the temporary formation of an enolate from an alcohol by borrowing hydrogen has not been exploited more widely in addition to other electrophiles.

One of the first examples of the use of a homogeneous catalyst for the coupling of two alcohols was reported by Cho and coworkers [9]. The reaction of 1-phenylethanol **12** and benzyl alcohol **13** led to the formation of the coupled alcohol product **14** (Scheme 7-4). In this early example, the presence of an excess of 1-dodecene was required to act as a hydrogen acceptor, and it was suggested that 1,4-dioxane acts as an oxidant. Subsequently, there have been several reports of related reactions, where these additives were not needed. For example, Crabtree and coworkers used the ruthenium terpyridine complex **15** to catalyze the same reaction using a 1:1 ratio of alcohols with 1 mol% catalyst in 1 h in toluene at reflux [10]. The use of $[Ru(dmso)_4Cl_2]$ has also been reported as a catalyst for the same transformation, although the reactions took up to 7 days to reach completion [11].

Fujita and coworkers used $[Cp^*IrCl_2]_2$, a complex that has been used very successfully in a range of borrowing hydrogen chemistry, for the coupling of alcohols [12]. It was more efficient than other catalysts for the coupling of aliphatic alcohols. For example, 1-phenylethanol **12** reacted with butanol **16** to give the coupled product **17** in good isolated yield when this catalyst was used (Scheme 7-4). The same catalyst was used by Ishii and coworkers for the dehydrative dimerization of primary alcohols [13]. For example, hexanol **18** was converted into the higher alcohol **19** in good yield using this catalyst (Scheme 7-4).

On the other hand, Allen and Crabtree have demonstrated that transition metal catalysts are not always required in order to achieve alcohol-coupling reactions [14]. They were able to couple 1-phenylethanol **12** and benzyl alcohol **13** using stoichiometric amounts of simple metal hydroxides such as NaOH or KOH.

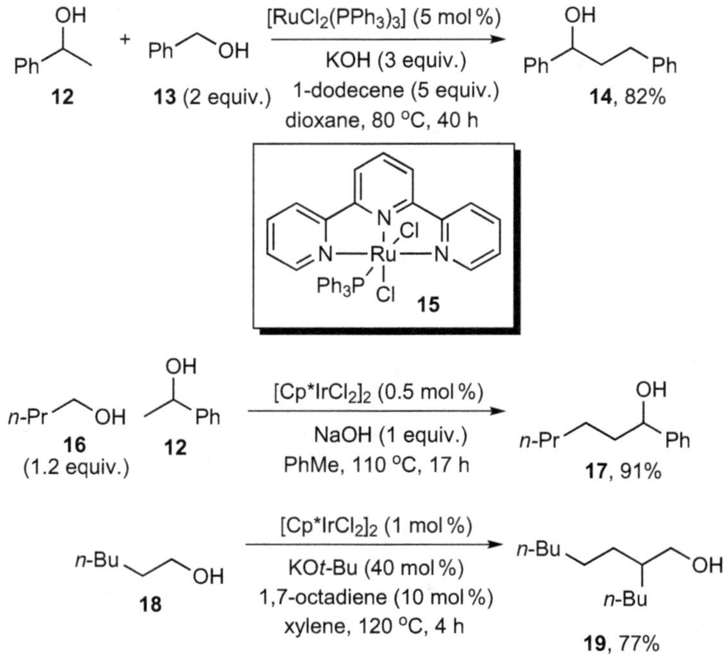

SCHEME 7-4. Ru- and Ir-catalyzed alcohol coupling reactions.

Primary alcohols can also be used as alkylating agents for ketones where the need to oxidize the secondary alcohol component is not required. Cho and coworkers demonstrated this for a range of ketones, although acetophenones and their derivatives gave the most successful results [15]. For example, tetralone **20** was alkylated with benzyl alcohol **13** to give the product **21** in good isolated yield (Scheme 7-5). 1-Dodecene acted as a hydrogen

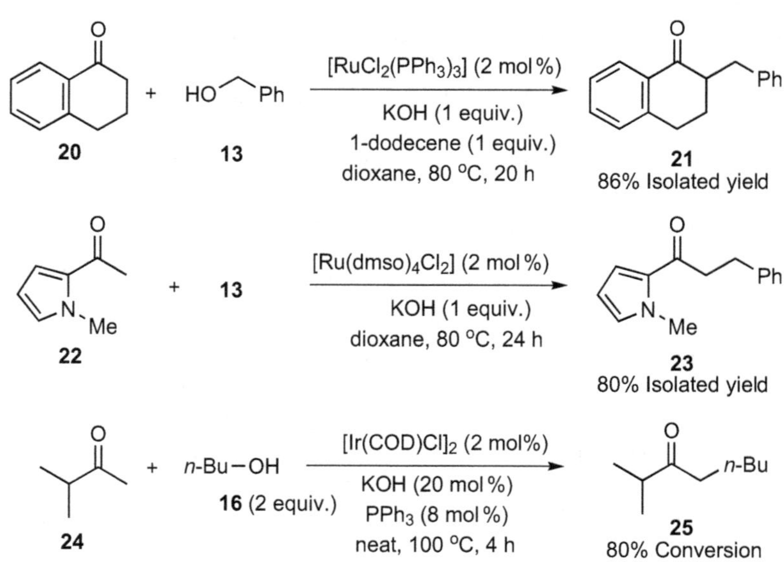

SCHEME 7-5. α-Alkylation of ketones with alcohols.

acceptor to favor the formation of the product at the ketone oxidation level. However, the authors also reported that in the absence of a hydrogen acceptor, the reaction pathway could be diverted to give the corresponding alcohol as the product [16]. Several other catalysts have been reported for the alkylation of ketones with alcohols. For example, $[RuCl_2(dmso)_4Cl_2]$ was used for the alkylation of ketone **22** with benzyl alcohol **13** to give the elaborated ketone **23** (Scheme 7-5) [17]. An iridium catalyst was found to be more effective when aliphatic alcohols were used as the alkylating agent [18]. The condensation of ketone **24** with butanol **16** was catalyzed in good conversion using this catalyst (Scheme 7-5).

Nitriles can also be alkylated with alcohols using the borrowing hydrogen strategy. The earliest examples of this were reported by Grigg in 1981, where arylacetonitriles were alkylated with a range of alcohols, catalyzed by Ru, Rh, or Ir complexes, including $[RuH_2(PPh_3)_4]$ [19]. Twenty-five years later, Grigg and coworkers reported the use of $[Cp{*}IrCl_2]_2$ as an efficient catalyst for the same transformation [20]. The parent reaction involves condensation of phenylacetonitrile **26** with benzyl alcohol **13** to give the alkylated product **27** (Scheme 7-6). However, a wide range of related substrates could also be coupled. For example, 3-pyridylacetonitrile **28** and 3-pyridylmethanol **29** underwent C–C bond formation in good yield to give product **30**. The reaction could be performed under

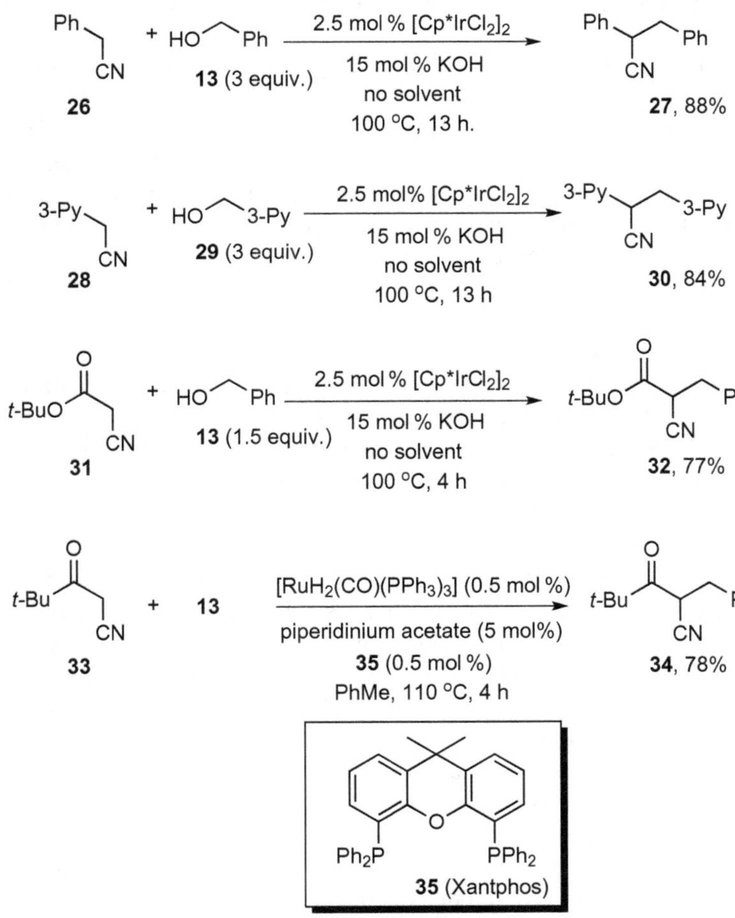

SCHEME 7-6. α-Alkylation of activated nitriles with alcohols.

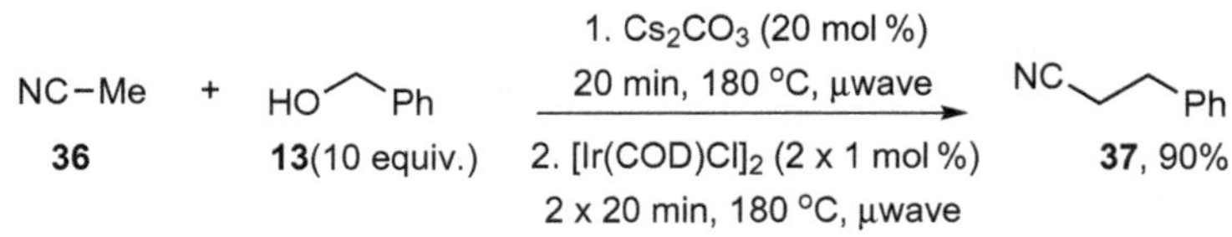

SCHEME 7-7. α-Alkylation of unactivated acetonitrile.

the conditions indicated in Scheme 7-6, or by using microwave irradiation at 110 °C for 10 min.

tert-Butyl cyanoacetate **31** has been alkylated with alcohols by several research groups [21]. $[Cp^*IrCl_2]_2$ and $[Ir(COD)Cl]_2/PPh_3$ have both been reported as effective catalysts for this transformation [22]. For example, the coupling of *tert*-butyl cyanoacetate **31** with benzyl alcohol **13** gave the alkylated product **32** in good yield (Scheme 7-6). Williams and coworkers used $[RuH_2(CO)(PPh_3)_3]$ as a catalyst for the coupling of the related ketonitrile **33** with benzyl alcohol **13** (Scheme 7-6) [23]. The presence of Xantphos **35** as a ligand accelerated the reaction considerably. When aliphatic alcohols were used as the alkylating agent, a higher catalyst loading of 5 mol% Ru was needed. The presence of piperidinium acetate was thought to catalyze the condensation of the intermediate benzaldehyde with the nitrile.

There has generally been less success in the alkylation of nonactivated nitriles, which is disappointing, given the synthetic versatility of the nitrile functional group. However, a recent report from Cossy and coworkers is very encouraging [24]. Under fairly forcing conditions, and using 10 equiv. of acetonitrile **36**, alkylation with benzyl alcohol **13** was achieved (Scheme 7-7). The best yields were obtained when starting materials were preheated for 20 min at 180 °C under microwave irradiation. Addition of iridium catalyst and further heating was performed, and the procedure was repeated. The reaction worked well for electron-rich benzyl alcohols but not for the electron-deficient variants. The reaction could be applied to alkylation using aliphatic alcohols but required microwave heating for 12 h at 180 °C. Sawaguchi and Obora recently reported that $[Ir(COD)(OH)]_2/PPh_3$ was a useful alternative to iridium catalyst for the alkylation of nitriles [25].

The formation of alkylated esters that do not contain additional electron-withdrawing groups has been achieved by several different approaches. First, an indirect Wittig reaction on alcohols using ruthenium *N*-heterocyclic carbene (NHC) complex **38** has been reported by Whittlesey and coworkers (Scheme 7-8) [26]. The reaction pathway was presumed to proceed by oxidation of benzyl alcohol **13** to give benzaldehyde that underwent a Wittig reaction to give an alkene, followed by return of hydrogen to give the ester product **40**. While the reaction was achieved under relatively mild conditions, the disadvantage of this chemistry is the formation of triphenylphosphine oxide as a stoichiometric by-product. Subsequently, Pridmore and Williams demonstrated that malonate half esters could be used successfully in borrowing hydrogen chemistry with decarboxylation occurring to give loss of the carboxylic acid group during the reaction [27]. Treatment of benzyl alcohol **13** with the malonate half ester **41** led to formation of the alkylated ester product **42** in good yield, but several additives were required in order to prevent the reaction stalling at the alkene stage (Scheme 7-8). The direct alkylation of *tert*-butyl acetate **43** with a range of aliphatic and benzylic alcohols has been achieved using an iridium catalyst, although a 10-fold excess of the ester was needed [28]. For example, reaction with benzyl alcohol **13** leads to formation of the alkylated product **44** in good conversion (Scheme 7-8).

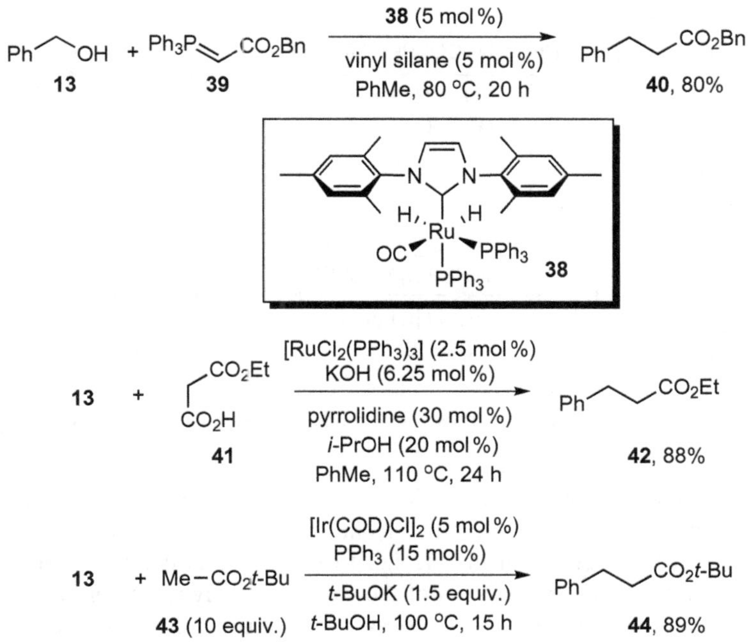

SCHEME 7-8. Strategies for the formation of elaborated esters.

Other classes of nucleophile have been *C*-alkylated by alcohols, including indoles [29], oxindoles [30,31] and an interesting elaboration of the methyl group of pyrimidine **45** [32]. Blank and Kempe reacted this substrate with alcohols, achieving C–C bond formation to give the alkylated product. For example, the reaction with benzyl alcohol **13** led to the formation of product **46** using an iridium catalyst in the presence of the *P,N*-ligand **47** (Scheme 7-9).

Krische and coworkers have developed a wide range of C–C bond forming reactions that involve catalytic hydrogen transfer using alcohols as starting materials (see Chapter 16). Their work involves the coupling of an unsaturated compound such as an alkyne or diene, which undergoes hydrometallation and couples with an aldehyde. The aldehyde may

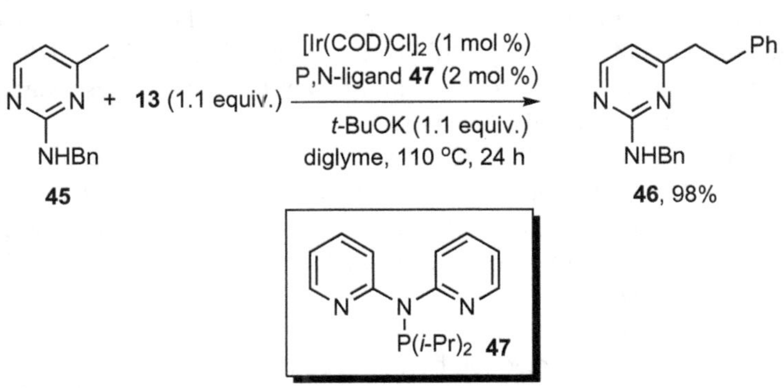

SCHEME 7-9. Alkylation of a methyl group on a heterocycle.

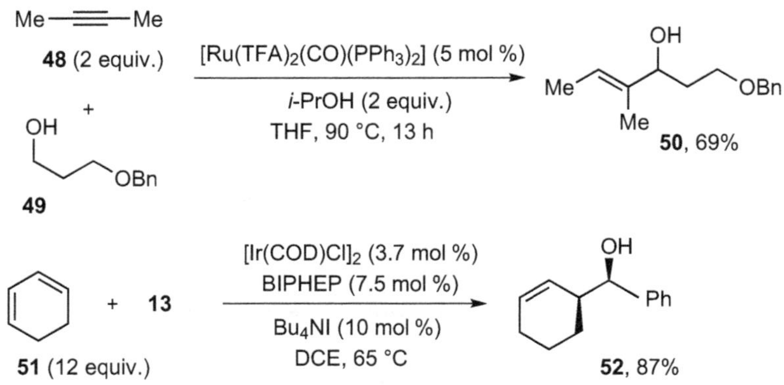

SCHEME 7-10. Examples of Krische's C–C bond forming reactions (see also Chapter 16).

be generated from an alcohol by catalytic removal of hydrogen. For example, alkyne **48** reacts with alcohol **49** to give the allylic alcohol **50** (Scheme 7-10) [33]. The ruthenium catalyst removes hydrogen from the alcohol to give an intermediate aldehyde. Hydrometallation of the alkyne and addition of the so-formed vinylruthenium to the aldehyde leads to product formation. Similarly, diene **51** acts as an alternative unsaturated species for hydrometallation, leading to the formation of product **52** with good diastereoselectivity and regioselectivity (Scheme 7-10) [34]. Many other unsaturated species can be used in these reactions including allenes, 1,3-enynes, and related compounds, some of which have been performed enantioselectively [35,36].

7.3 FORMATION OF C–N BONDS FROM ALCOHOLS

In the early 1980s, the first examples of homogeneously catalyzed alkylation of amines by alcohols were reported by Watanabe [37], Grigg [38], and Murahashi [39]. More recently, improved catalysts have been developed for these reactions, and many modifications of catalysts, usually ruthenium or iridium complexes, have been reported [2–8].

The combination of readily available $[RuCl_2(p\text{-cymene})]_2$ complexed *in situ* to diphosphines, for example, DPEphos, provides good reactivity for coupling primary alcohols with primary or secondary amines, although this catalyst is less efficient when using secondary alcohols [40]. Examples of its use include the reaction of alcohol **53** with *tert*-butylamine **54**, the monoalkylation of ethylene glycol **56** with *N*-benzylaniline **57**, and the subsequent conversion of the product **58** into one of the first antihistamine drugs, Antergan **59**, by coupling with dimethylamine **60** (Scheme 7-11).

An alternative ruthenium catalyst developed by Beller and coworkers is more reactive for the amination of secondary alcohols, although it also works well with primary alcohols [41]. In particular, ligand **61** works effectively with $[Ru_3(CO)_{12}]$ for a wide range of substrates. Examples of the use of this catalyst include condensation of alcohols **12** and **64** with hexylamine **63** (Scheme 7-12).

Perhaps the most widely used catalyst for amine/alcohol coupling reactions is $[Cp^*IrCl_2]_2$, first used for this reaction by Fujita and coworkers [42]. Aniline **66** can be coupled with alcohols including 2-phenylethanol **53**, and secondary amines such as pyrrolidine **68** can be coupled to benzyl alcohol **13** (Scheme 7-13). Cumpstey and

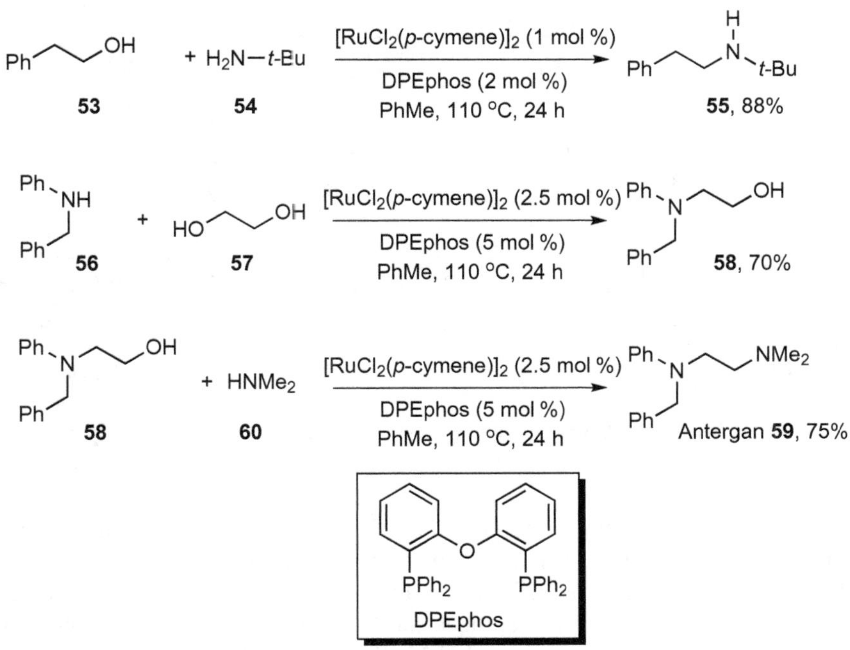

SCHEME 7-11. Amine alkylation reactions using $[RuCl_2(p\text{-cymene})]_2$ as catalyst.

coworkers have used the $[Cp^*IrCl_2]_2$ catalyst for the synthesis of amino sugars [43]. The reaction of amine **70** with alcohol **71** led to the formation of the pseudodisaccharide amine **72** in remarkably good yield, bearing in mind the complexity of the reacting partners (Scheme 7-13). The authors reported that other catalysts reported in the literature were either ineffective or less effective.

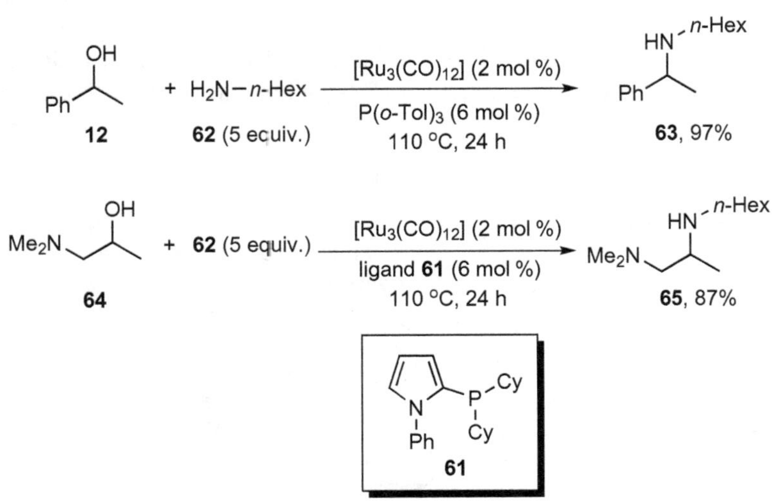

SCHEME 7-12. Amine alkylation reactions using Beller's ruthenium catalyst.

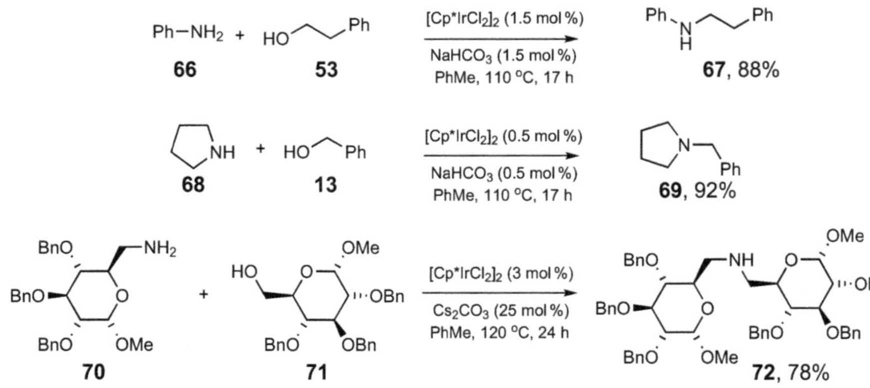

SCHEME 7-13. Amine alkylation reactions using $[Cp^*IrCl_2]_2$ as catalyst.

An alternative iridium catalyst developed by Kempe and coworkers is able to catalyze reactions at 70 °C, lower than the temperatures usually used with other catalysts [44]. The use of $[Ir(COD)Cl]_2$ with $Py_2NP(i\text{-}Pr)_2$ (**47**) forms the catalyst **73** *in situ.* The catalyst works well for the alkylation of anilines and 2-aminopyridines, represented by the alkylation of aniline **66** with benzyl alcohol **13** (Scheme 7-14). Even at 0.2 mol% loading of iridium, 71% yield of the product was obtained after 24 h at 70 °C.

Microwave irradiation has been used effectively as an alternative to conventional heating, with both iridium [45] and ruthenium [46] catalysts having been used successfully for amine alkylation reactions with alcohols. The majority of alkylation reactions have been performed in the absence of water, and some are run in the presence of molecular sieves in order to remove water as it is formed. However, water has been used as the reaction medium with iridium catalysts [47,48], clearly indicating that the reactions are not reversed by the presence of water. In fact, it would be useful if conditions could be found to reverse the alkylation reactions, to allow a synthesis of alcohols from amines, but this has not been reported. There have been many reports of amine alkylation reactions being performed in the absence of solvents, which often leads to shorter reaction times. One example from the group of Yamaguchi and Fujita involves the transformation of ammonium acetate **74** into a tertiary amine by a trialkylation reaction [49]. For example, the

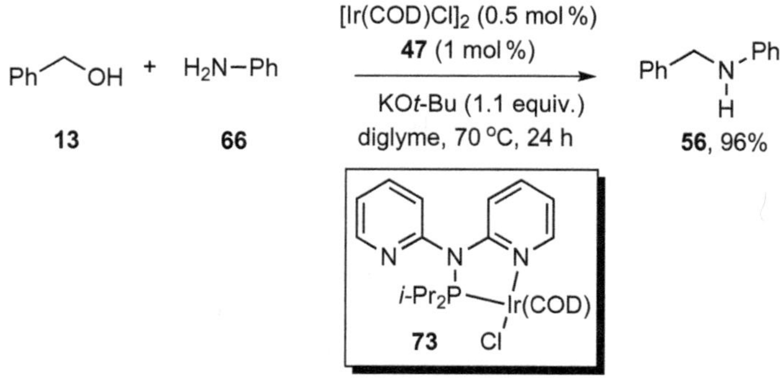

SCHEME 7-14. Lower temperature amine alkylation.

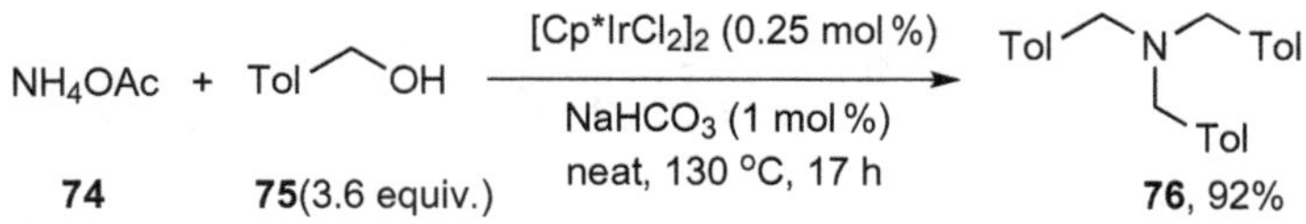

SCHEME 7-15. Conversion of an ammonium salt into a trisubstituted amine.

reaction with 3.6 equiv. of tolyl alcohol **75** under solvent-free conditions gives the tertiary amine **76** (Scheme 7-15). Interestingly, when ammonium tetrafluoroborate was used in place of ammonium acetate, the reaction was selective for the formation of secondary amines. Other ruthenium and iridium catalysts have also been reported for this reaction [50].

Catalytic amine alkylation reactions have been dominated by ruthenium and iridium catalysts, and research is active in the development of new catalysts for this transformation, including iridium pincer complexes and NHC complexes of both ruthenium and iridium [51–53]. Reports of the use of alternative metals, including manganese dioxide [54], palladium [55], and an iron/amino acid complex [56], are starting to appear, although there is no improvement in reactivity. In the case of the iron catalyst, the reaction is believed to proceed through an S_N2 mechanism rather than the borrowing hydrogen pathway.

Catalytic amine alkylation has been applied to the synthesis of hundreds of secondary and tertiary amines. In addition to this, research has been directed toward exploiting this methodology further, including applications in cyclization reactions, primary amine synthesis, and the use of other nitrogen-containing substrates. Cyclization reactions of diols with amines provide an interesting route for the formation of pyrrolidines, piperidines, and related *N*-heterocycles. $[Cp^*IrCl_2]_2$ has been used in a diastereoselective heterocyclization reaction of enantiomerically pure amine **77** with the racemic diol **78** (Scheme 7-16) [57]. The free piperidine could be formed by selective hydrogenolysis of the exocyclic benzyl group. This strategy was used by Trudell and coworkers in the synthesis of the natural product noranabasamine **80**, isolated from Colombian poison-dart frogs [58].

The direct conversion of alcohols into primary amines by condensation with ammonia is a significant synthetic challenge. The main problem is the need to prevent overalkylation, but significant progress has been made by the groups of Beller [59,60], Milstein [61], and Vogt [62]. Milstein's remarkable ruthenium complex **81** catalyzes the conversion of benzyl alcohol **13** into benzylamine **82** at low catalyst loading (Scheme 7-17) and was also successful with other alcohols, with benzylic alcohols giving the best yields. An alternative catalytic system **83** used by Beller was effective for the conversion of many primary and secondary alcohols into primary amines, including the conversion of 2-adamantanol **84**

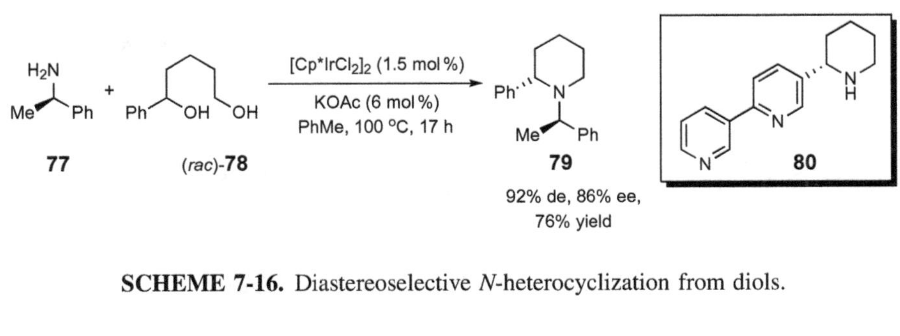

SCHEME 7-16. Diastereoselective *N*-heterocyclization from diols.

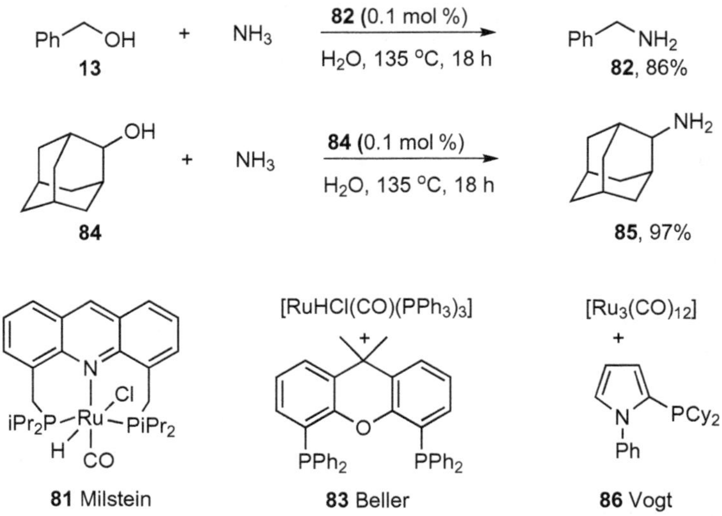

SCHEME 7-17. Catalysts for the conversion of alcohols into primary amines.

into the corresponding amine **85** (Scheme 7-17). Vogt's catalyst **86** (generated from a mixture of $[Ru_3(CO)_{12}]$ and ligand **61**), as well as a similar system reported by Beller, were also effective for primary amine formation, but selectivity and reactivity were somewhat lower.

The *N*-alkylation of sulfonamides with alcohols can be readily achieved using $[Cp^*IrCl_2]_2$ [63], and the reaction is effective even at low catalyst loadings. Benzyl alcohol **13** reacts with *p*-toluenesulfonamide **87** to give the alkylated product **88** even when only 0.05 mol% of catalyst is present (Scheme 7-18). The same reaction has also been reported to be catalyzed by 0.05 mol% of $Pd(OAc)_2$, although the reaction was run at 150 °C [64]. Copper(II) acetate has also been found to be effective for this reaction [65,66]. The reaction of sulfonamide **89** with benzyl alcohol **13** gives the alkylated product **90** in good yield when the reaction is run under air (Scheme 7-18). The conversion was reduced when the reaction was performed under argon. Sulfonamide alkylation has also been reported using 5 mol% iron(II) chloride as the catalyst [67].

Sulfonamides appear to be amongst the easiest substrates to alkylate, perhaps because they are less able to deactivate the catalyst by complexation. However, carboxamides and

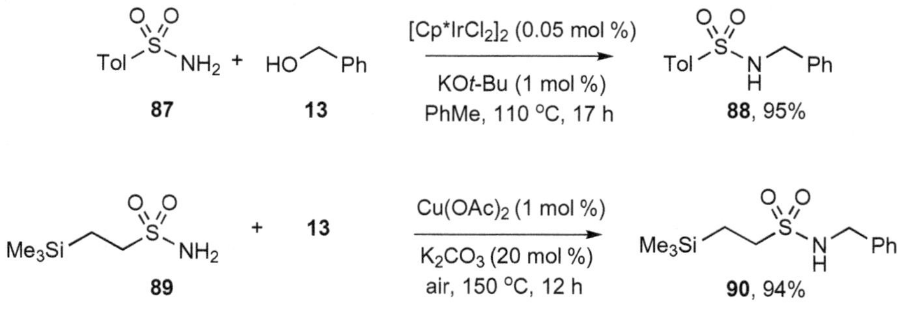

SCHEME 7-18. *N*-Alkylation of sulfonamides.

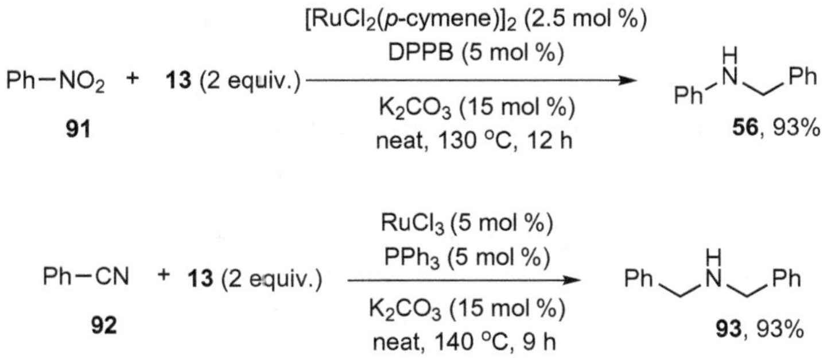

SCHEME 7-19. Reductive alkylation reactions.

carbamates usually require more forcing conditions, and these substrates are still rather challenging. For example, using $[Cp^*IrCl_2]_2$, temperatures of 130 °C are required for the alkylation of carboxamides and carbamates even in the absence of solvent [68]. The *N*-alkylation of indoles has been reported using a ruthenium catalyst [69].

Shi and coworkers have demonstrated that nitro and nitrile compounds can be reduced to amines, which then undergo alkylation, adding an interesting synthetic approach to this chemistry [70]. The reaction of nitrobenzene **91** with 2 equiv. of benzyl alcohol **13** in the presence of a ruthenium catalyst gives *N*-benzylaniline **56** in good isolated yield (Scheme 7-19). Similarly, benzonitrile **92** reacted with excess benzyl alcohol **13** to give dibenzylamine **93** using an alternative simple ruthenium catalyst. In both cases, an excess of alcohol is required to act as a reducing agent for the reaction.

In summary, the activation of alcohols by the temporary removal of hydrogen has attracted a lot of attention from many research groups. Although the reactions have shown considerable scope, substrates that act as good ligands for the catalyst tend to need higher catalyst loadings and temperatures in order to be successful. In general, relatively high temperatures and catalyst loading are required even for simple substrates. The identification of simple, cheap catalysts capable of achieving these reactions under milder conditions, ideally room temperature, remains a goal for using alcohols as alkylating agents.

REFERENCES

1. Mitsunobu O (1981). The use of diethyl azodicarboxylate and triphenylphosphine in synthesis and transformation of natural products. *Synthesis* 1–28.
2. Hamid MHSA, Slatford PA, Williams JMJ (2007). Borrowing hydrogen in the activation of alcohols. *Adv. Synth. Catal.* 349, 1555–1575.
3. Guillena G, Ramón DJ, Yus M (2007). Alcohols as electrophiles in C–C bond-forming reactions: the hydrogen autotransfer process. *Angew. Chem., Int. Ed.* 46, 2358–2364.
4. Nixon TD, Whittlesey MK, Williams JMJ (2009). Transition metal catalyzed reactions of alcohols using borrowing hydrogen methodology. *Dalton Trans.* 753–762.
5. Dobereiner GE, Crabtree RH (2010). Dehydrogenation as a substrate-activating strategy in homogeneous transition-metal catalysis. *Chem. Rev.* 110, 681–703.
6. Guillena G, Ramón DJ, Yus M (2010). Hydrogen autotransfer in the *N*-alkylation of amines and related compounds using alcohols and amines as electrophiles. *Chem. Rev.* 110, 1611–1641.

7. Watson AJA, Williams JMJ (2010). The give and take of alcohol activation. *Science* 329, 635–636.
8. Obora Y, Ishii Y (2011). Iridium-catalyzed reactions involving transfer hydrogenation, addition, *N*-heterocyclization and alkylation using alcohols and diols as key substrates. *Synlett* 30–51.
9. Cho CS, Kim BT, Kim HS, Kim TJ, Shim SC (2003). Ruthenium-catalyzed one-pot β-alkylation of secondary alcohols with primary alcohols. *Organometallics* 22, 3608–3610.
10. Gnanamgari D, Leung CH, Schley ND, Hilton ST, Crabtree RH (2008). Alcohol cross-coupling reactions catalyzed by Ru and Ir terpyridine complexes. *Org. Biol. Chem.* 6, 4442–4445.
11. Martínez R, Ramón DJ, Yus M (2006). $RuCl_2(dmso)_4$ catalyzes the β-alkylation of secondary alcohols with primary alcohols through a hydrogen autotransfer process. *Tetrahedron* 62, 8982–8987.
12. Fujita K, Asai C, Yamaguchi T, Hanasaka F, Yamaguchi R (2005). Direct β-alkylation of secondary alcohols with primary alcohols catalyzed by a Cp*Ir complex. *Org. Lett.* 7, 4017–4019.
13. Matsu-ura T, Sakaguchi S, Obora Y, Ishii Y (2006). Guerbet reaction of primary alcohols leading to β-alkylated dimer alcohols catalyzed by iridium complexes. *J. Org. Chem.* 71, 8306–8308.
14. Allen LJ, Crabtree RH (2010). Green alcohol couplings without transition metal catalysts: base-mediated β-alkylation of alcohols in aerobic conditions. *Green Chem.* 12, 1362–1364.
15. Cho CS, Kim BT, Kim TJ, Shim SC (2002). Ruthenium-catalyzed regioselective α-alkylation of ketones with primary alcohols. *Tetrahedron Lett.* 43, 7987–7989.
16. Cho CS, Kim BT, Kim TJ, Shim SC (2001). An unusual type of ruthenium-catalyzed transfer hydrogenation of ketones with alcohols accompanied by C–C coupling. *J. Org. Chem.* 66, 9020–9022.
17. Martínez R, Brand GJ, Ramón DJ, Yus M (2006). Easy α-alkylation of ketones with alcohols through a hydrogen autotransfer process catalyzed by $RuCl_2(dmso)_4$. *Tetrahedron* 62, 8988–9001.
18. Taguchi K, Nakagawa H, Hirabayashi T, Sakaguchi S, Ishii Y (2004). An efficient direct α-alkylation of ketones with primary alcohols catalyzed by $[Ir(COD)Cl]_2/PPh_3/KOH$ system without solvent. *J. Am. Chem. Soc.* 126, 72–73.
19. Grigg R, Mitchell TRB, Sutthivaiyakit S, Tongpenyai N (1981). Oxidation of alcohols by transition metal complexes part V. Selective catalytic monoalkylation of arylacetonitriles by alcohols. *Tetrahedron Lett.* 22, 4107–4110.
20. Löfberg C, Grigg R, Whittaker MA, Keep A, Derrick A (2006). Efficient solvent-free selective mono-alkylation of arylacetonitriles with mono-, bis-, and tris-primary alcohols catalyzed by a Cp*Ir complex. *J. Org. Chem.* 71, 8023–8027.
21. Grigg R, Löfberg C, Whitney S, Sridharan V, Keep A, Derrick A (2009). Iridium catalysed alkylation of *tert*-butyl cyanoacetate with alcohols under solvent free conditions. *Tetrahedron* 65, 849–854.
22. Iuchi Y, Hyotanishi M, Miller BE, Maeda K, Obora Y, Ishii Y (2010). Synthesis of ω-hydroxy carboxylic acids and α,ω-dimethyl ketones using α,ω-diols as alkylating agents. *J. Org. Chem.* 75, 1803–1806.
23. Slatford PA, Whittlesey MK, Williams JMJ (2006). C–C bond formation from alcohols using a xantphos ruthenium complex. *Tetrahedron Lett.* 47, 6787–6789.
24. Anxionnat B, Gomez-Pardo D, Ricci G, Cossy J. (2011). Monoalkylation of acetonitrile by primary alcohols catalyzed by iridium complexes. *Org. Lett.* 13, 4084–4087.
25. Sawaguchi T, Obora Y (2011). Iridium-catalyzed α-alkylation of acetonitrile with primary and secondary alcohols. *Chem. Lett.* 40, 1055–1057.
26. Burling S, Paine BM, Nama D, Brown VS, Mahon MF, Prior TJ, Pregosin PS, Whittlesey MK, Williams JMJ (2007). C–H activation reactions of ruthenium *N*-heterocyclic carbene

complexes: application in a catalytic tandem reaction involving C−C bond formation from alcohols. *J. Am. Chem. Soc.* 129, 1987–1995.

27. Pridmore SJ, Williams JMJ (2008). C−C bond formation from alcohols and malonate half esters using borrowing hydrogen methodology. *Tetrahedron Lett.* 49, 7413–7415.
28. Iuchi Y, Obora Y, Ishii Y (2010). Iridium-catalyzed α-alkylation of acetates with primary alcohols and diols. *J. Am. Chem. Soc.* 132, 2536–2537.
29. Whitney S, Grigg R, Derrick A, Keep A (2007). $[Cp^*IrCl_2]_2$-catalyzed indirect functionalization of alcohols: novel strategies for the synthesis of indoles. *Org. Lett.* 9, 3299–3302.
30. Grigg R, Whitney S, Sridharan V, Keep A, Derrick A (2009). Iridium catalyzed C-3 alkylation of oxindole with alcohols under solvent free thermal or microwave conditions. *Tetrahedron* 65, 4375–4383.
31. Jensen T, Madsen R (2009). Ruthenium-catalyzed alkylation of oxindole with alcohols. *J. Org. Chem.* 74, 3990–3992.
32. Blank B, Kempe R (2010). Catalytic alkylation of methyl-*N*-heteroaromatics with alcohols. *J. Am. Chem. Soc.* 132, 924–925.
33. Patman RL, Chaulagain MR, Williams VM, Krische MJ (2009). Direct vinylation of alcohols or aldehydes employing alkynes as vinyl donors: a ruthenium catalyzed C−C bond-forming transfer hydrogenation. *J. Am. Chem. Soc.* 131, 2066–2067.
34. Bower JF, Patman RL, Krische MJ (2008). Iridium-catalyzed C−C coupling via transfer hydrogenation: carbonyl addition from the alcohol or aldehyde oxidation level employing 1,3-cyclohexadiene. *Org. Lett.* 10, 1033–1035.
35. Ngai MY, Kong JR, Krische MJ (2007). Hydrogen-mediated C−C bond formation: a broad new concept in catalytic C−C coupling. *J. Org. Chem.* 72, 1063–1072.
36. Han SB, Kim IS, Han H, Krische MJ (2009). Enantioselective carbonyl reverse prenylation from the alcohol or aldehyde oxidation level employing 1,1-dimethylallene as the prenyl donor. *J. Am. Chem. Soc.* 131, 6916–6917.
37. Watanabe Y, Tsuji Y, Ohsugi Y (1981). The ruthenium catalyzed *N*-alkylation and *N*-heterocyclization of aniline using alcohols and aldehydes. *Tetrahedron Lett.* 22, 2667–2670.
38. Grigg R, Mitchell TRB, Sutthivaiyakit S, Tongpenyai N (1981). Transition metal-catalyzed *N*-alkylation of amines by alcohols. *J. Chem. Soc. Chem. Commun.* 611–612.
39. Murahashi SI, Kondo K, Hakata T (1982). Ruthenium catalyzed synthesis of secondary or tertiary amines from amines and alcohols. *Tetrahedron Lett.* 23, 229–232.
40. Hamid MHSA, Allen CL, Lamb GW, Maxwell AC, Maytum HC, Watson AJA, Williams JMJ (2009). Ruthenium-catalyzed *N*-alkylation of amines and sulfonamides using borrowing hydrogen methodology. *J. Am. Chem. Soc.* 131, 1766–1774.
41. Hollmann D, Tillack A, Michalik D, Jackstell R, Beller M (2007). An improved ruthenium catalyst for the environmentally benign amination of primary and secondary alcohols. *Chem. Asian J.* 2, 403–410.
42. Fujita K, Enoki Y, Yamaguchi R (2008). Cp^*Ir-catalyzed *N*-alkylation of amines with alcohols. A versatile and atom economical method for the synthesis of amines. *Tetrahedron* 64, 1943–1954.
43. Cumpstey I, Agrawal S, Borbas KE, Martín-Matute B (2011). Iridium-catalyzed condensation of alcohols and amines as a method for aminosugar synthesis. *Chem. Commun.* 47, 7827–7829.
44. Blank B, Michlik S, Kempe R (2009). Selective iridium-catalyzed alkylation of (hetero)aromatic amines and diamines with alcohols under mild reaction conditions. *Chem. Eur. J.* 15, 3790–3799.
45. Zhang W, Dong X, Zhao W (2011). Microwave-assisted solventless reaction of iridium-catalyzed alkylation of amines with alcohols in the absence of base. *Org. Lett.* 13, 5386–5389.

46. Watson AJA, Maxwell AC, Williams JMJ (2011). Borrowing hydrogen methodology for amine synthesis under solvent-free microwave conditions. *J. Org. Chem.* 76, 2328–2331.

47. Saidi O, Blacker AJ, Farah MM, Marsden SP, Williams JMJ (2010). Iridium-catalyzed amine alkylation with alcohols in water. *Chem. Commun.* 46, 1541–1543.

48. Kawahara R, Fujita K, Yamaguchi R (2011). *N*-Alkylation of amines with alcohols catalyzed by a water-soluble Cp*iridium complex: an efficient method for the synthesis of amines in aqueous media. *Adv. Synth. Catal.* 353, 1161–1168.

49. Yamaguchi R, Kawagoe S, Asai C, Fujita K (2008). Selective synthesis of secondary and tertiary amines by Cp*iridium-catalyzed multialkylation of ammonium salts with alcohols. *Org. Lett.* 10, 181–184.

50. Segarra C, Mas-Marzá E, Mata JA, Peris E (2011). Shvo's catalyst and [IrCp*Cl_2(amidine)] effectively catalyze the formation of tertiary amines from the reaction of primary amines and ammonium salts. *Adv. Synth. Catal.* 353, 2078–2084.

51. Andrushko N, Andrushko V, Roose P, Moonen K, Börner A (2010). Amination of aliphatic alcohols and diols with an iridium pincer catalyst. *ChemCatChem* 2, 640–643.

52. Gnanamgari D, Sauer ELO, Schley ND, Butler C, Incarvito CD, Crabtree RH (2009). Iridium and ruthenium complexes with chelating *N*-heterocyclic carbenes: efficient catalysts for transfer hydrogenation, β-alkylation of alcohols and *N*-alkylation of amines. *Organometallics* 28, 321–325.

53. Prades A, Corberán R, Poyatos M, Peris E (2008). [$IrCl_2$Cp*(NHC)] complexes as highly versatile efficient catalysts for the cross-coupling of alcohols and amines. *Chem. Eur. J.* 14, 11474–11479.

54. Yu X, Liu C, Jiang L, Xu Q (2011). Manganese dioxide-mediated *N*-alkylation of amides and amines with alcohols under air. *Org. Lett.* 13, 6184–6187.

55. Zhao Y, Foo SW, Saito S (2011). Iron/amino acid catalyzed direct *N*-alkylation of amines with alcohols. *Angew. Chem., Int. Ed.* 50, 3006–3009.

56. Zhang Y, Qi X, Cui X, Shi F, Deng Y (2011). Palladium catalyzed *N*-alkylation of amines with alcohols. *Tetrahedron Lett.* 52, 1334–1338.

57. Fujita K, Fujii T, Yamaguchi R (2004). Cp*Ir complex-catalyzed *N*-heterocyclization of primary amines with diols: a new catalytic system for environmentally benign synthesis of cyclic amines. *Org. Lett.* 6, 3525–3528.

58. Miao L, DiMaggio SC, Shu H, Trudell ML (2009). Enantioselective syntheses of both enantiomers of noranabasamine. *Org. Lett.* 11, 1579–1582.

59. Imm S, Bähn S, Zhang M, Neubert L, Neumann H, Klasovsky F, Pfeffer J, Haas T, Beller M (2010). Improved ruthenium-catalyzed amination of alcohols with ammonia: synthesis of diamines and amino esters. *Angew. Chem., Int. Ed.* 50, 7599–7603.

60. Imm S, Bähn S, Neubert L, Neumann H, Beller M (2010). An efficient and general synthesis of primary amines by ruthenium-catalyzed amination of secondary alcohols with ammonia. *Angew. Chem., Int. Ed.* 49, 8126–8129.

61. Gunanathan C, Milstein D (2008). Selective synthesis of primary amines directly from alcohols and ammonia. *Angew. Chem., Int. Ed.* 47, 8661–8664.

62. Pingen D, Müller C, Vogt D (2010). Direct amination of secondary alcohols using ammonia. *Angew. Chem., Int. Ed.* 49, 8130–8133.

63. Zhu M, Fujita K, Yamaguchi R (2010). Simple and versatile catalytic system for *N*-alkylation of sulfonamides with various alcohols. *Org. Lett.* 12, 1336–1339.

64. Martínez-Asencio A, Yus M, Ramón DJ (2011). Palladium(II) acetate as a catalyst for the *N*-alkylation of aromatic amines, sulfonamides and related nitrogenated compounds with alcohols by a hydrogen autotransfer process. *Synthesis* 3730–3740.

65. Shi F, Tse MK, Cui X, Gördes D, Michalik D, Thurow K, Deng Y, Beller M (2009). Copper-catalyzed alkylation of sulfonamides with alcohols. *Angew. Chem., Int. Ed.* 48, 5912–5915.

66. Martinez-Asenci A, Ramón DJ, Yus M (2011). *N*-Alkylation of poor nucleophilic amines and derivatives with alcohols by a hydrogen autotransfer process catalyzed by copper(II) acetate: scope and mechanistic considerations. *Tetrahedron* 67, 3140–3149.

67. Cui X, Shi F, Zhang Y, Deng Y (2010). Fe(II)-catalyzed *N*-alkylation of sulfonamides with benzylic alcohols. *Tetrahedron Lett.* 51, 2048–2051.

68. Fujita K, Komatsubara A, Yamaguchi R (2009). *N*-Alkylation of carbamates and amides with alcohols catalyzed by a Cp*Ir complex. *Tetrahedron* 65, 3624–3628.

69. Bähn S, Imm S, Mevius K, Neubert L, Tillack A, Williams JMJ, Beller M (2010). Selective ruthenium-catalyzed *N*-alkylation of indoles by using alcohols. *Chem. Eur. J.* 16, 3590–3593.

70. Cui X, Zhang Y, Shi F, Deng Y (2011). Ruthenium-catalyzed nitro and nitrile compounds coupling with alcohols: alternative route for *N*-substituted amine synthesis. *Chem. Eur. J.* 17, 2587–2591.

8

APPLICATION OF A REDOX-NEUTRAL ALCOHOL AMINATION IN THE KILOGRAM-SCALE SYNTHESIS OF A GLYT1 INHIBITOR

MARTIN A. BERLINER

Chemical Research and Development, Pfizer Inc., Groton, CT, USA

8.1 INTRODUCTION

In a recent review, Roughley and Jordan [1] analyzed 139 publications from selected pharmaceutical companies across several medicinal chemistry journals for the distribution of reactions used in the synthesis of drug targets. While most chemists casually acknowledge that C–N bond formation reactions are important, this survey indicated that almost 50% of all transformations in this selection involved the synthesis of a carbon–nitrogen bond. Within this grouping, more than one-quarter of all reactions were alkyl aminations, comprising both direct alkylations of amines and reductive aminations. These two authors commented that "Alkylation of nitrogen with alkyl halides, despite the contingent issues with overalkylation and side reactions, accouns for almost a quarter of transformations in this class. Reductive aminations, offering similar products with a greater degree of control over reactivity and the product thus obtained, are surprisingly only equally popular. These two methods together, however, account for half of all heteroatom derivatizations."

As compounds progress in development, one would expect that these proportions would change as the manufacturing synthesis of target molecules is refined to meet operational and costing objectives. A 2006 survey of 128 separate process syntheses of active pharmaceutical ingredients (APIs) [2] revealed that carbon–nitrogen bond formations decreased to approximately 23% of the total number of transformations. It is surprising that

Sustainable Catalysis: Challenges and Practices for the Pharmaceutical and Fine Chemical Industries, First Edition. Edited by Peter J. Dunn, K. K. (Mimi) Hii, Michael J. Krische, and Michael T. Williams.

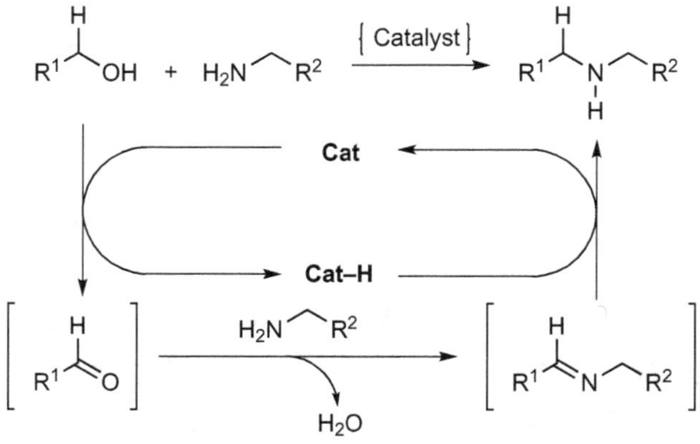

SCHEME 8-1. A prototypical redox–neutral alcohol amination.

in this class the relative proportion of alkyl aminations did not change significantly, with these reactions representing almost 40% of the total.

Thus, there is a significant opportunity across the synthetic chemistry community for the development of new methods for alkyl amination that do not carry the same liabilities. While still in its infancy, the redox–neutral amination of alcohols, which is covered in Chapter 7 and a recent review [3], has the potential to fulfill this role (Scheme 8-1). This type of process offers several practical advantages over the current best practices. Unlike the discrete alkylation or reductive amination reactions, the redox–neutral chemistry is comparatively atom efficient (water is the only by-product) and simple to run. It also combines the activation of an alcohol and alkylation in a single pot, which eliminates the need to separately prepare and handle potentially unstable and/or toxic electrophiles (alkyl halides/sulfonates and carbonyl compounds).

This chapter describes the application of redox–neutral alcohol amination chemistry to the synthesis of PF-03463275 (**1**), a glycine transporter type 1 (GlyT1) inhibitor. To the best of our knowledge, this work represents the first reported kilogram-scale use of this technology for an API [4].

8.2 BACKGROUND AND INITIAL SYNTHETIC WORK

PF-03463275 (**1**) is a GlyT1 inhibitor that was nominated as a potential therapy for schizophrenia and related illnesses [5]. The medicinal chemistry synthesis is shown in Scheme 8-2 and starts from the known *exo-N*-benzyl-3-azabicyclo[3.1.0]hexane-6-methanol **2** [6]. This molecule has been previously utilized by Pfizer in the first synthesis of the antibiotic trovafloxacin [7], and thus a significant amount of institutional knowledge had accumulated from those efforts. In particular, the chemists who worked on **2** and similar derivatives noted that *O*-derivatization chemistries were far more successful when the protecting group on the nitrogen was *not* an alkyl group[1]. The medicinal chemistry team hewed to this approach, and thus the *N*-benzylamine **2** was converted to its *N*-Boc derivative **4** using standard transformations in high overall yield via the water-soluble

[1] Private communication to the author from Katherine Brighty.

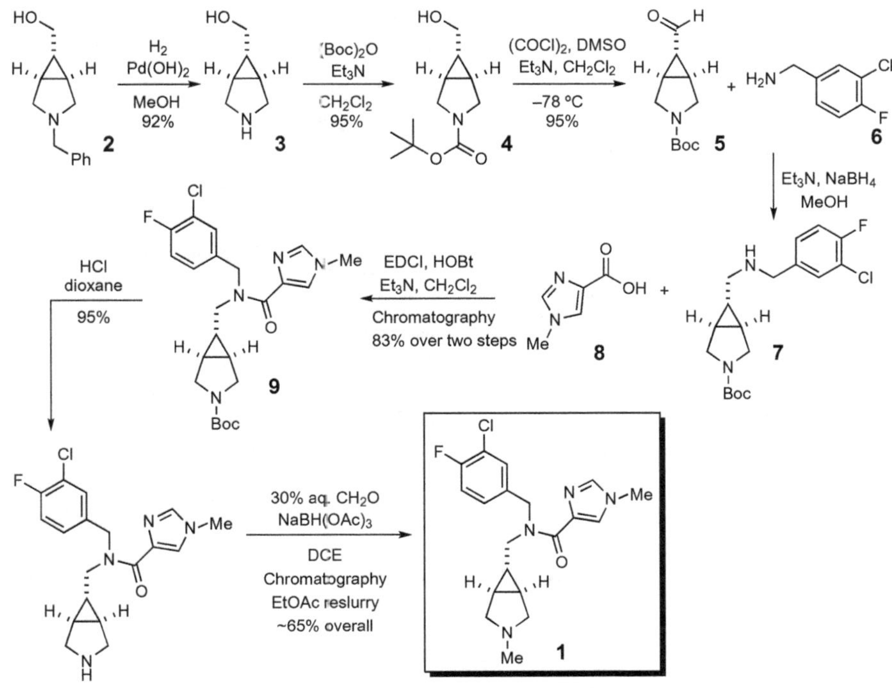

SCHEME 8-2. Medicinal chemistry synthesis of **1**.

amino alcohol **3**. After oxidation of the alcohol to aldehyde **5** using Swern conditions, a reductive amination with 3-chloro-4-fluorobenzylamine **6** was employed to generate the secondary amine **7**. Amidation between **7** and *N*-methyl-4-imidazolecarboxylic acid **8** using standard conditions gave **9**. Deprotection of the Boc group, followed by *N*-methylation of the pyrrolidine nitrogen with formaldehyde and sodium triacetoxyborohydride (STAB) in dichloroethane generated **1**, which was purified by chromatography and rendered crystalline by a reslurry in ethyl acetate. The overall efficiency of this sequence is good (43% over seven steps), with only two chromatographic purifications. Notably, all bond-formation reactions in this sequence involve the synthesis of carbon–nitrogen bonds (two reductive aminations, two acylations).

While sufficient to prepare small quantities of materials for preclinical toxicology, several areas were identified that required improvement or modification as the program progressed into regulatory studies. An initial focus was to move formaldehyde use earlier in the synthesis (or remove it altogether) to eliminate the need to develop methods for formaldehyde quantitation in the API. Other concerns included the use of two toxic solvents in the final two steps (dioxane, a potential carcinogen, and dichloroethane, which is an ICH Class 1 solvent), and certain operational procedures, including chromatography and the low-temperature Swern oxidation, which tend to be more difficult to execute on scale. Finally, eliminating the protecting group chemistry was a desirable but not essential improvement as it contributes two steps to the process.

From these criteria, an overall synthetic strategy was developed that targeted a final-step acylation between a new secondary amine intermediate **10** and carboxylic acid **8** (Scheme 8-3). A retrosynthesis of **10** identified two preferred strategies, the first

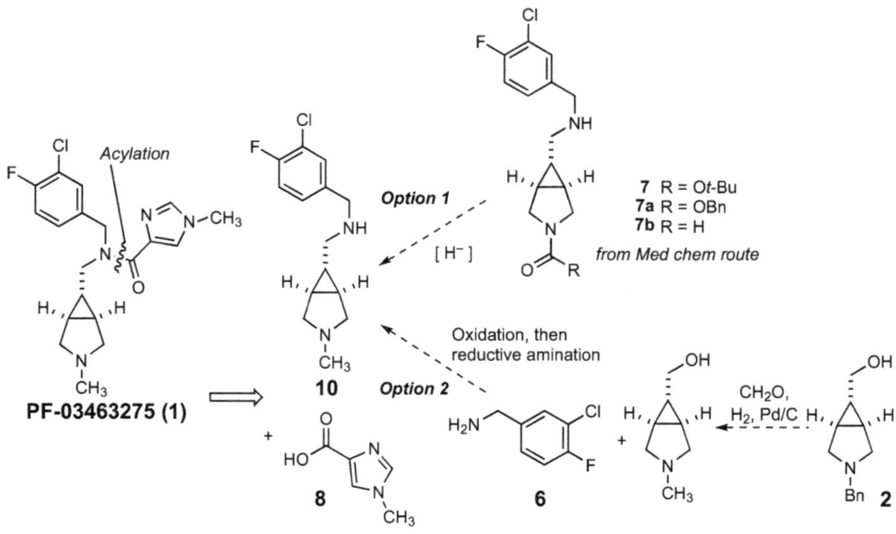

SCHEME 8-3. Process retrosynthesis.

involving reduction of a carbamate precursor such as **7** (option 1) and the second by a new route in which the pyrrolidine is *N*-methylated earlier in the synthesis and then coupled with amine **6** (option 2). Project timelines often demand that teams commit to a raw materials strategy before a route is fully established. Thus, to ensure the maximum flexibility in route design, the *N*-benzyl-azabicyclo[3.1.0]hexanol **2** was chosen as a key raw material, in addition to amine **6** and carboxylic acid **8**.

In order to gain experience with the chemistry, option 1 was investigated first. This approach had the additional benefit of paralleling the medicinal chemistry route for the first several steps. In this sequence, the debenzylation reaction to generate **3** proceeded nicely without modification (Scheme 8-4), and protection of 3 proved straightforward and the corresponding *N*-Boc **4**, *N*-CBz **4a** and *N*-formyl **4b** derivatives were prepared using typical conditions. Oxidation under Swern conditions was also uneventful, although work was discontinued on the water-soluble formamide **5b**, due to difficulties in its recovery from an aqueous workup. Without purification, the crude aldehydes, which were colorless oils, were converted to secondary amines **7** and **7a** by reductive amination with **6**. This was a good reaction, completed in minutes via the stable imine intermediates **11** and **11a**. The reduction of the carbamate functionality in amines **7** and **7a** proved a more serious obstacle. Aryl fluorides have a propensity to undergo significant dehalogenation (> 1 mol%) using aluminum-based reductants such as lithium aluminum hydride or Vitride[2], and these mixtures of fluoro- and desfluoro products tend to be difficult or impossible to separate. Borane has been shown to provide the desired chemoselectivity in a reduction of an amide, but in the case of carbamates **7** and **7a**, the use of borane did not result in any reaction, and

[2] Private communication to the author from Robert Singer. Given that the API has an aryl chloride we were correctly reluctant to utilize aluminum reagents for this step, as we later demonstrated that the *des*-chloro derivative of **10** and **1** could not be adequately purged at any step in the synthesis.

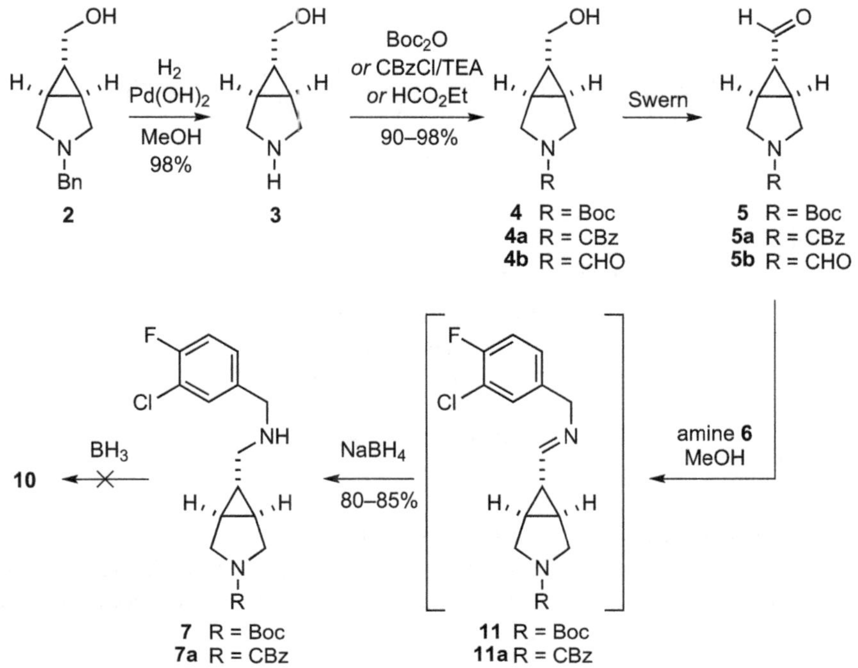

SCHEME 8-4. Initial attempts to synthesize **10**.

without spending any additional time optimizing the process, this particular synthetic approach was abandoned.

The approach described in option 2 proved more successful. Amino alcohol **3** participates in a facile reductive amination with aqueous formaldehyde in methanol in the presence of Pd/C under a hydrogen atmosphere to provide the water-soluble *exo-N*-methylazabicyclo[3.1.0]-6-hexanol **12** in excellent yield as a thick colorless oil after workup (Scheme 8-5). An attempt to develop conditions to telescope the debenzylation reaction and the reductive amination was unsuccessful, as formaldehyde suppresses the debenzylation in this particular substrate, even with very high loadings of catalyst. However, by delaying the charge of the aqueous formaldehyde solution until after the debenzylation was complete, it was possible to telescope both steps, and only a slight excess (1.1 equiv.) of formaldehyde was required for the reaction to proceed to completion. Here, the workup of the reaction proved critical. Since the product is water soluble, the extra formaldehyde cannot be extracted, and since it polymerizes in nonaqueous media, it cannot be removed during concentration of the filtrate of **12** under reduced pressure. Thus, removal of formaldehyde must occur during the hydrogenation, not afterward. The easiest solution, by reducing the formaldehyde charge to 1 equiv., led to incomplete conversion of **3**, which created another problem (purification of **12** from **3**) that was even more difficult to solve. In the end, a third hydrogenation stage was implemented. After the formation of **12** was complete, approximately 0.2 equiv. of an aq. dimethylamine solution was charged to the reactor. Hydrogenation was continued for another 4–8 h until hydrogen uptake had ceased. Analysis of the reaction mixture indicated no remaining formaldehyde, and the volatile by-products (dimethylamine and trimethylamine) were both easily removed during

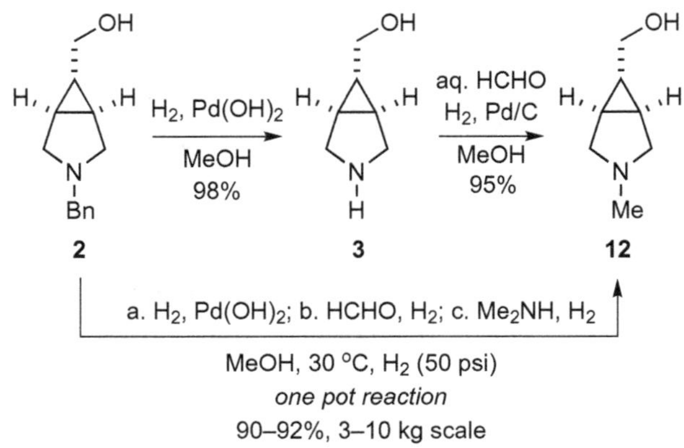

SCHEME 8-5. Synthesis of **12**.

the workup, which involved catalyst filtration and displacement into toluene to remove MeOH and water. As compared with other alternatives, including amine resin treatment, this proved inexpensive, efficient, and very easy to implement on both laboratory and plant scales.

8.3 FIRST-GENERATION SYNTHESIS OF 10

With **12** in hand, the next requirement was the development of a scalable synthesis of **10**. This molecule has three defining attributes that strongly influence its properties. First, it is very water soluble, and therefore the operational aspects of reaction workup must be adjusted to prevent loss through aqueous side streams. Second, the *N*-methylpyrrolidine functionality in **12** confounds workups for reactions that use exogenous bases as reagents (i.e., many oxidations) since typical methods for removing amine bases, such as extractions with acids, will also extract **12** and products generated from **12**. Finally, to make things worse, many of these molecules are oils, which preclude any direct crystallization methods to isolate intermediates.

The types of chemistry considered for preparation of **10** from alcohol **12** and amine **6** are shown in Scheme 8-6. All involve some type of activation of **12** to generate an electrophilic carbon, which is then utilized in an amination reaction[3]. Activation of the alcohol in **12** as a tosylate to provide **13** looked attractive on paper, but its stability was poor and this approach was abandoned. Somewhat greater success was achieved with oxidation of **12**, initially employing TPAP as catalyst with NMO as stoichiometric oxidant since no aqueous workup is required [8]. Isolation of **14** was problematic due to the challenge of removing Ru residues from the very polar product. Aldehyde **14** proved to be configurationally unstable and slowly isomerized to give a mixture of *exo* and *endo* aldehydes, but it was sufficiently stable to use immediately in the next transformation. From these materials, the first samples of diamine **10** were prepared.

[3] This choice should not be surprising given the statistics that are discussed in Section 8.1. Most organic chemists think about alkylation or reductive amination approaches when considering how to prepare alkyl-nitrogen bonds and try these methods first before moving on to more esoteric methods.

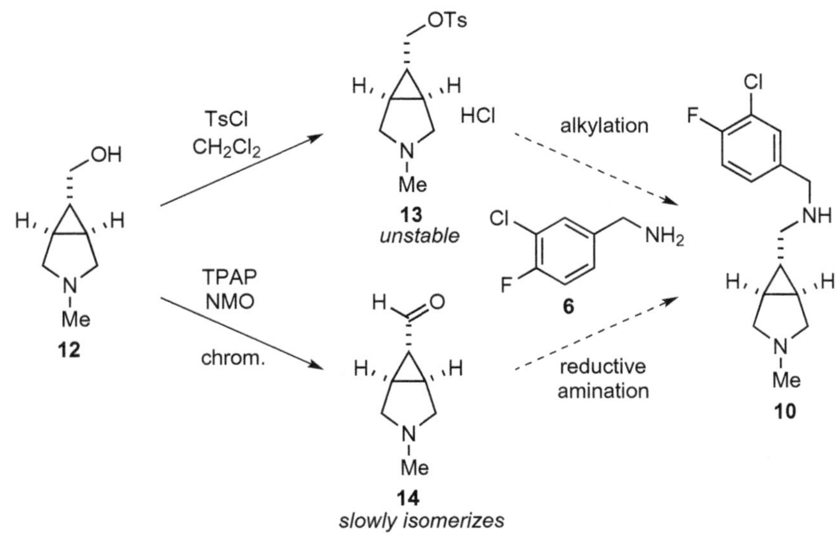

SCHEME 8-6. Routes and initial planned synthesis of **10**.

Encouraged by this result, additional screening identified activated dimethyl sulfoxide (DMSO) oxidations [9] as the best-performing methods. On a small scale, the three methods with the best overall profile proved to be the Corey–Kim [10], Swern [11], and the Parikh–Doering [12] reactions. All proceeded cleanly to completion without aldehyde isomerization. The water solubility of **14** precluded its isolation from the reaction matrix and attempts to form the bisulfite adduct of the aldehyde (which is frequently crystalline) failed. Therefore, once the oxidation was complete, the reaction mixtures were treated with benzyl amine **6** to form imine **15**. Reaction progress was more difficult to gauge with HPLC since this particular imine, unlike **11**, rapidly degrades in the presence of water. GCMS, FTIR, and especially NMR proved suitable for reaction monitoring, and NMR analysis of reaction aliquots was used almost exclusively for monitoring this transformation, since **12** and **14** are not UV-active. The solution of imine **15** was then diluted with methanol and a reductant (initially STAB, later $NaBH_4$) was added. This resulted in efficient formation of the desired secondary amine **10**, nicely demonstrated by *in situ* FTIR monitoring (Figure 8-1).

Pleasingly, amine **10** can be readily extracted from a basic aqueous reaction mixture with ethyl acetate and could be crystallized from 2-propanol as a nonhygroscopic bis (hydrochloride) salt. This greatly simplified product isolation provided a necessary cleanup opportunity before synthesizing the API, as a number of other compounds are formed during the oxidation–reductive amination process (Figure 8-2). Tertiary amine **16** results from reductive amination between **10** and **14** and was typically the major impurity (and was always accompanied by unconsumed primary amine **6**). To a varying extent, a pair of *N*-methylated impurities (**17** and **18**) was also formed. These arise from the known propensity of Me_2S-derived active oxidants to degrade via a Pummerer process [13], and their origin was established by running oxidation reactions using d_6-DMSO, which generated d_3-*N*-methylated isomers of **17** and **18**. While the proportions of **6** and **16** were fairly constant (10–20 mol%) for all three types of oxidations, the amounts of **17** and **18** formed in this reaction were condition and scale dependent. On a larger scale, the Corey–Kim oxidation

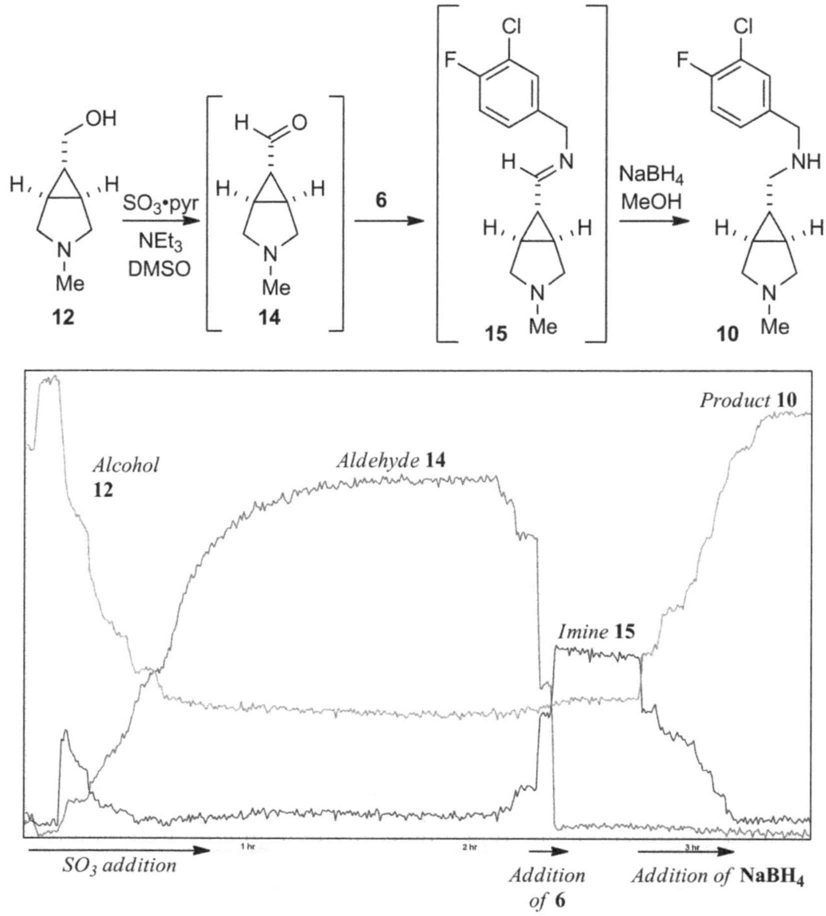

FIGURE 8-1. FTIR of the telescoped conversion of **12** to **10**.

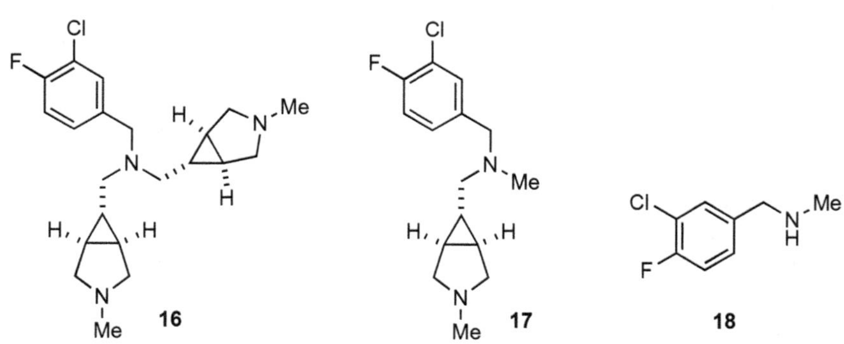

FIGURE 8-2. Impurities generated during the synthesis of **10**.

resulted in the formation of **17** as the major product as two equivalents of reagents (DMS and NCS) were required to achieve full conversion. The Swern reaction was significantly cleaner and had the highest yields (~50%) over two steps, but the use of cryogenic conditions precluded our ability to run this chemistry on scale. The Parikh–Doering conditions resulted in slightly lower overall yields (35–40%) but were the cleanest and could be run in standard reactors, so this process was selected for scale-up.

In moving this chemistry from the laboratory to the plant, only minor adjustments to the procedure were required. Notably, multiple charges of granular sodium borohydride were utilized to minimize rapid hydrogen evolution and to facilitate complete reduction of the imine. The workup, on the other hand, was extensively modified. On a small scale, it was possible to dilute the reaction mixture with ethyl acetate and conduct multiple water washes to remove the DMSO and ammonium salts. After concentration of the reaction mixture on the rotary evaporator, the crude product was coevaporated from 2-propanol to remove any residual pyridine and triethylamine and then acidified with commercial 5N HCl in 2-propanol to crystallize **10**·2HCl (Scheme 8-7). Attempts to isolate the freebase directly were complicated by the formation of stable diastereomeric boron complexes of **10** that made analysis more challenging.

When running on scale, addition of approximately 1.5 equiv. of aq. 2N NaOH served to neutralize all ammonium salts, break any boron-amine complexes, and precipitate borate salts from the reaction mixture. Triethylamine and pyridine were then distilled out at ambient pressure as their water azeotropes, leaving a solution of **10** and solid borate and sulfate salts in DMSO in the reactor. After cooling, the solids were removed by filtration and the liquid phase extracted once with ethyl acetate, which efficiently partitioned the desired product into the organic phase. The solvent was exchanged into 2-propanol via a constant-volume azeotropic displacement and the product crystallized as before with 5N HCl in 2-propanol. This crystallization proved very efficient at purging unreacted amino alcohol **12** and impurities **16**, **17**, and **18**. However, an unexpectedly large amount (10 mol%) of benzyl amine HCl salt (**6**·2HCl) was present, indicating that the purge factor was far less than originally expected. Overall, from 500 g of **12**, a total of 496 g of **10**·2HCl was isolated, representing a 37% combined yield for the two steps. While sufficient for the first clinical campaign, a new process was required for any future scale-up efforts. Surprisingly, the lead for this development came from even smaller scale work in the Isotope Synthesis Group.

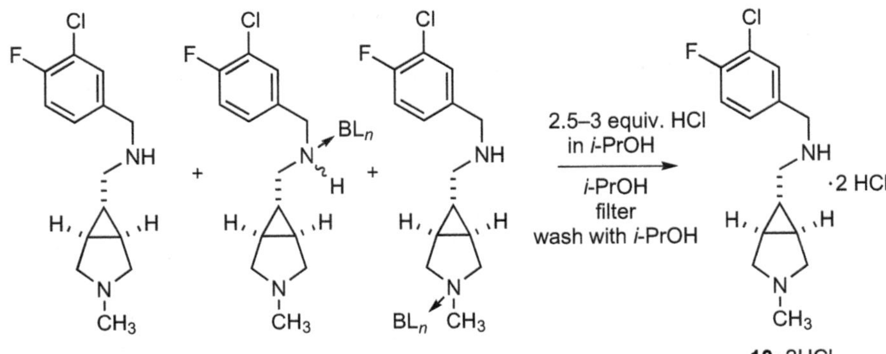

10 **10-Borate** Complexes **10**·2HCl

SCHEME 8-7. Crystallization of **10**·2HCl salt.

8.4 FIRST APPLICATION OF IR CHEMISTRY AND INITIAL PROCESS DEVELOPMENT EFFORTS

The oxidation/reductive amination process was rapidly deemed unsuitable for small-scale work during efforts to prepare an M + 5 isomer of **1** for drug safety studies. From **12** and benzyl amine **19** (the M + 4-labeled isomer of **6**, prepared as shown in Scheme 8-8), only milligram quantities of **20** (the desired M + 5 isomer of **10**) were obtained by using the first-generation process. The main problem was not so much a lack of reaction robustness (which there was) as a lack of good analytical methods to follow the reaction. More specifically, the imine was successfully formed but was not fully consumed when $NaBD_4$ was added, such that **12** and **19** were regenerated on reaction quench and workup.

At this juncture, it was recognized that a recently published method for the synthesis of amines from Fujita and Yamaguchi—an iridium-catalyzed redox-neutral coupling between alcohols and amines—might provide an ideal solution for this synthetic problem [14]. In the event, a mixture of **12** and **19** was treated with MeOD and D_2O to deuterate all exchangeable protons followed by approximately 5 mol% $[Cp^*IrCl_2]_2$ and 10 mol% K_2CO_3 in d_8-toluene at reflux overnight. After a simple workup, the desired M + 5 product was formed in 60–65% overall isolated yield (Scheme 8-9). The overall improvement was pleasing, and the formation of an M + 5 product was completely unexpected. It took many years and some amount of additional experimentation to fully appreciate the significance of this result and what it implied for the mechanism of the reaction, which will be discussed later (Section 8.5).

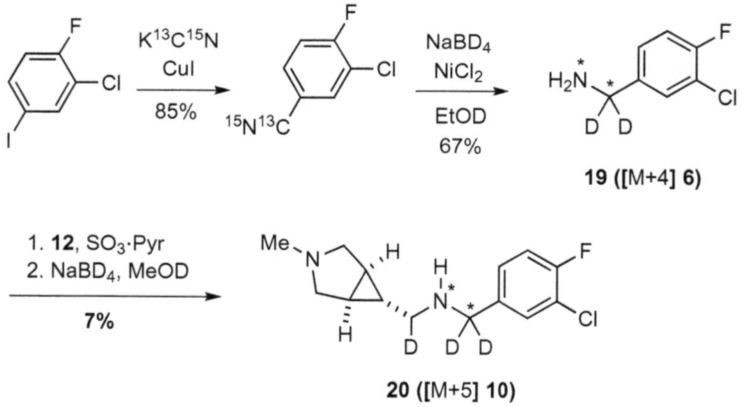

SCHEME 8-8. Unsuccessful synthesis of labeled compound **10**.

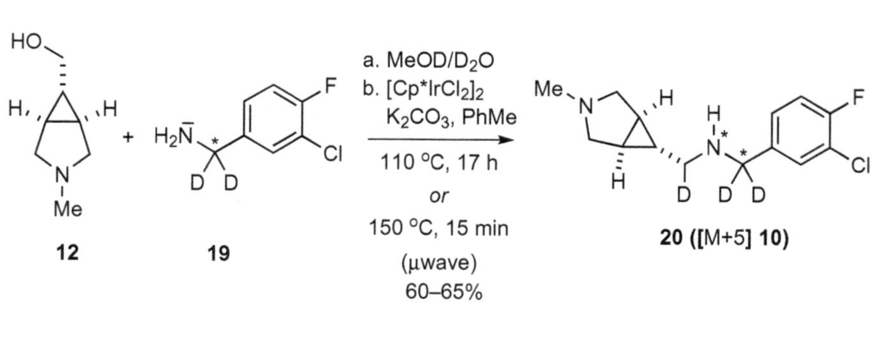

SCHEME 8-9. First Ir-catalyzed synthesis of **20**.

TABLE 8-1. Initial Screen of the Ir-Catalyzed Coupling Reaction

Run	Ir Catalyst (mol%)	H_2O (% wt Toluene)	Product Yield (%)
1	4	1.25	49
2	3	2.5	20
3	2	3.75	10
4	2	1.25	12
5	4	3.75	41
6	3	2.5	26

Filled with heady enthusiasm at this result, a quick optimization of the Ir-catalyzed coupling reaction was undertaken in an attempt to identify conditions that could be utilized for the next campaign. Screening this reaction against Ru and Ir catalysts revealed that only the $[Cp^*IrCl_2]_2$-catalyzed reactions proceeded to completion. With this catalyst, the effects of catalyst loading, base identity, and initial water content on reaction performance were assessed. Unfortunately, the reaction did not proceed to any reasonable extent with any less than 4 mol% Ir (2 mol% catalyst), and surprisingly, a small amount of water needed to be added to prevent catalyst deactivation (Table 8-1). The identity of added base did not seem to be important, although no reactions were attempted in the absence of base. The initial optimized conditions used 4.5 mol% Ir (2.25 mol% catalyst) and 5 mol% K_2CO_3 in 5 l kg^{-1} toluene with 3 vol% added water. Benzyl amine **6** was deliberately undercharged (0.9 equiv.) since the first campaign had demonstrated that crystallization could not adequately purge large amounts of **6**. In gram-scale experiments in the laboratory, this process typically ran to completion at 95–100 °C in 5 h or less, providing a mixture of desired product **10**, starting amine **6** and tertiary amine **16** as the only significant by-product. From this reaction, a 60–70% isolated yield of pure **10**·2HCl was obtained after workup and crystallization.

A new workup process was investigated prior to scale-up, focusing specifically on removal of as much iridium as possible. This was a critical issue because the catalyst remained soluble in the organic phase during the workup, and thus a dark red crude product solution containing >40,000 ppm Ir was taken into the crystallization. Unfortunately no improvements in iridium removal were realized using extractive methods or commercial scavenging reagents. This subject is discussed at greater length later (Section 8.6).

Upon scale-up of this reaction to 1.5 kg in the Kilo Lab, the reaction stalled at 65% conversion and could not be rescued by the addition of extra catalyst. Proceeding with workup and crystallization, a 2:1 ratio of **10**·2HCl and **6**·HCl was isolated with an adjusted yield of **10** of 55% over the process and high (2000 pppm) iridium. The failure of the reaction to proceed to completion was quickly ascribed to poor experimental design that did not adequately model the reactor, a 50 l Büchi glass-on-steel vessel with glass overheads and a descending condenser (see Figure 8-3). This is a lesson that every process chemist learns once—that reactions in round-bottom flasks with ascending condensers do not always behave the same as fixed reactors with descending condensers. In order to prevent vapor loss from the reactor, an inverse U-shaped vapor trap is installed between the condenser and the reactor. Typically used during a reflux operation, distillate is collected behind the vapor trap until it overflows and returns to the reactor. In this process, the amount of water added was small (<100 ml) and was efficiently distilled into the trap over the first hour of the reaction as a toluene/water azeotrope. The unintentional dehydration of the reaction (the vapor trap functioned as a Dean–Stark trap) led to the reaction stalling.

A second campaign afforded an opportunity to address these issues. An attempt to simulate the reactor configuration by using an addition funnel as an "in-line" Dean–Stark

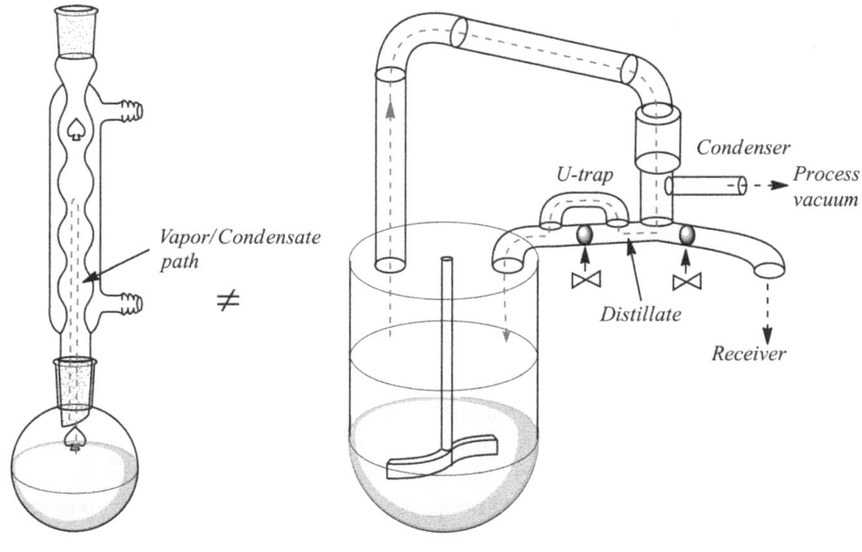

FIGURE 8-3. Reactor configuration during reflux operations.

trap was successful and revealed that semicontinuous return of water to the reactor prevented the reaction from stalling. More effort around understanding the crystallization was required to address efficiency and cake homogeneity issues encountered in the first campaign. The temperature of the crystallization had a greater influence on filtration efficiency than other factors. Adding the solution of freebase to a hot solution of HCl in 2-propanol led to a lesser propensity for the filter cake to crack, which was ascribed to a difference in final particle size. A larger wash volume was implemented since it resulted in a reduction of Ir in the final product by up to a factor of 2. On scale, the second campaign proceeded to 85% overall conversion (measured as consumption of **6**) before the reaction stalled, a better result than before but still not ideal. The improved crystallization and filtration conditions were highly successful, however. From 1.5 kg of **12**, a total of 2.45 kg of **10**·2HCl was isolated as a fine white solid in >98% purity with <1% residual **6**·HCl and only 360 ppm Ir. This represents a 125-fold reduction in Ir during the crystallization and a 57% overall yield. A comparison of the first two scale-up campaigns is presented in Table 8-2.

8.5 PROCESS OPTIMIZATION OF THE AMINATION REACTION

The need for another API manufacturing campaign led to a reinvestigation of this reaction. A review of new literature accounts on redox-neutral coupling reactions quickly identified a review article by Fujita and Yamaguchi in 2008 [15] as particularly interesting. Compared with their 2003 publication, this work had full experimental descriptions for different Cp*Ir(III)-catalyzed reactions and certain details stood out. Several examples were reported to proceed to completion with only 1 mol% of Ir rather than the 5 mol% utilized in the earlier work. Notably, reactions were run in sealed vials at 110 °C; that is, above the boiling point of the toluene/water azeotrope. Some reactions were also run at high concentrations (1 ml g^{-1} of toluene was typical). However, no mention in the text of the paper focused on experimental procedure, indicating that the authors had not investigated, or found the need to report on, the experimental aspects of this chemistry.

TABLE 8-2. First Ir-Catalyzed Scale-Up Runs

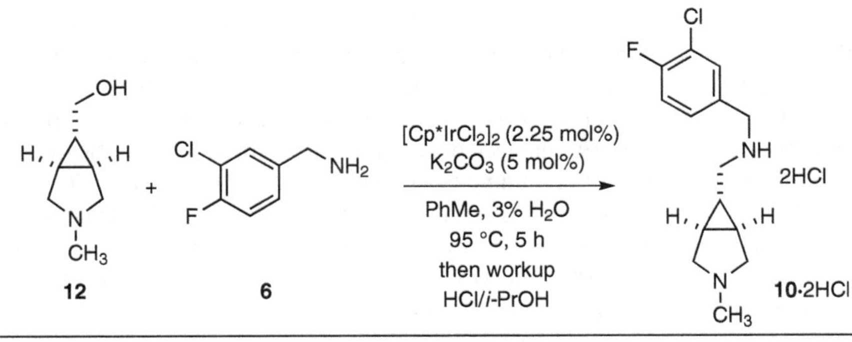

Run	**12** (kg)	Conversion[a] (%)	Purity[b] (%)	Yield[c] (%)	Ir (ppm)
1	1.5	60	66	55	2,000
2	1.5	85	>98	57	360

[a]Measured as a ratio of **10** to unreacted **6**.
[b]Calculated based on HPLC area%, UV detection.
[c]Isolated **10**·2HCl.

8.5.1 Reliability Optimization

Initial work focused on fixing the reliability of the transformation using the previously optimized conditions (4.5 mol% Ir, 5% K_2CO_3, 0.9 equiv. **6**, 3 vol% water, 5 volumes toluene, 95 °C, 5 h). Superb and reliable reaction performance (> 40:1 ratio of **10**:**6** in the crude reaction mixture) was obtained on gram-scale when reactions were conducted in screw-cap GL18 test tubes in a Mettler–Toledo EasyMax instrument (Table 8-3, entry 1). Scaling up to

TABLE 8-3. Ir-Catalyzed Reaction Performance in Sealed Vessels

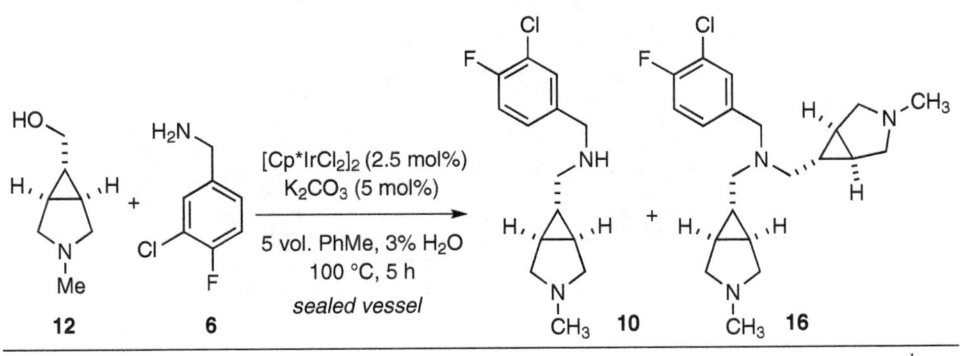

Entry	**12** (g)	Reactor Type	**6** (equiv.)	Ratio 10:16:6[a]	Yield[b] (%)
1	1	GL18	0.9	89:9:2	65
2	10	100 ml	0.9	60:0:40	N/A
3	10	Parr	0.9	82:18:0	60
4	10	Parr	1	88:7:5	74
5	1400	8 l Parr (2 runs)	1	88:7:5	78

[a]Reaction monitoring was via ^{1}H NMR analysis of samples; ratios are reported as molar ratios.
[b]Yield is of pure isolated **12**·2HCl salt after workup.

10 g in an EasyMax 100 mL reactor at 95 °C (N_2 atmosphere, with a standard condenser) resulted in the reaction stalling at 60% conversion, as was observed in the first campaign (entry 2). Adding more water resulted in only 20% additional conversion. Thinking the issue a problem of poor quality catalyst or reactant, the reaction was repeated in a GL18 test tube and proceeded to completion. In a pivotal experiment, the same batches of reactants and catalyst were run on 10 g scale in a 100 ml Parr pressure reactor. This reaction was the first to ever proceed to completion (*no* unreacted **12**) and generated an 82:18 ratio of desired secondary amine **10** to the tertiary amine by-product **16** and a 60% overall yield of pure **10**·2HCl after crystallization (entry 3). Simply changing the ratio of starting materials to 1:1 resulted in an increase in the isolated yield to 74% (entry 4). Run on scale, a total of 1.4 kg of **12** was converted to 2.82 kg of **10**·2HCl in two runs in an 8 l Hastelloy Parr vessel in an overall yield of 78%, the best overall performance to date for this process (entry 5).

8.5.2 Catalyst Loading Optimization

Continuing optimization in GL18 tubes, the reaction was investigated with lower levels of solvent and catalyst to determine the floor of reactivity for this particular substrate. As shown in Figure 8-4, the reaction continued to proceed to completion (assessed as <1% unreacted **12** by ^{1}H NMR analysis of crude reaction mixtures) through a change in catalyst loading from 5 to 0.03 mol% Ir, representing a reduction of *over two orders of magnitude.* Reaction temperature was kept constant at 110–115 °C at high catalyst loadings and increased to 125–130 °C at low loadings to maintain overall reaction efficiency (completion in 24 h or less). The total amount of solvent was also reduced to 0.2 ml g^{-1}**12**, with no obvious impact on product distribution or quality. At very low catalyst levels, the reaction needed to be run on 10 g or larger scale (in the 100 ml Parr reactor) to avoid inconsistent results from using extremely small amounts of catalyst ($\leq$ 3 mg catalyst/g of **12**). An essential control experiment, running the reaction in the Parr reactor in the absence of catalyst, resulted in no conversion and no degradation after 24 h at 115–120 °C. Simply

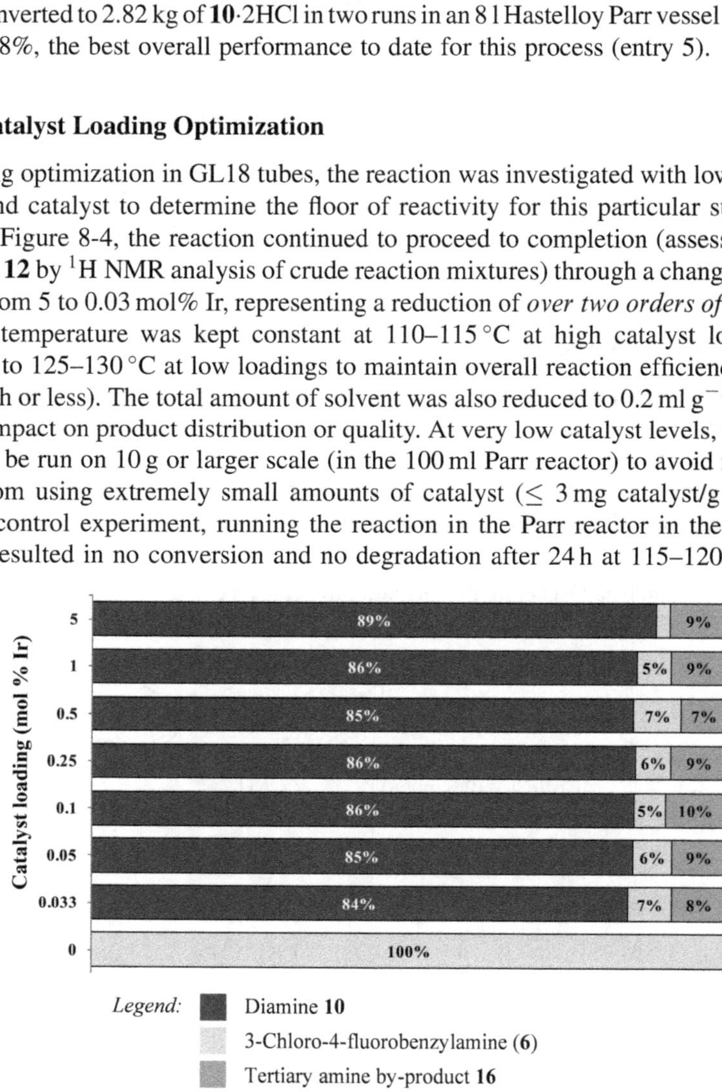

FIGURE 8-4. Catalyst loading optimization. Reported values are molar ratios as determined by proton NMR analysis of the crude reaction mixture. All reactions were conducted in toluene (0.2 or 1 l kg^{-1} of **12**) at 110 °C or 125–130 °C in sealed vessels.

adding catalyst to that mixture (corresponding to 0.05 mol% Ir) and heating to 115 °C overnight gave complete conversion. It is notable that the product distribution from each of these reactions is remarkably similar, indicating that catalyst loading does not significantly impact the course of the reaction. Finally, it is notable that the reaction appears to be amenable to running under flow conditions. In a single experiment, a solution of **6** and **12** in toluene (1 ml g^{-1}**12**) and 0.1 mol% Ir was recirculated through a 150 °C column with 20 min residence time. After running for 24 h, the mixture had proceeded to 75% conversion and had a similar product distribution to batch reactions.

8.5.3 Solvent Optimization and Additional Parameters

Additional reaction space exploration looked at the effect of different solvents on the reaction. Figure 8-5 shows that the best results were obtained in nonparticipating solvents and that the range of solvent compatibility (and by extension, functional group compatibility) is large. The use of solvents with Lewis basic character results in a decrease in overall reaction rate, as can be observed with DMAc, DMSO, and 2-methyltetrahydrofuran (2-MeTHF). The very slow reaction performance with 2-MeTHF and EtOAc is likely due to the solvent interacting with the metal and retarding turnover. Reaction temperatures from 85 to 130 °C were evaluated and little difference in reaction profile was observed, although lower temperatures resulted in significantly longer reaction times.

8.5.4 Kilogram-Scale Runs under Optimized Conditions

Two additional large-scale runs of this chemistry were executed. In the first run, 2.4 kg of **12** was utilized with 0.05 mol% Ir catalyst (3.5 g) and 1N aq. $NaHCO_3$ as base (0.1%, 9 ml).

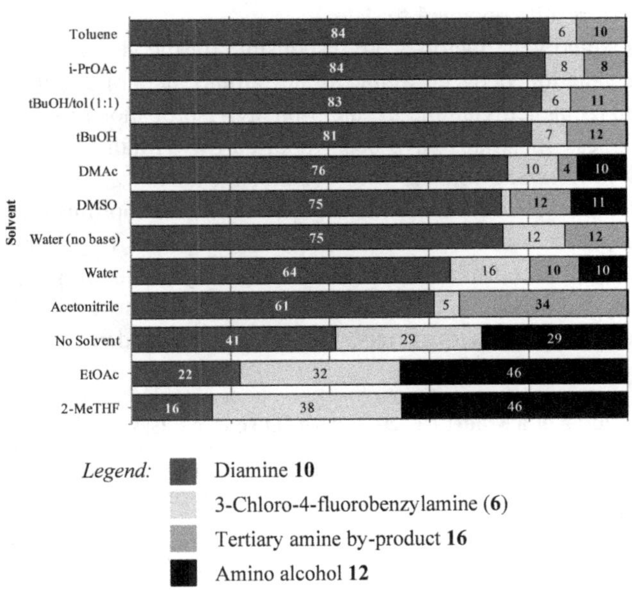

FIGURE 8-5. Solvent optimization screen. Reported values are molar ratios as determined by ^{1}H NMR analysis of the crude reaction mixture. All reactions were conducted with 0.25 mol% Ir and 1 l kg^{-1} solvent in GL18 threaded, capped test tubes at 110 °C for 24 h.

TABLE 8-4. Large-Scale Low-Catalyst Runs

Campaign	**12** (kg)	Ir used (mol%)	**10** (%)	**6** (%)	**16** (%)	**12** (%)	Yield (10·2HCl)	Ir (ppm)
1	2.4	0.0625	80.6	8.4	9.3	1.7	4.81 kg (76%)	9.2
2	2.4	0.0750	83.6	8.1	8.0	0.3	4.54 kg (69%)	6.9

The total reaction volume was 7 l, which allowed an 8 l Hastelloy pressure reactor to be deployed. After 24 h at 110 °C, the reaction was incomplete, so a small amount of excess catalyst was added to force it to completion overnight. After workup, a total of 4.81 kg of **10**·2HCl was isolated, representing a 76% overall yield (Table 8-4) The second run was the same scale, with a slightly larger initial catalyst charge (0.075 mol% Ir) to push the reaction to completion overnight, and very similar results were obtained. The direct translation of development runs on 10 g scale to plant runs on a 25 times higher scale with identical results proved very gratifying and indicated that for this substrate, redox-neutral chemistry was ideally suited for manufacturing use.

8.6 MECHANISTIC DISCUSSION

At the time of writing, the exact mechanism of this reaction was not known. It is not to say that the overall mechanistic profile is in question; every publication on redox-neutral alcohol aminations invokes a sequential process (Scheme 8-1) involving complexation of the alcohol with the catalyst followed by oxidation, imine formation, and reduction of the imine with the iridium hydride formed during the oxidation process. What happens within this mechanistic envelope is still very much in question, and this section is an attempt to make sense of the results obtained to date.

First, to understand what factors were responsible for the very high catalyst efficiency observed in the reaction between amine **6** and amino alcohol **12,** the Cp*Ir-catalyzed reaction between α-phenethylamine **21** and 3-phenyl-1-propanol **22** was investigated as a model system. These substrates have been reported by Williams and coworkers [16] to react sluggishly in refluxing toluene with 1–2 mol% catalyst.

Starting with "robust" conditions derived from the earlier optimization (0.25 mol% Ir, 1 ml g^{-1} toluene, sealed vessel at 110 °C), an initial screen of several aqueous bases addressed the thought that different sources of soluble hydroxide might positively impact reaction outcome. Surprisingly, this was not the case: little to no catalyst activity was observed (Table 8-5). More surprisingly, the reaction proceeded to completion when 1

TABLE 8-5. Screening of a Model Reaction with Inorganic Bases and Triethylamine[a]

Entry	Base (Amount)	Conversion[b] (%)	Conversion After NEt_3 Addition[c] (%)
1	1 N $NaHCO_3$ (0.25%)	36	95
2	1 N NaOH (1%)	0	>90
3	1 N Me_4NOH (1%)	0	>90

[a]All reactions conducted with 0.25 mol% Ir catalyst in 1 ml g^{-1} toluene at 110–115 °C in a capped test tube.
[b]Overall consumption of amine **21**. Molar ratios of products were measured by proton NMR analysis of the reaction mixtures and used to calculate conversions.
[c]1 equiv. of triethylamine was added and the reactions were continued for an additional 24 h at 110 °C.

equiv. of triethylamine was added and heating was continued. These data clearly indicated that carbonate or other inorganic salts were not beneficial in the reaction.

The next screen involved using triethylamine as the base with and without water present. All of these reactions proceeded to completion as measured by consumption of the starting alcohol **22**. One interesting result from these reactions was the impurity profile of the reaction mixture, first hinted by the ^{1}H NMR spectrum of the reaction product (Figure 8-6). LC/MS analysis revealed that the reaction mixture contained significant (>5%) quantities of compounds in which one or more *N*-ethyl groups were present. This result highlights one particularly interesting limitation of redox–neutral coupling reactions: the ligands on amines in the reaction mixture can participate in a similar redox–neutral *amine amination* pathway. This process has been recently documented by Williams and coworkers using trialkylamines as alkyl group donors [17], albeit at much higher reaction temperatures (>130 °C).

Screening several other bases revealed that facile alkyl group exchange occurs from amine bases containing *n*-alkyl groups and isopropyl groups (for example, Hünig's base and *N*-ethyl-pyrrolidine) but *not* from amines containing *N*-methyl groups and selected cyclic amines (see bottom NMR trace, Figure 8-6, also Figure 8-7). This behavior is consistent with the mechanism of the amine amination reaction in which oxidation of the tertiary amine generates an iminium ion that isomerizes to an enamine and hydrolyzes to generate a secondary amine and a carbonyl compound; bases that do not readily form enamines cannot participate in this reaction pathway. The amount of added tertiary amine base appears to not be critical to reaction performance, as similar results were obtained using catalytic (1–5 mol%) or stoichiometric (1 equiv., as reaction solvent) quantities. *N*-Methylpyrrolidine was employed in an additional round of optimization due to its high volatility and structural similarity to **12**. Successively larger scale reactions were conducted with and without toluene as a cosolvent and with and without water as an additive; optimum results were obtained in the absence of additional solvent and water seemed to not make a difference. Under these conditions, a 10 g scale reaction with 0.1% Cp*Ir catalyst proceeded to 87% overall conversion (870 turnovers) and provided the desired product in 65% yield after crystallization as the HCl salt. Notably, no alkyl group migration was observed in the LC/MS data of the crude reaction product.

From these results, a preliminary mechanistic portrait of this reaction can be proposed (Scheme 8-10). Dissolution of the catalyst is likely to provide an amine-complexed monomer (**I**)—similar to cationic complexes prepared and characterized by Kawahara and coworkers from the reaction of $[Cp^{*}IrCl_2]_2$ with ammonia [18]. Complex **I** oxidizes the alcohol to generate a Cp*Ir–H complex (**II**) and a carbonyl compound. As expected, the carbonyl compound reacts with the primary amine to generate an imine that is reduced by **II** to provide an iridium-product complex **III**. Decomplexation of the product from the metal is likely to be the turnover-limiting step[4]. The tertiary amine (or amines) that coordinates with **III** moderates the Lewis acidity of the metal and consequently increases

[4] Balcells and co-workers have conducted a theoretical mechanistic study of this reaction using carbonate as the base. While the work described in this Chapter appears to nullify many of their mechanistic assumptions, one of their calculations indicated that decomplexation of the product from iridium was likely to be a highly endothermic event. See Ref. 19.

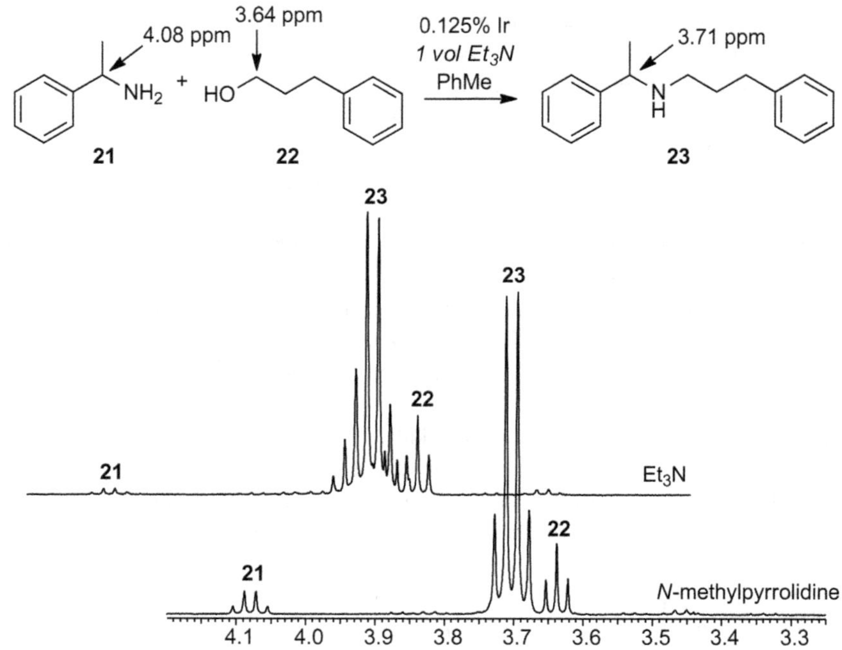

Reaction mixture component	m/z (ES$^+$), Area %	Proposed origin
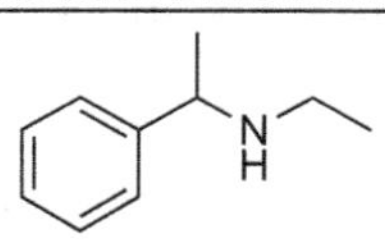	150.1 (13%)	Ethyl transfer from Et_3N to **21**
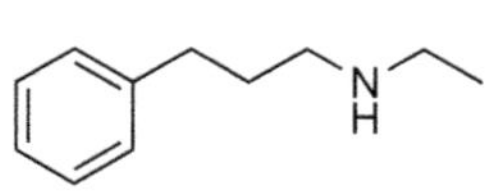	164.1 (2.6%)	Deethylation of m/z 192.1
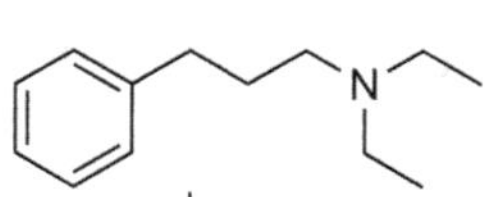	192.1 (16%)	Alkylation of diethylamine with **22**
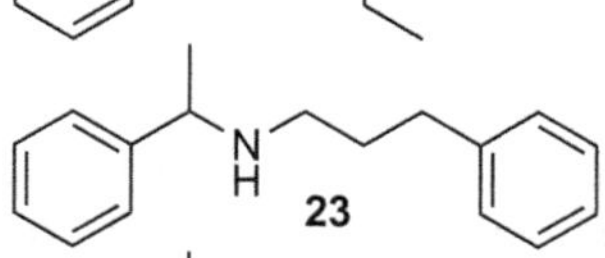	240.1 (41%)	Desired reaction product
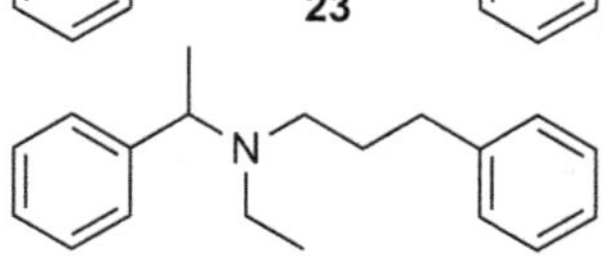	268.1 (1.3%)	Alkylation of **23** with Et_3N
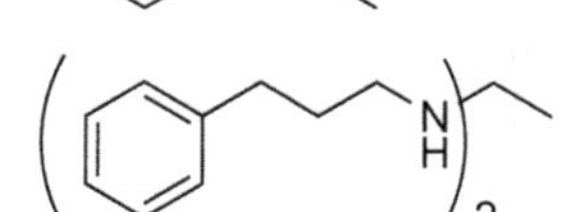	282.1 (6%)	Alkylation of m/z 164.1 with **22**
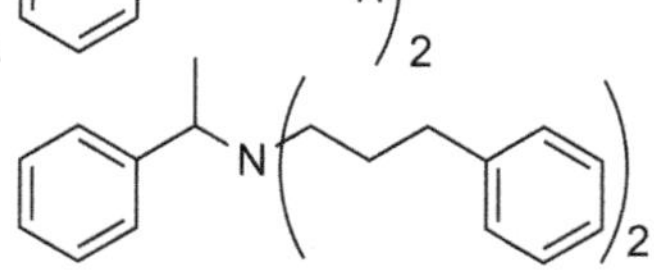	358.2 (4%)	Alkylation of **23** with **22**

FIGURE 8-6. Analysis and by-products from model reaction with triethylamine.

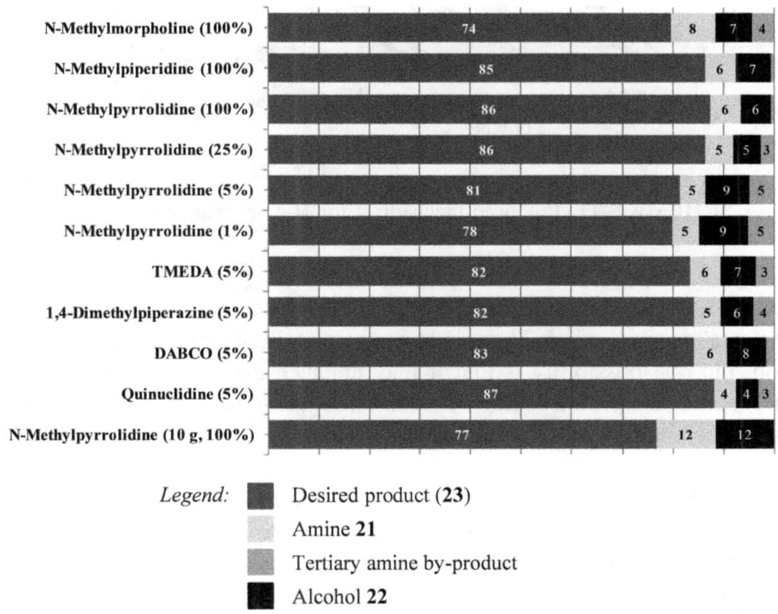

FIGURE 8-7. Reaction outcome using amine bases with nontransferrable ligands.

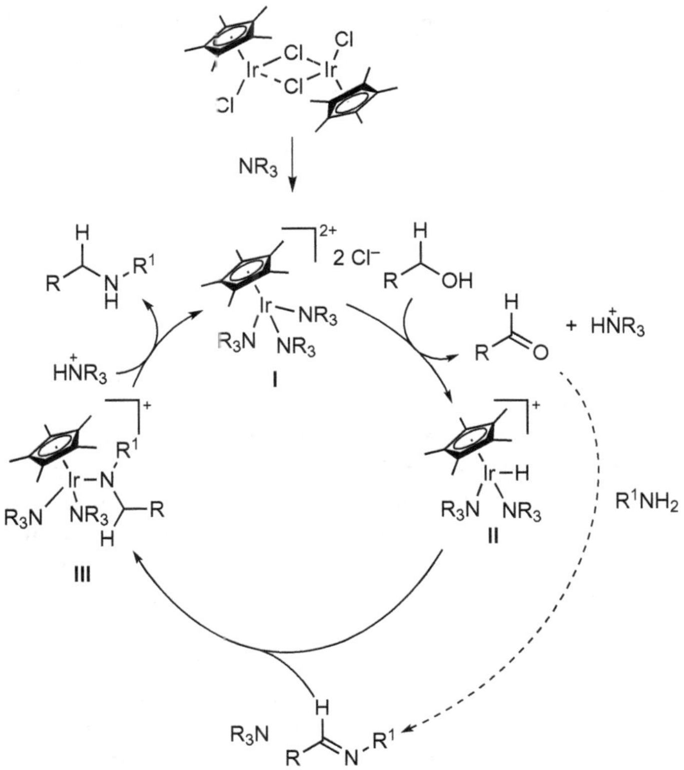

SCHEME 8-10. Proposed reaction mechanism.

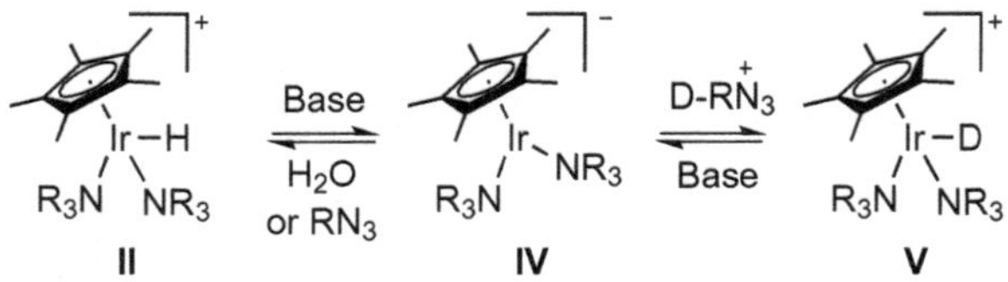

SCHEME 8-11. Proposed mechanism for H–D exchange on iridium.

the basicity of the secondary amine, which is protonated to release the product and regenerate catalyst monomer **I**. The role of water in facilitating catalyst turnover is not yet understood, but one possibility is that it may enhance the ability of trialkylammonium species to function as a proton shuttle between **I** and **III**. Obviously, a significant amount of additional experimentation is required to prove or disprove these proposals.

Another interesting result from these studies was the discovery of an unexpected deuteration at the site of imine reduction in the synthesis of the M + 5 isomer **20**. Additional experiments by the Isotope Synthesis Group revealed that the presence of D_2O at the beginning of the reaction was *required* for any deuteration to occur. In an experiment where the precursor components were concentrated multiple times from MeOD and not exposed to D_2O, the isolated product was the M + 4 isomer. It is not obvious what the role of D_2O is in this reaction, but since these experiments were conducted with solid K_2CO_3 as base, it is possible that the D_2O is simply facilitating dissolution of K_2CO_3 at the beginning of the reaction. The role of base in these reactions has been explored by Krische and coworkers who have demonstrated that Ir(III) hydrides are acidic and implicates these resulting Ir(I) anions in asymmetric alcohol allylation reactions [20] (see also Chapter 16). In this case, the presence of carbonate likely accelerates the equilibrium exchange of hydrogen for deuterium (**II** $\rightleftharpoons$ **V**) during the reaction by increasing the relative concentration of the intermediate Ir(I) anion **IV** (Scheme 8-11).

8.7 IRIDIUM CONTROL

Taken as a whole, the platinoid-group elements (transition metals, groups 8, 9, and 10) have been used to develop the most versatile catalysts currently in use in organic synthesis applications. Toxicological data to support their use in the synthesis of human medicines is much less well developed, and the ICH Guidelines for metal impurities (Q3D) are currently still in development. The most frequently cited regulatory guidance document is a draft guideline for the specification of residues from metal catalysts from the European Medicines Agency [21]. This document, formalized in 2007, divides frequently used catalytic metals into three classes on the basis of their potential risk to human health. Class I metals demonstrate carcinogenicity, are suspected carcinogens, or have other significant toxicities, and include all the platinum-group elements. Of these, platinum, palladium, iridium, rhodium, ruthenium, and osmium have the lowest permissible daily exposure (PDE) for oral medications of 100 $\mu g\,day^{-1}$ or 10 ppm in finished APIs. Molybdenum, nickel, chromium, and vanadium have slightly higher PDEs of 300 $\mu g\,day^{-1}$ or 30 ppm in APIs.

Given these requirements, any implementation of a process using a transition metal catalyst in pharmaceutical manufacturing requires advance planning on how to control residual metals and metal-containing by-products. A number of companies, including SiliCycle, Reaxa, and Johnson-Matthey, have exploited this opportunity to offer a variety of scavenging agents to reduce metal content in pharmaceutical applications. Thiol-derived silica gels work very well for platinum and palladium, but they do not efficiently remove iridium from reaction mixtures. One vendor claims that an amine-capped silica gel is effective for removing Ir^{3+} catalyst residues, but these types of scavengers were also found to be ineffective. A likely reason is that the starting materials and products from these redox–neutral reactions are amines, and the scavenger does not compete effectively with the product as a substrate for metal binding.

For a vast majority of substrates, the only effective iridium scavenger proved to be an activated carbon. As shown in Figure 8-8, a wide variety of carbons are capable of achieving a reasonable (85–90%) reduction in iridium levels while still minimizing overall substrate loss to the support. This performance pales in comparison of the best scavenging agents for palladium, which can reduce metal levels by a factor of 100 or more in typical circumstances.

Fortunately, the Cp*Ir–amine complexes formed in this reaction are readily destroyed in the presence of strong acid. In the case of the synthesis of **10**, crystallization of material containing 45,000 ppm Ir with excess HCl in 2-propanol led to isolation of bis(hydrochloride) salt with 360 ppm Ir, a 125-fold reduction in metal content. A beneficial effect was observed with higher crystallization temperatures and increased cake wash volumes, each of which resulted in a reduction of Ir content by a factor of 2. Of course, the greatest improvement in iridium control was reducing the overall catalyst charge (Figure 8-9), with very satisfactory results being obtained with the low loading conditions. As long as the product can tolerate strongly acidic treatment, this is by far the most effective and the least expensive method for removing iridium.

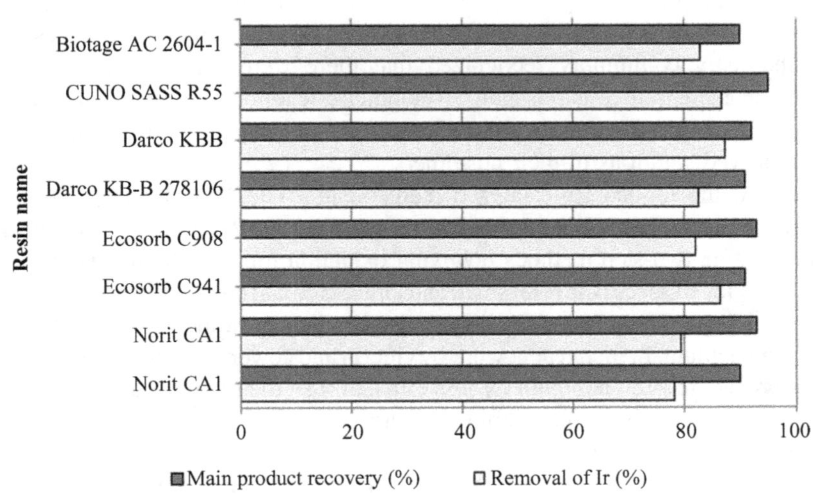

FIGURE 8-8. Iridium removal from **1** using activated carbons.

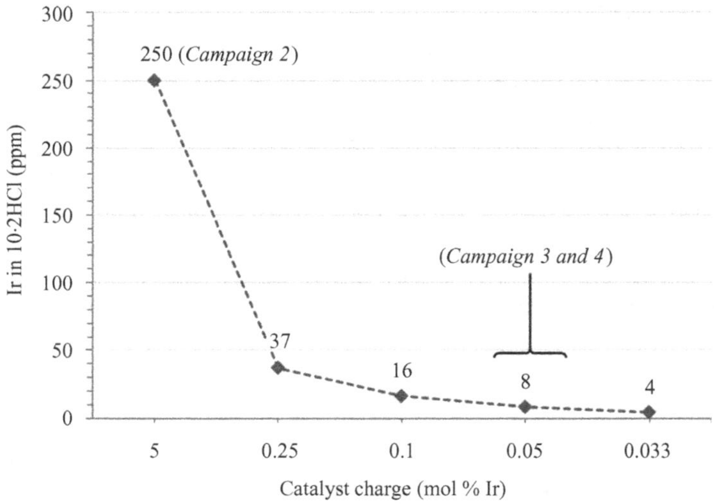

FIGURE 8-9. Catalyst loading versus iridium concentration in isolated **10**·2HCl.

8.8 FINAL COMMENTS

When asked to write a chapter for this book, I spent significant time thinking about what "sustainable catalysis" meant for redox-neutral amine coupling reactions of the type described in this chapter. I have come to the conclusion that these redox–neutral processes have to overcome a greater hurdle to achieve general utilization in the chemical community.

Transition metal-catalyzed C–C bond forming processes and aryl amination reactions are two examples of transformations that have revolutionized the construction of complex molecules by completely supplanting earlier methods with new, reliable, and predictable techniques. As synthetic organic chemists started to use these methods in their research, the greater chemistry community recognized that there were academic and commercial benefits from increased research and development, leading to the current environment where a number of groups in academia and industry are engaged in cutting-edge research to deliver improved solutions to their customers.

Conversely, the Ir- and Ru-catalyzed redox–neutral coupling chemistry is still in its infancy, with relatively little known about the scope and even less about the mechanism of the reaction. Existing two-step methods involving oxidation followed by reductive amination have been employed successfully on small and large scale and enjoy a high degree of comfort among practicing organic chemists, due in part to their prominent place in the existing chemistry curriculum. In fact, recent work by Guérin and coworkers extends this modality to a one-pot procedure [22]; although far less efficient than the redox–neutral chemistry, it is likely to be as frequently employed in the future due entirely to procedural comfort.

So how to change this situation? Of greatest importance is the need to understand the mechanism of this reaction, especially under realistic catalyst loadings (< 0.1 mol% metal; > 1000 turnovers). Industrial chemists will likely eschew this chemistry for less costly alternatives until catalyst loading is dramatically reduced. More work to establish the functional group compatibility of these methods is essential, and a better understanding of the limitations of the chemistry will help others to develop processes without needing to rediscover the limitations each time. It is my hope that the work described in this chapter will encourage those working in the field to achieve these goals, for this chemistry can positively impact how molecules are made.

ACKNOWLEDGMENTS

The work described in this chapter benefited from countless hours of assistance and collaboration by a large number of very dedicated scientists and technicians at Pfizer. In particular, I would like to recognize the efforts of Yinsheng Zhang of the Isotope Synthesis Group for having the prescience to attempt this chemistry for the synthesis of **1** which started us on a journey of discovery. Teresa Makowski, Barb Sitter, and Karl Ng contributed to reaction screening and optimization efforts preceding the first scale-up campaign. Thanks also go to current and former CRD colleagues Stephane Dubant, Kristin Price, Rob Singer, Al Pettman, Carrie Wager, Peter Thomson, Steve Hoagland, and the operators in the Groton Kilo Lab for assistance during the evolution of this project.

REFERENCES

1. Roughley SD, Jordan AM (2011). The medicinal chemist's toolbox: an analysis of reactions used in the pursuit of drug candidates. *J. Med. Chem.* 54, 3451–3479.
2. Carey JS, Laffan D, Thomson C, Williams MT (2006). Analysis of the reactions used for the preparation of drug candidate molecules. *Org. Biomol. Chem.* 4, 2337–2347.
3. For a recent review, see Guillena G, Ramón DJ, Yus M (2010). Hydrogen autotransfer in the *N*-alkylation of amines and related compounds using alcohols and amines as electrophiles. *Chem. Rev.* 110, 1611–1641.
4. Berliner MA, Dubant SPA, Makowski T, Ng K, Sitter B, Wager C, Zhang Y (2011). Use of an iridium-catalyzed redox-neutral alcohol-amine coupling on kilogram scale for the synthesis of a GlyT1 inhibitor. *Org. Process Res. Dev.* 15, 1052–1062.
5. (a) Lebel L, Tunucci D, Valentine J, Lowe III JA, Hou X, Schmidt C, Tingley III FD, McHardy S, Kalman M, DeNinno S, Sanner M, Ward K (2009). The discovery of a structurally novel class of inhibitors of the type 1 glycine transporter. *Bioorg. Med. Chem. Lett.* 19, 2974–2976. (b) For a review, see Wolkenberg S, Sur C (2010) Recent progress in the discovery of non-sarcosine based GlyT1 inhibitors. *Curr. Top. Med. Chem.* 10, 170–186.
6. Brighty K, Castaldi M (1996). Synthesis of (1α,5α,6α)-6-amino-3-azabicyclo[3.1.0]hexane, a novel achiral diamine. *Synlett* 1097–1099.
7. Brighty KE, Gootz TD (1997). The chemistry and biological profile of trovafloxacin. *J. Antimicrob. Chemother.* 39 (Suppl. 2), 1–14.
8. Ley SV, Norman J, Griffith WP, Marsden SP (1994). Tetrapropylammonium perruthenate, $Pr_4N^+RuO_4^-$, TPAP: a catalytic oxidant for organic synthesis. *Synthesis* 639–666.
9. (a) Mancuso A, Swern D (1981) Activated dimethyl sulfoxide: useful reagents for synthesis. *Synthesis* 165–185. (b) Tidwell T (1990). Oxidation of alcohols by activated dimethyl sulfoxide and related reactions: an update. *Synthesis* 10, 857–870.
10. Kim KS, Cho IH, Yoo BK, Song YH, Hahn CS (1984). Selective oxidation of primary and secondary alcohols using di-isopropyl sulphide–*N*-chlorosuccinimide. *J. Chem. Soc., Chem. Commun.* 762–763.

11. Mancuso AJ, Huang S, Swern D (1978). Oxidation of long-chain and related alcohols to carbonyls by dimethyl sulfoxide "activated" by oxalyl chloride. *J. Org. Chem.* 43, 2480–2482.
12. Parikh JR, Doering WVE (1967). Sulfur trioxide in the oxidation of alcohols by dimethyl sulfoxide. *J. Am. Chem. Soc.* 89, 5505–5507.
13. See, for example, van der Linden JJM, Hilberink PW, Kronenburg CMP, Kemperman GJ (2008). Investigation of the Moffatt–Swern oxidation in a continuous flow microreactor system. *Org. Process Res. Dev.* 12, 911–920.
14. Fujita K, Li Z, Ozeki N, Yamaguchi R (2003). *N*-Alkylation of amines with alcohols catalyzed by a Cp*Ir complex. *Tetrahedron Lett.* 44, 2687–2690.
15. Fujita K, Enoki Y, Yamaguchi R (2008). Cp*Ir-catalyzed *N*-alkylation of amines with alcohols. A versatile and atom economical method for the synthesis of amines. *Tetrahedron* 64, 1943–1954.
16. Saidi O, Blacker AJ, Lamb GW, Marsden SP, Taylor JE, Williams JMJ (2010). Borrowing hydrogen in water and ionic liquids: iridium-catalyzed alkylation of amines with alcohols. *Org. Process Res. Dev.* 14, 1046–1049.
17. Saidi O, Blacker AJ, Farah MM, Marsden SP, Williams JMJ (2009). Selective amine cross-coupling using iridium-catalyzed "borrowing hydrogen" methodology. *Angew. Chem., Int. Ed.* 48, 7375–7378.
18. Kawahara R, Fujita K, Yamaguchi R (2010). Multialkylation of aqueous ammonia with alcohols catalyzed by water-soluble Cp*Ir–ammine complexes. *J. Am. Chem. Soc.* 132, 15108–15111.
19. Balcells D, Nova A, Clot E, Gnanamgari D, Crabtree RH, Eisenstein O (2008). Mechanism of homogeneous iridium-catalyzed alkylation of amines with alcohols from a DFT study. *Organometallics* 27, 2529–2535.
20. Han SB, Kim IS, Krische MJ (2009). Enantioselective iridium-catalyzed carbonyl allylation from the alcohol oxidation level via transfer hydrogenation: minimizing pre-activation for synthetic efficiency. *Chem. Commun.* 47, 7278 and references therein.
21. http://www.ema.europa.eu/docs/en_GB/document_library/Scientific_guideline/2009/09/WC500003587.pdf. Last accessed January 14, 2013.
22. Guérin C, Bellosta V, Guillamot G, Cossy J (2011). Mild nonepimerizing *N*-alkylation of amines by alcohols without transition metals. *Org. Lett.* 13, 3534–3537.

9

OLEFIN METATHESIS: FROM ACADEMIC CONCEPTS TO COMMERCIAL CATALYSTS

JUSTYNA CZABAN,[1] CHRISTIAN TORBORG,[2] AND KAROL GRELA[1,2]

[1]*Institute of Organic Chemistry, Polish Academy of Sciences, Warsaw, Poland*
[2]*Department of Chemistry, Warsaw University, Warsaw, Poland*

9.1 INTRODUCTION

Catalysis tackles important problems chemists encounter in industrial processes, as introducing catalytic reactions can lead to shorter synthetic pathways and reduction of the reaction time, waste, and energy costs. Consequently, at least one catalytic process is included in the production of over 80% of today's chemical products [1]. The majority of these processes are heterogeneous catalytic ones, largely for the production of bulk chemicals. In contrast, the market share for homogeneous catalytic processes, with the catalyst types ranging from simple acids or bases to transition-metal complexes containing sophisticated ligands, is estimated to be 15%. Particularly for the construction of complex molecules such as potential pharmaceuticals or agrochemicals, ligand-bound transition-metal catalysts are often applied to the formation of carbon–carbon and carbon–nitrogen bonds, oxidations, reductions, and other transformations [2]. With respect to C−C bond formations, one of the most reliable methods is currently olefin metathesis [3]. The ongoing success of this reaction [4], which is being used in the fine chemical industry [5], in natural product synthesis [6], and also in polymerizations [7], culminated in the award of the 2005 Nobel Prize in Chemistry to three outstanding contributors to this field—Grubbs, Schrock, and Chauvin [8].

The generally accepted reaction mechanism of the olefin metathesis reaction was first proposed by Hérisson and Chauvin in 1971 [9], and confirmatory experimental evidence was subsequently found by various groups [10].

Sustainable Catalysis: Challenges and Practices for the Pharmaceutical and Fine Chemical Industries, First Edition. Edited by Peter J. Dunn, K. K. (Mimi) Hii, Michael J. Krische, and Michael T. Williams.

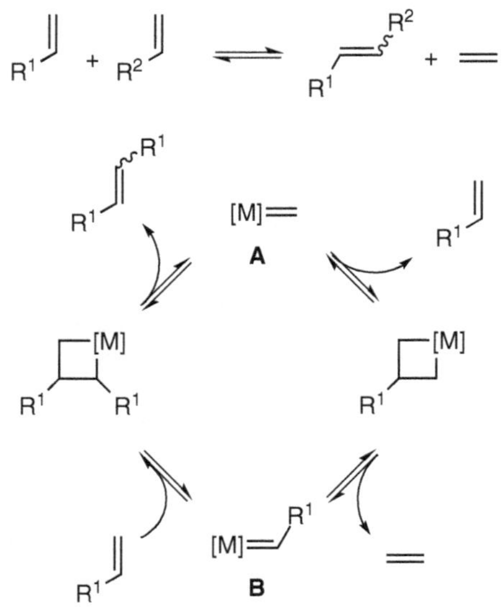

SCHEME 9-1. Mechanism of the alkene metathesis reaction.

A general scheme is depicted in Scheme 9-1: the reaction is initiated by a metal alkylidene of type **A**—the active catalyst—reacting with the olefin substrate. The metallocyclobutane is formed as an intermediate, in turn either producing ethylene and a new metal alkylidene **B** or reverting back to the initiating species. Intermediate **B** can subsequently react with a new olefin to give the final product and the alkylidene initiator, which restarts the cycle.

Depending of the nature of the substrate(s) applied, different categories of olefin metathesis transformations are known (Scheme 9-2). Careful choice of substrate features and reaction conditions can lead to high conversions within these reactions, even though olefin metathesis itself is a reversible process.

By refluxing the reaction mixture, for example, a volatile alkene product (e.g., ethylene) can be removed and the equilibrium shifted toward the product. In contrast, the mechanism of the ene–yne metathesis (Scheme 9-2) [11a] is considered to be more complex than metathesis reactions of olefins [11b]. In particular, the reversibility of the reaction steps of the ene–yne metathesis reaction is still under discussion [12].

Generally metathesis catalysts are based on tungsten, molybdenum, and ruthenium, with the last being considered as generally less active, but more stable toward air and moisture than early transition-metal complexes. Ruthenium-based catalysts were discovered about 40 years ago, when $RuCl_3$ hydrate was found to be a suitable precatalyst for the ring-opening polymerization (ROMP) [13]. The catalytic system underwent a renaissance in the late 1980s, when Grubbs et al. found that the addition of water is beneficial for the polymerization reaction [14]. Suspecting the active species to be an alkylidene complex [15], the synthesis of a defined ruthenium alkylidene precatalyst was achieved in 1992, with the isolation of the first metathesis-active ruthenium alkylidene complex **3** (Scheme 9-3) [16].

This was the starting point for the development of well-defined ruthenium precatalysts, a research area that continues to this day. Table 9-1 gives an overview of commercially available Ru-based defined olefin metathesis precatalysts.

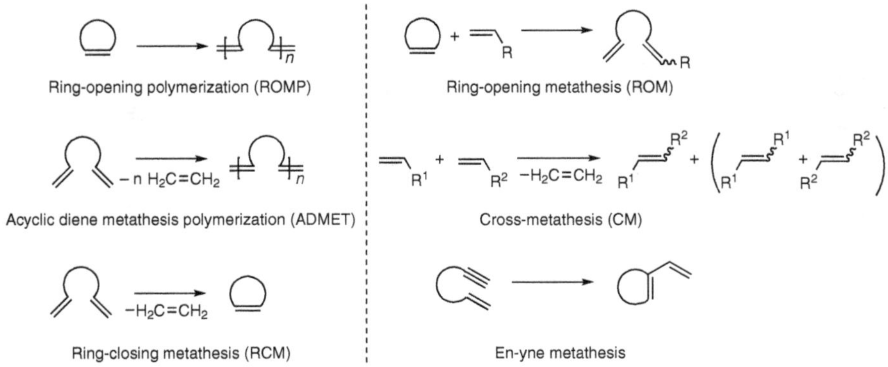

SCHEME 9-2. Types of alkene metathesis.

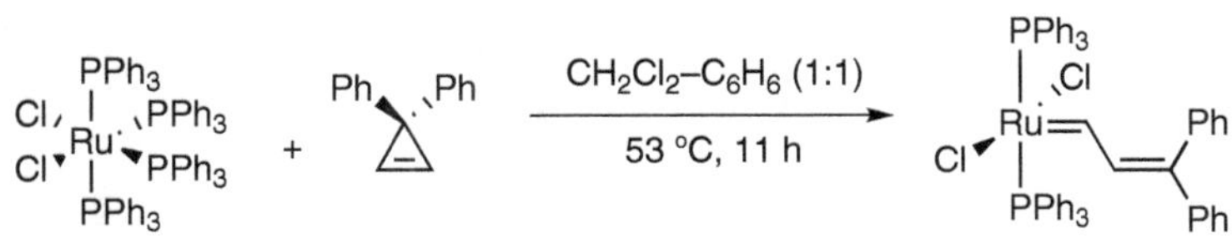

1 2 3 > 99 %

SCHEME 9-3. Synthesis of **3** as reported by Grubbs et al.

The most striking advantage of well-defined precatalysts over ill-defined ones is their higher efficiency, for example, in alkyne metathesis [17]. In contrast, a large amount of metal is thought to be inactive in ill-defined systems [18], and the corresponding reactions often suffer from long initiation periods. The substitution of the PPh_3 ligand with the more sterically demanding and basic tricyclohexylphosphine ligand, as well as the incorporation of the readily initiating benzylidene moiety, led to the discovery of catalyst **4**, nowadays referred to Grubbs' first-generation catalyst [19]. **4** constitutes a highly active catalyst tolerating most functional groups, although its limited stability and low activity toward substituted double bonds are major drawbacks. However, complex **4** has been intensively used in ROMP reactions and the synthesis of natural products and still represents an alternative to more active but also more expensive catalysts [20]. The introduction of *N*-heterocyclic carbene (NHC) ligands to Ru-metathesis catalysts [21] in 1998 by Herrmann et al. [22] led to further important progress. The initially synthesized complexes **55** and **56** (Figure 9-1) bearing two NHC ligands turned out to be active in both ring-closing metathesis (RCM) and ROMP, but have still been outperformed by the first-generation catalyst **4**.

Considering this, groups of Herrmann [23] and Fürstner [24], Grubbs [25], and Nolan [26] independently, and almost simultaneously, combined the stable, strongly donating NHC ligand and the labile phosphine group within one complex. These compounds, subsequently known as Grubbs' second-generation catalysts, present higher activity in overall metathesis reactions than their parent bis(NHC) and bis(phosphine) complexes. The most striking example is the (SIMes)-substituted **5** [27] that exhibits an activity comparable to most powerful molybdenum-based catalysts, high stability toward air and

TABLE 9-1. Commercially Available Ruthenium Precatalysts[a]

Catalyst	Name/Description	CAS Number	Supplier
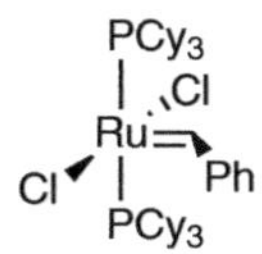 **4**	Grubbs catalyst, first generation	172222-30-9	Aldrich, Materia
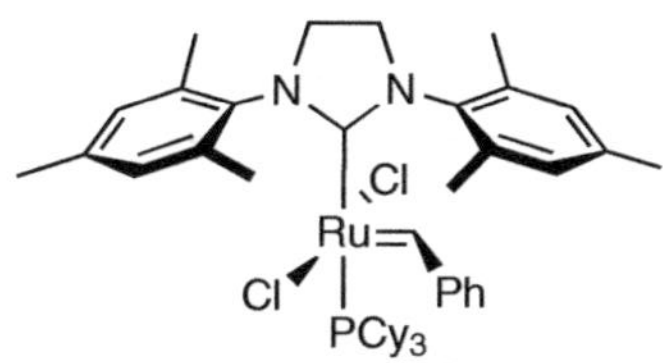 **5**	Grubbs catalyst, second generation, G2	246047-72-3	Aldrich, Materia
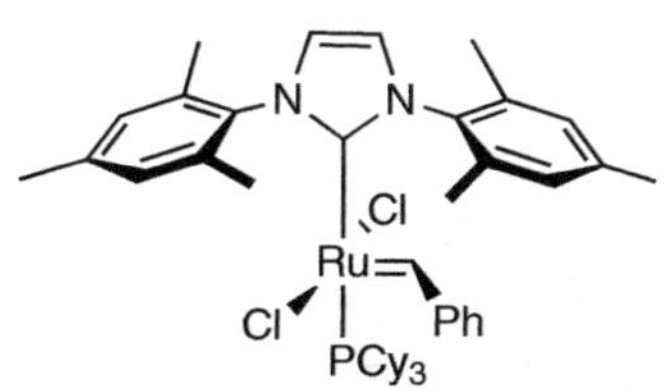 **6**	Ruthenium, dichloro[1,3-bis(2,4,6-trimethylphenyl)-imidazol-2-ylidene](phenylmethylene)(tricyclohexylphosphine)	223415-64-3	Shanghai Chem
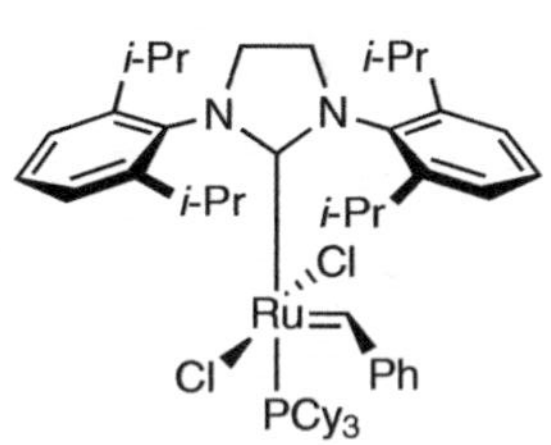 **7**	Dichloro[4,5-dihydro-1,3-bis (2,6-isopropylphenyl)-2-imidazolidinylidene](benzylidene)(tricyclohexylphosphine)ruthenium(II)	373640-75-6	Aldrich, Materia

Structure	Name	CAS	Supplier
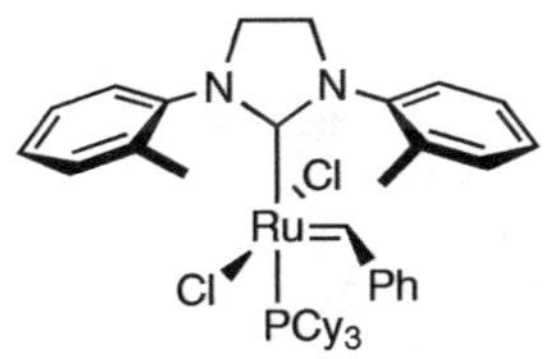 8	1,3-bis(2-methylphenyl)-2-imidazolidinylidene] dichloro-(benzylidene)(tricyclohexylphosphine) ruthenium(II)	927429-60-5	Aldrich, Materia
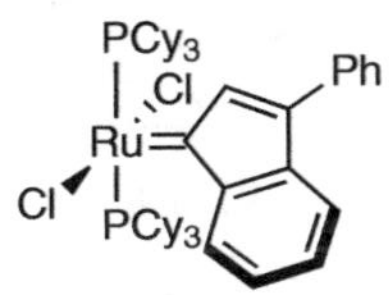 9	M1	250220-36-1	Strem, Umicore
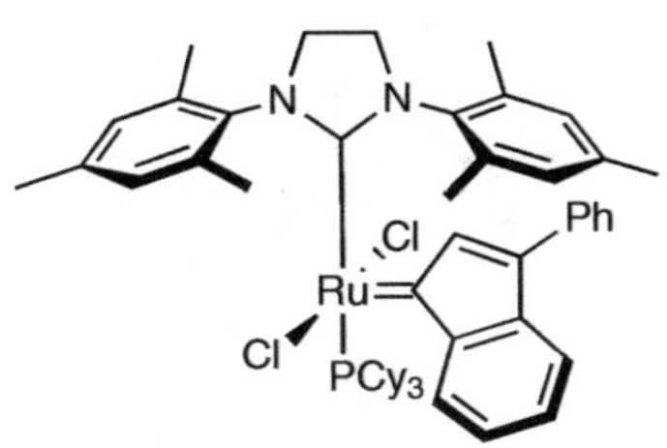 10	M2	536724-67-1	Strem, Umicore
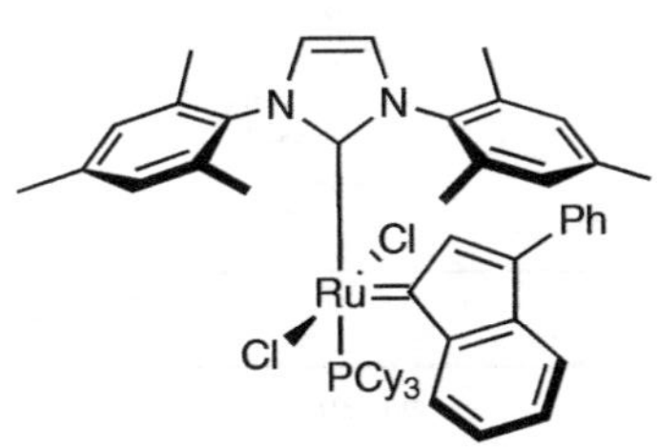 11	catMETium RF1	254972-49-1	Strem, Evonik-Degussa

(*continued*)

TABLE 9-1. (*Continued*)

Catalyst	Name/Description	CAS Number	Supplier
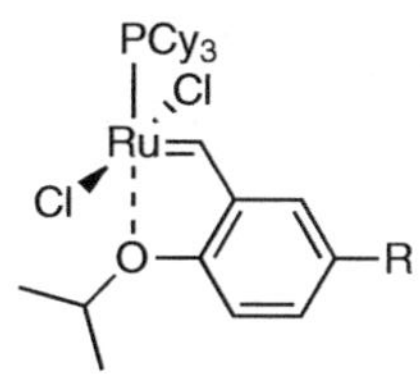 **12, 13**	**12**: Hoveyda–Grubbs catalyst, first generation (R = H) **13**: Zhan catalyst 1C (R = SO_2NMe_2)	203714-71-0 918871-44-0	Aldrich, Materia Strem, Zannan Pharma
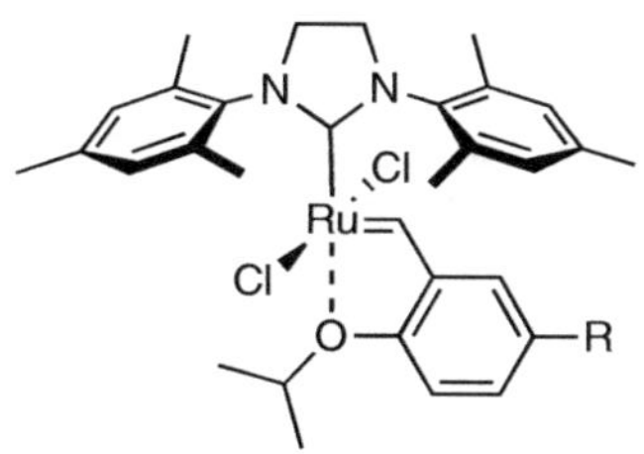 **14–16**	**14**: Hoveyda–Grubbs catalyst, second generation **15**: Nitro catalyst (R = NO_2) **16**: Zhan catalyst 1B (R = SO_2NMe_2)	301224-40-8 502964-52-5 918870-76-5	Aldrich, Materia Apeiron Catalysts Strem, Zannan Pharma
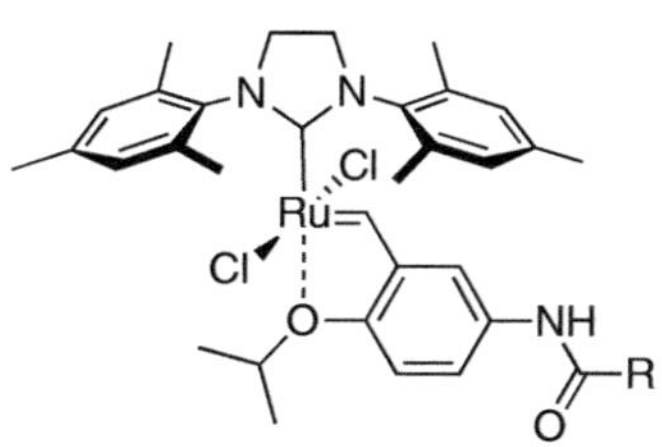 **17–20**	**17**: $M7_1$-SIMes (R = CF_3) **18**: $M7_2$-SIMes (R = C_6F_5) **19**: $M7_3$-SIMes (R = O*i*-Bu) **20**: $M7_4$-SIMes (CO_2Et)	1025728-56-6 1030618-02-0 1025718-57-7 1030618-11-1	Omega Cat System

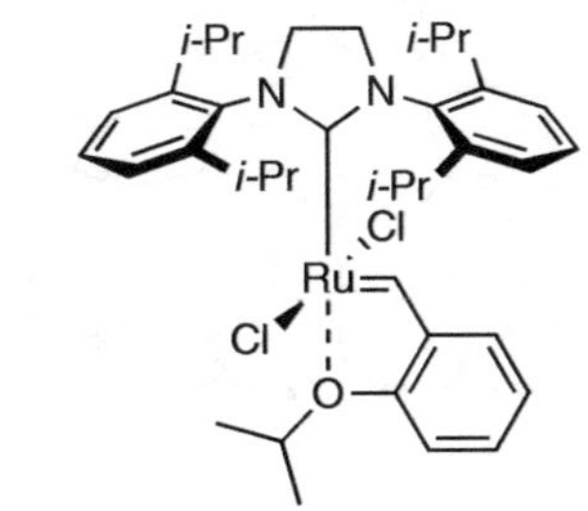 **21**	Dichloro[1,3-bis(2,6-isopropylphenyl)-2-imidazolidinylidene] (2-isopropoxyphenylmethylene) ruthenium(II)	635679-24-2	Aldrich, Materia
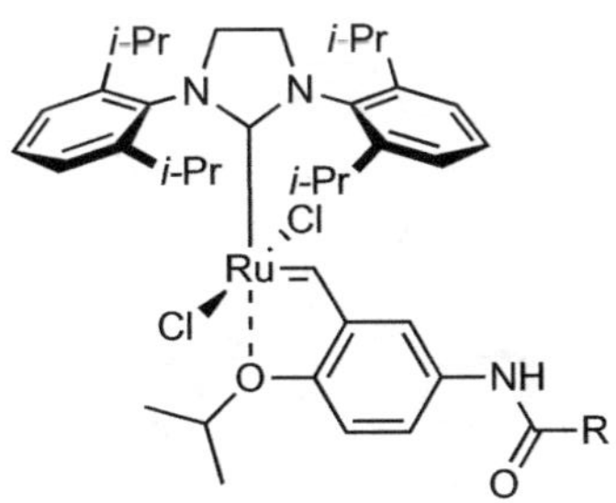 **22–25**	**22**: $M7_1$-SIPr (R = CF_3) **23**: $M7_2$-SIPr (R = C_6F_5) **24**: $M7_3$-SIPr (R = O*i*-Bu) **25**: $M7_4$-SIPr (R = CO_2Et)	1212008-99-5 – 1212009-05-6 1212009-08-9	Omega Cat System
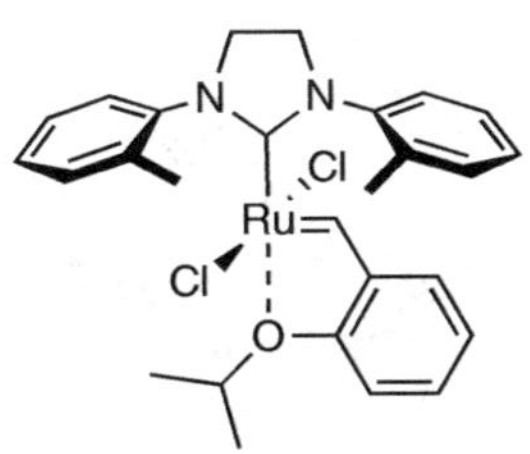 **26**	Dichloro[1,3-bis(2-methylphenyl)-2-imidazolidinylidene](2-isopropoxyphenylmethylene) ruthenium(II)	927429-61-6	Aldrich, Materia

(continued)

TABLE 9-1. (*Continued*)

Catalyst	Name/Description	CAS Number	Supplier
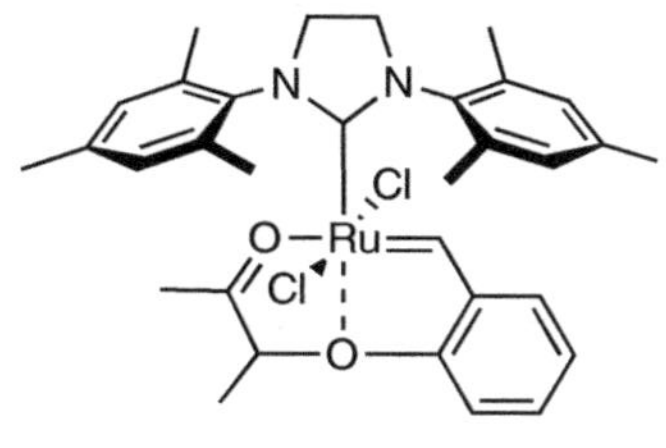 **27**	Umicore $M5_1$	1031262-71-1	Umicore
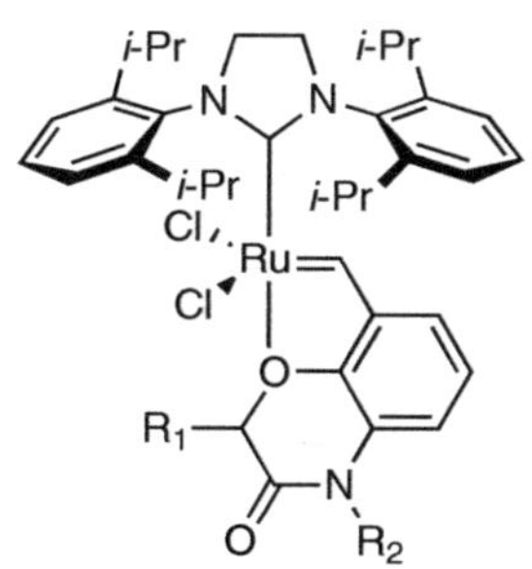 **28–31**	**28**: $M8_{13}$-SIPr ($R^1 = Me$, $R^2 = C(O)C_6F_5$) **29**: $M8_{31}$-SIPr ($R^1 = Et$, $R^2 = H$) **30**: $M8_{32}$-SIPr ($R^1 = Et$, $R^2 = C(O)O$*i*-Bu) **31**: $M8_{43}$-SIPr ($R^1 = Et$, $R^2 = C(O)C_6F_5$)	1025728-56-6 – – –	Omega Cat System
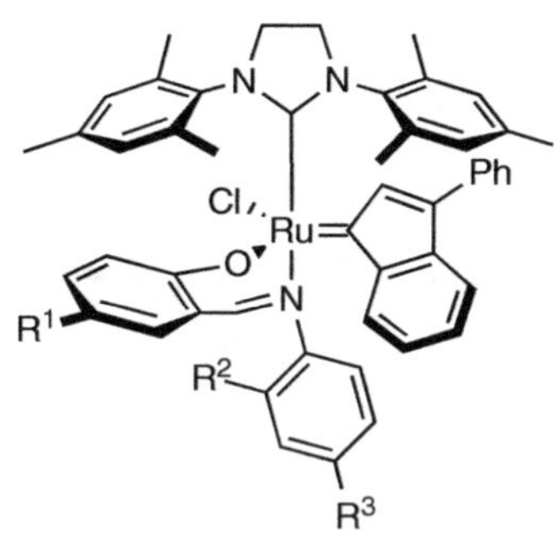 **32, 33**	**32**: $M4_1$ ($R^1 = NO_2$, $R^2 = H$, $R^3 = Me$) **32**: $M4_1$ ($R^1 = NO_2$, $R^2 = H$, $R^3 = Me$) **33**: $M4_2$ ($R^1 = H$, $R^2 = Me$, $R^3 = H$)	934538-04-2 934538-04-2 934538-12-2	Strem, Umicore Strem, Umicore

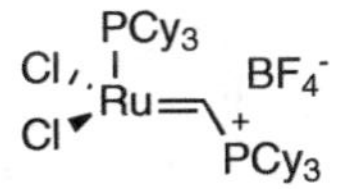 34	Piers-type catalyst Dichloro(tricyclohexylphosphine) [(tricyclohexylphosphoranyl) methylidene]ruthenium (II) tetrafluoroborate	1020085-61-3	Aldrich, Materia
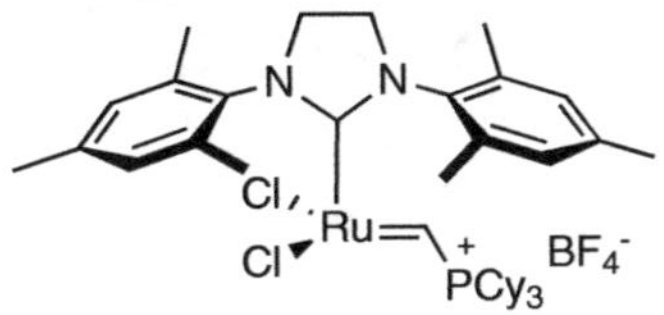 35	Piers-type catalyst dichloro[1,3-bis(2,4,6-trimethylphenyl)-2-imidazolidinylidene] [(tricyclohexyl-phosphoranyl) methylidene] ruthenium(II) tetrafluoroborate	832146-68-6	Aldrich, Materia
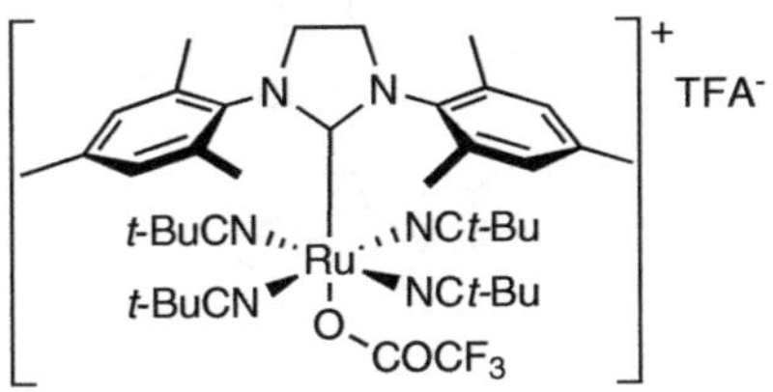 36	Trifluoroacetato[1,3-bis(2,4,6-trimethylphenyl) imidazol-2-ylidene]tetra(2,2-dimethylpropanenitrile) ruthenium(II) trifluoroacetate	1034198-65-6	Strem
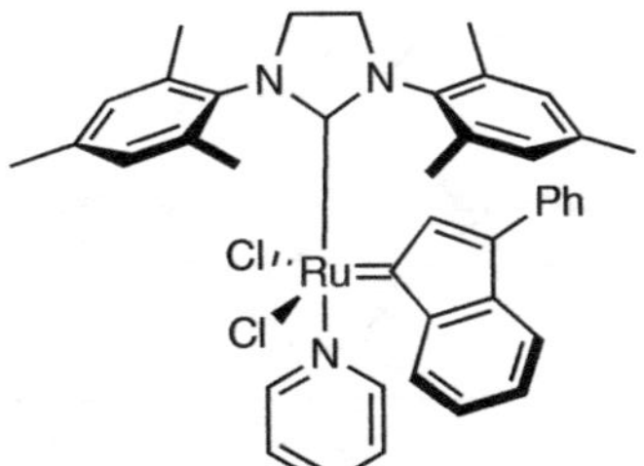 37	Umicore $M3_1$	1031262-76-6	Umicore

(*continued*)

TABLE 9-1. (*Continued*)

Catalyst	Name/Description	CAS Number	Supplier
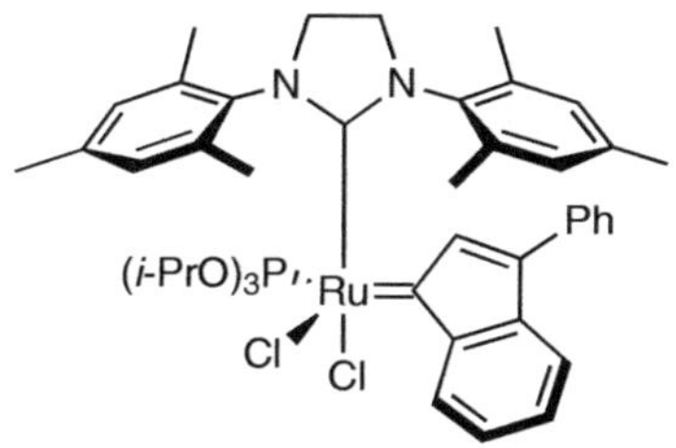 **38**	*cis*-Caz-1	1255536-61-8	Strem
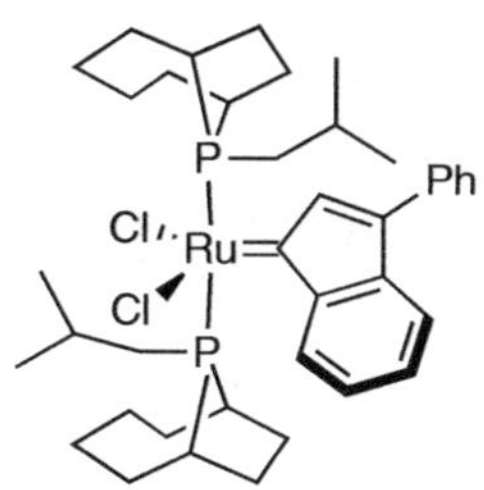 **39**	$M1_1$	894423-99-5	Strem, Umicore
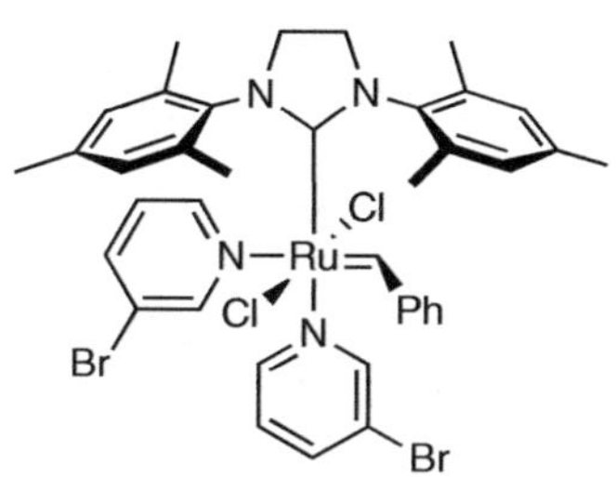 **40**	Dichloro[4,5-dihydro-1,3-bis(2,4,6-trimethylphenyl)-2-imidazolidinylidene] (benzylidene)bis(3-bromopyridine)-ruthenium(II)	900169-53-1	Aldrich, Materia

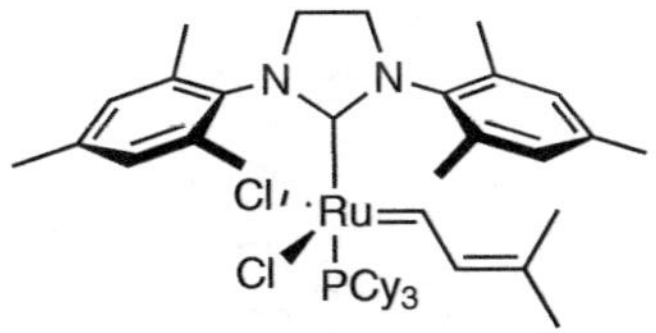 **41**	Dichloro[1,3-bis(2,4,6-trimethylphenyl)-2-imidazolidinylidene] (3-methyl-2-butenylidene) (tricyclohexylphosphine)-ruthenium(II)	253688-91-4	Aldrich, Materia
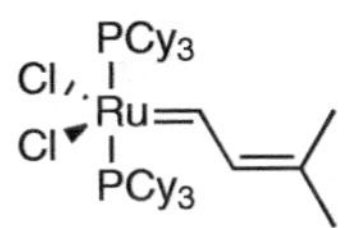 **42**	Dichloro(3-methyl-2-butenylidene)bis (tricyclohexylphosphine)-ruthenium(II)	194659-03-9	Aldrich, Materia
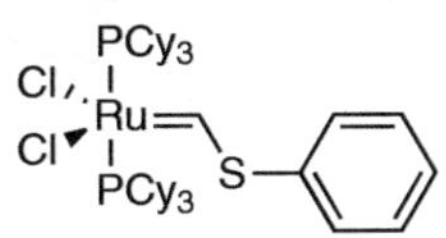 **43**	Bis(tricyclohexylphosphine) [(phenylthio)methylene] ruthenium(II) dichloride	219770-99-7	Strem
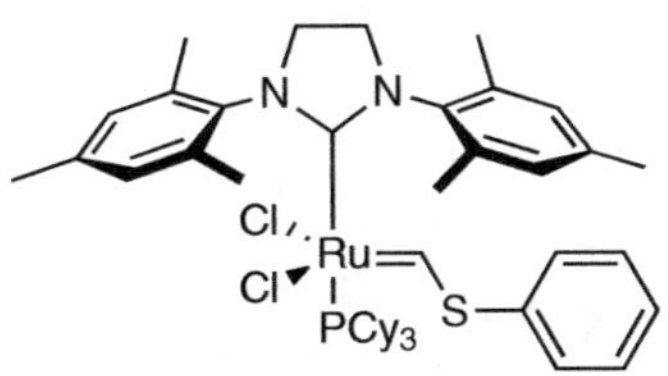 **44**	Tricyclohexylphosphine[1,3-bis(2,4,6-trimethylphenyl)-2-imidazolidinylidene][(phenylthio)methylene] ruthenium(II) dichloride	1155422-69-7	Strem

(*continued*)

TABLE 9-1. (*Continued*)

Catalyst	Name/Description	CAS Number	Supplier
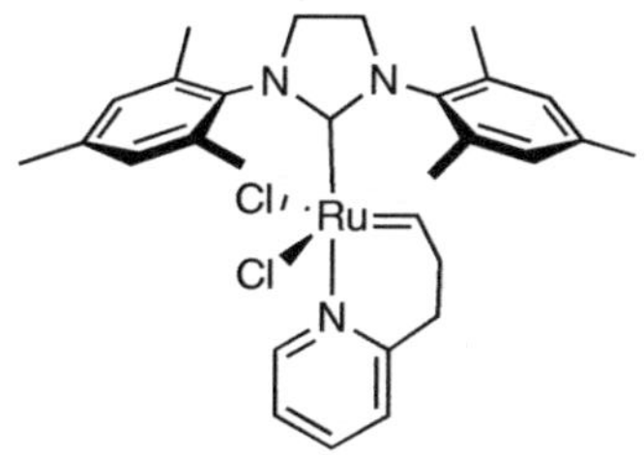 **45**	Dichloro[1,3-bis(2,4,6-trimethylphenyl)-2-imidazolidinylidene] [3-(2-pyridinyl)propylidene] ruthenium(II)	802912-44-3	Aldrich, Materia
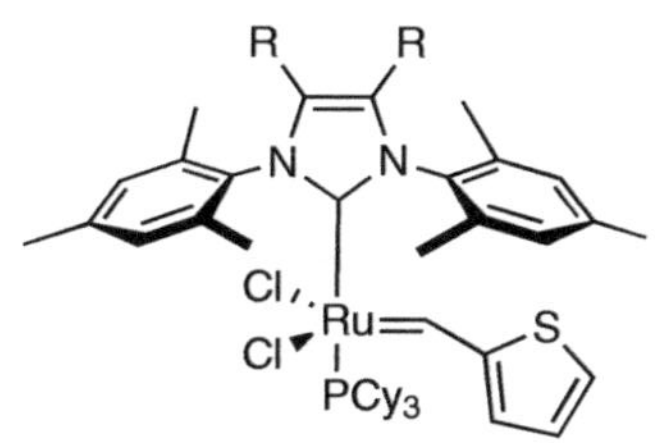 **46, 47**	**46**: catMETium RF 2 (R = H) **47**: catMETium RF 3 (R = Me)	1190427-49-6 1190427-50-9	Strem, Evonik-Degussa
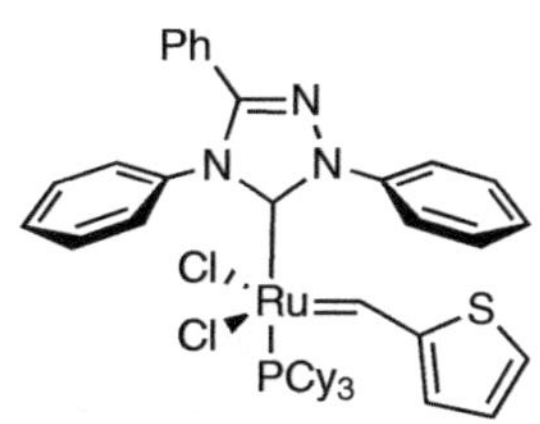 **48**	catMETium RF 4	1190427-51-0	Strem, Evonik-Degussa

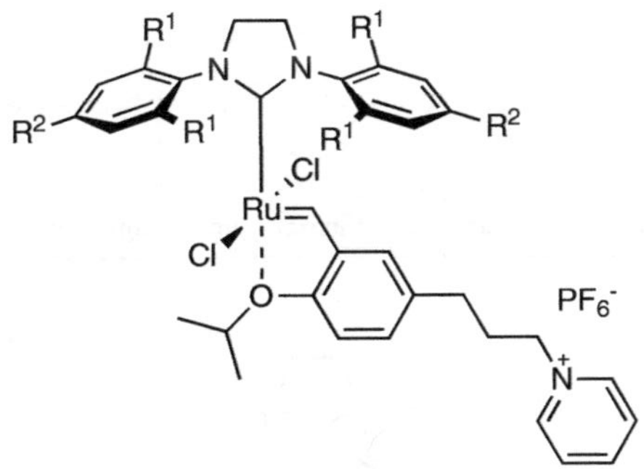 **49, 50**	**49**: Omega$_{31}$-SIMes ($R^1 = Me$, $R^2 = Me$) **50**: Omega$_{31}$-SIPr ($R^1 = i$-Pr, $R^2 = H$)	921194-52-7 1227263-67-3	Omega Cat System
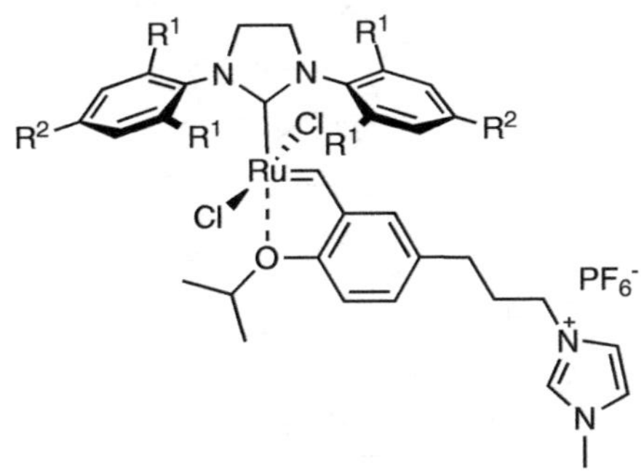 **51, 52**	**51**: Omega$_{32}$-SIMes ($R^1 = R^2 = Me$) **52**: Omega$_{32}$-SIPr ($R^1 = i$-Pr, $R^2 = H$)	805260-26-8 1227263-65-1	Omega Cat System

(*continued*)

TABLE 9-1. (*Continued*)

Catalyst	Name/Description	CAS Number	Supplier
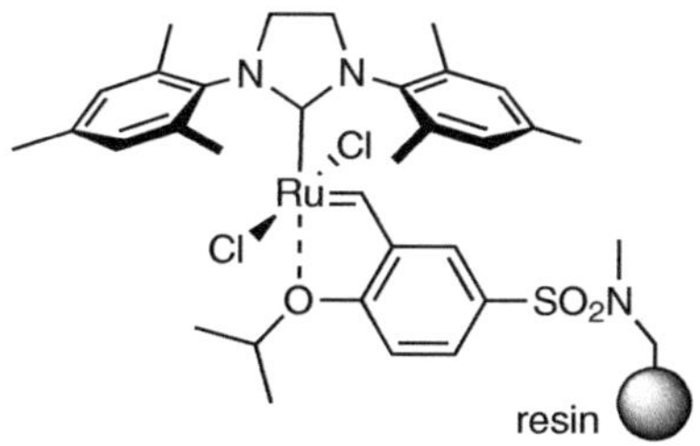 **53**	Zhan catalyst II	–	Strem, Zannan Pharma
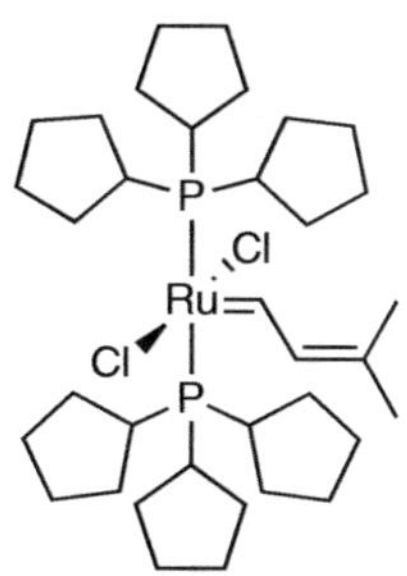 **54**	Dichloro(3-methyl-2-butenylidene) bis (tricyclopentylphosphine)ruthenium(II)	220883-08-9	Aldrich

[a]According to an Internet search, conducted in February 2012.

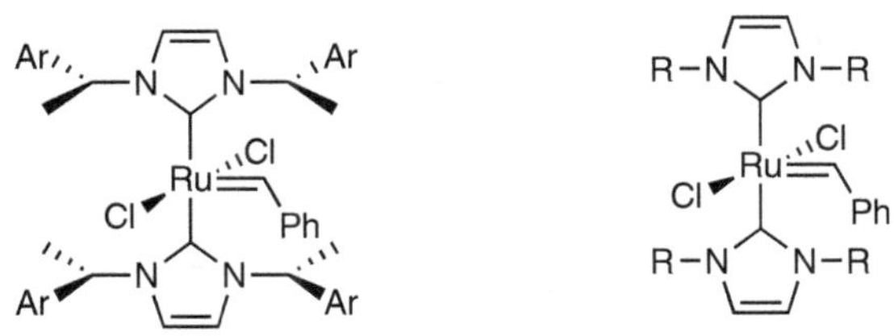

(*R*, *R*)-**55a** Ar = Ph
(*R*, *R*)-**55b** Ar = naphthyl

56a R = Cy
56b R = *i*-Pr

FIGURE 9-1. Herrmann's bis-NHC Ru complexes.

moisture, improved thermal stability, and functional group tolerance. Nowadays, complex **5** is one of the most often applied metathesis catalysts; it is highly active in benchmark RCM and ROMP reactions and enables RCM of tri- and tetrasubstituted olefins [25]. It has been found to be the most efficient in the ene–yne cycloisomerization of challenging substrates [28]. Moreover, the first example of a cross metathesis (CM) leading to a trisubstituted olefin was demonstrated by the use of **5** [29]. Its wide applicability was also proven in the syntheses of natural products and potential pharmaceuticals [30]. The general differences in reactivity between bis(phosphine) complexes (first generation) and mixed NHC/phosphine complexes have been thoroughly investigated. Despite the slower phosphine dissociation ("initiation") from Grubbs' second-generation catalysts compared to the first-generation catalyst, **4** is mostly outperformed by NHC-containing catalysts in more challenging metathesis reactions. Ru-3-phenylindenylidene complexes [31], such as the so-called M2 catalyst **10**, were developed as an alternative to benzylidene complexes, exhibiting even higher thermal stability and longevity [32]. On the other hand, **10** and similar catalysts are slow to initiate and exhibit low activity at ambient temperature. For instance, in RCM reactions leading to di- and trisubstituted alkenes, only low conversions can be observed even after prolonged reaction times (days). At elevated temperatures (70 °C), however, they are comparable with other state-of-the-art catalysts in most RCM and CM model reactions. A further crucial discovery in the field of olefin metathesis catalysts was made in 1998 by Hoveyda et al., that complex **12** (Hoveyda (I)), in which the benzylidene moiety is replaced by a styrenyl ether moiety, shows remarkable stability [33]. The corresponding derivative **14**, known today as Hoveyda or Hoveyda–Grubbs (II) catalyst, was independently reported by Blechert and coworkers [34] and Hoveyda [35]. The SIMes-substituted complex shows high activity in the CM of electron-deficient [36] as well as demanding alkenes, simultaneously remaining most reliable in the RCM at tetrasubstituted alkenes. Being remarkably stable toward moisture and air, it offers the possibility for immobilization and catalyst recovery. The outstanding performance of the Hoveyda–Grubbs catalysts enables it to engage in the so-called release-return (or "boomerang") mechanism (Scheme 9-4).

It was assumed that the initiation of Grubbs-type and Hoveyda–Grubbs-type precatalysts leads to the active 14-electron species **C** (release) [37]. After a few catalytic cycles, the precatalyst may be regenerated by the reaction with *o*-isopropoxystyrene (return). The release-return mechanism was supported by results of deuterium-exchange experiments, and also by the fact that significant amounts of precatalyst can be recovered after the reaction. However, Plenio et al. [38] recently questioned the importance of the described

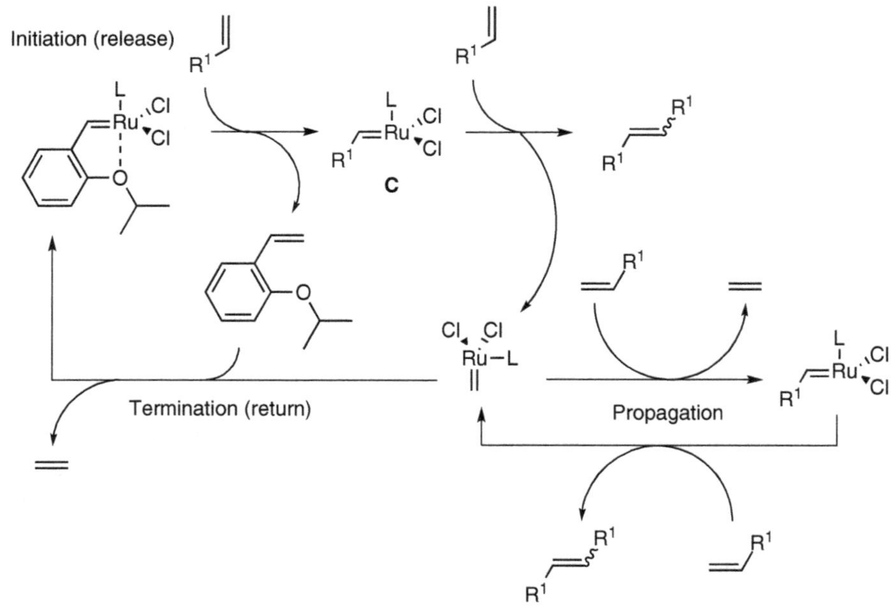

SCHEME 9-4. Suggested boomerang mechanism of a Hoveyda–Grubbs-type catalyst.

mechanism, as they found no evidence for a release-return pathway when conducting olefin metathesis with a fluorophore-tagged Hoveyda–Grubbs (II)-type catalyst. Despite these uncertainties, the Hoveyda–Grubbs (II) catalyst has been widely applied in the construction of complex molecules [39].

Given that the formation of the active 14-electron species involves the cleavage of the Ru· · ·O chelation, it was reasonably assumed that weakening this bond, by the introduction of electron-withdrawing groups to the aromatic ring of the styrene moiety, may lead to faster initiation. In 2002, Grela et al. reported the synthesis of *nitro*-substituted Hoveyda–Grubbs (II)-type catalyst **15** [40], which showed high reactivity toward different challenging CM and RCM substrates and outperforms other state-of-the-art catalysts in straightforward RCM reactions. Consequently, **15** has been successfully applied in the construction of natural products, pharmaceuticals, and organic materials [41].

9.2 RECOVERY AND REUSE OF RU-BASED METATHESIS CATALYSTS: THE ACADEMICS' VIEW

The removal of heavy-metal impurities is an important task in pharmaceutical and fine chemical production [42], and it remains rather challenging, since the acceptable ruthenium content should not exceed 10 ppm in active pharmaceutical ingredients (APIs) [43]. Furthermore, the applicability and the cost-efficiency of the purification method are also of high importance.

On laboratory scale, the most common way to purify metathesis products is column chromatography (CC). However, protocols that include silica gel CC, or even filtration through silica, are expensive and difficult to implement on an industrial scale. Moreover, in some cases, CC may not be sufficient if product and ruthenium residues are of similar polarity. Even if purification via CC turns out to be successful, the remaining concentration of ruthenium in the product is generally too high for pharmaceutical applications (around 100 ppm). Consequently, other purification methods have to be sought, if a Ru-catalyzed metathesis protocol is to be utilized in industry. Various industrial chemists and academic research groups have worked on this issue and have presented several solutions for the purification of products and the recovery of the catalysts. Within this section, recent results are presented and compared with earlier work.

Some published strategies for removing residues of commercially available Ru complexes are shown in Table 9-2. One of them is the double purification of the products on silica gel, combined with activated carbon treatment. With this method, the ruthenium contamination can be decreased to levels of 60 ppm (Table 9-2, entry 4) [44]. Another method is the utilization of ruthenium scavengers such as dimethyl sulfoxide or triphenylphosphine oxide, followed by filtration through silica gel, which led to reduced Ru levels of 240 and 360 ppm, respectively [45]. Similar levels of ruthenium (240 ppm) were obtained by using lead tetraacetate as scavenger [46], but because of the high oxidizing power and toxicity of this reagent, the usage of this method is limited [47]. Grubbs et al. reported that stirring crude product with a water-soluble tris(hydroxymethyl)phosphine and silica gel, followed by filtration, reduced the level of ruthenium to 200 ppm [48]. However, this process required 89 equiv. of the phosphine scavenger and 2 equiv. of triethylamine, and for large-scale application the mixture has to be stirred at least for 12 h [49]. Another method is a sequestration of the ruthenium contamination from the product by Fe_xO_y and 3-mercaptopropionic acid, which provided a 91% decrease of contamination (Table 9-2, entry 9) [50]. Recently, special polymers have been developed and commercialized for the removal of heavy metals [51]. Treatment of the metathesis product with amine-modified silica lowered ruthenium contamination to <2000 ppm (Table 9-2, entry 10) [52]. Recently, Diver proposed a simple method by using a polar isocyanide, which decreased the ruthenium contamination to 120 ppm (Table 9-2, entry 11) [53]. In all cases, a large excess of the scavenger (around 80 equiv.) is necessary to obtain low levels of ruthenium in the product. Hence, although scavenging methods give valid results, they are still not practical, from environmental and economic points of view.

TABLE 9-2. Purification Methods for Metathesis Products

Entry	Method	Ru Residue (ppm)	References
1	None	14,316	[44,48]
2	Column chromatography (CC)	1912	[44]
3	CC purification + activated carbon	578	[44]
4	Double CC purification + activated carbon	60	[44]
5	Dimethyl sulfoxide	240	[45]
6	Triphenylphosphine oxide	360	[45]
7	Lead tetraacetate	240	[46,47]
8	Tris(hydroxymethyl)phosphine and silica gel	200	[48,49]
9	Fe_xO_y and 3-mercaptopropionic acid	1250	[50]
10	Amine-modified silica	2000	[51,52]
11	Isocyanide	120	[53]

Large-scale transformations under supercritical (sc) conditions have been used in industry for most of the twentieth century [54], and the application of sc fluids in catalytic reactions is slowly emerging [55]. Usage of sc fluids as reaction media has the potential to solve the problem of catalyst recovery. Having verified that CO_2 did not participate in the reaction, $scCO_2$ was first used as a reaction medium for the metathesis reaction in 1996 by DeSimone [56], and since then, $scCO_2$ has been applied in CM [57], RCM [58], and ROMP [57b,59]. Unfortunately, in all cases, the issues of residual Ru levels and catalytic recycling were not addressed by the authors. Recently, it was reported that RCM products could be separated from Ru contaminants by extraction with $scCO_2$ [60]. Also catalysts immobilized on solid support were used in RCM in $scCO_2$ [61], an approach that allowed an efficient recovery of the catalyst up to six cycles, without loss of activity. More remarkably, only a low level of ruthenium leaching was observed (18–21 ppm).

To enhance its capability of being recycled, modifications of a basic state-of-the-art metathesis catalyst were undertaken. Figure 9-2 shows the number of possibilities for attaching different functionalities to the ruthenium complex.

Tags can be connected through nonlabile ionic ligands and NHCs, or through labile ones, for example, benzylidene moiety, phosphine-, or pyridine-based ligands. Ru catalysts have been bound to different tags, including: (i) insoluble polymers; (ii) inorganic materials (e.g., silica gel); (iii) ionic liquid (IL) functionalities; (iv) perfluorinated hydrocarbons; (v) soluble polymers or small-molecule functionalities. In an ideal case, modified catalysts can be easily retrieved from reaction mixture by either filtration (nanofiltration) when considering (i), (ii), and (v) above or extraction into IL in case (iii), fluorous phase in case (iv), and nonpolar solvent in case (v).

The immobilization of ruthenium catalysts on different resins or silica was first proposed as a modification aiming to facilitate purification, but this also led to an improvement of the catalyst stability, simultaneously preventing undesirable bimolecular decomposition pathways by inhibition of intermolecular interaction [62]. This field has been very well described in a recent series of reviews [43a,63]. Examples are summarized in Table 9-3.

Solid support has been attached to the catalysts through all possible positions described above in Figure 9-2. Unfortunately, in most cases, catalysts have lower catalytic activity compared to homogeneous analogs. Nevertheless, **57** could be reused up to three times [67,70]. When the solid support was attached through phosphine ligands, a high decrease of catalytic activity was observed. With **58** (Table 9-3, entry 2), CM is 150 times slower, compared to its homogeneous analog [65]. The monolithic catalyst **59** (entry 3) efficiently

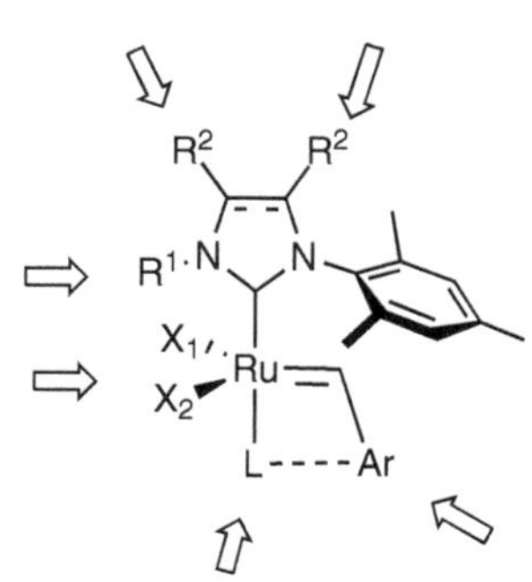

FIGURE 9-2. Possible positions for tagging.

TABLE 9-3. Solid-Supported Ruthenium Catalysts[a]

Entry	Catalyst	Substrate, Conditions	Cycles	Conversion (%)	Ru Residue (ppm)	Reference
1	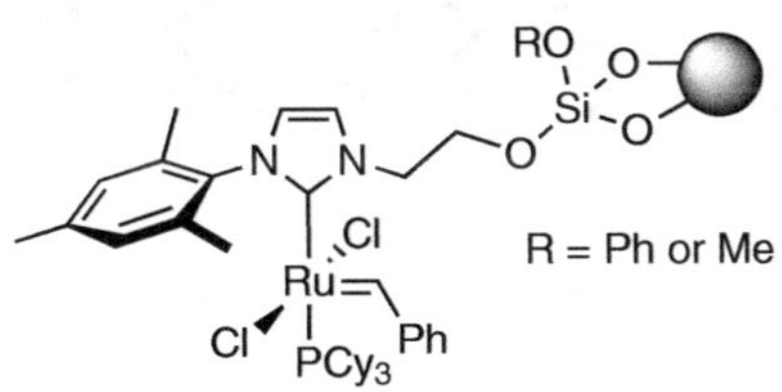 **57**	DEDAM, 5 mol% Ru, CH_2Cl_2, 40 °C, 24 h, then filtration	3	98	250	[64]
2	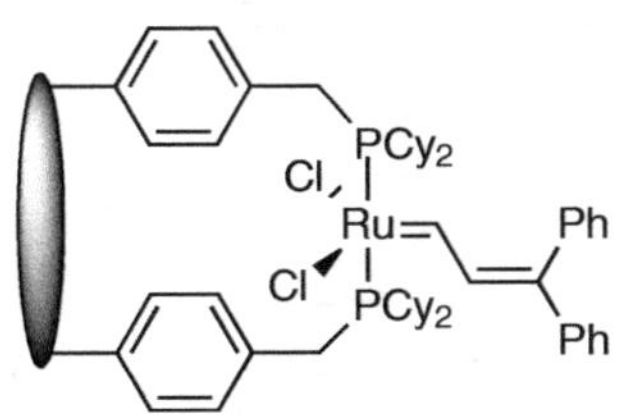 **58**	*cis*-2-Pentene, 1 mol% Ru, C_6D_6, rt, 10 days, filtration	1 2 3	100 80 60	No data	[65]

(*continued*)

TABLE 9-3. (*Continued*)

Entry	Catalyst	Substrate, Conditions	Cycles	Conversion (%)	Ru Residue (ppm)	Reference
3	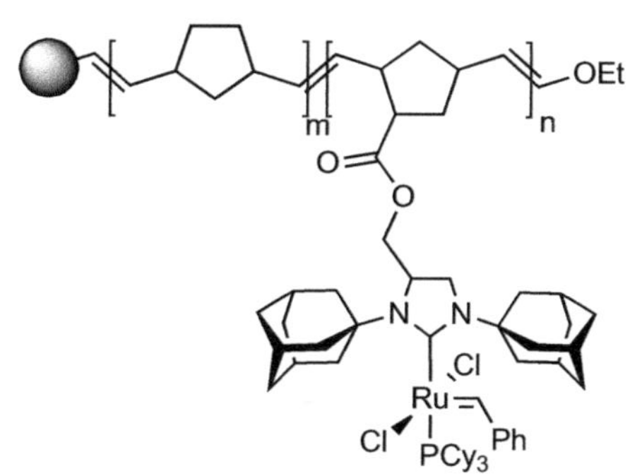 **59**	DEDAM, cartridge with Ru, 56 °C			70	[66]
		50 min	1	100		
		5 min	2	92		
4	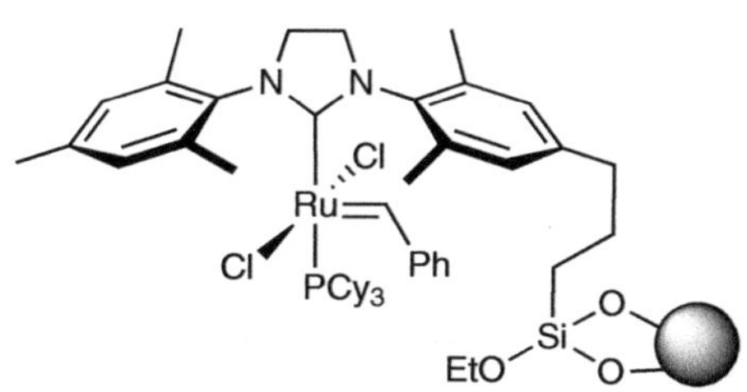 **60**	DEDAM, 0.75 mol% Ru, C_6H_6, 30 °C, 2 h	8	60–82	5×10^{-3}	[67]

5	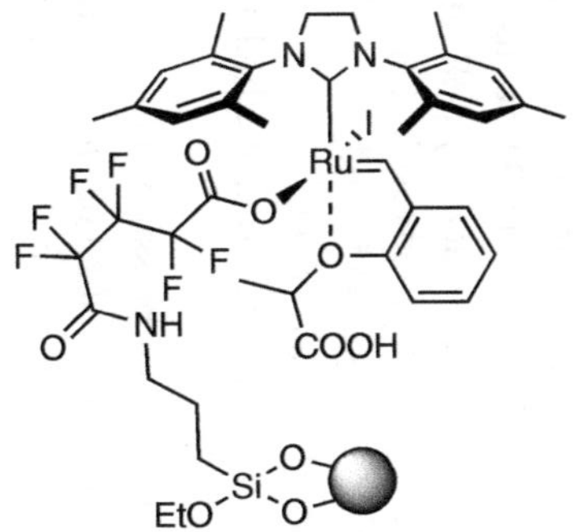 **61**	DEDAM, 0.2 mol% Ru, PhMe, 60 °C, 1 h	1	99.9	17	[68]
			2	92		
6	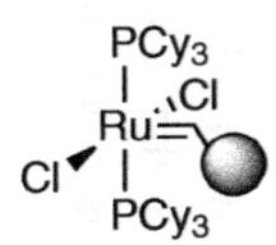 **62**	DEDAM, 3 mol% Ru, CH_2Cl_2, rt, 40 min	1	>98		[69]
		Filtration			500	
		Filtration + CC			55	

[a]DEDAM, diallyl diethyl malonate.

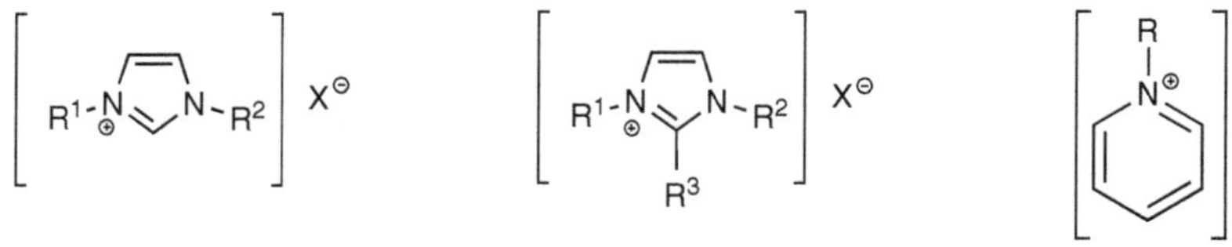

[emim] = R^1= Me, R^2= Et
[bmim]= R^1= Me, R^2= *n*-Bu
[hmim] = R^1= Me, R^2= *n*-Hex
[omim] = R^1= Me, R^2= *n*-Oct

[bdmim]= R^1 = Me, R^2 = *n*-Bu, R^3= Me

X= BF_4, PF_6, OAc,TfO, NTf_2

FIGURE 9-3. IL with possible counterions.

promoted RCM and can be deployed as a fixed bed in a cartridge for applications in combinatorial or flow chemistry; the ruthenium contamination in the product was found to be approximately 70 ppm [66]. On the other hand, **60** (entry 4) can be recycled multiple times very efficiently without leaching ruthenium; ICP-MS analysis of the reaction solution shows <5 ppb of ruthenium contamination after each cycle [67]. Tagging through an anionic ligand was developed by Mol [71] and optimized by Buchmeiser [72]. This approach was also employed by groups of Grela and Hynek, who showed that **61** (entry 5) can be recycled in only two cycles, but with ruthenium contamination level in the product was low as 17 ppm, while the activity of **61** was comparable to its homogenous analog [68,73]. Tagging through the alkylidene ligand was first reported by Barrett and coworkers **62** (entry 6) [69]. There are various catalysts based on this strategy synthesized by Nolan [74], Blechert [37a,75], Hoveyda [76], Dowden [77], Grela [78], and Yao [79]. This type of catalyst could be reused for up to 20 cycles without a significant loss of activity, but the ruthenium contamination in the product is high in most cases (above 300 ppm). An alternative purification and recycling strategy exists, employing magnetic nanoparticles. Examples are summarized in Table 9-4.

The simplicity of the separation by applying a magnetic field led to increased attention to this research field [82]. The use of magnetically retrievable metathesis catalysts was reported independently by Zhu [81] and Che [80]. Notably, **63** (Table 9-4, entry 1) was used in RCM reactions and was recycled up to 22 cycles with almost no loss of activity [80]. On the other hand, **64** (entry 2) was used in oleo metathesis reactions with good results [81]. It could be recycled up to five cycles with TOFs of 488, 484, 487, 485, and 486 h^{-1}, respectively, and the Ru content found in the product was as low as 3 ppm.

Utilizing ILs as solvent in metathesis reactions represents an alternative method to simplify the purification of the product (Figure 9-3) [55d,83].

ILs are nonflammable, recoverable solvents with negligible vapor pressure and can be used as "green" alternatives for conventional organic solvents [84]. The first usage of an IL was in a biphasic system, for metathesis reactions with **65** (Table 9-5, entry 1), reported by Dixneuf and coworkers [85]. To improve the solubility in the IL, catalysts containing an ionic moiety have been designed by the groups of Mauduit (**66**) [86] and Yao (**67**) [87]. These systems exhibit excellent reactivity, and the catalyst could be recovered, up to 10 cycles in a biphasic CH_2Cl_2/[bmim]PF_6 medium. In the case of **66** (entry 2), very low ruthenium contamination was detected in the product; up to 22 ppm was detected after eight cycles.

TABLE 9-4. Catalysts Immobilized Onto Magnetic Nanoparticles

Entry	Catalyst	Substrate, Conditions	Cycles	Conversion (%)	Ru Residue (ppm)	Reference
1	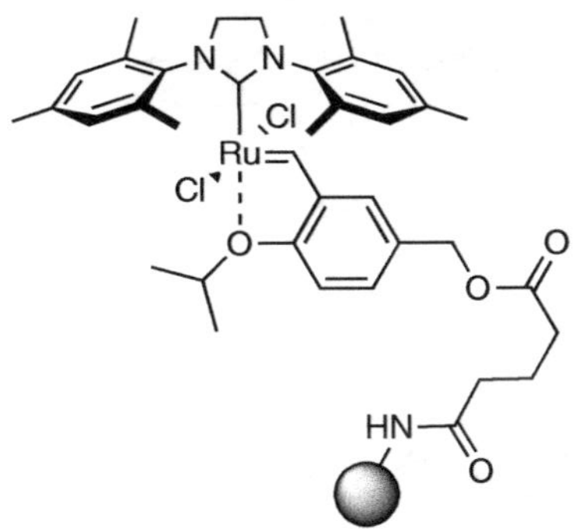 **63**	DATA[a], 2.5 mol% Ru, CH_2Cl_2, 40 °C, 2 h (1–13)[b], 4 h (14–18)[b], 10 h (19–22)[b]	22	>95–98	No data	[80]
2	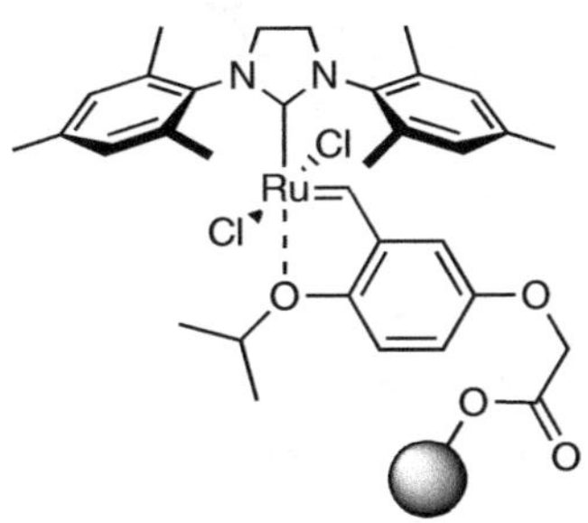 **64**	Methyl oleate, 0.05 mol% Ru, neat, 50 °C, 3 h	5	75	3	[81]

[a]DATA, *N,N*-diallyltosylamine.
[b]Values in parentheses refer to cycle numbers.

TABLE 9-5. Ionic Tagged Catalysts[a]

Entry	Catalyst	Substrate, Conditions	Cycles	Conversion (%)	Ru Residue (ppm)	Reference
1	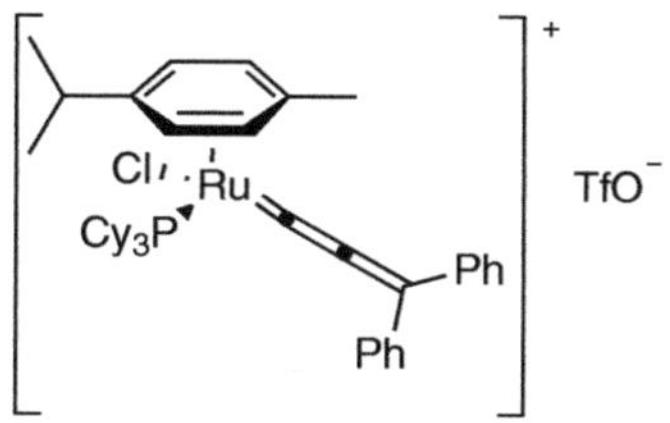 **65**	Norbornene, 0.3 mol% Ru, 40 °C $[bdmim][PF_6]$, 2 h	2	61–98	No data	[85]
		$[bdmim][PF_6]$/PhMe, 0.5 h	4	87–98		
2	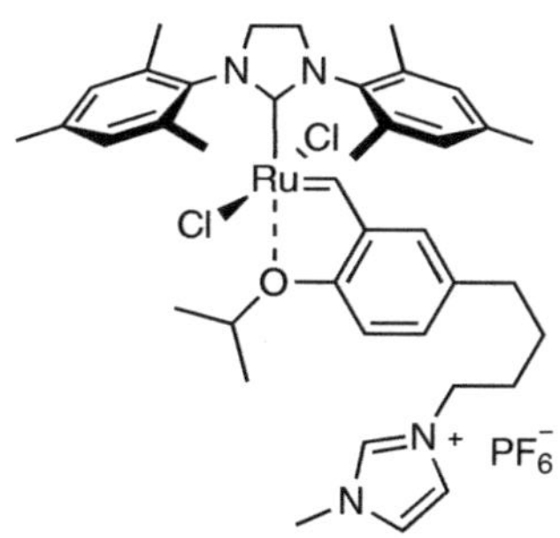 **66**	Batch DATA, 2.5 mol% Ru, $[bmim][PF_6]$, 60 °C, 45 min, extraction with toluene	10	95	1.2–22	[86]
		SILP-SCF[b] system methyl oleate, 0.02% Ru, CO_2, 50 °C, 3 h, 0.05 $cm^3\ min^{-1}$	cont.	67	8	[88]
3	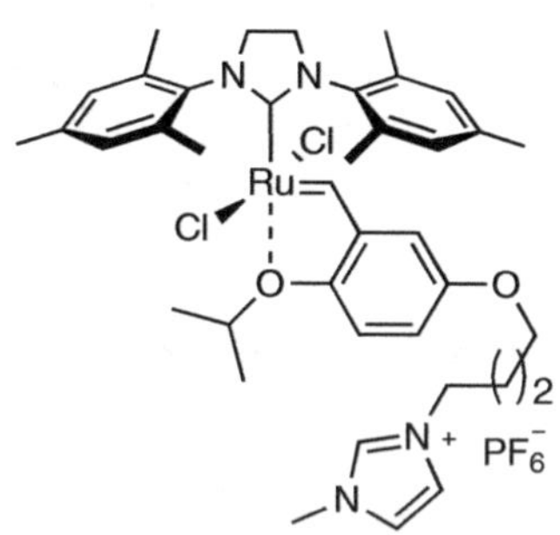 **67**	DEDAM, 1 mol% Ru, [bmim] $[PF_6]$/ CH_2Cl_2, 45 °C, 1 h	10	90–98	No data	[87]

4	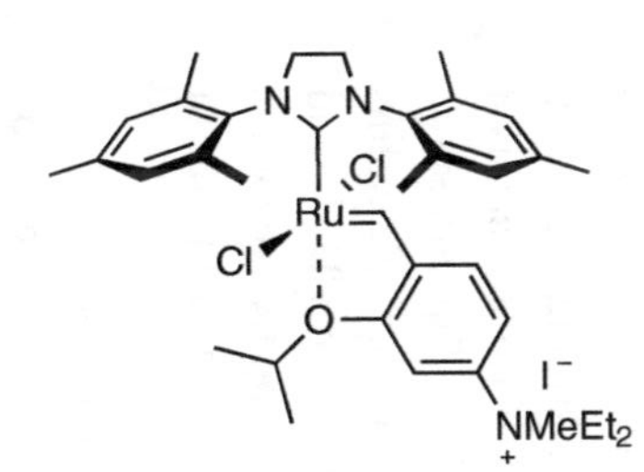 **68**	Batch DEDAM, 5 mol% Ru, CH_2Cl_2, rt, 60 min	1	98	12	[89]
		Raschig rings, AMATA, 5 mol% Ru, CH_2Cl_2, rt, 24 h	6	56–85	100	
		PASS*flow* AMATA, 5 mol% Ru, CH_2Cl_2, 40 °C, 5 h	2	79–100	No data	
5	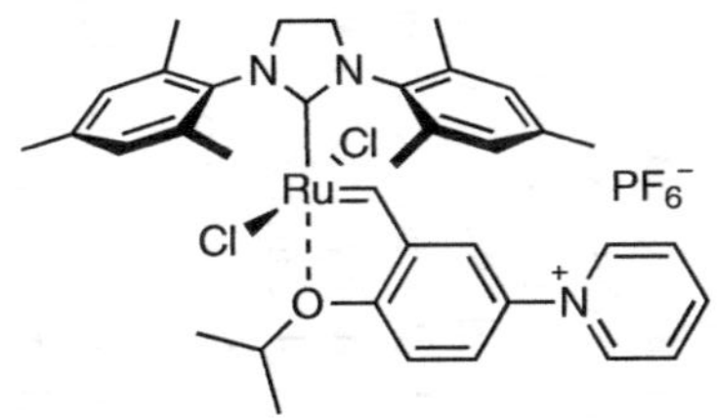 **69**	AMATA, 5 mol% Ru, [bmim][PF_6]/ PhMe, 25 °C, 3 h	3	65–89	No data	[90a]
6	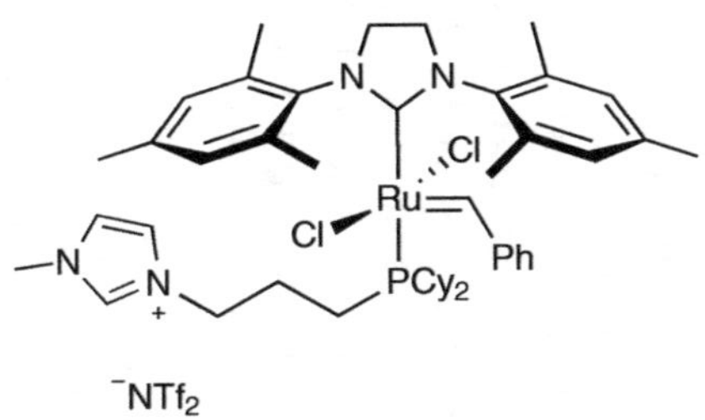 **70**	1.7 octadiene, 0.25 mol% Ru, [bmim][PF_6]/PhMe, 45 °C, 40–90 min	8	88–98	<2	[90b]

[a]DEDAM, diallyl diethyl malonate; DATA, *N,N*-diallyltosylamine, AMATA, 2-allyl-2-methylallyltosylamine.
[b]SILP–SCF system, supported IL phase with compressed CO_2 as the flowing medium.

The groups of Grela [89] and Mauduit [90a] reported a new concept of ionic tagged ruthenium catalysts, whereby an electron-withdrawing group was attached to the benzylidene ligand; these pyridinium and ammonium moieties are responsible for an electronic activation of the catalysts, for their high solubility in the ionic phase. Leaching of ruthenium was very low, and after simple filtration through a pad of silica, Ru contents of 12–59 ppm for **68** [89a,89b] and 25–173 ppm for **69** [90a] were determined. However, these catalysts are efficient only in the first cycles, with significant loss of activity in subsequent runs. This type of catalyst **68** (Table 9-5, entry 4) was also immobilized onto ionic resins [89b,91]. Catalysts attached onto Raschig rings can be reused six times without a significant loss of activity and leaching (21–102 ppm). Reactions under continuous-flow conditions were investigated, using a catalyst attached onto an ionic polymer inside a PASS*flow* reactor [92]. The reactor can be reused twice, but no product was observed in the third run [93]. A similar concept, with **68** supported in an IL (entry 2), was used in the continuous-flow homogeneous alkene metathesis with CO_2 as a transport vector [88]. This system was used in RCM and CM reactions, with only slight loss of activity and leaching of the ruthenium (1–15 ppm). In 2008, Consorti and Dupont presented an ionophilic Ru complex **70** (entry 6), with an IL tagged to a phosphine ligand [90b]. This approach enabled efficient recovery and reuse of the catalyst up to eight cycles, with <2 ppm of Ru contamination in the product at 0.25 mol% loading. Unfortunately, the catalyst is suitable only for simple substrates.

A similar purification method utilizes fluorous biphasic chemistry [55d,83,94]. This approach has become popular because of its potential as a new paradigm of green chemistry [95]. Purification of the desired product can be achieved either via filtration through a short pad of fluorous-phase silica or by fluorous-phase extraction. Only a few fluorous-tagged catalysts have so far been prepared, including **71** (Table 9-6, entry 1) among others [96]. The supported catalyst is remarkably active and very stable and can be reused up to 20 cycles without loss of activity. Unfortunately, no Ru contamination data are available. **72** (entry 2) synthesized by Curran [97] is also very active and can be reused with up to seven cycles. In each cycle, only 65–88% of the catalyst could be recovered, so after seventh cycle, only 10% of the original catalyst was retrieved. Ru contamination in the product was estimated at around 500 ppm.

Catalyst **73** (entry 3) could be recycled up to three times without significant loss of activity [98]. It could be used under both monophasic and biphasic conditions, although the catalyst is much more active under the latter conditions. Interestingly, a simple extraction by fluorous/organic liquid was used as a recycling protocol, but the recovery of the catalyst was not quantitative (first cycle: 90%, second: 85%, and third: 57%). Subsequently, Matsugi proposed a few other fluorous-tagged catalysts [99], where the fluorous side chains are connected directly to the alkylidene ligand. By performing reactions in $CDCl_3$ and CH_2Cl_2, catalyst **74** (entry 4) was active and stable and could be recycled up to five cycles with sustained activity (95–100% conversion). After the fifth cycle, 84% of the catalyst can be recovered.

A similar purification strategy was used by Bazzi and Bergbreiter [100], who presented a catalyst attached to a heptane-soluble polymer, which could be easily recovered by simple liquid/liquid or liquid/solid separation. Based on previous work [101], nonpolar catalyst **75** was prepared [100c] (Table 9-7, entry 1). This particular catalyst is very active and stable, could be recycled up to five cycles without significant loss of activity, and the Ru contamination in the product was low, ca. 20 ppm. The reaction was carried out in heptane and then the product was extracted with acetonitrile. Within catalyst **76**, a nonpolar

TABLE 9-6. Fluorous-Tagged Complexes[a]

Entry	Catalyst	Substrate, Conditions	Cycles	Conversion (%)	Ru Residue (ppm)	Reference
1	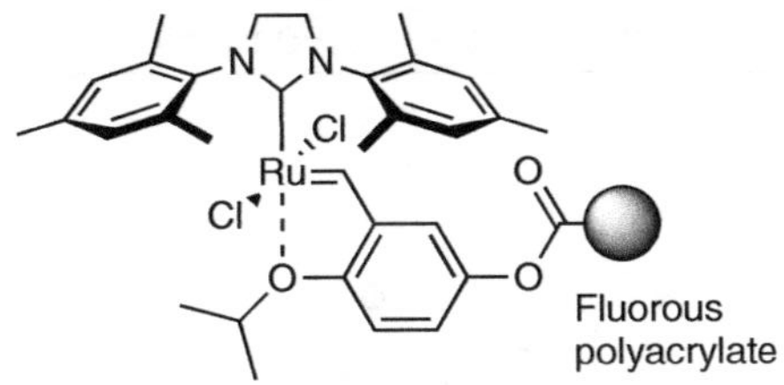 **71**	DATA, 1 mol% Ru, $PhCF_3/CH_2Cl_2$, 50 °C, 2 h	20	95–98	No data	[96]
2	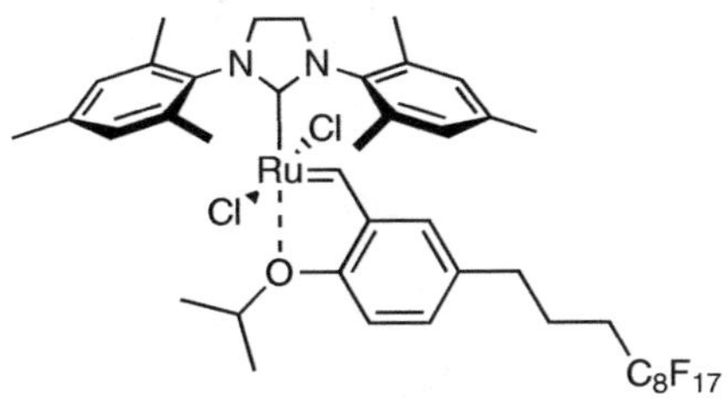 **72**	DATA, 5 mol% Ru, CH_2Cl_2, 55 °C, 2 h, fluorous solid-phase extraction	7	96–98	500	[97]

(continued)

TABLE 9-6. (*Continued*)

Entry	Catalyst	Substrate, Conditions	Cycles	Conversion (%)	Ru Residue (ppm)	Reference
3	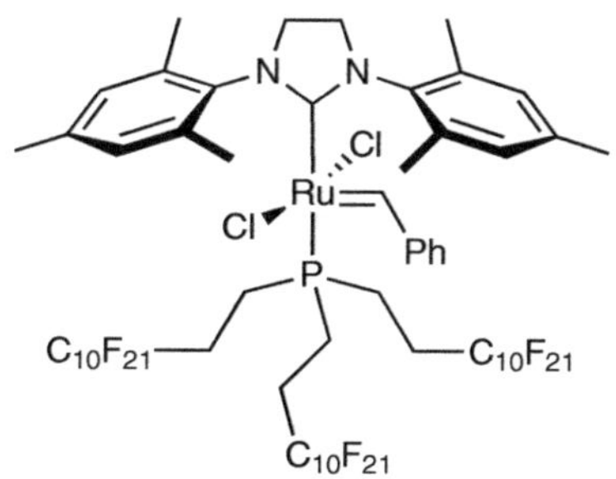 **73**	DEDAM, 2.6 mol% Ru, $CH_2Cl_2/CF_3C_6F_{11}$, rt, extraction	1 2 3	79 78 79	246 369 1157	[98]
4	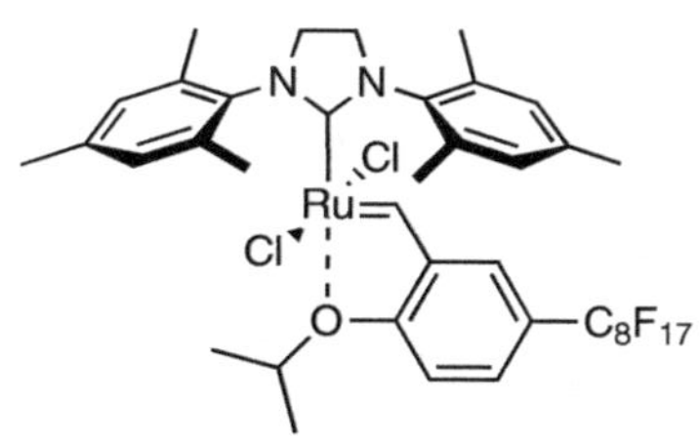 **74**	DEDAM, 5 mol% Ru, CH_2Cl_2, rt, 1 h, FSPE[b]	5	82–100	No data	[99]

[a]DATA, *N,N*-diallyltosylamine; DEDAM, diallyl diethyl malonate.
[b]FSPE, fluorous solid-phase extraction.

TABLE 9-7. Ruthenium Catalysts with a Nonpolar Tag[a]

Entry	Catalyst	Substrate, Conditions	Cycles	Conversion (%)	Ru Residue (ppm)	Reference
1	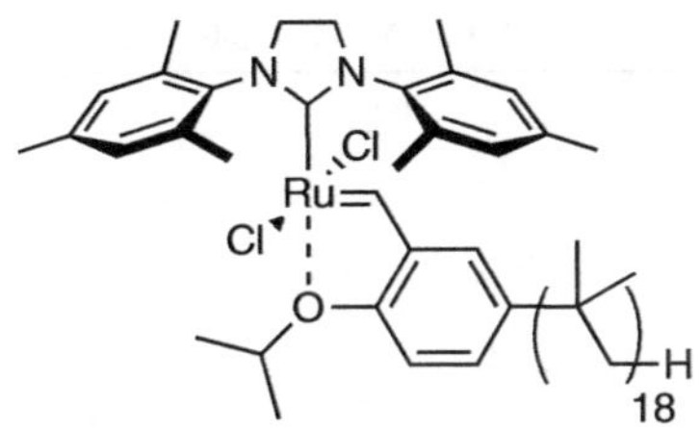 **75**	DEDAM, 5 mol% Ru, heptane, rt, 1 h	5	66–99	20	[100c]
2	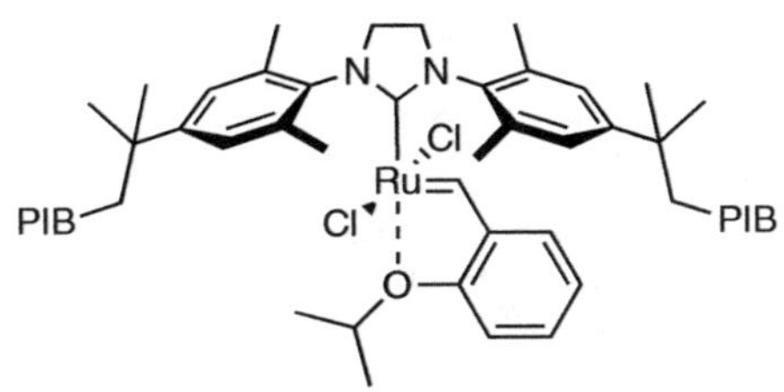 **76**	DATA, 5 mol% Ru, heptane, rt, 2 h	20	72–99	17.6–30	[100b]

[a]DATA, *N,N*-diallyltosylamine; DEDAM, diallyl diethyl malonate; PIB, polyisobutylene.

TABLE 9-8. Temperature-Dependent Catalysts[a]

Entry	Catalyst	Conditions	Cycles	Conversion (%)	Ru Residue (ppm)	Reference
1	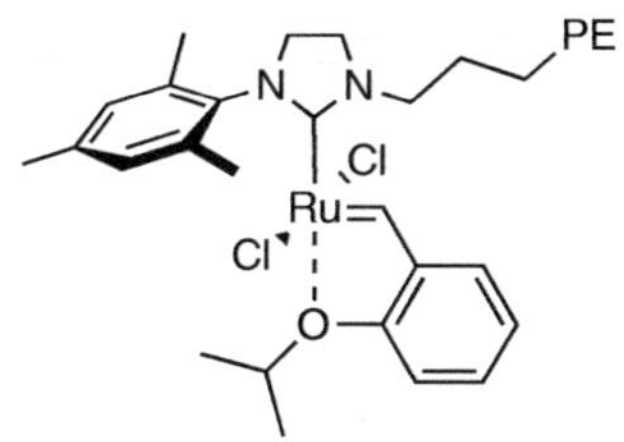 **77**	DEDAM, 5 mol% Ru, PhMe, 65 °C, 5 min	5	92–98	77	[103]
2	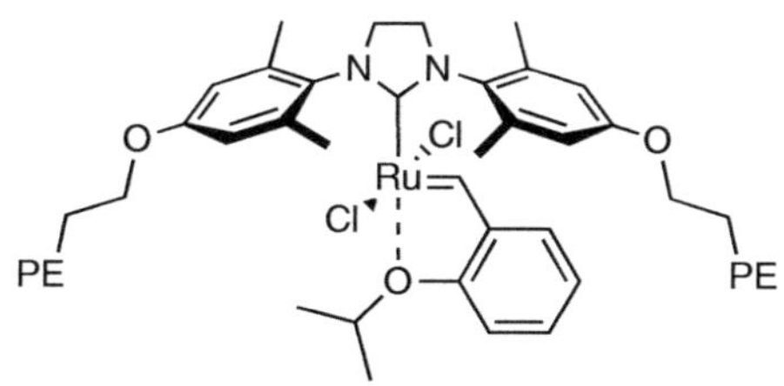 **78**	AMATA, 2,5 mol% Ru, PhMe, 65 °C, 5 min	8	81–99	No data	[103]

[a]DEDAM, diallyl diethyl malonate; AMATA, 2-allyl-2-methylallyltosylamine; PE, polyethylene oligomer.

tag (polyisobutylene, PIB, entry 2) was attached to the NHC ligand [102b]. This approach led to a 10-fold reduction in Ru contamination in comparison to **75** (0.37–0.63% of the starting Ru), and the catalyst could be recycled up to 20 times.

An approach proposed by Bazzi and Bergbreiter utilizes the idea of attaching a recently described [102] thermomorphic polyethylene (PE) tag to the Ru complex [103]. Recovery of the catalyst is very simple: decreasing the temperature after reaction led to precipitation of the catalyst. **77** (Table 9.8, entry 1) could be recovered up to five times with sustained activity (92–98 %) and **78** even up to eight cycles. Ru contamination in the product was as low as 77 ppm.

The most recent method to purify postmetathesis reaction mixtures utilizes nanofiltration. The uses of membranes in separation processes are well known [104] and are used on the industrial scale in aqueous media [105] for water treatment and in the food industry [106]. Organic solvent nanofiltration (OSNF) is one of the most promising methods, allowing separation of organic mixtures at the molecular level [55d,107]. Separation of transition-metal complexes by membranes was described in the early 1970s [108]. OSNF has already been used to recycle different types of catalysts, such as Co-Jacobsen catalyst (four cycles) [109], Suzuki catalyst (up to 10 cycles) [110], hydrogenation catalyst [111], and many others [112]. The type of membranes used in OSNF can be divided into polymeric (organic) [113] and ceramic (inorganic) membranes [114], both of which are commercially available in a range of molecular weight cut-off (MWCO). MWCO is defined as the molecular weight at which 90% of the solutes are retained by the membrane, but in many cases reaction conditions, such as temperature and solvent, have a great influence on the pore size [115]. Modifications of the catalyst include attaching a soluble molecular weight enlargement (MWE) unit, which can be soluble polymers, hyperbranched polymers, dendronized polymers, dendrimers, dendrones, or polyhedral oligomeric silsesquioxanes (POSSs) [106a,116]. OSNF was first used in a metathesis reaction by van Koten [117], who reported the performance of an immobilized catalyst in the RCM reaction of DEDAM, carried out in an NF reactor (SelRO nanofiltration membrane MPS-60; MWCO = 400). At 20% conversion, the catalyst was found to decompose, forming a black precipitate in the vessel. It was initially thought that decomposition of **79** (Table 9-9, entry 1) was facilitated by the membrane surface, even though leaching of the catalyst through membrane did not apparently occur (the permeate was clear and colorless). Later, the experiment was repeated by Vosloo, who concluded that low conversion and problems with reusing **79** were most probably due to the short lifetime of the catalyst, instead of the lack of stability on the membrane [118].

Rabiller-Baudry et al. prepared a variety of catalysts of different molecular weights, ranging from 625 to 2195 $g\,mol^{-1}$ [119], in which the MWE unit was attached to the benzylidene ligand of the standard Hoveyda–Grubbs' (II) catalyst. Initial nanofiltration tests showed that Ru fouling increased slightly with the catalyst's molecular mass. This result can be explained by the adsorption of the heaviest molecules on the membrane surface [122]. Further RCM experiments with catalyst **80** showed that dimethyl carbonate (DMC) is a much better solvent for the RCM reaction than toluene, as the complex was more stable and, subsequently, could be recycled five times without compromising catalytic activity, compared to three cycles in toluene (Table 9-9 entry 2). An MWE attached to the NHC ligand was recently described by Plenio [120]. Catalysts tagged in this way turned out to be more easily recovered than their alkylidene-tag counterparts. Complex **81** has been utilized in RCM and CM reactions, and 4 ppm Ru content was detected after nanofiltration. Moreover, RCM of DEDAM in a continuous

TABLE 9-9. Ruthenium Catalysts Used in Nanofiltration[a]

Entry	Catalyst	Conditions	Cycles	Conversion (%)	Ru Residue	Reference
1	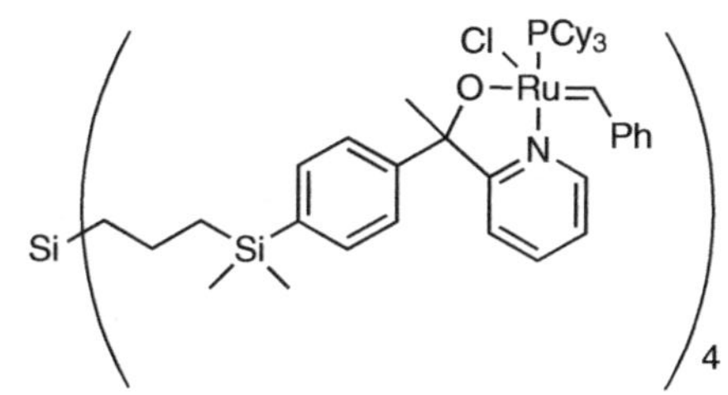 **79**, MW = 3245 g mol^{-1}	DEDAM, 1 mol% Ru, DCE, 80 °C, SelRo membrane MPS60 MWCO = 400	1	20	No data	[117]
2	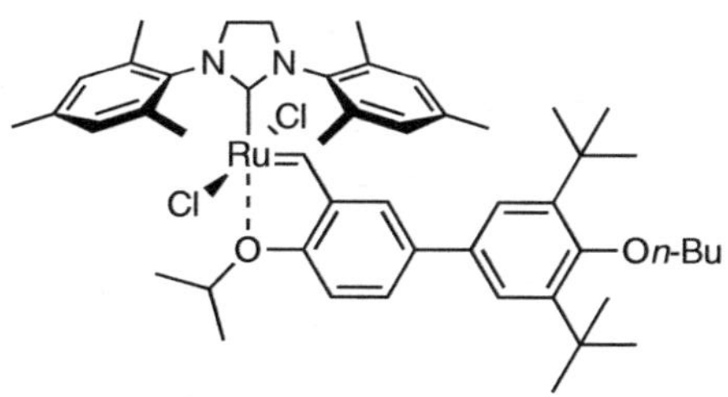 **80**, MW = 887 g mol^{-1}	DATA, 2,5 mol% Ru, DMC, rt Starmem 228	5	98–100	220	[119]

3	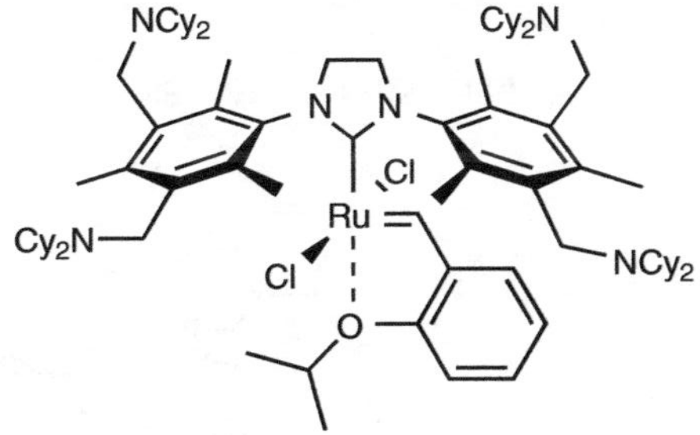 **81**, MW = 1401 g mol^{-1}	Batch DEDAM, 0.1 mol% Ru, CH_2Cl_2, 40 °C, 5 min PDMS membrane	1	100	4	[120]
		Continuous 500 min PDMS membrane	Cont.	37	7	
4	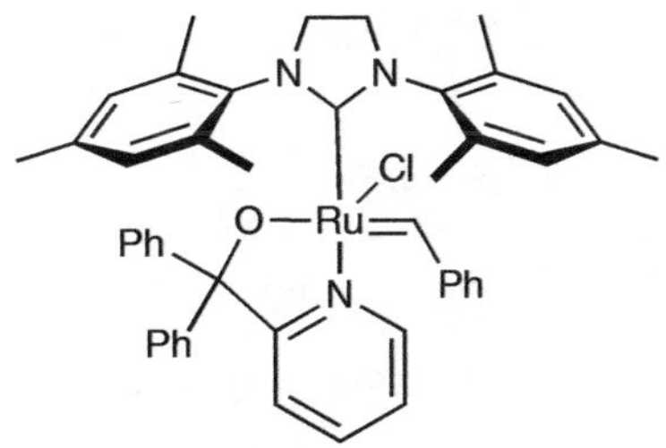 **82**, MW = 794 g mol^{-1}	1-Octene, 0.014 mol% Ru, neat, 80 °C, 400 min, Starmem 228	3	73	9	[118]
5	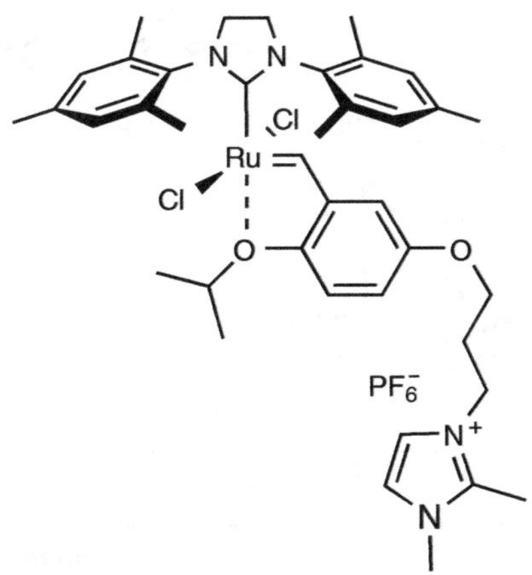 **83**, MW = 926 g mol^{-1}	DATA, 2.5 mol% Ru on IL membrane, toluene, 35 °C, 1 h MWCO = 228	3	52–99	No data	[121]

(*continued*)

TABLE 9-9. (*Continued*)

Entry	Catalyst	Conditions	Cycles	Conversion (%)	Ru Residue	Reference
6	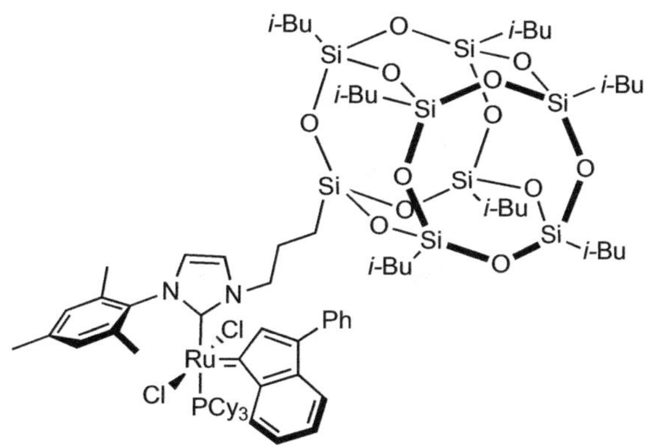 **84**, MW = 1687 g mol^{-1}	DEDAM, 1 mol% Ru, toluene, 60 °C, overnight, Starmem 280	1	99	3.65	[b]
			2	90	2.3	
			3	73	0.35	
7	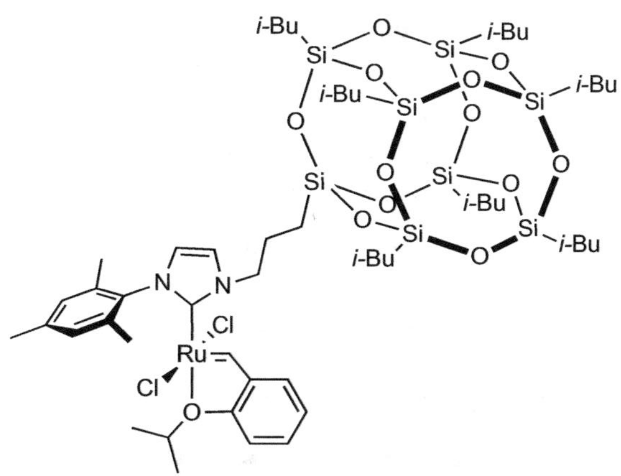 **85**, MW = 1365 g mol^{-1}	DEDAM, 1 mol% Ru, toluene, 60 °C, overnight, Starmem 280	1	66	2	[b]
			2	55	3	
			3	40	2	

8	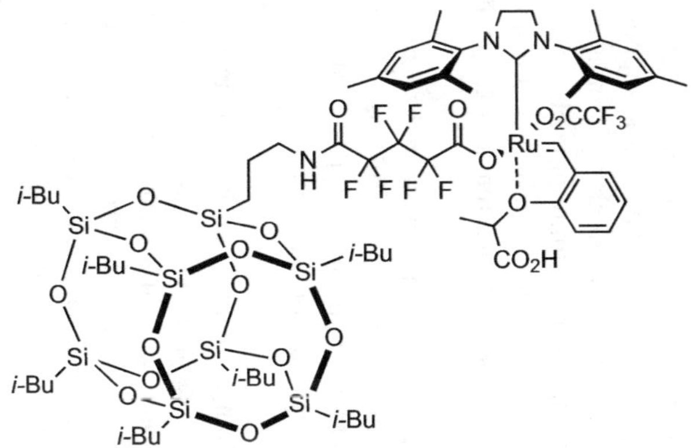 **86**, MW = 1795 g mol^{-1}	DEDAM, 1 mol% Ru, toluene, 60 °C, 5 h, Starmem 280	1	69	0.09	[b]
			2	63	0.34	
			3	62	0.29	
		DEDAM, 1 mol% Ru, CH_2Cl_2, rt, 5 h, APTS 240	1	92	0.005	
			2	91	0.09	
			3	89	6.89	

[a]DATA, *N,N*-diallyltosylamine; DEDAM, diallyl diethyl malonate.

[b]J. Czaban, A. Kajetanowicz, K. Leszczyńska, H. Gerittsen, H. Abbenhuis, H. Siddique, L. Peeva, A.G. Livingstone, and K. Grela, unpublished results.

system has been conducted. The experiment was carried out for 500 min, after which samples were collected to determine the Ru contamination, which was 7 ppm in all cases (entry 3). However, the conversion was only moderate, around 37% after 120 min. Recently, OSNF has also been applied to purify postreaction mixtures of 1-octene self-metathesis [118]. The separation was achieved with the commercially available Starmem™ 228 membrane achieving high retention, with up to 99% of the catalysts retained in the membrane. Complex **82** was recycled four times without significant loss of activity (entry 4), but it deactivated completely after 60 h (overall TON = 5500). Another novel approach was reported by Fischemeister and Rabiller-Baudry, whereby the catalyst was immobilized on the membrane [121]. The modification was achieved in two steps, first by the absorption of the IL ([bmim][PF_6]) and then dispersion of **83** onto the IL membrane. The permeability of the modified membrane was two times higher than an untreated membrane (0.35—Starmem 228 compared to 0.67—IL Starmem 228). The activity of the system was subsequently evaluated by performing RCM of *N,N*-diallyltosylamine (DATA); the catalyst could be recycled three times before any decreasing of activity was observed (entry 5). On the other hand, Grela et al. found that by attaching a polyhedral oligosilsesquioxane (POSS) as an MWE to the NHC and the ionic ligand [123], complex **84** becomes more active than commercially available catMETium® RF1 (**11**) (J. Czaban, et al., unpublished). It could be recovered in three cycles before its activity decreased, and the Ru content of the product was 4 ppm (entry 6). In comparison, complex **85** is less active compared to its symmetrical counterpart, even though it can be activated at higher temperature, and recovered three times, before any loss of activity (entry 7); the Ru contamination in the permeate remains <3 ppm. Last but not the least, the complex **86** had a similar activity to its untagged counterpart [73] and could be recovered twice before any loss of activity, and in the permeate the residual Ru content is < 0.1 ppm (J. Czaban, et al., unpublished), entry 8.

In conclusion, numerous techniques have been devised to purify metathesis products and to recycle the catalyst. Most of them require modification of the catalyst's structure, by attaching various tags, or by anchoring it to organic or inorganic supports. Despite these, there are only a few strategies that are currently suitable for industrial applications. However, the situation is rapidly changing, as more immobilized catalysts are being developed and patented [100]. A number of OSNF membranes are now commercially available, though there may be many more avenues that remain to be explored.

Applications of metathesis that do not require a purification step and utilize low to very low amounts of the catalyst are crucial to make this technology useful for low-cost production of bulk chemicals [124]. This strategy could be, for example, very useful in the conversion of Fischer–Tropsch olefin fractions and vegetable oils to various commodity chemicals and biodiesel components.

Recent research on metathesis shows that this transformation can be used in many "alternative" media. It is now possible to conduct metathesis reactions in aqueous media [125] under both homogeneous and heterogeneous conditions (in water/organic mixtures [126], in neat water using acoustic emulsification [127], and in the presence of various surfactants [128]). There are also specially modified catalysts soluble in neat water [129]. Different enabling techniques have also been explored by the scientists, including: solvent effects [58a,130], microwave [42b,131], thermal activation, concentration effects, ultrasound [127], pressure, and additives (co-catalysts) [132].

9.3 APPLICATION OF RUTHENIUM METATHESIS CATALYSTS IN WATER

The utilization of water as a medium for organic reactions is accompanied by various advantages over the use of common organic solvents [133]. Water is a widely available, cost-efficient, nontoxic, nonflammable liquid owning a high heat capacity and therefore predestined for the application as a solvent. Moreover, modifications of biomolecules such as proteins are possible in water [134]. However, many transition-metal catalysts either are deactivated by water [135] or possess poor solubility. Therefore, till recently, conducting transformation of advanced molecules by olefin metathesis in water still remained a challenge [136]. Two approaches are commonly made when conducting olefin metathesis in water: first, an insoluble catalyst is applied in combination with a surfactant [137] or ultrasound [138]. Second, a catalyst is modified by the attachment of tags in order to increase its solubility in water. Tags can be connected through nonlabile ionic ligands and NHCs, or through labile ones, for example the benzylidene moiety, phosphine-, or pyridine-based ligands. Figure 9-2 shows the number of possibilities for attaching different functionalities to the ruthenium complex.

Hydrophilic tags commonly used are poly(ethylene) glycol chains (Figure 9-4, **87**) [139] or quaternary ammonium groups (Figure 9-4, **88–90**) [140]; furthermore, water-soluble catalysts owning tertiary phosphine [141] and sulfone groups [142] are known. A more sophisticated approach is the incorporation of tagged Ru catalysts into peptides to give artificial metalloenzymes as water-soluble RCM catalysts [143].

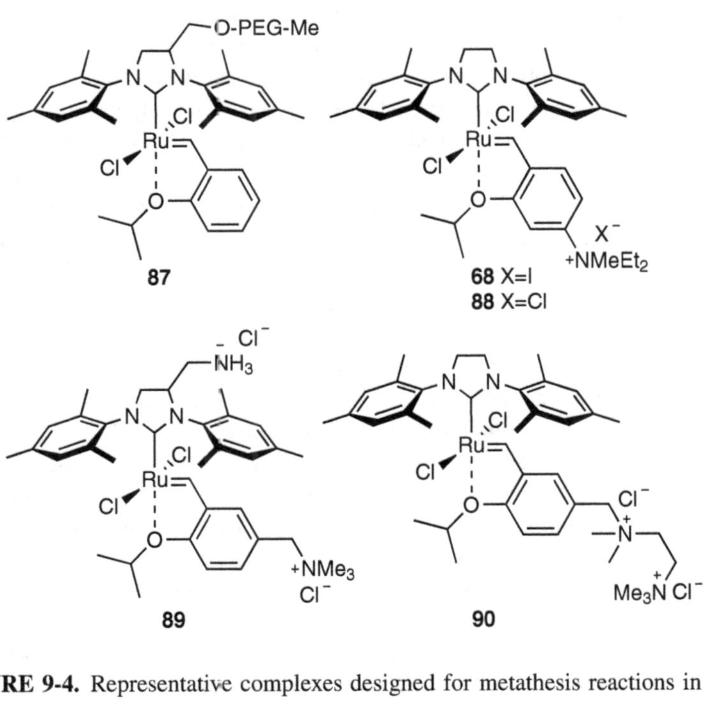

FIGURE 9-4. Representative complexes designed for metathesis reactions in water.

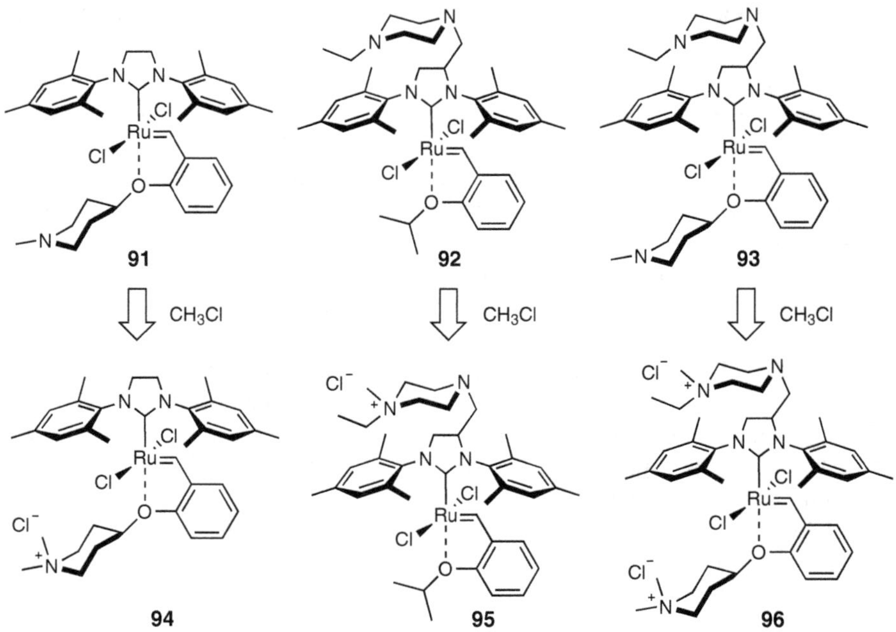

FIGURE 9-5. Catalysts bearing a quaternary ammonium fragment.

Recently, a group of researchers from an industrial company, together with their colleagues from the university, jointly reported the first incorporation of a quaternary ammonium fragment in the NHC ligand of Ru-based metathesis catalysts (Figure 9-5, **95**, **96**) [144]. Initially, the synthesis of this kind of catalysts was found to be troublesome. Direct generation of a free carbene from a tagged imidazolium salt and subsequent reaction with the first-generation Hoveyda catalyst could not be achieved due to the highly basic conditions required for this transformation [144]. On the other hand, the purification of complexes prepared via benzylidene exchange route proved to be difficult in the presence of preformed quaternary ammonium tag. Therefore, the group established a new method for the synthesis and purification of ammonium-tagged olefin metathesis catalysts, in which the quaternary ammonium groups were formed at the very end of the synthesis (Figure 9-5).

Catalyst **94** was synthesized in three steps, while catalysts bearing the quaternary ammonium group in the backbone of the NHC were both prepared in seven steps. In all cases, quaternization was accomplished in high yield by heating of the free base (**91**, **92**, and **93**) with methyl chloride in a pressurized vessel. Interestingly, catalysts **94** and **95** showed low solubility in water (2 and 3 mg ml^{-1}, respectively), while the incorporation of a second quaternary ammonium group dramatically increased the solubility of **96** (to 35 mg ml^{-1}). The catalytic activity of the new catalysts was tested in model CM, RCM, and ene–yne metathesis reactions with water-soluble substrates in neat D_2O. In addition, the RCM reaction of water-insoluble DEDAM was successfully performed with catalyst **96** in DCM, and after the completion of the reaction the catalyst extracted with D_2O. To this extract, another substrate, (*Z*)-but-2-ene-1,4-diol, was added, and the reaction mixture was subsequently analyzed by ^{1}H-NMR. After 1 h, the isomerization product (*E*)-but-2-ene-1,4-diol was formed in 94% yield, proving the ability of catalyst to be recycled. Similarly, quaternary ammonium salt-tagged catalysts were also shown to be easily removed from postreaction mixtures, either via extraction or filtration through silica gel, providing metathesis products with remarkable low Ru contents [144c].

9.4 SUMMARY AND OUTLOOK

Despite its power and applicability, a wider application of the metathesis reaction is prevented by certain limitations, ranging from low stability and/or low activity of some metal complexes, and problems with removing and recycling of catalysts after the reaction, to unfavorable patent situations and licensing strategies. However, further development of new metathesis catalysts and strategies should hopefully eliminate or help to bypass many of these issues. The variety of metathesis catalysts now available on the market as well as the number of commercial players (including the growing number of academic spin-off companies) are increasing rapidly. In Chapter 10, a case history is presented to demonstrate how the limitations in the metathesis methodology can be tackled during the scale-up of an industrial application. Coupled with the wider appreciation of the industrial value of the metathesis methodology, more commercial applications will surely follow.

Yet another of many possible improvements in olefin metathesis catalyst production and application is the use of microreactors. For example, the use of continuous-flow chemistry for the preparation of a ligand precursor for the successful nitro metathesis catalyst **15** [145] has been described jointly by university and industrial chemists [146]. Using a microreactor, it was possible to design a selective, efficient, and reproducible nitration of 2-isopropoxybenzaldehyde to produce the 2-isopropoxy-5-nitrobenzaldehyde building block for the preparation of **15**. The challenging and hazardous nitration process was performed with red fuming HNO_3 using a flow-through silicon–glass microreactor equipped with a set of temperature sensors. The productivity of $13\,g\,h^{-1}$ enabled a reproducible chemical process that was amenable for the production of sufficient quantities of 2-isopropoxy-5-nitrobenzaldehyde for ongoing large-scale synthesis of **15** [146b]. In our opinion, numerous examples of successful University-Industry cooperation cited in this chapter show that at least some of the existing problems and limitations can be soon solved, and that olefin metathesis is now a sufficiently mature technology to be used in large-scale processes.

ACKNOWLEDGMENTS

JC thanks the Foundation for Polish Science for financial support. The project "Ventures/2011–7/3" is realized within the "Ventures" program of Foundation for Polish Science, cofinanced by European Union, Regional Development Fund. CT and KG thank the Foundation for Polish Science "Team Programme" cofinanced by the EU European Regional Development Fund—Operational Program Innovative Economy 2007–2013.

REFERENCES

1. Hagen J, *Industrial Catalysis: A Practical Approach*, 2nd edition. Wiley-VCH, Weinheim; 2006.
2. Magano J, Dunetz JR (2011). Large-scale applications of transition metal-catalyzed couplings for the synthesis of pharmaceuticals. *Chem. Rev.* 111, 2177–2250.
3. Grubbs RH, *Handbook of Metathesis*, Vols. 1–3. Wiley-VCH, Weinheim; 2003.
4. Hoveyda AH, Zhugralin AR (2007). The remarkable metal-catalyzed olefin metathesis reaction. *Nature* 450, 243–251.
5. Požgan F, Dixneuf PH. Recent applications of alkene metathesis for fine chemical and supramolecular system synthesis. In: Imamoglu Y, Dragutan V, Karabulut, S, editors. *Metathesis Chemistry: From Nanostructure Design to Synthesis of Advanced Materials*. Springer, Dordrecht; 2007, pp 195–222.
6. (a) Nicolaou KC, Bulger PG, Sarlah D (2005). Metathesis reactions in total synthesis. *Angew. Chem., Int. Ed.* 44, 4490–4527. (b) Fürstner A (2011). Metathesis in total synthesis. *Chem. Commun.* 6505–6511.
7. (a) Leitgeb A, Wappel J, Slugovc C (2010). The ROMP toolbox upgraded. *Polymer* 51, 2927–2946. (b) Bielawski CW, Grubbs RH (2007). Living ring-opening metathesis polymerization. *Prog. Polym. Sci.* 32, 1–29. (c) Mol JC (2004). Industrial applications of olefin metathesis. *J. Mol. Catal. A: Chem.* 213, 39–45.
8. The Nobel Prize in Chemistry last access 2013.01.12; http://www.nobelprize.org/nobel_prizes/chemistry/laureates/2005/.
9. Hérisson PJL, Chauvin Y (1971). Catalyze de transformation des oléfines par les complexes du tungstène. II. télomérisation des oléfines cycliques en présence d'oléfines acycliques. *Die Makromol. Chem.* 141, 161–176.
10. (a) Katz TJ, McGinnis J (1977). Metathesis of cyclic and acyclic olefins. *J. Am. Chem. Soc.* 99, 1903–1912. (b) Grubbs RH, Carr DD, Hoppin C, Burk PL (1976). Consideration of the mechanism of the metal catalyzed olefin metathesis reaction. *J. Am. Chem. Soc.* 98, 3478–3483. (c) Katz TJ, Rothchild R (1976). Mechanism of the olefin metathesis of 2,2′-divinylbiphenyl. *J. Am. Chem. Soc.* 98, 2519–2526. (d) Katz TJ, McGinnis J (1975). Mechanism of the olefin metathesis reaction. *J. Am. Chem. Soc.* 97, 1592–1594. (e) Grubbs RH, Burk PL, Carr DD (1975). Mechanism of the olefin metathesis reaction. *J. Am. Chem. Soc.* 97, 3265–3267.
11. (a) Diver ST, Giessert AJ (2004). Enyne metathesis (enyne bond reorganization). *Chem. Rev.* 104, 1317–1382. (b) Galan BR, Giessert AJ, Keister JB, Diver ST (2005). Studies on the mechanism of intermolecular enyne metathesis: kinetic method and alkyne substituent effects. *J. Am. Chem. Soc.* 127, 5762–5763.
12. Maifeld SV, Lee D (2005). Group-selective ring-closing enyne metathesis. *Chem. Eur. J.* 11, 6118–6126.
13. (a) Michelotti FW, Keaveney WP (1965). Coordinated polymerization of the bicyclo-[2.2.1]-heptene-2 ring system (norbornene) in polar media. *J. Polym. Sci., Part A: Gen. Pap.* 3, 895–905. (b) Rinehart RE, Smith HP (1965). The emulsion polymerization of the norbornene ring system catalyzed by noble metal compounds. *J. Polym. Sci., Part B: Polym. Lett.* 3, 1049–1052.
14. (a) Hillmyer MA, Lepetit C, McGrath DV, Novak BM, Grubbs RH (1992). Aqueous ring-opening metathesis polymerization of carboximide-functionalized 7-oxanorbornenes. *Macromolecules* 25, 3345–3350. (b) Novak BM, Grubbs RH (1988). Catalytic organometallic chemistry in water: the aqueous ring-opening metathesis polymerization of 7-oxanorbornene derivatives. *J. Am. Chem. Soc.* 110, 7542–7543.
15. (a) McGrath DV, Grubbs RH, Ziller JW (1991). Aqueous ruthenium(II) complexes of functionalized olefins: the X-ray structure of $Ru(H_2O)_2(\eta^1(O){:}\eta^2(C,C')-OCOCH_2CH{=}CHCH_3)_2$.

J. Am. Chem. Soc. 113, 3611–3613. (b) France MB, Grubbs RH, McGrath DV, Paciello RA (1993). Chain transfer during the aqueous ring-opening metathesis polymerization of 7-oxa-norbornene derivatives. *Macromolecules* 26, 4742–4747.

16. Nguyen ST, Johnson LK, Grubbs RH, Ziller JW (1992). Ring-opening metathesis polymerization (ROMP) of norbornene by a group VIII carbene complex in protic media. *J. Am. Chem. Soc.* 114, 3974–3975.
17. (a) Wengrovius JH, Sancho J, Schrock RR (1981). Metathesis of acetylenes by tungsten(VI)-alkylidyne complexes. *J. Am. Chem. Soc.* 103, 3932–3934. (b) Fürstner A, Guth O, Rumbo A, Seidel G (1999). Ring closing alkyne metathesis. comparative investigation of two different catalyst systems and application to the stereoselective synthesis of olfactory lactones, azamacrolides, and the macrocyclic perimeter of the marine alkaloid nakadomarin A. *J. Am. Chem. Soc.* 121, 11108–11113. (c) Fürstner A, Grela K, Mathes C, Lehmann CW (2000). Novel and flexible entries into prostaglandins and analogues based on ring closing alkyne metathesis or alkyne cross metathesis. *J. Am. Chem. Soc.* 122, 11799–11805.
18. Ivin KJ, Mol JC. The metal carbene/metallacyclobutane mechanism. In: *Olefin Metathesis and Metathesis Polymerization II*. Academic Press, London; 1997, pp 50–81.
19. Nguyen ST, Grubbs RH, Ziller JW (1993). Syntheses and activities of new single-component, ruthenium-based olefin metathesis catalysts. *J. Am. Chem. Soc.* 115, 9858–9859.
20. Trnka TM, Grubbs RH (2001). The development of $L_2X_2Ru{=}CHR$ olefin metathesis catalysts: an organometallic success story. *Acc. Chem. Res.* 34, 18–29.
21. Samojłowicz C, Bieniek M, Grela K (2009). Ruthenium-based olefin metathesis catalysts bearing *N*-heterocyclic carbene ligands. *Chem. Rev.* 109, 3708–3742.
22. Weskamp T, Schattenmann WC, Spiegler M, Herrmann WA (1998). A novel class of ruthenium catalysts for olefin metathesis. *Angew. Chem., Int. Ed.* 37, 2490–2493.
23. Frenzel U, Weskamp T, Kohl FJ, Schattenmann WC, Nuyken O, Herrmann WA (1999). *N*-heterocyclic carbenes: application of ruthenium-alkylidene complexes in ring-opening metathesis polymerization. *J. Organomet. Chem.* 586, 263–265.
24. Ackermann L, Fürstner A, Weskamp T, Kohl FJ, Herrmann WA (1999). Ruthenium carbene complexes with imidazolin-2-ylidene ligands allow the formation of tetrasubstituted cycloalkenes by RCM. *Tetrahedron Lett.* 40, 4787–4790.
25. Scholl M, Trnka TM, Morgan JP, Grubbs RH (1999). Increased ring closing metathesis activity of ruthenium-based olefin metathesis catalysts coordinated with imidazolin-2-ylidene ligands. *Tetrahedron Lett.* 40, 2247–2250.
26. Huang J, Stevens ED, Petersen JL, Nolan SP (1999). Olefin metathesis-active ruthenium complexes bearing a nucleophilic carbene ligand. *J. Am. Chem. Soc.* 121, 2674–2678.
27. Scholl M, Ding S, Lee CW, Grubbs RH (1999). Synthesis and activity of a new generation of ruthenium-based olefin metathesis catalysts coordinated with 1,3-dimesityl-4,5-dihydroimidazol-2-ylidene ligands. *Org. Lett.* 1, 953–956.
28. Bieniek M, Michrowska A, Usanov DL, Grela K (2008). In an attempt to provide a user's guide to the galaxy of benzylidene, alkoxybenzylidene, and indenylidene ruthenium olefin metathesis catalysts. *Chem. Eur. J.* 14, 806–818.
29. Chatterjee AK, Grubbs RH (1999). Synthesis of trisubstituted alkenes via olefin cross-metathesis. *Org. Lett.* 1, 1751–1753.
30. Rivkin A, Yoshimura F, Gabarda AE, Chou TC, Dong H, Tong WP, Danishefsky SJ (2003). Complex target-oriented total synthesis in the drug discovery process: the discovery of a highly promising family of second generation epothilones. *J. Am. Chem. Soc.* 125, 2899–2901.
31. Dragutan V, Dragutan I, Verpoort F (2005). Ruthenium indenylidene complexes. *Platinum Met. Rev.* 49, 33–40.

32. Jafarpour L, Schanz HJ, Stevens ED, Nolan SP (1999). Indenylidene–imidazolylidene complexes of ruthenium as ring-closing metathesis catalysts. *Organometallics* 18, 5416–5419.

33. Harrity JPA, La DS, Cefalo DR, Visser MS, Hoveyda AH (1998). Chromenes through metal-catalyzed reactions of styrenyl ethers. Mechanism and utility in synthesis. *J. Am. Chem. Soc.* 120, 2343–2351.

34. Gessler S, Randl S, Blechert S (2000). Synthesis and metathesis reactions of a phosphine-free dihydroimidazole carbene ruthenium complex. *Tetrahedron Lett.* 41, 9973–9976.

35. Garber SB, Kingsbury JS, Gray BL, Hoveyda AH (2000). Efficient and recyclable monomeric and dendritic Ru-based metathesis catalysts. *J. Am. Chem. Soc.* 122, 8168–8179.

36. (a) Randl S, Buschmann N, Connon SJ, Blechert S (2001). Highly efficient and recyclable polymer-bound catalyst for olefin metathesis reactions. *Synlett* 1547–1550. (b) Imhof S, Randl S, Blechert S (2001). Ruthenium catalyzed cross metathesis with fluorinated olefins. *Chem. Commun.* 1692–1693. (c) Demchuk OM, Pietrusiewicz KM, Michrowska A, Grela K (2003). Synthesis of substituted P-stereogenic vinylphosphine oxides by olefin cross-metathesis. *Org. Lett.* 5, 3217–3220. (d) Vinokurov N, Michrowska A, Szmigielska A, Drzazga Z, Wójciuk G, Demchuk OM, Grela K, Pietrusiewicz KM, Butenschön H (2006). Homo- and cross-olefin metathesis coupling of vinylphosphane oxides and electron-poor alkenes: access to P-stereogenic dienophiles. *Adv. Synth. Catal.* 348, 931–938. (e) Grela K, Bieniek M (2001). Highly selective cross-metathesis with phenyl vinyl sulphone using the 'second generation' Grubbs' catalyst. *Tetrahedron Lett.* 42, 6425–6428.

37. (a) Love JA, Morgan JP, Trnka TM, Grubbs RH (2002). A practical and highly active ruthenium-based catalyst that effects the cross metathesis of acrylonitrile. *Angew. Chem., Int. Ed.* 41, 4035–4037. (b) Kingsbury JS, Hoveyda AH (2005). Regarding the mechanism of olefin metathesis with sol–gel-supported Ru-based complexes bearing a bidentate carbene ligand. Spectroscopic evidence for return of the propagating Ru carbene. *J. Am. Chem. Soc.* 127, 4510–4517.

38. Vorfalt T, Wannowius KJ, Thiel V, Plenio H (2010). How important is the release–return mechanism in olefin metathesis? *Chem. Eur. J.* 16, 12312–12315.

39. (a) Wang H, Goodman SN, Dai QY, Stockdale GW, Clark Jr., WM (2008). Development of a robust ring-closing metathesis reaction in the synthesis of SB-462795, a cathepsin K inhibitor. *Org. Process Res. Dev.* 12, 226–234. (b) Wang H, Matsuhashi H, Doan BD, Goodman SN, Ouyang X, Clark Jr., WM (2009). Large-scale synthesis of SB-462795, a cathepsin K inhibitor: the RCM-based approaches. *Tetrahedron* 65, 6291–6303.

40. (a) Grela K, Harutyunyan S, Michrowska A (2002). A highly efficient ruthenium catalyst for metathesis reactions. *Angew. Chem., Int. Ed.* 41, 4038–4040. (b) Michrowska A, Bujok R, Harutyunyan S, Sashuk V, Dolgonos G, Grela K (2004). Nitro-substituted Hoveyda–Grubbs ruthenium carbenes: enhancement of catalyst activity through electronic activation. *J. Am. Chem. Soc.* 126, 9318–9325. (c) Grela K, inventor; Boehringer Ingelheim Int GmbH, assignee. Ruthenium complexes as (Pre)catalysts for metathesis reactions. World Patent WO 2004/035596. April 29, 2004. (d) Apeiron Catalysts http://www.apeiron-catalysts.com/.

41. (a) Honda T, Namiki H, Kaneda K, Mizutani H (2004). First diastereoselective chiral synthesis of (−)-securinine. *Org. Lett.* 6, 87–89. (b) Rosyk PJ, inventor; Boehringer Ingelheim Pharma GmbH, assignee. Process for the preparation of a tetrahydro-1*H*-azepines. World Patent WO 2008/009671. January 24, 2008. (c) Michrowska A, List B (2009). Concise synthesis of ricciocarpin A and discovery of a more potent analogue. *Nat. Chem.* 1, 225–228. (d) Ostrowski S, Mikus A (2003). A new approach to the synthesis of porphyrin-fullerene dyads. *Mol. Divers.* 6, 315–321. (e) Albert BJ, Sivaramakrishnan A, Naka T, Koide K (2006). Total synthesis of FR901464, an antitumor agent that regulates the transcription of oncogenes and tumor suppressor genes. *J. Am. Chem. Soc.* 128, 2792–2793. (f) Seiser T, Kamena F, Cramer N (2008). Synthesis and biological activity of largazole and derivatives. *Angew. Chem., Int. Ed.* 47, 6483–6485. (g) Weychardt H, Plenio H (2008). Acyclic diene metathesis polymerization

of divinylarenes and divinylferrocenes with Grubbs-type olefin metathesis catalysts. *Organometallics* 27, 1479–1485. (h) Stellfeld T, Bhatt U, Kalesse M (2004). Synthesis of the A,B, C-ring system of hexacyclinic acid. *Org. Lett.* 6, 3889–3892. (i) Honda T, Namiki H, Watanabe M, Mizutani H (2004). First total synthesis of (+)-viroallosecurinine. *Tetrahedron Lett.* 45, 5211–5213.

42. (a) Clavier H, Grela K, Kirschning A, Mauduit M, Nolan SP (2007). Sustainable concepts in olefin metathesis. *Angew. Chem., Int. Ed.* 46, 6786–6801. (b) Goldup SM, Pilkington CJ, White AJP, Burton A, Barrett AGM (2006). A simple, short, and flexible synthesis of viridiofungin derivatives. *J. Org. Chem.* 71, 6185–6191.
43. (a) The European Agency for the Evaluation of Medicinal Products, http://www.ema.europa.eu/docs/en_GB/document_library/Scientific_guideline/2009/09/WC500003586.pdf (last access 2013.01.13). (b) Nicola T, Brenner M, Donsbach K, Kreye P (2005). First scale-up to production scale of a ring closing metathesis reaction forming a 15-membered macrocycle as a precursor of an active pharmaceutical ingredient. *Org. Process Res. Dev.* 9, 513–515.
44. Cho JH, Kim M (2003). An efficient method for removal of ruthenium by-products from olefin metathesis reactions. *Org. Lett.* 5, 531–533.
45. Ahn YM, Yang K, Georg GI (2001). A convenient method for the efficient removal of ruthenium by-products generated during olefin metathesis reactions. *Org. Lett.* 3, 1411–1413.
46. Mendez-Andino J, Paquette LA (2000). Tandem deployment of iridium-, ruthenium-, and lead-promoted reactions. four-carbon intercalation between the carbonyl group of open-chain and cyclic α-diketones. *Org. Lett.* 2, 1263–1265.
47. Balskus EP, Méndez-Andino J, Arbit RM, Paquette LA (2001). Intercalation of multiple carbon atoms between the carbonyls of α-diketones. *J. Org. Chem.* 66, 6695–6704.
48. Maynard HD, Grubbs RH (1999). Purification technique for the removal of ruthenium from olefin metathesis reaction products. *Tetrahedron Lett.* 40, 4137–4140.
49. Grubbs RH, Ferguson ML, O'Leary DJ (2003). Ring-closing metathesis synthesis of *N*-Boc-3-pyrroline (1*H*-pyrrole-1-carboxylic acid, 2,5-dihydro-, 1,1-dimethylethyl ester). *Org. Synth.* 80, 85–92.
50. Macdonald JE, Veinot JGC (2008). Removal of residual metal catalysts with iron/iron oxide nanoparticles from coordinating environments. *Langmuir* 24, 7169–7177.
51. (a) SiliCycle Inc., www.silicycle.com. (b) S. AMIT http://www.s-amit.com/ (last access 2013.01.13).
52. McEleney K, Allen DP, Holliday AE, Crudden CM (2006). Functionalized mesoporous silicates for the removal of ruthenium from reaction mixtures. *Org. Lett.* 8, 2663–2666.
53. Galan BR, Kalbarczyk KP, Szczepankiewicz S, Keister JB, Diver ST (2007). A rapid and simple cleanup procedure for metathesis reactions. *Org. Lett.* 9, 1203–1206.
54. (a) Skerget M, Knez Z, Knez-Hrncic M (2011). Solubility of solids in sub- and supercritical fluids: a review. *J. Chem. Eng. Data* 56, 694–719. (b) Bhardwaj L, Sharma PK, Visht S, Garg VK, Kumar N (2010). A review on methodology and application of supercritical fluid technology in pharmaceutical industry. *Pharm. Sin.* 1, 183–194. (c) Cummings S, Trickett K, Enick R, Eastoe J (2011). CO_2: a wild solvent, tamed. *Phys. Chem. Chem. Phys.* 13, 1276–1289. (d) Brunner G (2010). Applications of supercritical fluids. *Annu. Rev. Chem. Biomol. Eng.* 1, 321–342.
55. (a) Jessop PG, Ikariya T, Noyori R (1999). Homogeneous catalysis in supercritical fluids. *Chem. Rev.* 99, 475–494. (b) Prajapati D, Gohain M (2004). Recent advances in the application of supercritical fluids for carbon–carbon bond formation in organic synthesis. *Tetrahedron* 60, 815–833. (c) Hintermair U, Francio G, Leitner W (2011). Continuous flow organometallic catalysis: new wind in old sails. *Chem. Commun.* 47, 3691–3701. (d) Gläser, R. Heterogeneous catalysis. In: *Handbook of Green Chemistry*. Wiley-VCH, Weinheim; 2010.

56. (a) Mistele CD, Thorp HH, Desimone JM (1996). Ring-opening metathesis polymerizations in carbon dioxide. *J. Macromol. Sci., Part A* 33, 953–960. (b) Hamilton JG, Rooney JJ, DeSimone JM, Mistele C (1998). Stereochemistry of ring-opened metathesis polymers prepared in liquid CO_2 at high pressure using $Ru(H_2O)_6(Tos)_2$ as catalyst. *Macromolecules* 31, 4387–4389.

57. (a) Fabris M, Aquino C, Ward AJ, Perosa A, Maschmeyer T, Selva M (2009). Self-metathesis of 1-octene using alumina-supported Re_2O_7 in supercritical CO_2. *Top. Catal.* 52, 315–321. (b) Selva M, Perosa A, Fabris M, Canton P (2009). The metathesis of α-olefins over supported Re-catalysts in supercritical CO_2. *Green Chem.* 11, 229–238. (c) Song J, Hou M, Liu G, Zhang J, Han B, Yang G (2009). Effect of phase behavior on the ethenolysis of ethyl oleate in compressed CO_2. *J. Phys. Chem. B* 113, 2810–2814.

58. (a) Fürstner A, Ackermann L, Beck K, Hori H, Koch D, Langemann K, Liebl M, Six C, Leitner W (2001). Olefin metathesis in supercritical carbon dioxide. *J. Am. Chem. Soc.* 123, 9000–9006. (b) Fürstner A, Koch D, Langemann K, Leitner W, Six C (1997). Olefin metathesis in compressed carbon dioxide. *Angew. Chem., Int. Ed.* 36, 2466–2469.

59. Hu X, Blanda MT, Venumbaka SR, Cassidy PE (2005). Ring-opening metathesis polymerization (ROMP) of norbornene in supercritical carbon dioxide using well-defined metal carbene catalysts. *Polym. Adv. Technol.* 16, 146–149.

60. Gallou F, Saim S, Yee NK, inventors; Boehringer Ingelheim International, GmbH, assignee. Removal of ruthenium catalyst residue by supercritical fluid processing. World Patent WO2005/056182. June 23, 2005.

61. Michalek F, Mädge D, Rühe J, Bannwarth W (2006). Ring-closure metathesis in supercritical carbon dioxide as sole solvent with use of covalently immobilized ruthenium catalysts. *Eur. J. Org. Chem.* 577–581.

62. (a) Collman JP, Belmont JA, Brauman JI (1983). A silica-supported rhodium hydroformylation catalyst: evidence for dinuclear elimination. *J. Am. Chem. Soc.* 105, 7288–7294. (b) Drago RS, Pribich DC (1985). A new method for enhancing site isolation on silica gel and for improving the lifetime of site-isolated catalysts. *Inorg. Chem.* 24, 1983–1985.

63. (a) Vougioukalakis GC, Grubbs RH (2010). Ruthenium-based heterocyclic carbene-coordinated olefin metathesis catalysts. *Chem. Rev.* 110, 1746–1787. (b) Buchmeiser MR (2009). Polymer-supported well-defined metathesis catalysts. *Chem. Rev.* 109, 303–321. (c) Buchmeiser MR (2004). Recent advances in the synthesis of supported metathesis catalysts. *New J. Chem.* 28, 549–557. (d) Sommer WJ, Weck M (2006). Supported *N*-heterocyclic carbene complexes in catalysis. *Coord. Chem. Rev.* 251, 860–873. (e) Coperet C, Basset JM (2007). Strategies to immobilize well-defined olefin metathesis catalysts: supported homogeneous catalysis vs. surface organometallic chemistry. *Adv. Synth. Catal.* 349, 78–92.

64. Prühs S, Lehmann CW, Fürstner A (2003). Preparation, reactivity, and structural peculiarities of hydroxyalkyl-functionalized "second-generation" ruthenium carbene complexes. *Organometallics* 23, 280–287.

65. Nguyen SBT, Grubbs RH (1995). The syntheses and activities of polystyrene-supported olefin metathesis catalysts based on $Cl_2(PR_3)_2Ru{=}CH{-}CH{=}CPh_2$. *J. Organomet. Chem.* 497, 195–200.

66. Mayr M, Mayr B, Buchmeiser MR (2001). Monolithic materials: new high-performance supports for permanently immobilized metathesis catalysts. *Angew. Chem., Int. Ed.* 40, 3839–3842.

67. Allen DP, Van Wingerden MM, Grubbs RH (2009). Well-defined silica-supported olefin metathesis catalysts. *Org. Lett.* 11, 1261–1264.

68. Bek D, Gawin R, Grela K, Balcar H (2012). Mesoporous molecular sieves SBA-15 immobilized ruthenium metathesis catalyst bearing chelating carboxylate ligand. *Catal. Commun.* 21, 42–45.

69. Ahmed M, Barrett AGM, Braddock DC, Cramp SM, Procopiou PA (1999). A recyclable 'boomerang' polymer-supported ruthenium catalyst for olefin metathesis. *Tetrahedron Lett.* 40, 8657–8662.

70. Vehlow K, Maechling S, Blechert S (2005). Ruthenium metathesis catalysts with saturated unsymmetrical *N*-heterocyclic carbene ligands. *Organometallics* 25, 25–28.

71. Nieczypor P, Buchowicz W, Meester WJN, Rutjes FPJT, Mol JC (2001). Synthesis and application of a new polystyrene-supported ruthenium carbene catalyst for alkene metathesis. *Tetrahedron Lett.* 42, 7103–7105.

72. (a) Krause JO, Lubbad SH, Nuyken O, Buchmeiser MR (2003). Heterogenization of a modified Grubbs–Hoveyda catalyst on a ROMP-derived monolithic support. *Macromol. Rapid Commun.* 24, 875–878. (b) Yang L, Mayr M, Wurst K, Buchmeiser MR (2004). Novel metathesis catalysts based on ruthenium 1,3-dimesityl-3,4,5,6-tetrahydropyrimidin-2-ylidenes: synthesis, structure, immobilization, and catalytic activity. *Chem. Eur. J.* 10, 5761–5770. (c) Krause JO, Nuyken O, Wurst K, Buchmeiser MR (2004). Synthesis and reactivity of homogeneous and heterogeneous ruthenium-based metathesis catalysts containing electron-withdrawing ligands. *Chem. Eur. J.* 10, 777–784. (d) Halbach TS, Mix S, Fischer D, Maechling S, Krause JO, Sievers C, Blechert S, Nuyken O, Buchmeiser MR (2005). Novel ruthenium-based metathesis catalysts containing electron-withdrawing ligands: synthesis, immobilization, and reactivity. *J. Org. Chem.* 70, 4687–4694.

73. Gawin R, Makal A, Woźniak K, Mauduit M, Grela K (2007). A dormant ruthenium catalyst bearing a chelating carboxylate ligand: in-situ activation and application in metathesis reactions. *Angew. Chem., Int. Ed.* 46, 7206–7209.

74. (a) Jafarpour L, Nolan SP (2000). Simply assembled and recyclable polymer-supported olefin metathesis catalysts. *Org. Lett.* 2, 4075–4078. (b) Jafarpour L, Heck MP, Baylon C, Lee HM, Mioskowski C, Nolan SP (2002). Preparation and activity of recyclable polymer-supported ruthenium olefin metathesis catalysts. *Organometallics* 21, 671–679.

75. (a) Connon SJ, Blechrt S (2002). A solid-supported phosphine-free ruthenium alkylidene for olefin metathesis in methanol and water. *Bioorg. Med. Chem. Lett.* 12, 1873–1876. (b) Connon SJ, Dunne AM, Blechert S (2002). A self-generating, highly active, and recyclable olefin-metathesis catalyst. *Angew. Chem., Int. Ed.* 41, 3835–3838.

76. (a) Hoveyda AH, Gillingham DG, Van Veldhuizen JJ, Kataoka O, Garber SB, Kingsbury JS, Harrity JPA (2004). Ru complexes bearing bidentate carbenes: from innocent curiosity to uniquely effective catalysts for olefin metathesis. *Org. Biomol. Chem.* 2, 8–23. (b) Kingsbury JS, Garber SB, Giftos JM, Gray BL, Okamoto MM, Farrer RA, Fourkas JT, Hoveyda AH (2001). Immobilization of olefin metathesis catalysts on monolithic sol–gel: practical, efficient, and easily recyclable catalysts for organic and combinatorial synthesis. *Angew. Chem., Int. Ed.* 40, 4251–4256.

77. Dowden J, Savovic J (2001). Olefin metathesis in non-degassed solvent using a recyclable, polymer supported alkylideneruthenium. *Chem. Commun.* 37–38.

78. Grela K, Tryznowski M, Bieniek M (2002). A PS-DES immobilized ruthenium carbene: a robust and easily recyclable catalyst for olefin metathesis. *Tetrahedron Lett.* 43, 9055–9059.

79. (a) Yao QW (2000). A soluble polymer-bound ruthenium carbene complex: a robust and reusable catalyst for ring-closing olefin metathesis. *Angew. Chem., Int. Ed.* 39, 3896–3898. (b) Yao QW, Motta AR (2004). Immobilization of the Grubbs second-generation ruthenium-carbene complex on poly(ethylene glycol): a highly reactive and recyclable catalyst for ring-closing and cross-metathesis. *Tetrahedron Lett.* 45, 2447–2451.

80. Che C, Li W, Lin S, Chen J, Zheng J, Wu JC, Zheng Q, Zhang G, Yang Z, Jiang B (2009). Magnetic nanoparticle-supported Hoveyda–Grubbs catalysts for ring-closing metathesis reactions. *Chem. Commun.* 5990–5992.

81. Zhu Y, Loo K, Ng H, Li C, Stubbs LP, Siong CF, Tan M, Ship CP (2009). Magnetic nanoparticle supported second generation Hoveyda–Grubbs catalyst for metathesis of unsaturated fatty acid esters. *Adv. Synth. Catal.* 351, 2650–2656.

82. (a) Lu AH, Schmidt W, Matoussevitch N, Bönnemann H, Spliethoff B, Tesche B, Bill E, Kiefer W, Schüth F (2004). Nanoengineering of a magnetically separable hydrogenation catalyst. *Angew. Chem., Int. Ed.* 43, 4303–4306. (b) Hu AG, Yee GT, Lin WB (2005). Magnetically recoverable chiral catalysts immobilized on magnetite nanoparticles for asymmetric hydrogenation of aromatic ketones. *J. Am. Chem. Soc.* 127, 12486–12487. (c) Zheng Y, Stevens PD, Gao Y (2006). Magnetic nanoparticles as an orthogonal support of polymer resins: applications to solid-phase Suzuki cross-coupling reactions. *J. Org. Chem.* 71, 537–542. (d) Lee D, Lee J, Lee H, Jin S, Hyeon T, Kim BM (2006). Filtration-free recyclable catalytic asymmetric dihydroxylation using a ligand immobilized on magnetic mesocellular mesoporous silica. *Adv. Synth. Catal.* 348, 41–46. (e) Phan NTS, Gill CS, Nguyen JV, Zhang ZJ, Jones CW (2006). Expanding the utility of one-pot multistep reaction networks through compartmentation and recovery of the catalyst. *Angew. Chem., Int. Ed.* 45, 2209–2212. (f) Kawamura M, Sato K (2006). Magnetically separable phase-transfer catalysts. *Chem. Commun.* 4718–4719. (g) Luo S, Zheng X, Cheng JP (2008). Asymmetric bifunctional primary aminocatalysis on magnetic nanoparticles. *Chem. Commun.* 5719–5721. (h) Jun CH, Park YJ, Yeon YR, Choi JR, Lee WR, Ko SJ, Cheon J (2006). Demonstration of a magnetic and catalytic Co@Pt nanoparticle as a dual-function nanoplatform. *Chem. Commun.* 1619–1621. (i) Gleeson O, Tekoriute R, Gun'ko YK, Connon SJ (2009). The first magnetic nanoparticle-supported chiral DMAP analogue: highly enantioselective acylation and excellent recyclability. *Chem. Eur. J.* 15, 5669–5673. (j) Claesson EM, Mehendale NC, Gebbink RJMK, van Koten G, Philipse AP (2007). Magnetic silica colloids for catalysis. *J. Magn. Magn. Mater.* 311, 41–45.

83. DeSimone JM (2002). Practical approaches to green solvents. *Science* 297, 799–803.

84. (a) Wasserscheid P, Keim W (2000). Ionic liquids–new "solutions" for transition metal catalysis. *Angew. Chem., Int. Ed.* 39, 3772–3789. (b) Dupont J, de Souza RF, Suarez PAZ (2002). Ionic liquid (molten salt) phase organometallic catalysis. *Chem. Rev.* 102, 3667–3692. (c) Smiglak M, Reichert WM, Holbrey JD, Wilkes JS, Sun L, Thrasher JS, Kirichenko K, Singh S, Katritzky AR, Rogers RD (2006). Combustible ionic liquids by design: is laboratory safety another ionic liquid myth? *Chem. Commun.* 2554–2556.

85. Csihony S, Fischmeister C, Bruneau C, Horvath IT, Dixneuf PH (2002). First ring-opening metathesis polymerization in an ionic liquid. Efficient recycling of a catalyst generated from a cationic ruthenium allenylidene complex. *New J. Chem.* 26, 1667–1670.

86. (a) Audic N, Clavier H, Mauduit M, Guillemin JC (2003). An ionic liquid-supported ruthenium carbene complex: a robust and recyclable catalyst for ring-closing olefin metathesis in ionic liquids. *J. Am. Chem. Soc.* 125, 9248–9249. (b) Clavier H, Audic N, Mauduit M, Guillemin JC (2004). Ring-closing metathesis in biphasic [BMI][PF_6] ionic liquid/toluene medium: a powerful recyclable and environmentally friendly process. *Chem. Commun.* 2282–2283. (c) Clavier H, Audic N, Guillemin JC, Mauduit M (2005). Olefin metathesis in room temperature ionic liquids using imidazolium-tagged ruthenium complexes. *J. Organomet. Chem.* 690, 3585–3599.

87. (a) Yao Q, Zhang Y (2003). Olefin metathesis in the ionic liquid 1-butyl-3-methylimidazolium hexafluorophosphate using a recyclable Ru catalyst: remarkable effect of a designer ionic tag. *Angew. Chem., Int. Ed.* 42, 3395–3398. (b) Yao Q, Sheets M (2005). An ionic liquid-tagged second generation Hoveyda–Grubbs ruthenium carbene complex as highly reactive and recyclable catalyst for ring-closing metathesis of di-, tri- and tetrasubstituted dienes. *J. Organomet. Chem.* 690, 3577–3584.

88. Duque R, Ochsner E, Clavier H, Caijo F, Nolan SP, Mauduit M, Cole-Hamilton DJ (2011). Continuous flow homogeneous alkene metathesis with built-in catalyst separation. *Green Chem.* 13, 1187–1195.

89. (a) Michrowska A, Gułajski Ł, Kaczmarska Z, Mennecke K, Kirschning A, Grela K (2006). A green catalyst for green chemistry: synthesis and application of an olefin metathesis catalyst bearing a quaternary ammonium group. *Green Chem.* 8, 685–688. (b) Gułajski Ł, Michrowska A, Bujok R, Grela K (2006). New tunable catalysts for olefin metathesis: controlling the initiation through electronic factors. *J. Mol. Catal. A: Chem.* 254, 118–123. (c) Gułajski Ł, Grela K The influence of the anionic counter-ion on the activity of ammonium substituted Hoveyda-type olefin metathesis catalysts in aqueous media. In: Dragutan V, Demonceau A, Dragutan I, Finkelshtein ES, editors. *Green Metathesis Chemistry*. Springer, Dordrecht; 2010, pp 49–56. (d) Gułajski Ł, Mauduit M, Grela K (2009). Onium-tagged Ru complexes as universal catalysts for olefin metathesis reactions in various media. *Pure Appl. Chem.* 81, 2001–2012.

90. (a) Rix D, Clavier H, Coutard Y, Gułajski Ł, Grela K, Mauduit M (2006). Activated pyridinium-tagged ruthenium complexes as efficient catalysts for ring-closing metathesis. *J. Organomet. Chem.* 691, 5397–5405. (b) Consorti CS, Aydos GLP, Ebeling G, Dupont J (2007). Ionophilic phosphines: versatile ligands for ionic liquid biphasic catalysis. *Org. Lett.* 10, 237–240.

91. Michrowska A, Mennecke K, Kunz U, Kirschning A, Grela K (2006). A new concept for the noncovalent binding of a ruthenium-based olefin metathesis catalyst to polymeric phases: preparation of a catalyst on Raschig rings. *J. Am. Chem. Soc.* 128, 13261–13267.

92. (a) Jas G, Kirschning A (2003). Continuous flow techniques in organic synthesis. *Chem. Eur. J.* 9, 5708–5723. (b) Kirschning A, Solodenko W, Mennecke K (2006). Combining enabling techniques in organic synthesis: continuous flow processes with heterogenized catalysts. *Chem. Eur. J.* 12, 5972–5990. (c) Ceylan S, Kirschning A. Organic synthesis with mini flow reactors using immobilised catalysts. In: Benaglia M editor. *Recoverable and Recyclable Catalysts*. Wiley, Chichester; 2009, pp 379–410. (d) Wegner J, Ceylan S, Kirschning A (2011). Ten key issues in modern flow chemistry. *Chem. Commun.* 47, 4583–4592. (e) Kirschning A (2009). Chemistry in flow systems. *Beilstein J. Org. Chem.* 5, 15.

93. (a) Kirschning A, Altwicker C, Dräger G, Harders J, Hoffmann N, Hoffmann U, Schönfeld H, Solodenko W, Kunz U (2001). PASSflow syntheses using functionalized monolithic polymer-/glass composites in flow-through microreactors. *Angew. Chem., Int. Ed.* 40, 3995–3998. (b) Kunz U, Schönfeld H, Kirschning A, Solodenko W (2003). Polymer/carrier composites as materials and reactors for organic synthesis. *J. Chromatogr. A* 1006, 241–249.

94. (a) Horváth IT, Rábai J (1994). Facile catalyst separation without water: fluorous biphase hydroformylation of olefins. *Science* 266, 72–75. (b) Studer A, Hadida S, Ferritto R, Kim SY, Jeger P, Wipf P, Curran DP (1997). Fluorous synthesis: a fluorous-phase strategy for improving separation efficiency in organic synthesis. *Science* 275, 823–826.

95. (a) Horváth IT (1998). Fluorous biphase chemistry. *Acc. Chem. Res.* 31, 641–650. (b) Curran DP (1998). Strategy-level separations in organic synthesis: from planning to practice. *Angew. Chem., Int. Ed.* 37, 1174–1196. (c) de Wolf E, van Koten G, Deelman BJ (1999). Fluorous phase separation techniques in catalysis. *Chem. Soc. Rev.* 28, 37–41. (d) Fish RH (1999). Fluorous biphasic catalysis: a new paradigm for the separation of homogeneous catalysts from their reaction substrates and products. *Chem. Eur. J.* 5, 1677–1680. (e) Dobbs AP, Kimberley MR (2002). Fluorous phase chemistry: a new industrial technology. *J. Fluorine Chem.* 118, 3–17. (f) Hope EG, Stuart AM (1999). Fluorous biphase catalysis. *J. Fluorine Chem.* 100, 75–83.

96. Yao QW, Zhang YL (2004). Poly(fluoroalkyl acrylate)-bound ruthenium carbene complex: a fluorous and recyclable catalyst for ring-closing olefin metathesis. *J. Am. Chem. Soc.* 126, 74–75.

97. Matsugi M, Curran DP (2005). Synthesis, reaction, and recycle of light fluorous Grubbs–Hoveyda catalysts for alkene metathesis. *J. Org. Chem.* 70, 1636–1642.

98. (a) da Costa RC, Gladysz JA (2006). Fluorous phase-transfer activation of catalysts: application of a new rate-enhancement strategy to alkene metathesis. *Chem. Commun.* 2619–2621. (b) da

Costa RC, Gladysz JA (2007). Syntheses and reactivity of analogues of Grubbs' second generation metathesis catalyst with fluorous phosphines: a new phase-transfer strategy for catalyst activation. *Adv. Synth. Catal.* 349, 243–254.

99. Matsugi M, Kobayashi Y, Suzumura N, Tsuchiya Y, Shioiri T (2010). Synthesis and RCM reactions using a recyclable Grubbs–Hoveyda metathesis catalyst activated by a light fluorous tag. *J. Org. Chem.* 75, 7905–7908.

100. (a) Bergbreiter DE, Bazzi HS, Hongfa C, inventor; Howrey LLP-HN, assignee. Nonpolar phase-soluble metathesis catalysts. US Patent Appl. US2009/0203860. August 1, 2009. (b) Hongfa C, Su HL, Bazzi HS, Bergbreiter DE (2009). Polyisobutylene-anchored *N*-heterocyclic carbene ligands. *Org. Lett.* 11, 665–667. (c) Hongfa C, Tian J, Bazzi HS, Bergbreiter DE (2007). Heptane-soluble ring-closing metathesis catalysts. *Org. Lett.* 9, 3259–3261.

101. (a) Bergbreiter DE, Osburn PL, Smith T, Li C, Frels JD (2003). Using soluble polymers in latent biphasic systems. *J. Am. Chem. Soc.* 125, 6254–6260. (b) Bergbreiter DE, Liu YS, Osburn PL (1998). Thermomorphic rhodium(I) and palladium(0) catalysts. *J. Am. Chem. Soc.* 120, 4250–4251.

102. Grady MC, Kristjansdottir S, Tam W, Older CM, Ritter JC. Development of commercially viable thermomorphic catalysts for controlled free radical polymerization. In: Prunier ML, editor. *Catalysis of Organic Reactions*. CRC Press, Boca Raton, FL; 2008, pp 319–328.

103. Hobbs C, Yang YC, Ling J, Nicola S, Su HL, Bazzi HS, Bergbreiter DE (2011). Thermomorphic polyethylene-supported olefin metathesis catalysts. *Org. Lett.* 13, 3904–3907.

104. Janssen M, Muller C, Vogt D (2011). Recent advances in the recycling of homogeneous catalysts using membrane separation. *Green Chem.* 13, 2247–2257.

105. Fane AG, Schaefer A, Waite TD. *Nanofiltration Principles and Applications*. Elsevier, Amsterdam; 2004, p 560.

106. (a) Vandezande P, Gevers LEM, Vankelecom IFJ (2008). Solvent resistant nanofiltration: separating on a molecular level. *Chem. Soc. Rev.* 37, 365–405. (b) Li X, Vandezande P, Vankelecom IFJ (2008). Polypyrrole modified solvent resistant nanofiltration membranes. *J. Membr. Sci.* 320, 143–150.

107. Dijkstra HP, van Klink GPM, van Koten G (2002). The use of ultra- and nanofiltration techniques in homogeneous catalyst recycling. *Acc. Chem. Res.* 35, 798–810.

108. (a) Goldup A, Westaway MT, Walker G, inventors; British Petroleum Co. Ltd, assignee. Separation of organometallic catalysts by silicon latex membranes under pressure. DE 1953641 A. 1971; US 3645891 A. 1972. (b) Gosser LW, inventor; du Pont de Nemours, E. I., and Co., assignee. Reversible osmosis in the separation of metal catalysts from nitriles, DE 2336763 A1. 1974.

109. Aerts S, Weyten H, Buekenhoudt A, Gevers LEM, Vankelecom IFJ, Jacobs PA (2004). Recycling of the homogeneous co-Jacobsen catalyst through solvent-resistant nanofiltration (SRNF). *Chem. Commun.* 710–711.

110. (a) Nair D, Scarpello JT, White LS, dos Santos LMF, Vankelecom IFJ, Livingston AG (2001). Semi-continuous nanofiltration-coupled Heck reactions as a new approach to improve productivity of homogeneous catalysts. *Tetrahedron Lett.* 42, 8219–8222. (b) Nair D, Scarpello JT, Vankelecom IFJ, dos Santos LMF, White LS, Kloetzing RJ, Welton T, Livingston AG (2002). Increased catalytic productivity for nanofiltration-coupled Heck reactions using highly stable catalyst systems. *Green Chem.* 4, 319–324. (c) Datta A, Ebert K, Plenio H (2003). Nanofiltration for homogeneous catalysis separation: soluble polymer-supported palladium catalysts for Heck, Sonogashira, and Suzuki coupling of aryl halides. *Organometallics* 22, 4685–4691. (d) Pink CJ, Wong HT, Ferreira FC, Livingston AG (2008). Organic solvent nanofiltration and adsorbents; a hybrid approach to achieve ultra low palladium contamination of post coupling reaction products. *Org. Process Res. Dev.* 12, 589–595.

111. (a) Wong HT, See-Toh YH, Ferreira FC, Crook R, Livingston AG (2006). Organic solvent nanofiltration in asymmetric hydrogenation: enhancement of enantioselectivity and catalyst stability by ionic liquids. *Chem. Commun.* 2063–2065. (b) Nair D, Wong HT, Han S, Vankelecom IFJ, White LS, Livingston AG, Boam AT (2009). Extending Ru-BINAP catalyst life and separating products from catalyst using membrane recycling. *Org. Process Res. Dev.* 13, 863–869.

112. Chavan SA, Maes W, Gevers LEM, Wahlen J, Vankelecom IFJ, Jacobs PA, Dehaen W, De Vos DE (2005). Porphyrin-functionalized dendrimers: synthesis and application as recyclable photocatalysts in a nanofiltration membrane reactor. *Chem. Eur. J.* 11, 6754–6762.

113. (a) Koch Membrane Systems, Wilmington, USA, www.kochmembrane.com, (last access 2013.01.13). (b) Evonik MET Ltd., London, U.K., http://duramem.evonik.com/product/duramem-puramem/en/Pages/default.aspx (last accessed 2012.02.19).

114. Inopor, Veisdorf, Germany http://www.inopor.com/en/index.html (last access 2013.01.13).

115. (a) Müller C, Nijkamp MG, Vogt D (2005). Continuous homogeneous catalysis. *Eur. J. Inorg. Chem.* 2005, 4011–4021. (b) Darvishmanesh S, Degrève J, Van der Bruggen B (2010). Physicochemical characterization of transport in nanosized membrane structures. *ChemPhysChem* 11, 404–411.

116. (a) Janssen M, Muller C, Vogt D (2010). Molecular weight enlargement—a molecular approach to continuous homogeneous catalysis. *Dalton Trans.* 39, 8403–8411. (b) van de Coevering R, Gebbink RJMK, van Koten G (2005). Soluble organic supports for the non-covalent immobilization of homogeneous catalysts; modular approaches towards sustainable catalysts. *Prog. Polym. Sci.* 30, 474–490. (c) Hovestad NJ, Ford A, Jastrzebski JTBH, van Koten G (2000). Functionalized carbosilane dendritic species as soluble supports in organic synthesis. *J. Org. Chem.* 65, 6338–6344. (d) Dijkstra HP, Kruithof CA, Ronde N, van de Coevering R, Ramón DJ, Vogt D, van Klink GPM, van Koten G (2002). Shape-persistent nanosize organometallic complexes: synthesis and application in a nanofiltration membrane reactor. *J. Org. Chem.* 68, 675–685. (e) van Klink GPM, Dijkstra HP, van Koten G (2003). Recyclable nanosize homogeneous catalysts. *C. R. Chim.* 6, 1079–1085.

117. Wijkens P, Jastrzebski JTBH, van der Schaaf PA, Kolly R, Hafner A, van Koten G. (2000). Synthesis of periphery-functionalized dendritic molecules using polylithiated dendrimers as starting material. *Org. Lett.* 2, 1621–1624.

118. van der Gryp P, Barnard A, Cronje JP, de Vlieger D, Marx S, Vosloo HCM (2010). Separation of different metathesis Grubbs-type catalysts using organic solvent nanofiltration. *J. Membr. Sci.* 353, 70–77.

119. Keraani A, Renouard T, Fischmeister C, Bruneau C, Rabiller-Baudry M (2008). Recovery of enlarged olefin metathesis catalysts by nanofiltration in an eco-friendly solvent. *ChemSusChem* 1, 927–933.

120. Schoeps D, Buhr K, Dijkstra M, Ebert K, Plenio H (2009). Batchwise and continuous organophilic nanofiltration of Grubbs-type olefin metathesis catalysts. *Chem. Eur. J.* 15, 2960–2965.

121. Keraani A, Rabiller-Baudry M, Fischmeister C, Bruneau C (2010). Immobilisation of an ionically tagged Hoveyda catalyst on a supported ionic liquid membrane: an innovative approach for metathesis reactions in a catalytic membrane reactor. *Catal. Today* 156, 268–275.

122. White LS (2002). Transport properties of a polyimide solvent resistant nanofiltration membrane. *J. Membr. Sci.* 205, 191–202.

123. (a) Azap C, Wolf D, Abbenhuis HCL, Gerritsen G, Grela K, Wilting J, Leszczyńska K, Czaban J, Wojtasiewicz A, inventors; Evonik Degussa GmbH, assignee. Polyhedral oligomeric silesquioxane (poss)-linked ligands. World Patent WO/2011/107417. September 9, 2011. (b) Azap C, Wolf D, Abbenhuis HCL, Gerritsen G, Grela K, Wilting J, Leszczyńska K,

Czaban J, Wojtasiewicz A, inventors; Evonik Degussa GmbH, assignee. Polysilsesquioxane (POSS)-linked imidazole-based carbene and phosphine ligands for transition metal catalysts. European Patent EP2368894. September 28, 2011.

124. (a) Bantreil X, Randall RAM, Slawin AMZ, Nolan SP (2010). Ruthenium complexes bearing two *N*-heterocyclic carbene ligands in low catalyst loading olefin metathesis reactions. *Organometallics* 29, 3007–3011. (b) Kuhn KM, Champagne TM, Hong SH, Wei WH, Nickel A, Lee CW, Virgil SC, Grubbs RH, Pederson RL (2010). Low catalyst loadings in olefin metathesis: synthesis of nitrogen heterocycles by ring-closing metathesis. *Org. Lett.* 12, 984–987. (c) Abbas M, Slugovc C (2011). As low as reasonably achievable catalyst loadings in the cross metathesis of olefins with ethyl acrylate. *Tetrahedron Lett.* 52, 2560–2562. (d) Schmid TE, Bantreil X, Citadelle CA, Slawin AMZ, Cazin CSJ (2011). Phosphites as ligands in ruthenium-benzylidene catalysts for olefin metathesis. *Chem. Commun.* 47, 7060–7062.

125. (a) Burtscher D, Grela K (2009). Aqueous olefin metathesis. *Angew. Chem., Int. Ed.* 48, 442–454. (b) Torborg C, Samojlowicz C, Grela K. Metathesis reaction. In: Kobayashi S, editor. *Water in Organic Synthesis*. Aqueous Media: Reaction of C–C Multiple Bonds, Vol. 2011/5, Science of Synthesis: Methods of Molecular Transformations. Georg Thieme Verlag KG, Stuttgart; 2012, pp 225–256.

126. (a) Binder JB, Blank JJ, Raines RT (2007). Olefin metathesis in homogeneous aqueous media catalyzed by conventional ruthenium catalysts. *Org. Lett.* 9, 4885–4888. (b) Lin YA, Chalker JM, Floyd N, Bernardes GJL, Davis BG (2008). Allyl sulfides are privileged substrates in aqueous cross-metathesis: application to site-selective protein modification. *J. Am. Chem. Soc.* 130, 9642–9643. (c) Connon SJ, Rivard M, Zaja M, Blechert S (2003). Practical olefin metathesis in protic media under an air atmosphere. *Adv. Synth. Catal.* 345, 572–575. (d) Hunter L, Condie GC, Harding MM (2010). Reversible aqueous metathesis reactions for potential application in dynamic combinatorial chemistry. *Tetrahedron Lett.* 51, 5064–5067.

127. Gułajski Ł, Sledz P, Lupa A, Grela K (2008). Olefin metathesis in water using acoustic emulsification. *Green Chem.* 10, 271–274.

128. (a) Lipshutz BH, Ghorai S, Abela AR, Moser R, Nishikata T, Duplais C, Krasovskiy A, Gaston RD, Gadwood RC (2011). TPGS-750-M: a second-generation amphiphile for metal-catalyzed cross-couplings in water at room temperature. *J. Org. Chem.* 76, 4379–4391. (b) Voigtritter K, Ghorai S, Lipshutz BH (2011). Rate enhanced olefin cross-metathesis reactions: the copper iodide effect. *J. Org. Chem.* 76, 4697–4702.

129. (a) Gawin R, Czarnecka P, Grela K (2010). Ruthenium catalysts bearing chelating carboxylate ligands: application to metathesis reactions in water. *Tetrahedron* 66, 1051–1056. (b) Lipshutz BH, Ghorai S (2010). PQS-2: ring-closing- and cross-metathesis reactions on lipophilic substrates; in water only at room temperature, with in-flask catalyst recycling. *Tetrahedron* 66, 1057–1063. (c) Gallivan JP, Jordan JP, Grubbs RH (2005). A neutral, water-soluble olefin metathesis catalyst based on an *N*-heterocyclic carbene ligand. *Tetrahedron Lett.* 46, 2577–2580. (d) Samanta D, Kratz K, Zhang X, Emrick T (2008). A synthesis of PEG- and phosphorylcholine-substituted pyridines to afford water-soluble ruthenium benzylidene metathesis catalysts. *Macromolecules* 41, 530–532. (e) Hong SH, Grubbs RH (2006). Highly active water-soluble olefin metathesis catalyst. *J. Am. Chem. Soc.* 128, 3508–3509.

130. (a) Adjiman CS, Clarke AJ, Cooper G, Taylor PC (2008). Solvents for ring-closing metathesis reactions. *Chem. Commun.* 2806–2808. (b) Samojlowicz C, Bieniek M, Zarecki A, Kadyrov R, Grela K (2008). The doping effect of fluorinated aromatic hydrocarbon solvents on the performance of common olefin metathesis catalysts: application in the preparation of biologically active compounds. *Chem. Commun.* 6282–6284. (c) Reichardt C, Welton T. Solvents and green chemistry. In: *Solvents and Solvent Effects in Organic Chemistry*. Wiley-VCH, Weinheim; 2010, pp 509–548. (d) Gladysz JA, Emnet C. Fluorous solvents and related media. In: *Handbook of Fluorous Chemistry*. Wiley-VCH, Weinheim; 2005, pp 11–23.

131. (a) Debleds O, Campagne JM (2008). 1, 5-Enyne metathesis. *J. Am. Chem. Soc.* 130, 1562–1563. (b) Fustero S, Jiménez D, Sánchez-Roselló M, del Pozo C (2007). Microwave-assisted tandem cross metathesis intramolecular aza-Michael reaction: an easy entry to cyclic β-amino carbonyl derivatives. *J. Am. Chem. Soc.* 129, 6700–6701. (c) Coquerel Y, Rodriguez J (2008). Microwave-assisted olefin metathesis. *Eur. J. Org. Chem.* 1125–1132. (d) Gebauer J, Arseniyadis S, Cossy J (2008). Total synthesis of cystothiazole A by microwave-assisted olefin cross-metathesis. *Eur. J. Org. Chem.* 2701–2704. (e) Grela K (2008). The joy and challenge of small rings metathesis. *Angew. Chem., Int. Ed.* 47, 5504–5507. (f) Kappe CO (2008). Microwave dielectric heating in synthetic organic chemistry. *Chem. Soc. Rev.* 37, 1127–1139. (g) Razzaq T, Kappe CO (2008). On the energy efficiency of microwave-assisted organic reactions. *ChemSusChem* 1, 123–132. (h) Ettari R, Nizi E, Di Francesco ME, Dude MA, Pradel G, Vicík R, Schirmeister T, Micale N, Grasso S, Zappalà M (2008). Development of peptidomimetics with a vinyl sulfone warhead as irreversible falcipain-2 inhibitors. *J. Med. Chem.* 51, 988–996. (i) Boddaert T, Coquerel Y, Rodriguez J (2009). Microwave-assisted cross-metathesis of acrylonitrile. *C. R. Chim.* 12, 872–875. (j) Cros F, Pelotier B, Piva O (2010). Microwave-assisted cross-metathesis of unsaturated thiocyanates: application to the synthesis of thiocyanatins A and B and analogues. *Synthesis* 233–238. (k) Bakhrou N, Lamaty F, Martinez J, Colacino E (2010). Ring-closing metathesis in glycerol under microwave activation. *Tetrahedron Lett.* 51, 3935–3937. (l) Samojłowicz C, Borré E, Mauduit M, Grela K (2011). Microwave-assisted ruthenium-catalyzed olefin metathesis in fluorinated aromatic hydrocarbons: a beneficial combination. *Adv. Synth. Catal.* 353, 1993–2002.

132. (a) Fürstner A, Langemann K (1997). Total syntheses of (+)-ricinelaidic acid lactone and of (−)-gloeosporone based on transition-metal-catalyzed C–C bond formations. *J. Am. Chem. Soc.* 119, 9130–9136. (b) Vedrenne E, Dupont H, Oualef S, Elkaïm L, Grimaud L (2005). Dramatic effect of boron-based Lewis acids in cross-metathesis reactions. *Synlett* 670–672.

133. (a) Kobayashi S. *Water in Organic Synthesis*. Science of Synthesis, Vol. 2011/7, Georg Thieme; 2012. (b) Lindstrom M. *Organic Reactions in Water: Principles, Strategies and Applications*. Blackwell Publishing; 2007.

134. (a) Lin YA, Chalker JM, Davis BG (2009). Olefin metathesis for site-selective protein modification. *ChemBioChem* 10, 959–969. (b) Chalker JM, Bernardes GJL, Lin YA, Davis BG (2009). Chemical modification of proteins at cysteine: opportunities in chemistry and biology. *Chem. Asian J.* 4, 630–640.

135. Dinger MB, Mol JC (2003). Degradation of the first-generation Grubbs metathesis catalyst with primary alcohols, water, and oxygen. Formation and catalytic activity of ruthenium(II) monocarbonyl species. *Organometallics* 22, 1089–1095.

136. (a) Burtscher D, Grela K (2009). Aqueous olefin metathesis. *Angew. Chem., Int. Ed.* 48, 442–454. (b) Zaman S, Curnow OJ, Abell AD (2009). Development of aqueous olefin metathesis catalysts. *Aust. J. Chem.* 62, 91–100. (c) Torborg C, Samojłowicz C, Grela K. Metathesis reactions in science of synthesis: methods of molecular transformations. In: Kobayashi, S, editor. *Water in Organic Synthesis*, Vol. 2011/5. Georg Thieme Verlag KG; 2012, Section 3.6, pp 225–256.

137. Lipshutz BH, Ghorai S (2008). Transition-metal-catalyzed cross couplings going green: in water at room temperature. *Aldrichim. Acta* 41, 59–72.

138. Gułajski Ł, Śledź P, Lupa A, Grela K (2008). Olefin metathesis in water using acoustic emulsification. *Green Chem.* 10, 271–282.

139. (a) Gallivan JP, Jordan JP, Grubbs RH (2005). A neutral, water-soluble olefin metathesis catalyst based on an *N*-heterocyclic carbene ligand. *Tetrahedron Lett.* 46, 2577–2580. (b) Hong SH, Grubbs RH (2006). Highly active water-soluble olefin metathesis catalyst. *J. Am. Chem. Soc.* 128, 3508–3509.

140. (a) Michrowska A, Gułajski Ł, Kaczmarska Z, Mennecke K, Kirschning A, Grela K (2006). A green catalyst for green chemistry: synthesis and application of an olefin metathesis catalyst bearing a quaternary ammonium group. *Green Chem.* 8, 685–688. (b) Gułajski Ł, Michrowska A, Narożnik J, Kaczmarska Z, Rupnicki L, Grela K (2008). A highly active aqueous olefin metathesis catalyst bearing a quaternary ammonium group. *ChemSusChem* 1, 103–109. (c) Jordan JP, Grubbs RH (2007). Small-molecule *N*-heterocyclic-carbene-containing olefin-metathesis catalysts for use in water. *Angew. Chem., Int. Ed.* 46, 5152–5155.

141. Roberts AN, Cochran AC, Rankin DA, Lowe AB, Schanz HJ (2007). Benzylidene-functionalized ruthenium-based olefin metathesis catalysts for ring-opening metathesis polymerization in organic and aqueous media. *Organometallics* 26, 6515–6518.

142. Rölle T, Grubbs RH (2002). Ring closing metathesis in protic media by means of a neutral and polar ruthenium benzylidene complex. *Chem. Commun.* 1070–1071.

143. (a) Lo C, Ringenberg MR, Gnandt D, Wilson Y, Ward TR (2011). Artificial metalloenzymes for olefin metathesis based on the biotin-(strept)avidin technology. *Chem. Commun.* 47, 12065–12067. (b) Mayer C, Gillingham DG, Ward TR, Hilvert D (2011). An artificial metalloenzyme for olefin metathesis. *Chem. Commun.* 47, 12068–12070.

144. (a) Apeiron Synthesis Sp. z o.o., http://www.apeiron-synthesis.com/. (b) Skowerski K, Szczepaniak G, Wierzbicka C, Gułajski Ł, Bieniek M, Grela K (2012). Highly active catalysts for olefin metathesis in water. *Catal. Sci. Technol.* 2, 2424–2427. (c) Skowerski K, Wierzbicka C, Szczepaniak G, Bieniek M, Grela K (2012). Easily removable olefin metathesis catalysts. *Green Chem.* 14, 3264–3268.

145. Bieniek M, Michrowska A, Gułajski Ł, Grela K (2007). A practical larger scale preparation of second-generation Hoveyda-type catalysts. *Organometallics* 26, 1096–1099.

146. (a) Apeiron Synthesis Sp. z o.o., http://www.apeiron-synthesis.com/. (b) Knapkiewicz P, Skowerski K, Jaskólska D, Barbasiewicz M, Olszewski TK (2012). Nitration under continuous flow conditions: convenient synthesis of 2-isopropoxy-5-nitrobenzaldehyde, an important building block in the preparation of nitro-substituted Hoveyda–Grubbs metathesis catalyst. *Org. Process Res. Dev.* 16, 1430–1435.

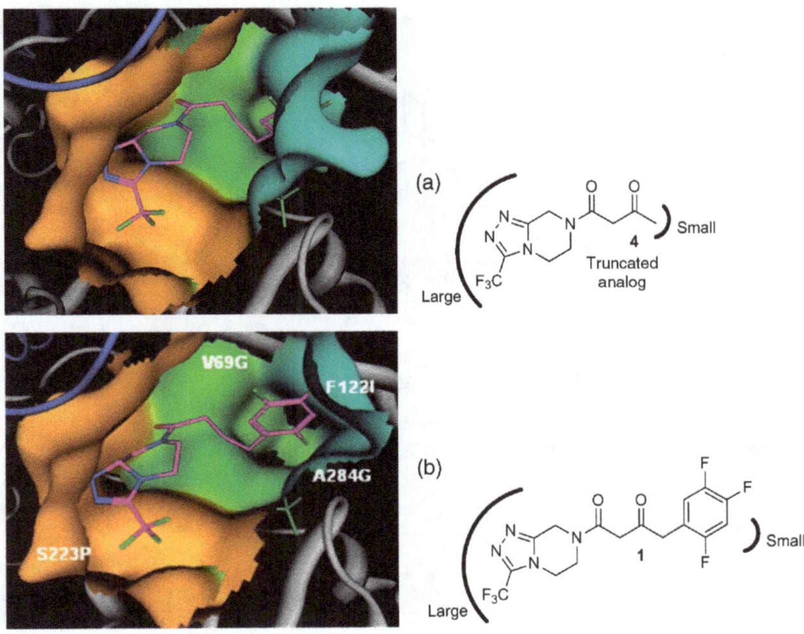

FIGURE 4-1. (a) Wild-type (ATA-117) homology model. (b) Model with V69G, F122I, A284G, and S223P. Reprinted with permission from Ref. 4.

Sustainable Catalysis: Challenges and Practices for the Pharmaceutical and Fine Chemical Industries, First Edition. Edited by Peter J. Dunn, K. K. (Mimi) Hii, Michael J. Krische, and Michael T. Williams.

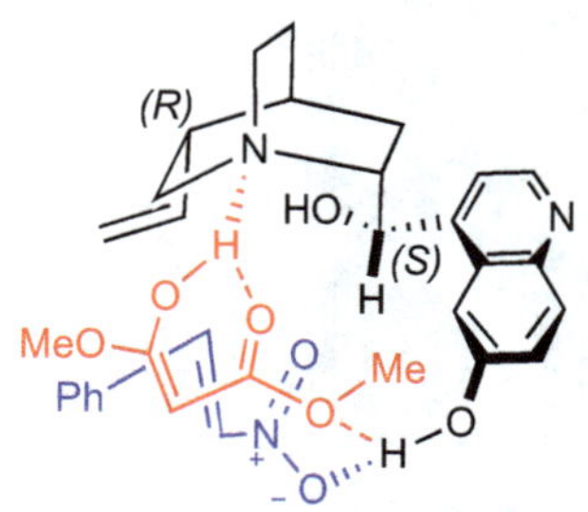

39

FIGURE 14-3. Stereochemical model for the Michael addition reaction catalyzed by DMQ **37**.

10

CHALLENGE AND OPPORTUNITY IN SCALING-UP METATHESIS REACTION: SYNTHESIS OF CILUPREVIR (BILN 2061)

NATHAN YEE, XUDONG WEI, AND CHRIS SENANAYAKE
Chemical Development, Boehringer Ingelheim Pharmaceutical, Inc., Ridgefield, CT, USA

10.1 INTRODUCTION

Over the last decade, considerable effort has been devoted to the design and synthesis of substrate-based peptidomimetic inhibitors that target specifically the NS3 serine protease enzyme of the hepatitis C virus (HCV) [1]. HCV is a severe medical problem, a global infection afflicting more than 170 million people that can cause cirrhosis of the liver and liver cancer [2,3]. Therefore, disrupting the catalytic function of the essential NS3 protease enzyme could provide a treatment option of significant socio-economic benefit.

A number of pharmaceutical companies have invested major efforts in discovery and development of substrate-based, small-molecule inhibitors targeting the HCV NS3 serine protease [1]. Boehringer Ingelheim discovered a new class of macrocyclic peptidomimetics that are significantly superior as drug candidates to their acyclic precursors. In 2002, the first human proof-of-concept for a direct acting anti-HCV drug was achieved with the Boehringer Ingelheim NS3 protease inhibitor ciluprevir (BILN 2061, Figure 10-1) [4,5], which is a 15-membered macrocyclic tripeptide composed of three unnatural amino acid residues with five chiral centers. The design of macrocyclic drugs which are peptidic in nature is undoubtedly inspired by nature's nonribosomally formed secondary metabolites, which are often composed of unusual amino acid fragments and are constrained into cyclic peptides with unique bioactive conformations; well-known pharmaceuticals include cyclosporine, vancomycin, and chloropeptin. However, unlike the production of natural products via fermentation, the chemical manufacturing of structurally complex macrocyclic peptide, such as ciluprevir, provides a unique synthetic challenge for process chemists to develop a scalable process for commercialization. This chapter describes the

Sustainable Catalysis: Challenges and Practices for the Pharmaceutical and Fine Chemical Industries, First Edition. Edited by Peter J. Dunn, K. K. (Mimi) Hii, Michael J. Krische, and Michael T. Williams.

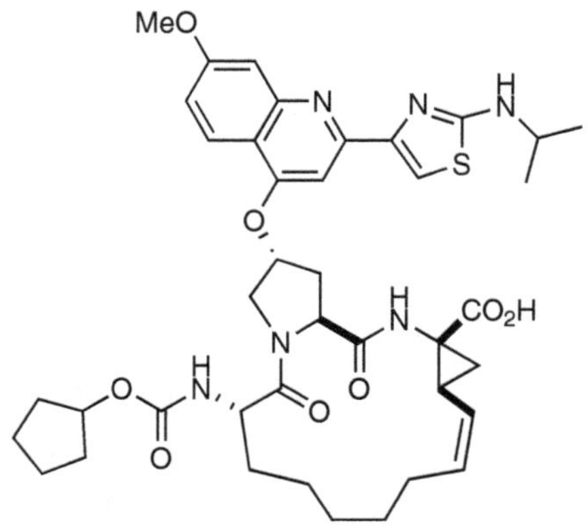

FIGURE 10-1. Ciluprevir (BILN 2061).

challenges and opportunities involved in developing a practical and efficient chemical process for ciluprevir through ring-closing metathesis (RCM) [6]. In particular, we highlight our efforts in developing the first scale-up of RCM reaction for potential commercialization.

10.2 SYNTHESIS OF CILUPREVIR (BILN 2061) AND CRITICAL CHALLENGES

Retrosynthetically, ciluprevir can be assembled from three amino acid building blocks (**1**, **2**, **3**) and a quinoline fragment **4** (Scheme 10-1). Among these building blocks, only fragment **2** (*trans*-(2*S*,4*R*)-hydroxyproline) is readily available from commercial sources. The strategic disconnections were focused on amide linkages and on ether bond, and the macrocyclization would be achieved by RCM. Despite significant advances in academia, discovery syntheses, and material science, no application of the RCM reaction has been reported on a large scale at the time of this project. This limitation is associated with the typical high dilution necessary to prevent dimerization, which limits productivity due to

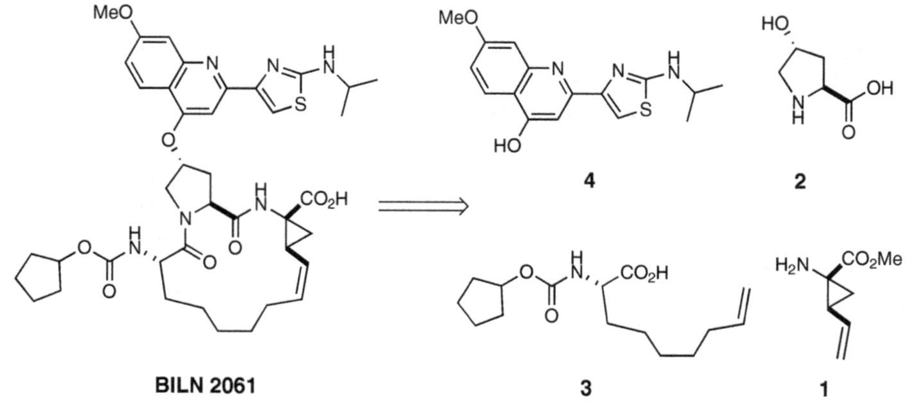

SCHEME 10-1. Retrosynthesis of ciluprevir (BILN 2061).

increasing the volume–time factor for production. Accordingly, key optimization of the reaction conditions and substrate would be necessary to obtain a process amenable for scale-up.

10.3 PREPARATIONS OF BUILDING BLOCKS

The building block (1*R*,2*S*)-1-amino-2-vinylcyclopropanecarboxylic acid methyl ester **1** was initially synthesized by a sequential S_N2–S_N2' dialkylation of protected glycine anions with *trans*-1,4-dibromo-2-butene **7** (Scheme 10-2) [7]. The resulting *cis*-ester isomer **8** was then hydrolyzed and converted into its Boc derivative (±)-**10**, followed by an enzymatic chiral resolution to afford the chiral Boc-protected building block (1*R*,2*S*)-**11** in >98% ee. This chemistry was further improved by the process chemists in terms of yield, efficiency, and operation time for execution in a pilot plant. Methyl glycinate and benzaldehyde were condensed in almost quantitative yield using trimethyl orthoformate as the dehydrating agent to afford the imine **6** which was employed directly for the cyclopropanation reaction with *trans*-1,4-dibromo-2-butene using lithium *tert*-butoxide as the base to provide the racemic intermediate **8** with the vinyl *cis* to the ester group. It should be noted that a systematic investigation was performed using a range of bases such as KOH, CsOH, LiHMDS, NaHMDS, KHMDS, *t*-BuONa, and *t*-BuOK in order to improve the *cis:trans* ratio. The system of *t*-BuOLi and LiHMDS eventually provided the optimal results (>40:1 ratio), presumably due to the formation of a more favorable *s-trans* configuration of the enolate in the transition state, aided by the use of the strongly coordinating Li^+ ion in nonpolar solvents. This procedure proved to be suitable for scale-up. After hydrolysis and treatment with Boc_2O, the racemic Boc-protected vinyl cyclopropane amino ester **10** was obtained in 65–70% overall yield from glycine methyl ester and *trans*-1,4-dibromo-2-butene. For subsequent enzymatic resolution, the optimal procedure utilized a phosphate buffer system to maintain the pH of between 8.1 and 8.2 (adjusted by the periodic addition of 1 N NaOH). The reaction was run until the enantiomeric purity of the remaining ester reached about 99% ee. The ester **11** was back-extracted and converted into the tosylate salt by treatment with *p*-toluenesulfonic acid. Ultimately, the six-step

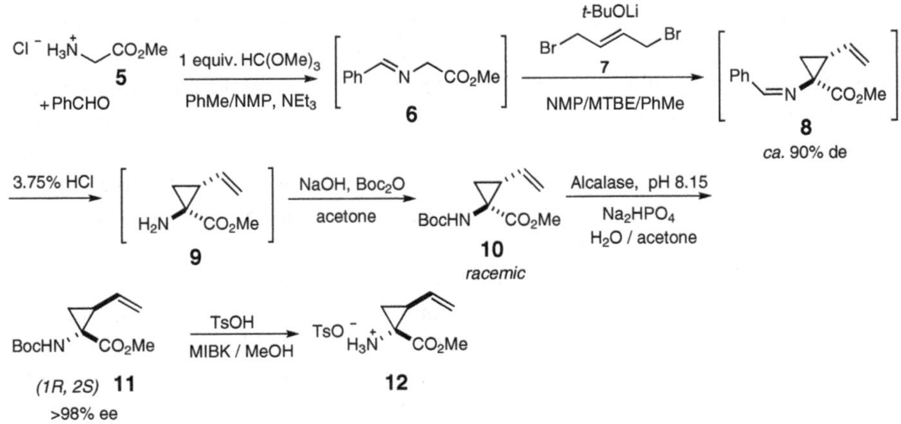

SCHEME 10-2. Synthesis of building block **1** (as salt **12**).

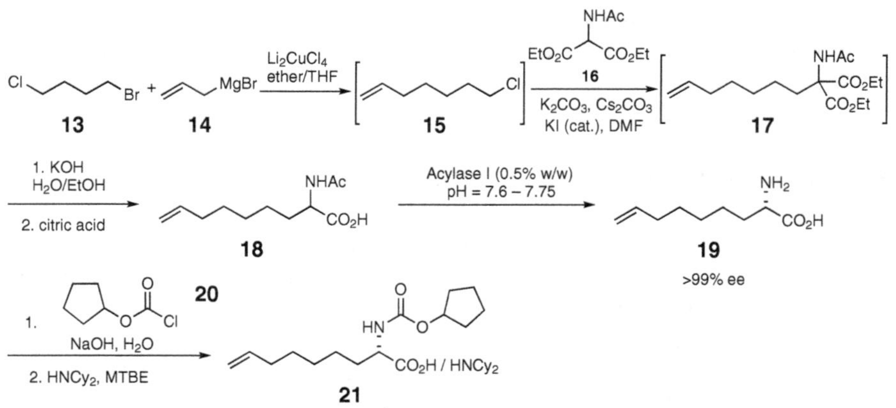

SCHEME 10-3. Synthesis of fragment **3** (as salt **21**).

sequence was streamlined and successfully performed on kilogram scale in only three operations, including a final purification by crystallization to afford the fragment **1** in high purity.

Although (*S*)-2-(*tert*-butyloxycarbonyl)amino-8-nonenoic acid **3** can be prepared by catalytic asymmetric hydrogenation [8], a kinetic resolution was adopted in the scale-up process (Scheme 10-3), due to the increased number of steps required to generate the hydrogenation substrate and associated protection group manipulation [9].

The synthesis began with the preparation of 7-chloro-1-heptene **15** by a Kochi cross coupling of 1-bromo-4-chlorobutane **13** with allylmagnesium bromide **14** in the presence of dilithium tetrachlorocuprate in THF. Alkylation of diethyl acetamidomalonate with the crude 7-chloro-1-heptene **15** was accomplished in high yield by heating the mixture in DMF at 80 °C with a mixture of 4:1 mixture of K_2CO_3 and Cs_2CO_3 bases and a catalytic amount of potassium iodide. Subsequent saponification and decarboxylation of the acetyl malonate derivative **17**, by treatment with KOH in aq. ethanol, followed by neutralization with citric acid, afforded the racemic *N*-acetyl nonenoic acid **18**. This racemic material was subjected to kinetic resolution with 0.5% (w/w) acylase I in water at pH 7.5–8.0. At 37 °C, the (*S*)-amino acid **19** precipitated out smoothly and the solid product was obtained simply by filtration to afford the chiral amino acid in ~45% yield with >99% ee. After acylation under Schotten–Baumann conditions, fragment **3** was produced quantitatively as an oil. For convenient handling and long-term storage purposes, a crystalline salt **21** was made by treatment with dicyclohexylamine in methyl *tert*-butyl ether (MTBE). The (*R*)-enantiomer of the amino acid derivative was easily recycled by racemization of (*R*)-**18** by Ac_2O in water. This process was successfully scaled up on multikilogram scale.

The building block thiazolyl quinolone **4** was synthesized in a comparatively straightforward manner. Isopropylthiourea **22** was condensed with 3-bromopyruvic acid **23** to form the crystalline 2-isopropylaminothiazole-4-carboxylic acid hydrobromide salt **24** in excellent yield (Scheme 10-4) [10]. The carboxylic acid **24** was coupled with 1-(2-amino-4-methoxyphenyl)ethanone **26** through the acid chloride intermediate **25** generated from oxalyl chloride in the presence of catalytic amount of DMF. Base-promoted cyclization in *tert*-butanol furnished the quinolone fragment **4**.

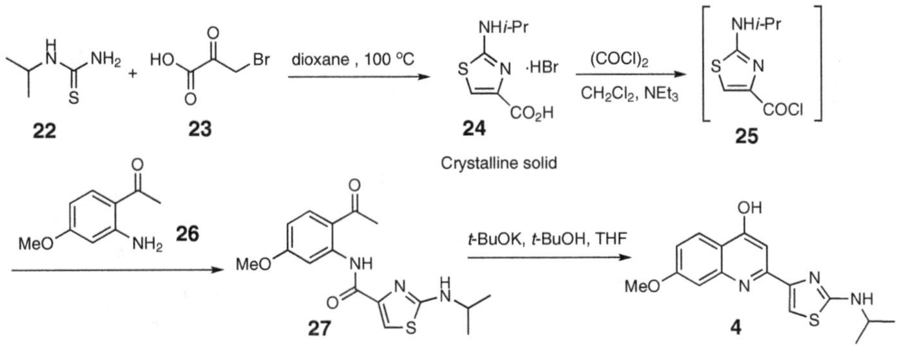

SCHEME 10-4. Synthesis of fragment **4**.

10.4 THE FIRST GENERATION CILUPREVIR (BILN 2061) PROCESS

In the discovery/medicinal chemistry program, ciluprevir was assembled in eight steps by a rational design approach, involving the use of traditional parallel synthesis, which allowed the late-stage modifications of the fragment **3** and heterocycle sites to facilitate structure–activity-relationship (SAR) studies (Scheme 10-5) [8]. Although this route was lengthy and not optimally convergent, it was capable of providing gram-scale quantities of structurally diverse drug candidates to support early-stage SAR, toxicology, and related discovery studies.

A very similar strategy was adopted for the first scale-up campaign, where the coupling of the quinolone building block **4** with the hydroxyproline unit **43** was achieved by an S_N2 substitution. Although the overall number of steps remained the same, improved convergence and efficiency were attained by pursuing two strategies [11,12]. First, the macrocycle was synthesized with the cyclopentyl carbamate unit **3** already in place, thereby circumventing the previous Boc deprotection and acylation operations. Secondly, the coupling of the quinoline unit to the macrocycle was performed after derivatizing the quinoline with the thiazole moiety (Scheme 10-6).

The synthesis commenced with a standard peptide coupling of **12** and **28** units utilizing EDCI to afford the dipeptide intermediate **29**. The stereochemistry of the hydroxy substituent of the proline moiety was inverted with *p*-nitrobenzoic acid through a Mitsunobu reaction, before the effective coupling of the proline amino acid portion of the dipeptide **40** with **3** to furnish the RCM precursor **41** with high yield. The subsequent RCM proceeded with good yield under the initial conditions (0.01 M, 5 mol% Hoveyda's catalyst **33**). After hydrolysis and treatment with *p*-Br-benzenesulfonyl chloride, the brosylate **43** was coupled with the quinolone **4** using cesium carbonate as the base and NMP as the solvent. Hydrolysis of the penultimate ester **38** completed the synthesis of ciluprevir. This process was used to produce the first kilogram batch of the ciluprevir.

This initial scale-up process required further refinement to improve the cost and volume efficiency. Two opportunities were identified for the second approach to reduce the raw material cost. First, delaying the introduction of the relatively high cost vinylcyclopropane fragment into the synthesis. Secondly, the removal of the Mitsunobu substitution (for inverting the hydroxy proline stereocenter) to eliminate the associated low atom economy reagents from the process. These two objects were implemented into the process by

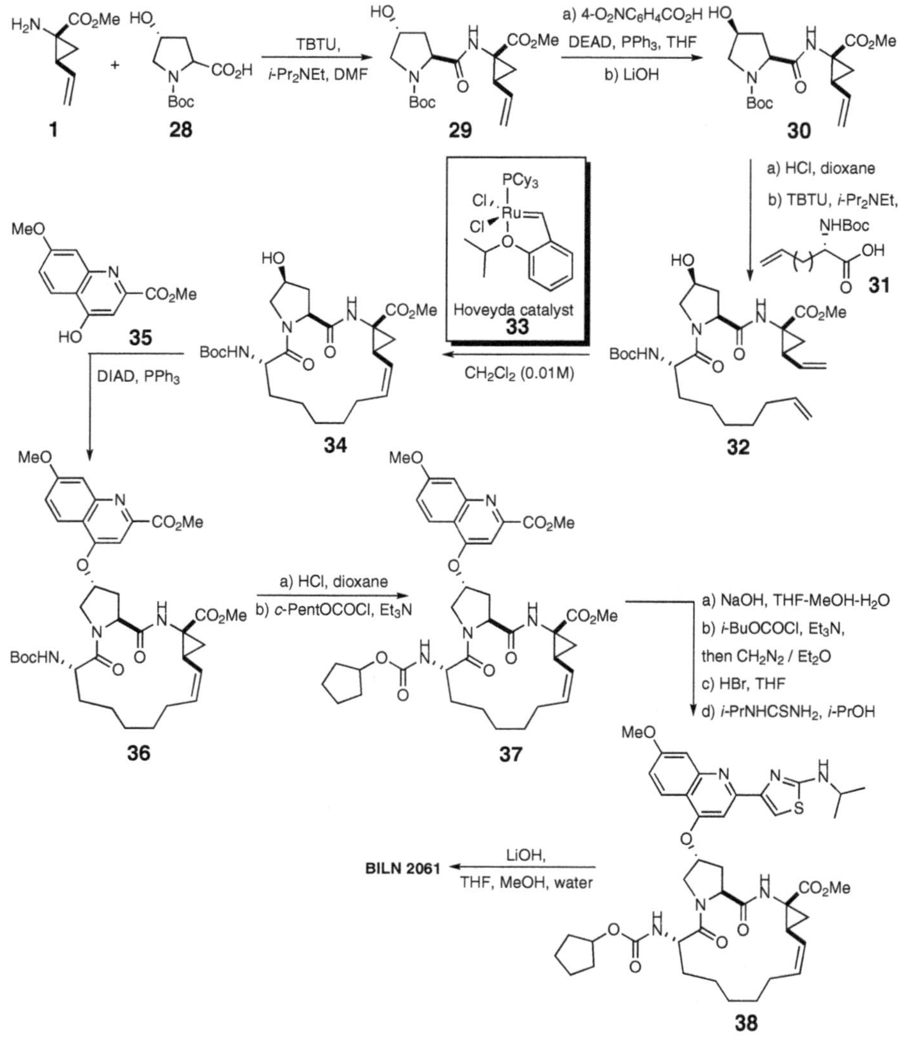

SCHEME 10-5. Synthesis of ciluprevir (BILN 2061) for the discovery program.

converting the *trans*-hydroxyproline segment **28** into the azabicyclo derivative **44**, which was coupled with the optically pure amino acid **3**, prior to coupling with the vinylcyclopropane unit **1** (Scheme 10-7) [11,12].

Treating **28** with methanesulfonyl chloride formed a mesylate which cyclized *in situ* to the azabicyclo intermediate **44,** which was simply purified by direct crystallization from the reaction mixture by addition of water to give the desired product with good yield and excellent quality (>98% purity). This lactonization provided the required inversion of configuration at the hydroxyl moiety of hydroxyproline in high diastereomeric excess. Boc cleavage proceeded smoothly using methanesulfonic acid in MeOAc to directly generate the crystalline mesylate **45** in excellent yield and quality. The first peptide coupling with the (*S*)-amino nonenoic acid derivative **3** was efficiently performed utilizing EDCI as the

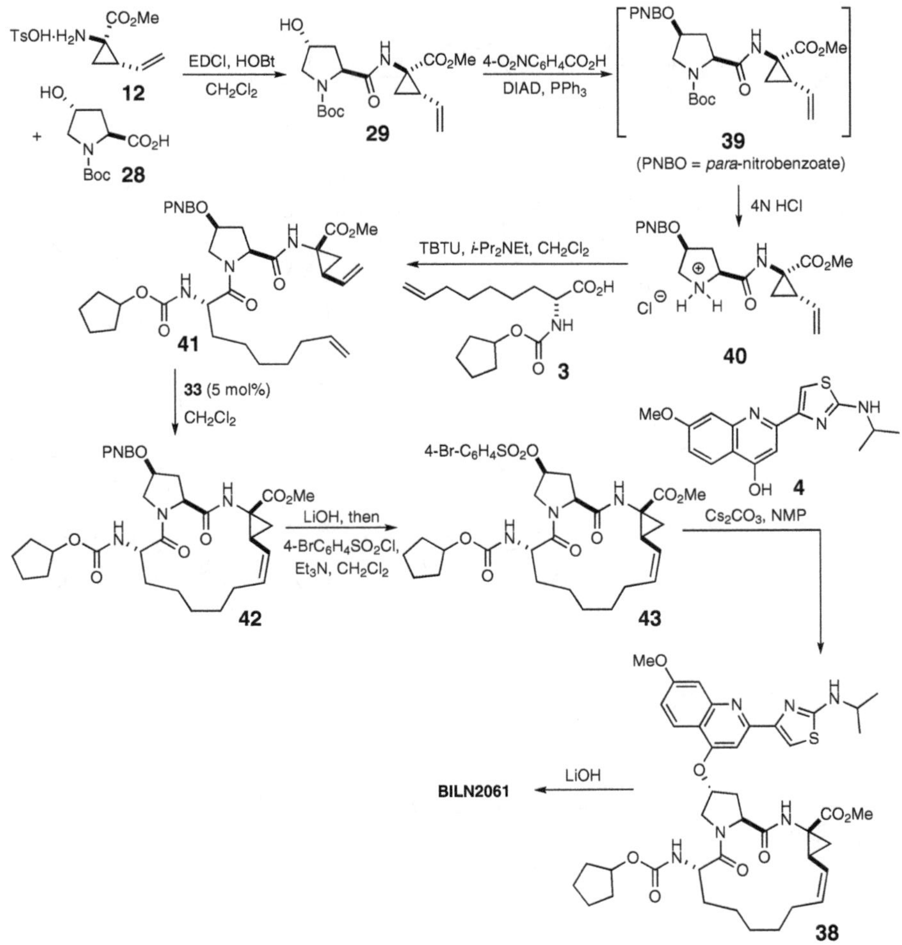

SCHEME 10-6. First scale-up synthesis of ciluprevir (BILN 2061).

more preferred coupling reagent over TBTU. Epimerization during the peptide formation was minimized by conducting the reaction at room temperature, and the resulting dipeptide **46** was recrystallized to acceptable purity from toluene/heptane. The coupling of dipeptide **46** with **1** occurred simply by treating the reaction mixture with sodium 2-ethyl-hexanoate as the base in water. Evidently, the strain of the lactone ring provided sufficient activation to enable the direct acylation and proceeded with a high yield.

Hence, the key intermediate **44** provided three favorable features to the synthesis: (i) it introduced the correct C-4 stereochemistry at a very early stage when no other amino acid was present in the substrate; (ii) it served as a protecting group in the first peptide coupling step; and (iii) it served as an activating group in the second coupling to eliminate a coupling reagent from one step in the process.

Subsequently, the hydroxy diene **47** was converted into the corresponding brosylate **48** with *p*-Br-benzenesulfonyl chloride. The RCM reaction of this brosylate was found to be very comparable to that previously achieved with the PNB-protected diene **41** (Scheme 10-6). Finally, S_N2 displacement with the quinolone and hydrolysis gave

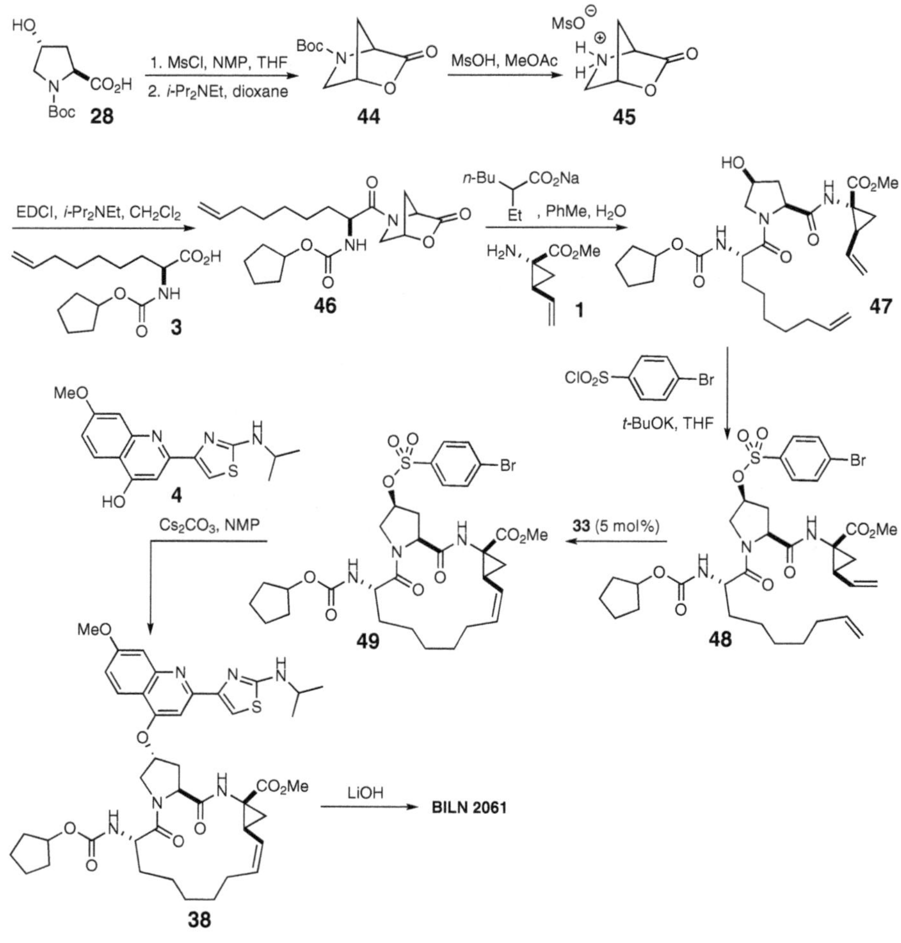

SCHEME 10-7. Second scale-up synthesis of ciluprevir (BILN 2061).

ciluprevir. This process was successfully scaled up to produce >100 kg of ciluprevir to support the early preclinical and clinical studies.

10.5 CHALLENGES IN SCALING UP THE RCM REACTION

While the second scale-up synthesis of ciluprevir (Scheme 10-7) was able to provide sufficient active pharmaceutical ingredient (API) to support development activities, there were still several unmet challenges which required innovative solutions for its further scale-up and commercialization, particularly for the RCM reaction. First, the high dilution of the RCM reaction (10 mM) had to be overcome; this problem is typical of macrocyclizations, including all RCM macrocyclizations known to date. In fact, a high-concentration RCM macrocyclization was unprecedented in the literature at the outset of our work, and our first-generation process was already operating at the high end of the

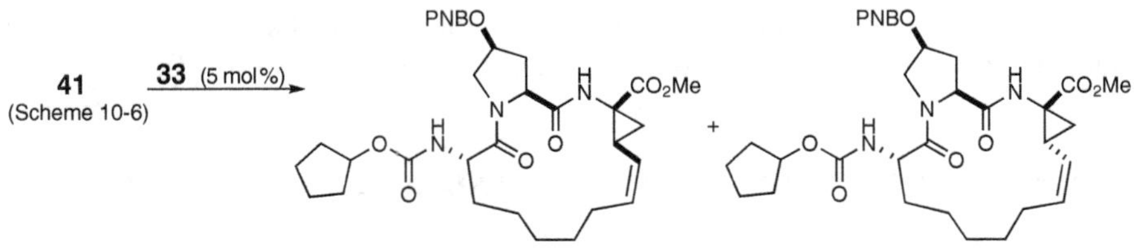

SCHEME 10-8. Epimerization of vinylcyclopropane.

concentration range used in the published work [13]. The reasons for this problem are easily understood: at higher concentration, intermolecular processes start to become competitive with ring closure, thus lowering the yield of the desired cyclization product. In practice, one is then forced to compromise between high yield and high throughput.

A second problem with the first-generation process was the high catalyst loading. In the early runs, 5–7 mol% of the expensive catalyst **33** was routinely used. On a pilot-plant scale, with the starting diene **48** in high purity, it was possible to reduce the catalytic loading to about 3 mol%, without compromising high yields of **49**. This limitation is due to the intrinsically low stability and low efficiency of catalyst **33**. On a commercial scale, this is impractical, due not only to the high cost but also to the extreme difficulty of removing the high levels of Ru metal from the API. Therefore, a better catalyst was required.

A third problem was the long reaction times, often in excess of 24 h and in some cases, reaching 40 h. In a highly streamlined commercial process, where all downstream manipulations have been optimized, this step would be the bottleneck, further reducing the already meager throughput.

A fourth problem we have encountered from time to time is the epimerization of the vinyl-bearing cyclopropane (P1) carbon atom (see Scheme 10-8), due to trace impurities of amines or phosphines in the reaction medium. Although we have clarified some of the mechanistic details of the epimerization [14], this troublesome side reaction requires scrupulous quality control in the reaction solvent as well as the starting diene, thus requiring extensive analytical work. Ideally, we would prefer a robust RCM reaction that can be realized under a more permissive range of conditions, thus improving ease of operation.

Second-generation catalysts, such as Grela catalyst **51** (Figure 10-2) [15], address the second and third problems rather successfully, due to their better TONs and TOFs.

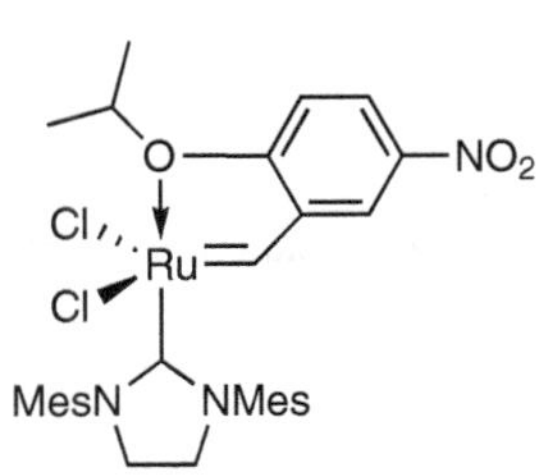

51

FIGURE 10-2. Grela catalyst **51**.

However, they operate under thermodynamic conditions instead of kinetic ones [11], which produce more unwanted products from intermolecular metathesis [16,17].

10.6 DEVELOPMENT OF A PRACTICAL AND SCALABLE RCM PROCESS

The development of a more practical RCM process started from various initial observations on our early RCM process. First, it was noticed that the remote substituent at the C-4 position of the hydroxyproline moiety had a small but detectable effect on the RCM rate and therefore on the effective molarity [18] ($EM = k_{intra}/k_{inter}$). This small effect was tentatively ascribed to subtle conformational factors. Second, when the initiation of the reaction was monitored using a stoichiometric amount of Grubbs' catalyst **52**, carbene transfer occurred to a large extent (96%) at the vinylcyclopropane moiety (Scheme 10-9), where the Ru may be stabilized by chelation. Such stabilization, in turn, may reduce the concentration of the active Ru catalyst in the reaction and negatively affect the rate of the RCM reaction [14].

Guided by these observations, a number of derivatives were subsequently prepared, in which the amide bond was protected with various removable groups [19,20]. We expected that such substitution might interrupt the coordinative stabilization by the ester group through steric interaction ($A^{1,3}$ strain) and therefore divert the catalyst to initiate at the nonenoic acid moiety, which may be beneficial to the RCM [21,22]. Gratifyingly, our postulation was experimentally verified (Scheme 10-10).

The resting state of the Grubbs' catalyst **52** was readily determined by ^{1}H NMR spectroscopy as already reported [14]. Indeed, whereas *N*-benzyl substitution (**55c**) had a relatively minor effect on the site of initiation, acylation of the *N* atom led to inhibition of carbene transfer to that position. Instead, carbene transfer took place completely (>98%) at the nonenoic acid moiety (57b and 57d were produced in >90% conversion).

With these results in hand, the RCM of substrate **55b** was attempted under standard conditions (10 mM, toluene, 60 °C), but employing a second-generation Ru catalyst **51** [15].[1] The desired RCM took place with an initial rate that was three to four times faster than that for substrate **55a** under identical conditions, leading to the desired product in

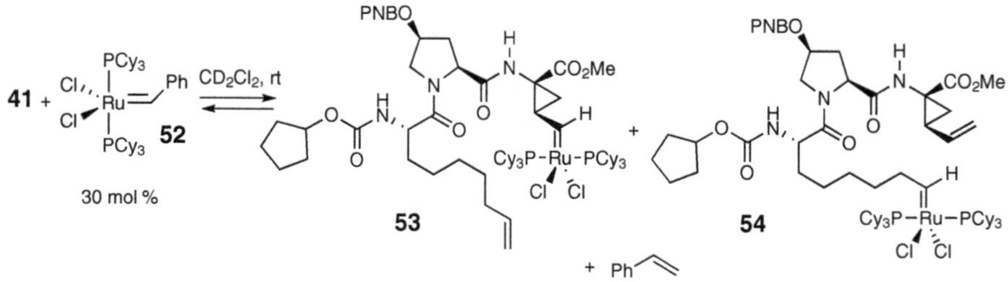

SCHEME 10-9. Initiation of RCM reaction.

[1] See Ref. 15. We are fully aware of the fact that initiation with the Grela catalyst may be different from that with catalyst **33** or **52**. On the other hand, only catalysts containing two strong donor ligands give rise to observable intermediates, and those with IMes and SIMes ligands cyclize too quickly to make the carbene transfer products observable.

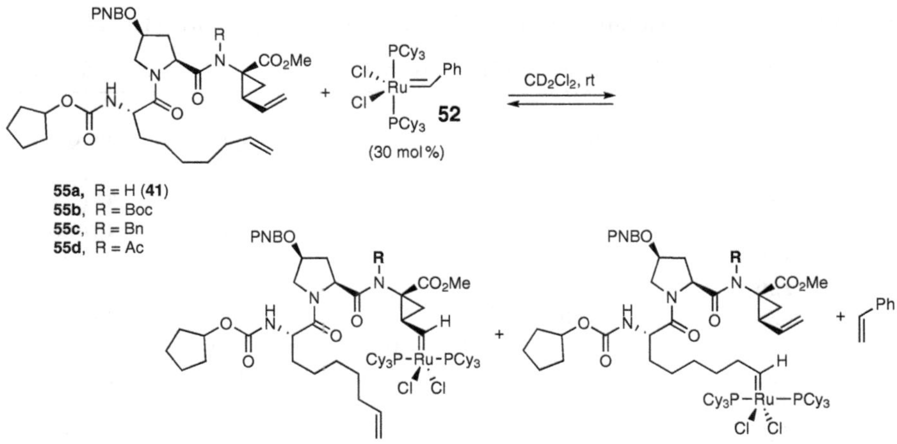

55a, R = H (**41**)
55b, R = Boc
55c, R = Bn
55d, R = Ac

56a, R = H, 96% (**53**)
56b, R = Boc, 0%
56c, R = Bn, 52%
56d, R = Ac, 0%

57a, R = H, 4% (**54**)
57b, R = Boc, 100%
57c, R = Bn, 48%
57d, R = Ac, 100%

SCHEME 10-10. Effect of *N*-protecting group on metathesis initiation.

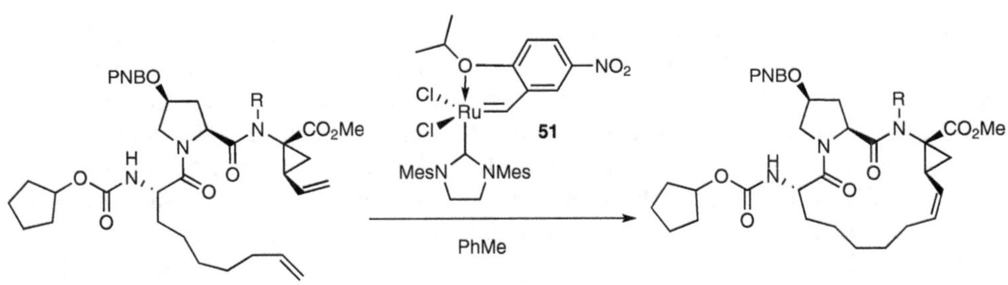

55a, R = H (**41**)
55b, R = Boc
55c, R = Bn
55d, R = Ac

58a, R = H (**42**)
58b, R = Boc
58c, R = Bn
58d, R = Ac

SCHEME 10-11. RCM reaction of modified substrates catalyzed by Grela catalyst **51**.

quantitative yield (>98%), without formation of dimers. This is a significant improvement over the RCM reaction of **55a** (vide infra). Reactions with our key substrates were then tested, presumably under thermodynamic conditions[2] on the basis of our previous work (Scheme 10-11 and Table 10-1).

As entries 1–4 show, in the case of the unprotected **55a**, the yield of the RCM product **58a** decreases as the substrate concentration increases. The *N*-Boc substrate, **55b**, on the other hand, operating at a 10-fold higher concentration (0.10 M, entry 8), provided essentially the same yield as the best result obtained with **55a**. Furthermore, by running

[2] Our previous studies on diene **41/55a** showed the reaction to be readily reversible under these conditions. See ref 11.

TABLE 10-1. Effect of Concentration on RCM Reactions of 55a–d Catalyzed by 51.[a]

Entry	Substrate	Concentration (M)	**51** (mol%)	T (°C)	Yield of **58a–d**[b] (%)
1	**55a (41)**	0.01	1	60	82 (95)
2	**55a (41)**	0.02	1	60	70 (91)
3	**55a (41)**	0.05	1	60	52 (80)
4	**55a (41)**	0.10	1	60	35 (72)
5	**55b**	0.01	1	60	98 (98)
6	**55b**	0.02	1	60	97 (98)
7	**55b**	0.05	1	60	87 (92)
8	**55b**	0.10	1	60	80 (87)
9	**55b**	0.05	0.1	110	97 (100)
10	**55b**	0.10	0.1	110	95 (100)
11	**55b**	0.20	0.1	110	93 (100)
12	**55b**	0.40	0.1	110	80 (99)
13	**55c**	0.01	1	60	85 (95)
14	**55c**	0.10	1	110	48 (90)
15	**55d**	0.01	1	60	97 (99)
16	**55d**	0.10	0.1	110	89 (98)

[a]Reactions were run on 1 mmol scale. Yields were determined by the quantitative HPLC assay.
[b]Values in parenthesis refer to starting material consumed as determined by the HPLC assay.

the reaction at higher temperatures [23], both the yield and the EM increased, such that catalyst loading can be lowered to 0.1%. Finally, entry 11 demonstrates that the goal of operating with $\leq$0.1 mol% catalyst can be achieved under standard concentrations ($\geq$0.2 M). At even higher concentration (0.4 M, entry 12), up to 20% yield of dimer was detected, therefore limiting the yield of the RCM[3]. In comparison, the *N*-benzyl derivative **55c** behaved similarly to **55a**, whereas the results obtained using the acetyl derivative **55d** mirrored that of **55b**. These data indicate that the increase of EM by *N*-substitution with an electron-withdrawing group was closely related to the shift of initiation from **56** to **57** [24].

Taking into account the reversible nature of the metathesis reaction and high reactivity of **51** suggests that at least part of the effect of the electron-withdrawing *N*-substituent is to increase the thermodynamic EM, possibly by reducing the ring strain. Indeed, both amide and the cyclopropane ring represent trans elements, which strain the ring system in **58a**. It is possible that the removal of the enforced planarity in **58b** by *N*-Boc substitution may reduce ring strain and improve the thermodynamic EM. Thus, a theoretical analysis was carried out to (1) calculate the strain energy of the macrocycle and (2) gain an understanding of the conformational characteristics of the open chain and cyclized molecules.

The strain energy content of the macrocycle was calculated by first determining the conformational energy change between the open chain molecules with and without Boc substitution, **55a** and **55b**, and then comparing with the energy change between cyclized molecules **58a** and **58b** with the same chemical modifications. The difference between these two energy changes, that is ΔE, is the contribution of Boc substitution to the strain energy of the molecule. Because we were only interested in the strain energy of the core structure, the PNB group was omitted from our calculations for simplicity. The conformational energy of the open chain and cyclized molecules was calculated by molecular

[3] The main cyclic dimeric product was isolated, and separately converted to **58b** under standard RCM conditions.

TABLE 10-2. Calculated Reduction of Strain Energy by Boc Substitution on Molecule 58b in Comparison to 58a

Method	OPLS01[a]	MM3[a]	MMFFs[a]	DFT/B3LYP[b]
Energy change (kcal mol^{-1})	−3.33	−1.99	−1.10	−2.18

[a]Calculated from Boltzmann average of the energy minima of the chain and ring molecules sampled by the Macromodel program in water. Solvation energy was calculated by the generalized Born model provided in the Macromodel program.
[b]Calculated from full geometry optimization of a single structure for each molecule in gas phase with the Jaguar program.[27] The basis set used in the calculation was 6-31G*. The calculated gas-phase strain energy change is −4.26 kcal mol^{-1}. Calculation with the Jaguar program using the continuous solvation model gave a difference of solvation energy of 2.08 kcal mol^{-1}. The sum of gas-phase strain energy and solvation energy is shown in the table.

mechanics (MM) and quantum chemical (QM) DFT methods. MM methods are computationally efficient. Using the powerful torsion-scan/low-mode search [25] MC algorithm from the Macromodel program [26], the unique energy minima and the lowest energy conformations of the relevant molecules can be identified. However, since the parametrizations of current force fields are not fully validated for the complex, polar-lipophilic macrocycles like the present system, the calculated energies obtained with the MM method may not reflect the true strain energy well. Therefore, QM DFT methods were further used to calculate the energies of the most representative conformation for the cyclized and open chain molecules.

Table 10-2 summarizes the calculated strain energy reduction in the macrocycle due to Boc substitution. The calculations using a number of theoretical methods invariably indicate that Boc substitution at the P1/P2 *N*-atom reduces the strain energy of the ring molecule by ~2 kcal mol^{-1}, which is consistent with the experimental results. By analyzing the QM DFT optimized structures, it was determined that the zero-point energy of the molecules does not make a net contribution to the ΔG change in the Boc-substituted macrocycle, thus minimizing entropic contributions.

The origins of the "*N*-Boc effect"[4] seem to be grounded in favorable kinetic and thermodynamic effects. We have shown that strategic introduction of removable groups on the RCM linker can direct the initiation site and have a remarkable effect on the RCM, thus complementing the known relay strategy [21,22] and can also affect the strain content in the product, thus dramatically increasing the thermodynamic EM.

Table 10-3 highlights the key features of the first-generation process (using either catalyst) versus the second-generation process [28]. For the first-generation process,

TABLE 10-3 Results of the First- versus Second-Generation RCM Process

	Process with Hoveyda Catalyst **33**	Process with Grela Catalyst **51**	Second-Generation Process with **51**
Catalyst loading	3–5 mol%	0.3–0.4 mol%	0.05–0.1 mol%
Reaction time	24–40 h	≤ 1 h	30 min
Initial diene concentration	0.01 M	0.01 M	0.1–0.2 M
RCM yield (HPLC assay)	92%	85%	93%

[4] We prefer the Boc group over other acyl groups like the acetyl because Boc groups are easily removed from the product, thus making the *N*-Boc a "protecting-activating moiety" in this case.

although the Grela catalyst (**51**) improves TON and TOF, the formation of large amounts of dimers compromised the isolated yield (ca. 75% after two recrystallizations, representing a further 10% yield loss), and therefore, its use, *per se*, does not result in a practical process improvement. The use of catalyst **51** in conjunction with modified substrate **58a**, on the other hand, leads to a major improvement in all areas and meets our original goals of an ideally efficient, green process. The following considerations hold regarding the new RCM process (from **55b** to **58b** with Grela catalyst **51**) shown, in comparison with our first-generation manufacturing process (from **55a** to **58a** with Hoveyda catalyst **33**):

(a) *Catalyst efficiency (TON and TOF)*: Second-generation catalysts react, in the process shown in Scheme 10-1, about three orders of magnitude more rapidly than first-generation ones like **52** (R.D. Simpson, unpublished observations). In addition, metathesis of **55b** is about three times faster than that of **55a** under typical conditions (C. Shu, unpublished observations). This allows the reaction to be complete in minutes instead of days (as required in the original process), thus removing the bottleneck. The much higher TOF achieved with **51** is complemented by a higher TON, which is 50–100 times higher than that of **33** under identical conditions. This introduces major savings in catalyst use. Furthermore, it was also observed that the reaction of substrates **55a–58a** requires scrupulous degassing for optimum results with Hoveyda's catalyst **33**, whereas the reaction of **55b–58b** requires only a brief boil-out before adding the Grela catalyst **51**. The superior robustness of **51** allowed us to dispense with oxygen sensors and oxygen specifications in the system, rendering the RCM much easier to execute on scale.

(b) *Epimerization reaction at P1*: Although the epimerization reaction in the RCM shown in Scheme 10-8 can be avoided by ensuring that all solvents used in the RCM are acid-washed, the new RCM protocol requires no such special precautions. The reasons can be summarized as follows [20]: changing the resting state, the Ru catalyst resides at the vinylcyclopropane moiety rather than at the nonenoic acid moiety. Thus epimerization at P1 cannot occur even in the presence of amines or phosphines. This feature allowed simplification of manipulations, quality control of solvents and intermediates, and makes the RCM reaction easier to execute.

(c) *Solvent consumption*: While the original process utilized as much as 150,000 l solvent to process 1 MT of diene **55a**, the same amount of diene **55b** can now be processed with only 7500 l using the new process. This has important repercussions on the type of equipment used to carry out the reaction (simple 2000 or 4000 l vessels are now adequate for production). Also importantly, the new process does not require the installation of equipment for rapid evaporation of organic solvents (such as wiped-film evaporators).

(d) *Ruthenium removal*: In the original process [11], quenching of the catalyst and extractions required excess 2-mercaptonicotinic acid (MNA, 2 kg for 1 kg diene) and large volumes of bicarbonate to remove some of the solubilized Ru. Silica filtration was then employed in order to further reduce the Ru level after the RCM reaction to 100–200 ppm. Subsequent chemical steps had to be coupled with charcoal filtrations in order to reduce the Ru levels in the API to below 10 ppm. These laborious operations could now be spared using the new RCM reaction: given the much lower catalyst load, the amount of MNA used was 50-fold lower. No silica pad or charcoal filtrations were necessary, and the Ru content of the RCM product was typically <50 ppm, yielding an API with <5 ppm Ru without further specific removal steps.

(e) *Overall reaction efficiency*: Of all parameters that have been proposed to evaluate the efficiency of a reaction, the *E*-factor, introduced by Sheldon, is the easiest to use [29]. This parameter refers to the amount of waste (e.g., in kg) for each unit (e.g., 1 kg) of useful product obtained, and includes all reagents and solvents. We have calculated the *E*-factors for both processes of **55a–58a** and **55b–58b**. The new process using **55b** includes the introduction and the removal of the Boc group. The first-generation RCM using 55a is accompanied by an *E*-factor of 370, whereas the second-generation process using **55b** has an *E*-factor of 52. The suitability and efficiency of solvent recycles have not yet been evaluated, although they would improve the *E*-factor of both processes. In conclusion, the second-generation RCM process improves the greenness of the process by about one order of magnitude.

10.7 THE SECOND GENERATION CILUPREVIR (BILN 2061) PROCESS

Since the major breakthrough for scaling up the key RCM reaction was achieved through the investigation of the kinetic and thermodynamic effective molarities, EM(kin) and EM (therm), the development of the second-generation process of ciluprevir was focused on the further improvement of its assembly convergence. The higher level of assembly convergence called for the direct installation of the quinoline heterocycle to the macrocycle through a S_NAr reaction, which avoided the double inversions of the stereocenter at the C-4 position of the hydroxyproline moiety.

The second-generation process for ciluprevir used the same building blocks as described above but employed a direct S_NAr reaction to efficiently attach the quinoline moiety to the macrocyclic core (Scheme 10-12) [30,31]. The building block (*S*)-2-(cyclopentyloxycarbonyl)-amino-8-nonenoic acid **21** was coupled with *trans*-hydroxyproline ester **59** using cyanuric chloride as the peptide-coupling reagent. After hydrolysis, the crystalline dipeptide **60** was activated as the mixed anhydride and reacted with building block **12** to form a tripeptide. In the same reaction vessel, the hydroxyl group of the intermediate was acetylated with acetic anhydride and the amide-NH was protected with Boc_2O to give highly crystalline intermediate **61** with high purity (>99% by HPLC). A solution of **61** (0.1 M in toluene) was heated to reflux and a solution of Grela catalyst **51** (0.1 mol%) was added. The RCM reaction was rapidly completed with the release of ethylene gas to furnish the desired RCM product **62** in excellent yield. Again, in the same vessel, the crude RCM product **62** was treated with methanesulfonic acid followed by NaOH to sequentially deprotect the Boc group, acetyl group, and ester, and provide the hydroxy acid intermediate **63**, which was also a highly crystalline compound. The quinolinyl heterocycle was directly attached to the macrocyclic core **63** through the S_NAr reaction with quinolinyl sulfone **64** (prepared from the corresponding quinolone **4**) [30,31] to complete the synthesis of ciluprevir.

While this final process represented the optimal convergent strategy for the assembly of ciluprevir, the most worthy of the accomplishments in this process was Boehringer Ingelheim's innovative solution to their scale-up of the RCM reaction from both academic and industrial perspectives.

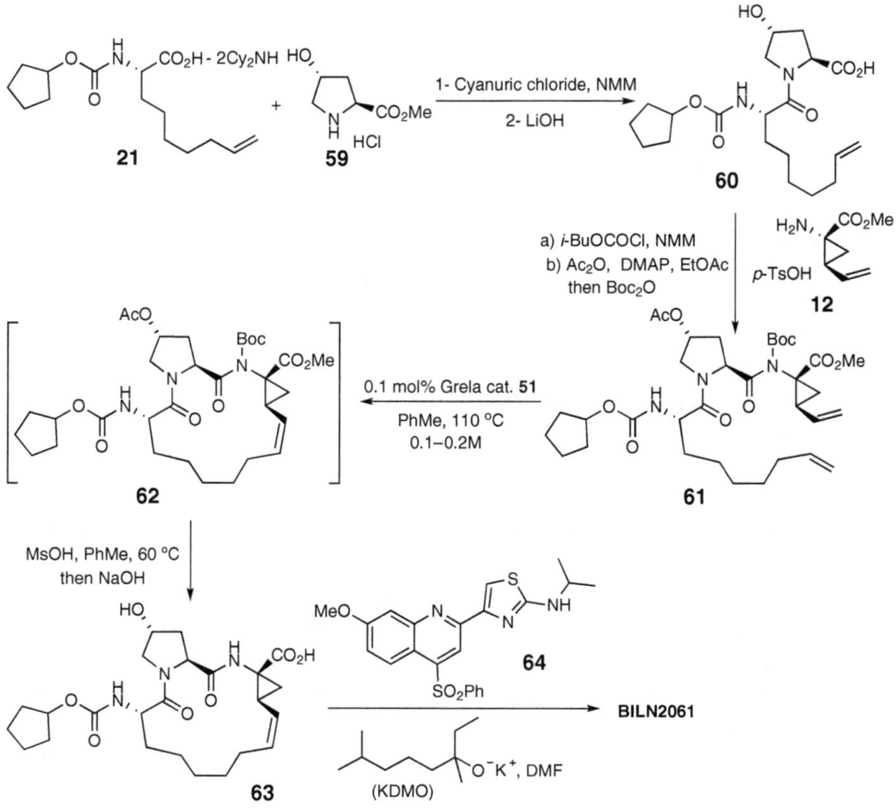

SCHEME 10-12. Second-generation synthesis of ciluprevir (BILN 2061).

10.8 CONCLUSION

Development of a scalable and cost-effective process for ciluprevir presents unique synthetic challenges for process chemists, particularly for the first scale-up of the RCM reaction for the macrocycle formation. During the course of these studies, a number of critical parameters for the RCM reaction, such as the nature of the precatalysts, initiation site, and conformations of substrate/products, were studied which led to the successful development of a practical process for scaling up the RCM reaction. Also, the second-generation ciluprevir process represents the optimal convergent strategy for the assembly of ciluprevir for its potential commercialization.

REFERENCES

1. Chen KX, Njoroge FG (2009). A review of HCV protease inhibitors. *Curr. Opin. Invest. Drugs* 10, 821–837.
2. Lavanchy D (2009). The global burden of hepatitis C. *Liver Int.* 29 (Suppl. 1), 74–81.
3. Brown RS (2005). Hepatitis C and liver transplantation. *Nature* 436, 973–978.

4. Lamarre D, Anderson PC, Bailey M, Beaulieu P, Bolge G, Bonneau P, Bos M, Cameron DR, Cartier M, Cordingley MG, Faucher AM, Goudreau N, Kawai SH, Kukolj G, Lagace L, LaPlante SR, Narjes H, Poupart MA, Rancourt J, Sentjens RE, St. George R, Simoneau B, Steinmann G, Thibeault D, Tsantrizos YS, Weldon SM, Yong CL, Llinas-Brunet M (2003). An NS3 protease inhibitor with antiviral effects in humans infected with hepatitis C virus. *Nature* 426, 186–189.
5. Hinrichsen H, Benhamou Y, Wedemeyer H, Reiser M, Sentjens RE, Calleja JL, Forns X, Erhardt A, Cronlein J, Chaves RL, Yong CL, Nehmiz G, Steinmann GG (2004). Short-term antiviral efficacy of BILN 2061, a hepatitis C virus serine protease inhibitor, in hepatitis C genotype 1 patients. *Gastroenterology* 127, 1347–1355.
6. Grubbs RH *Handbook of Metathesis*, Wiley-VCH, Weinheim; 2003.
7. Beaulieu PL, Gillard J, Bailey MD, Boucher C, Ducepee JS, Simoneau B, Wang XJ, Zhang L, Grozinger K, Houpis I, Farina V, Heimroth H, Krueger T, Schnaubelt J (2005). Synthesis of (1*R*,2*S*)-1-amino-2-vinylcyclopropanecarboxylic acid vinyl-ACCA derivatives: key intermediates for the preparation of inhibitors of the hepatitis C virus NS3 protease. *J. Org. Chem.* 70, 5869–5879.
8. Faucher AM, Bailey MD, Beaulieu PL, Brochu C, Duceppe JS, Ferland JM, Ghiro E, Gorys V, Halmos T, Kawai SH, Poirier M, Simoneau B, Tsantrizos YS, Llinàs-Brunet M (2004). Synthesis of BILN 2061, a HCV NS3 protease inhibitor with proven antiviral effect in humans. *Org. Lett.* 6, 2901–2904.
9. Wang XJ, Zhang L, Smith-Keenan LL, Houpis IN, Farina V (2007). Efficient synthesis of (*S*)-2-(cyclopentyloxycarbonyl)-amino-8-nonenoic acid: key building block for BILN 2061, an HVC NS3 protease inhibitor. *Org. Process Res. Dev.* 11, 60–63.
10. Frutos RP, Haddad N, Houpis IN, Johnson M, Smith-Keenan LL, Fuchs V, Yee NK, Farina V, Faucher AM, Brochu C, Hache B, Duceppe JS, Beaulieu P (2006). Practical synthesis of 2-(2-isopropylaminothiazol-4-yl)-7-methoxy-1*H*-quinoline-4-one: key intermediate for synthesis of potent HCV NS3 protease inhibitor BILN 2061. *Synthesis* 2563–2567.
11. Yee NK, Farina V, Houpis IN, Haddad N, Frutos RP, Gallou F, Wang XJ, Wei X, Simpson RD, Feng X, Fuchs V, Xu Y, Tan J, Zhang L, Xu J, Smith-Keenan LL, Vitous J, Ridges MD, Spinelli EM, Johnson M, Donsbach K, Nicola T, Brenner M, Winter E, Kreye P, Samstag W (2006). Efficient large-scale synthesis of BILN 2061, a potent HCV protease inhibitor, by a convergent approach based on ring-closing metathesis. *J. Org. Chem.* 71, 7133–7145.
12. Nicola T, Brenner M, Donsbach K, Kreye P (2005). First scale-up to production scale of a ring closing metathesis reaction forming a 15-membered macrocycle as a precursor of an active pharmaceutical ingredient. *Org. Process Res. Dev.* 9, 513–515.
13. Gradillas A, Péerez-Castells J (2006). Macrocyclization by ring-closing metathesis in the total synthesis of natural products: reaction conditions and limitations. *Angew. Chem., Int. Ed.* 45, 6086–6101.
14. Zeng X, Wei X, Farina V, Napolitano E, Xu Y, Zhang L, Haddad N, Yee NK, Grinberg N, Shen S, Senanayake CH (2006). Epimerization reaction of a substitued vinylcyclopropane catalyzed by ruthenium carbenes: mechanistic analysis. *J. Org. Chem.* 7, 8864–8875.
15. Michrowska A, Bujok R, Harutyunyan S, Sashuk V, Dolgonos G, Grela K (2004). Nitro-substituted Hoveyda–Grubbs ruthenium carbenes: enhancement of catalyst activity through electronic activation. *J. Am. Chem. Soc.* 126, 9318–9325.
16. Miller SJ, Kim SH, Chen ZR, Grubbs RH (1995). Catalytic ring-closing metathesis of dienes: application to the synthesis of eight-membered ring. *J. Am. Chem. Soc.* 117, 2108–2109.
17. Conrad JC, Ellman MD, Duarte Silva JA, Monfette S, Parnas HH, Snelgrove JL, Fogg DE (2007). Oligomers as intermediates in ring-closing metathesis. *J. Am. Chem. Soc.* 29, 1024–1025.
18. Mitchell L, Parkinson JA, Percy JM, Singh K (2008). A novel and versatile entry to asymmetrically substituted pyrazines. *J. Org. Chem.* 73, 2389–2395.

19. Shu C, inventor; Boehringer Ingelheim International GmbH, assignee. Ruthenium catalyzed ring-closing metathesis process for the preparation of macrocyclic peptides as antiviral agents. World Patent WO2007/030656. March 15, 2007.
20. Shu C, Zeng X, Hao MH, Wei X, Yee NK, Busacca CA, Han Z., Farina V, Senanayake CH (2008). RCM macrocyclization made practical: an efficient synthesis of HCV protease inhibitor BILN2061. *Org. Lett.* 10, 1303–1306.
21. Change of initiation site is known to affect RCM. See Wallace D (2005) Relay ring-closing metathesis: a strategy for achieving reactivity and selectivity in metathesis chemistry. *Angew. Chem., Int. Ed.* 44, 1912–1915.
22. Hoye TR, Jeffrey CS, Tennakoon MA, Wang J, Zhao H (2004). Relay ring-closing metathesis (RRCM): a strategy for directing metal movement throughout olefin metathesis sequence. *J. Am. Chem. Soc.* 126, 10210–10211.
23. Yamamoto K, Biswas K, Gaul C, Danishefsky SJ (2003). Effects of temperature and concentration in some ring-closing metathesis reactions. *Tetrahedron Lett.* 44, 3297–3299.
24. Visser MS, Heron NM, Didiuk MT, Sagal JF, Hoveyda AH (1996). Catalytic and enantioselective route to medium-ring heterocycles. Asymmetric zirconium catalyzed ethylmagnesation of seven and eight membered rings. *J. Am. Chem. Soc.* 118, 4291–4298.
25. Kolossvary I, Guida WC (1999). Low mode conformational search elucidated: application to $C_{39}H_{80}$ and flexible docking of 9-deazaguanine inhibitors into PNP. *J. Comput. Chem.* 20, 1671–1684.
26. Macromodel, v 9.1, Schrödinger, Inc., New York, 2006.
27. Jaguar, v 6.5, Schrödinger, Inc., New York, 2006.
28. Farina V, Shu C, Zeng X, Wei X, Han Z, Yee NK, Senanayake CH (2009). Second-generation process for the HCV protease inhitors BILN 2061: a greener approach to Ru-catalyzed ring-closing metathesis. *Org. Process Res. Dev.* 13, 250–254.
29. Sheldon R (2007). The *E*-factor: fifteen years on. *Green Chem.* 9, 1273–1283.
30. Busacca CA, Gallou F, Haddad N, Hossain A, Kapadia S, Liu J, Senanayake CH, Wei X, Yee N, inventors; Boehringer Ingelheim International GmbH, assignee. Process for preparing macrocyclic compounds. World Patent WO2006/096652. September 14, 2006.
31. Busacca CA, Frutos R P, Haddad N, Kapadia SR, Lorenz JC, Saha A, Senanayake CH, Wei X, inventors; Boehringer Ingelheim International GmbH, assignee. Process for the preparation of 4-(4-propyloxy)quinoline derivatives as HCV protease inhibitors. US Patent Application US2005/0267151. December 1, 2005.

11

C–H ACTIVATION OF HETEROAROMATICS

Koji Hirano and Masahiro Miura
Division of Applied Chemistry, Graduate School of Engineering, Osaka University, Suita, Osaka, Japan

11.1 INTRODUCTION

Selective C–C bond forming reaction has been one of the long-standing central topics in synthetic chemistry. As seen in the award of the Nobel Prize to Heck, Negishi, and Suzuki in 2010, the discovery of the palladium-catalyzed cross-coupling has brought about a revolution in the construction of C–C skeletons and now provides a most powerful and reliable access to various target structures [1]. On the other hand, metal-mediated C–H activation has recently received significant attention in modern organic synthesis because it can obviate at least one of the preactivation steps (halogenation or stoichiometric metalation) of starting materials [2] necessary for the application of "classical" cross-coupling methodologies. This feature is very beneficial from the viewpoint of atom-economy and green chemistry. This chapter focuses on the recent advances in the metal-mediated C–C bond formation on biologically and physically interesting heteroaromatics via C–H bond cleavage: direct arylation, alkenylation, alkynylation, and alkylation are described, based on reports published up to early 2011.

Although the detailed mechanisms of heteroaromatic C–H bond cleavage are still underdeveloped, they can be broadly categorized into four types (Scheme 11-1): (i) electrophilic metalation by the action of MX (SEAr mechanism) [3]; (ii) carboxylate-ligand-promoted concerted metalation–deprotonation (CMD mechanism) [4]; (iii) simple base-assisted metalation [5]; and (iv) oxidative addition of C–H to the metal center [6]. Electron-rich systems such as thiophene and pyrrole generally undergo the metalation by MX of electrophilic nature (i). In the reaction involving SEAr, the presence of electron-donating substitutions tends to accelerate the rate and increase the overall yield. In this

Sustainable Catalysis: Challenges and Practices for the Pharmaceutical and Fine Chemical Industries, First Edition. Edited by Peter J. Dunn, K. K. (Mimi) Hii, Michael J. Krische, and Michael T. Williams.

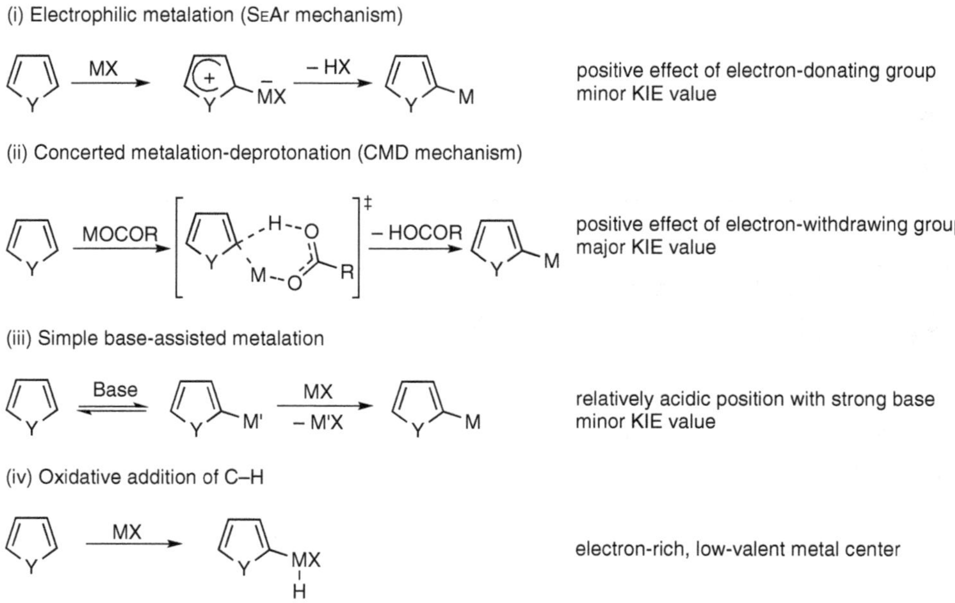

SCHEME 11-1. Four major mechanisms for C–H activation.

case, the C–H cleavage step is not involved in the rate-determining step, and thus a minor value of kinetic isotope effect (KIE) is often observed (<2.0). In contrast, the CMD mechanism (ii) shows the opposite trends: the reaction occurs more rapidly when an electron-withdrawing group is present, and a major KIE value is often recorded, in line with the rate-determining C–H bond scission. The appropriate introduction of electron-deficient substituents or changes in the reaction parameters, such as the nature of the ligand, can sometimes induce a switch between these two mechanisms. The third mechanism (iii), a simple base-assisted process, is proposed in the reaction of acidic C–Hs, including the C-2 position of 1,3-azoles, especially when a strong base is employed, such as LiO*t*Bu. During the course of the reaction, rapid H/D exchange could proceed prior to the C–C bond formation, leading to a small KIE of ~1.0. Finally, some C–H bonds can undergo oxidative addition to the metal center (iv), particularly when electron-rich, low-valent transition metals, such as Ni(0), Rh(I), or Ru(0), are employed.

In addition, there are other reaction pathways not categorized above. Particularly notable are catalytic reactions promoted by acetate anions: in certain rhodium- and iridium-catalyzed reactions, C–H activation is thought to proceed via an electrophilic deprotonation mechanism, resulting in an enhancement of reaction rate by electron-donating groups [7], and in certain palladium-mediated reactions, where σ-bond metathesis occurs with minor substitution effect but exhibits a large KIE value [8]. Thus, C–H bond cleavage is highly dependent upon the electronic nature of nucleus, substituents, catalysts, and reaction conditions.

11.2 DIRECT ARYLATION

Since the biaryl structure containing heterocyclic cores occurs frequently in many pharmaceuticals, biologically active compounds, and functional materials [9], the direct arylation reaction has been widely explored by many synthetic chemists. In this section, the arylation reactions are divided into three types of C–H/C–X, C–H/C–M, and C–H/C–H couplings [X = (pseudo)halogen, M = metal], and the scope and limitation are summarized.

11.2.1 C–H/ C–X Coupling

Since the pioneering work by Sakai and Ohta [10], the scope of the palladium-catalyzed direct arylation of heteroarenes with aryl halides has been expanded considerably, to include a variety of electron-rich heterocycles such as thiophene, indole, oxazole, and thiazole. However, the arylating reagent is somewhat restricted to relatively reactive aryl iodides, bromides, and triflates. This can be overcome by using the bulky and electron-rich biarylphosphine ligands, as reported independently by Daugulis [11] and Ackermann [12] (Scheme 11-2). The catalysts allow the less reactive but inexpensive aryl chlorides and sulfonates to be adopted as arylating reagents.

In addition to palladium, the catalytic activities of several other transition metals have also been developed: group 9 elements rhodium [13] and iridium [14] catalyze the direct arylation of N–H benzimidazoles and thiophenes, respectively (Scheme 11-3). In terms of

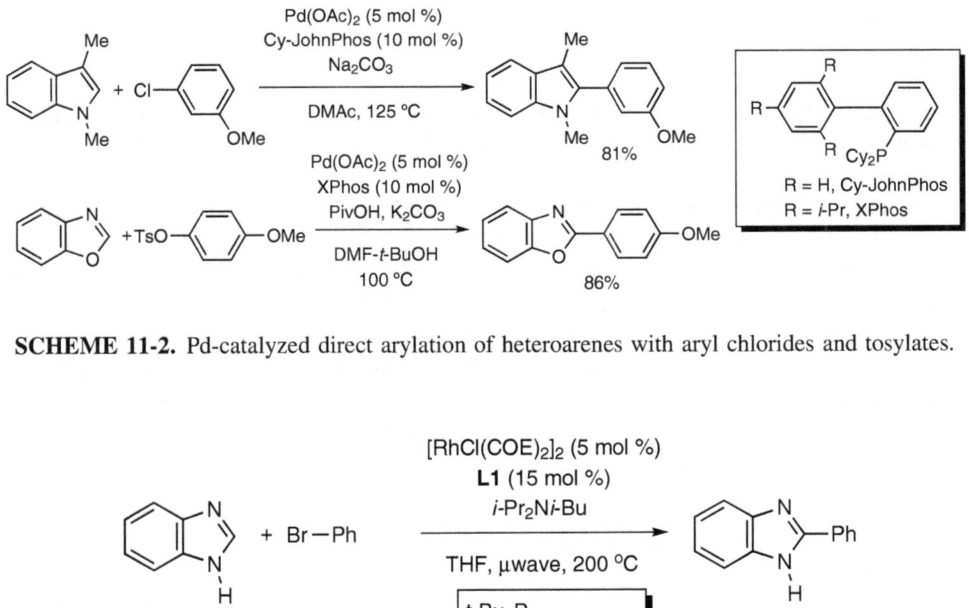

SCHEME 11-2. Pd-catalyzed direct arylation of heteroarenes with aryl chlorides and tosylates.

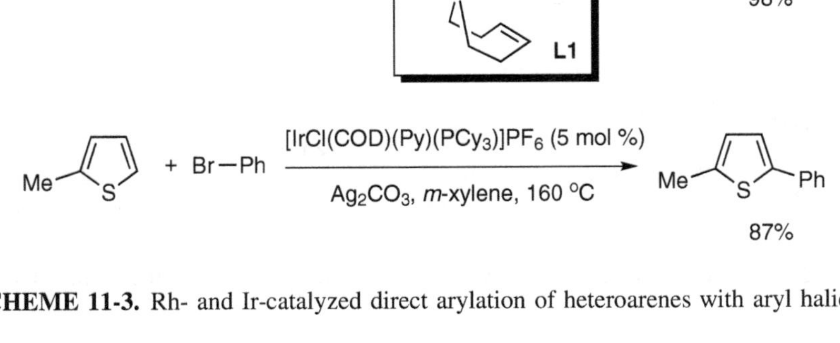

SCHEME 11-3. Rh- and Ir-catalyzed direct arylation of heteroarenes with aryl halides.

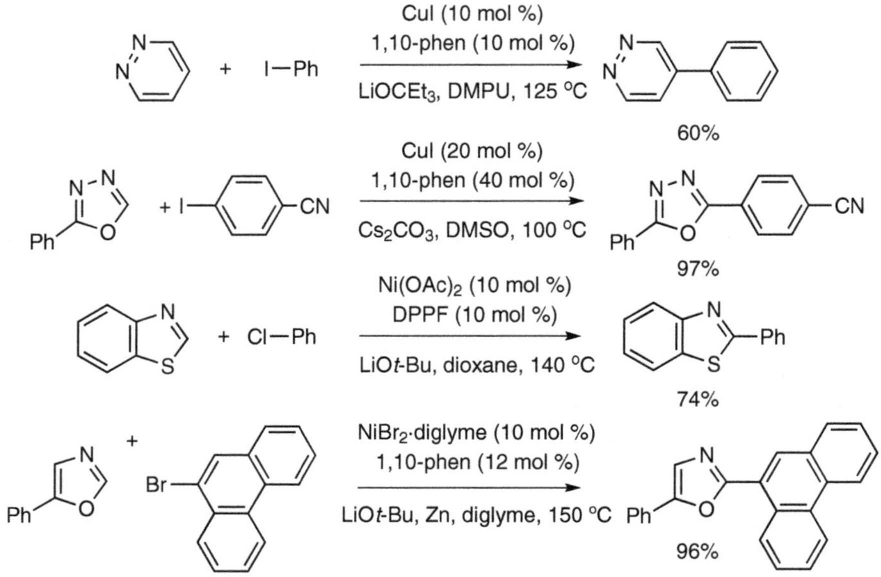

SCHEME 11-4. Cu- and Ni-catalyzed direct arylation of azoles with aryl halides.

cost and tractability, copper and nickel catalysts are good alternatives to palladium catalysts for the arylation of relatively acidic C–H bonds in azoles (Scheme 11-4) [5,15,16,17].

Among the heteroarenes, pyridine is considered as an unreactive substrate, largely due to its electron deficiency. To address the problem, suitable activating groups can be installed at the nitrogen atom to facilitate the cleavage of the proximal C–H bond (Scheme 11-5), which can be removed after the arylation to furnish the corresponding arylated pyridines in good overall yields [18].

On the other hand, *ortho*-arylation of unactivated pyridine derivatives can be achieved using a rhodium catalyst, as reported by Bergman and Ellman [19]. However, a sterically demanding substituent at C-2 is essential for the good conversion, so as to discourage the formation of the *N*-bound Rh intermediate, in favor of the less stable but reactive C–H

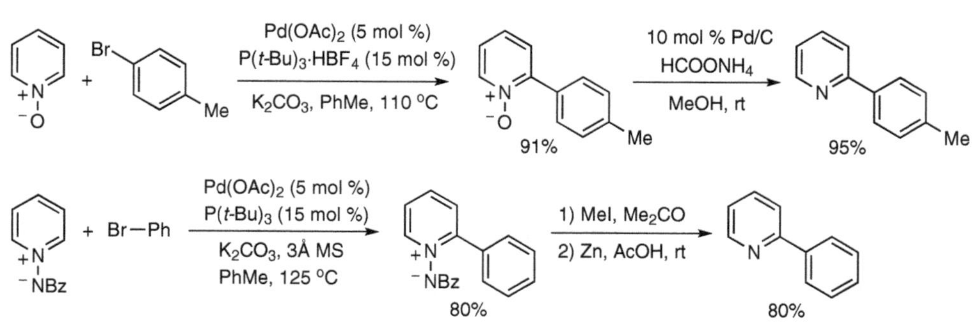

SCHEME 11-5. Pd-catalyzed direct C2-arylation of activated pyridines for the synthesis of arylated pyridines.

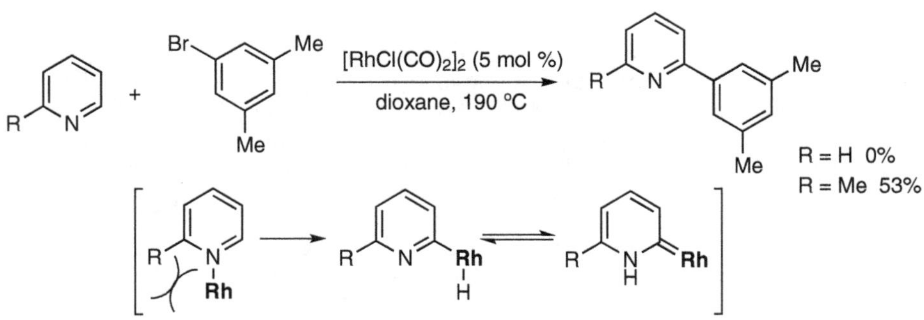

SCHEME 11-6. Rh-catalyzed direct arylation of unactivated pyridines and the reactivity profile of the rhodium-pyridine complex.

activated species (Scheme 11-6). *ortho*-Substituents at the aryl bromide substrate are also not tolerated.

The direct arylation protocol is easily applicable to multiple arylation, such that the facile construction of highly π-extended heterocyclic cores is possible in one synthetic operation. Thiophenes bearing activating substituents such as cyano [20] and carboxyl groups [21] undergo the palladium-catalyzed multiple arylation in the presence of excess of aryl bromides (Scheme 11-7). The presence of such functional groups works as a directing group to facilitate the proximal C–H bond cleavage. It is worth noting that the carboxyl group can also act as a leaving group for *ipso*-substitution.

In contrast to the above multiple reactions, site-selective monoarylations have also been developed. Fagnou reported the highly selective synthesis of triarylated thiazoles through the sequential C-2, C-5, and C-4 arylations of thiazole *N*-oxide as a platform. The key to

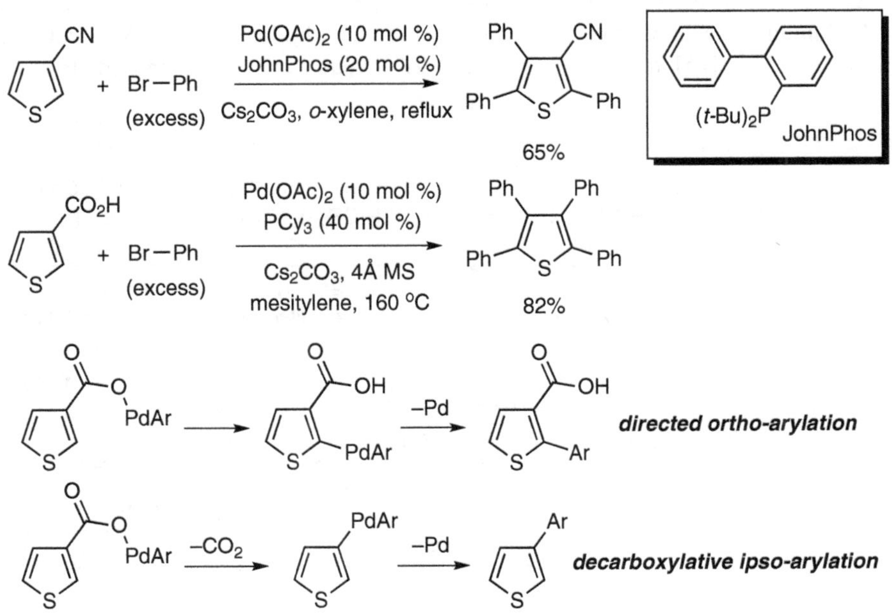

SCHEME 11-7. Pd-catalyzed arylation of thiophenes.

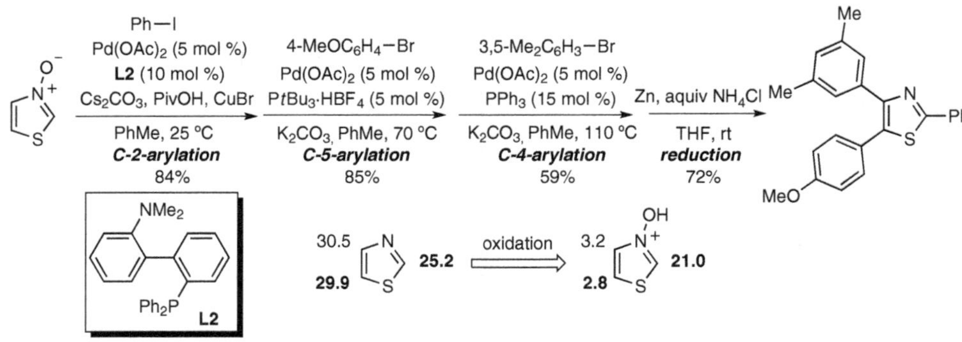

SCHEME 11-8. Highly selective synthesis of triarylthiazole through site-selective C–H arylation sequence of thiazole *N*-oxide, and atomic contribution (%) of HOMO in thiazole and thiazole *N*-oxide calculated by DFT at the B3LYP/TZVP level of theory.

success is the introduction of an *N*-oxide group to change the weak bias between C-2 and C-5 positions into a reliable C-2 > C-5 reactivity profile, which is supported by DFT calculations (Scheme 11-8) [22]. Itami also described the programmed synthesis of tetraarylthiophenes from 3-methoxythiophene, by a catalyst-controlled site-selective arylation sequence (Scheme 11-9) [23]. While the Rh/P[OCH(CF_3)$_2$]$_3$ complex activates selectively at C-2, the corresponding Pd/P[OCH(CF_3)$_2$]$_3$ catalyst shows unusual C-4-selectivity, even in the presence of an unsubstituted C-5 position.

Whereas catalytic direct arylation generally requires more forceful conditions compared to conventional cross-coupling using arylmetals, room temperature arylation of heteroarenes can be realized under certain conditions (Scheme 11-10). The intermolecular C–H arylation of imidazolinone proceeds at 23 °C in the presence of Pd(OAc)$_2$ and NaOAc·3H_2O as a catalyst and a base, respectively [24]. The use of highly electrophilic diaryliodonium salts allows the C-2 arylation of *N*-methylindole under [(IMes)Pd(OAc)$_2$] catalysis at 25 °C [25] for which a Pd(II)/Pd(IV) redox cycle is proposed. Instead of

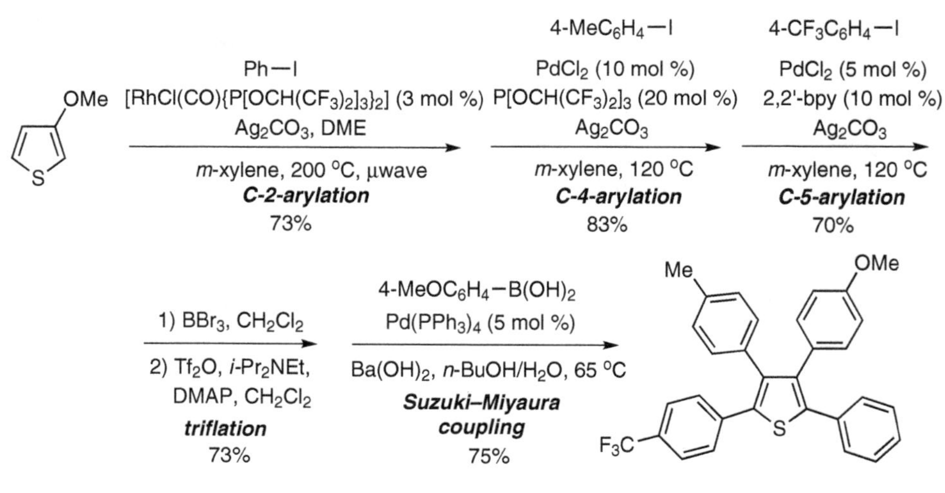

SCHEME 11-9. Programmed synthesis of tetra-arylthiophenes by catalyst-controlled C–H arylation of 3-methoxythiophene.

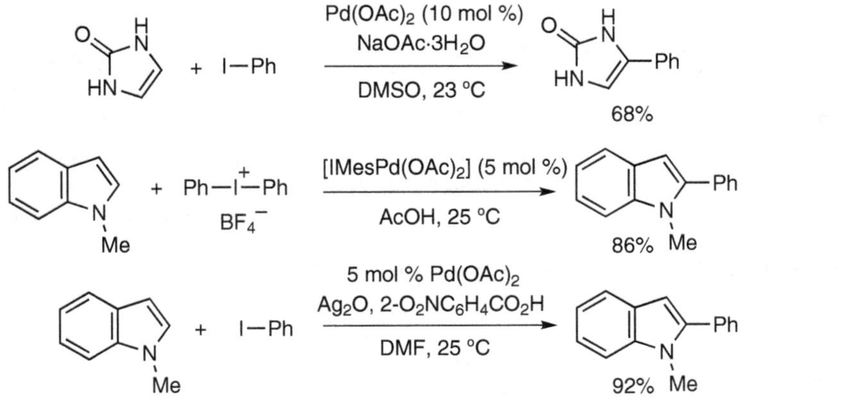

SCHEME 11-10. Room temperature Pd-catalyzed direct arylation of imidazolines and indoles.

hypervalent iodonium reagents, similar reactivity can be obtained by a combination of aryl iodide and silver carboxylate, using unligated $Pd(OAc)_2$ as a catalyst [26].

Significant enhancement in turnover number (TON) can also be achieved by modification of the catalyst system (Scheme 11-11). By using a ferrocene-based triphosphine ligand, a catalyst loading of 0.5 mol% can be deployed for the direct arylation of thiophene with aryl chloride [27]. On the other hand, microwave irradiation dramatically accelerates the arylation of 1,2,3-triazoles; conversion was complete within 15 min [28].

The utility of the methodology is also demonstrated by a highly efficient dehydrohalogenative polycondensation of 2-bromo-3-hexylthiophene for the synthesis of head-to-tail poly(3-hexylthiophene) (Scheme 11-12) [29]. An extensive catalyst screening identified a combination of Herrmann's palladacycle and $P(o\text{-}Me_2NC_6H_4)_3$ to be optimal; the corresponding polymer was obtained in 99% yield with good M_n and M_w/M_n of 30,600 and 1.60, respectively (M_n = the number average of molecular weight; M_w = the weight average of molecular weight).

Thus far, the processes described rely on a metal catalyst in the C–H bond activation and/or C–C bond-forming steps. However, it is important to note that potassium butoxide is able to promote the direct arylation of azines with aryl iodides, even in the absence of a

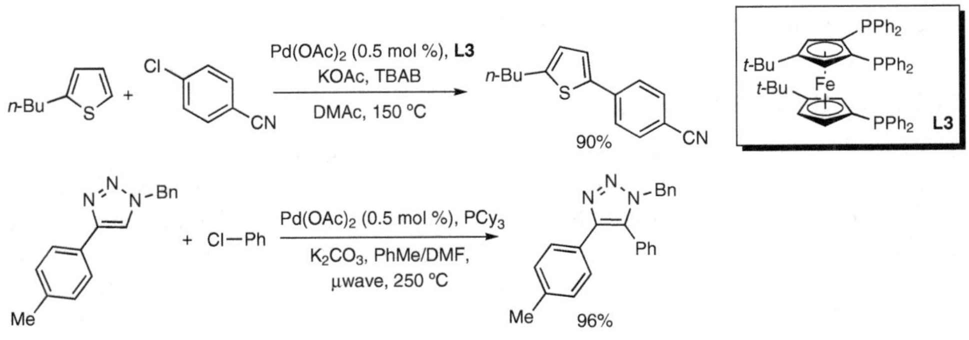

SCHEME 11-11. Direct arylation of thiophenes and triazoles with aryl chlorides under low catalyst loading of palladium.

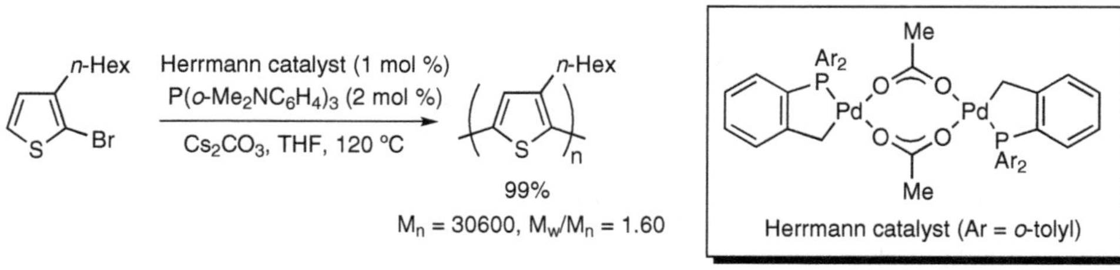

SCHEME 11-12. Pd-catalyzed dehydrohalogenative polycondensation of 2-bromo-3-hexylthiophene.

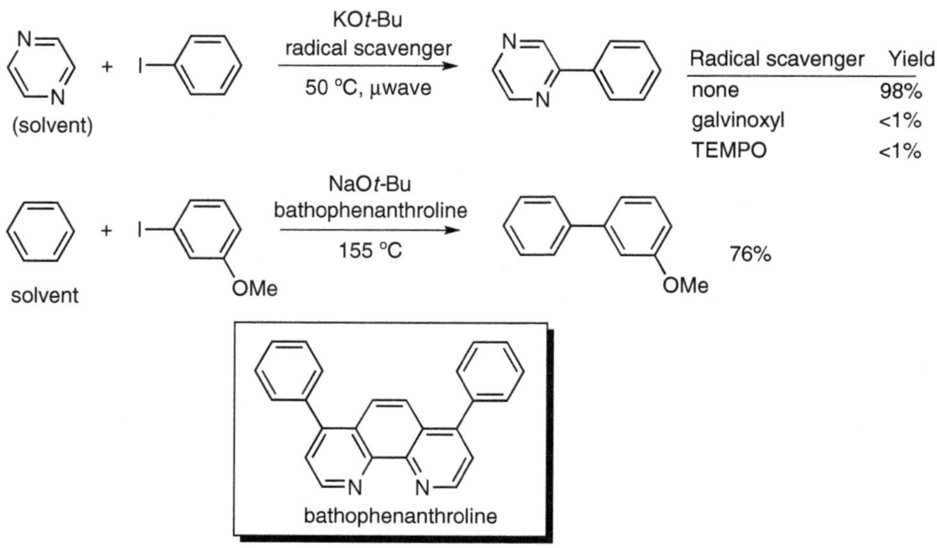

SCHEME 11-13. *tert*-Butoxide-mediated direct arylation of azines and benzenes.

metal catalyst (Scheme 11-13) [30]. The distribution of regioisomers, and reaction outcomes in the presence of radical inhibitors such as galvinoxyl and TEMPO, implicates the operation of a single-electron-transfer mechanism. Other metal-free systems have since been discovered, which include the arylation of benzene itself, enhanced by the presence of a diamine or phenanthroline derivative [31]. A representative example is shown.

11.2.2 C–H/C–M Coupling

The catalytic direct C–H arylation with arylmetals in the presence of an appropriate oxidant can complement the arylation strategy with aryl halides mentioned in Section 11.2.1, and allow greater access to the biaryl structure. Thus, the development of oxidative arylation reactions has grown rapidly in modern organic chemistry. In particular, air-stable and easy-to-handle arylboronic acid derivatives are mainly employed for this type of transformation (Scheme 11-14). Compared to the arylation with aryl halides, these systems have simpler experimental procedures that can be easily scaled up [32]. It can also afford exquisite C-4 selectivity in the arylation of thiazole, which is otherwise difficult to achieve [33].

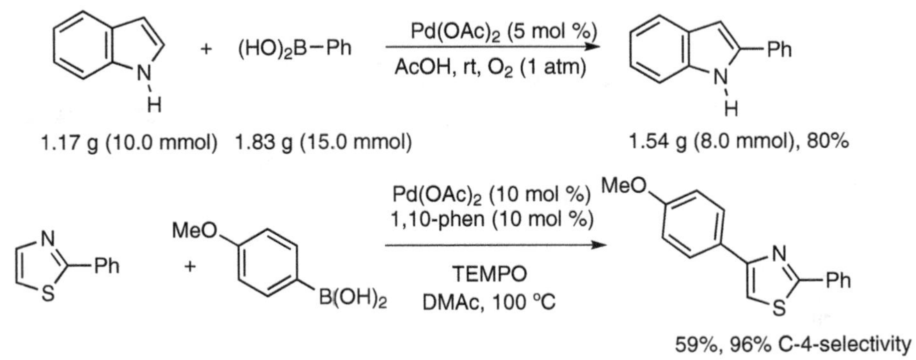

SCHEME 11-14. Pd-catalyzed direct arylation of heteroarenes with arylboronic acids.

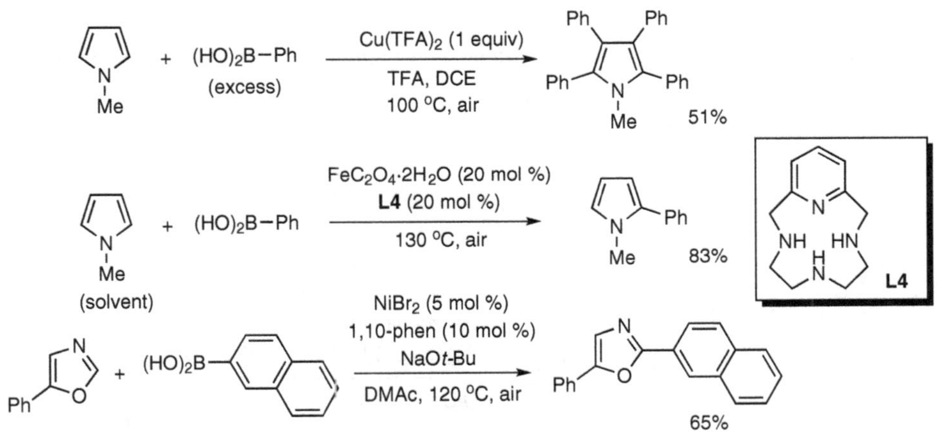

SCHEME 11-15. Cu-, Fe-, and Ni-promoted direct arylation of heteroarenes with arylboronic acids.

In some cases, the precious palladium complexes can be replaced with more cost-effective copper [34], iron [35], and nickel [36] catalysts (Scheme 11-15). All these reactions can be carried out in air.

Recently, Baran has expanded the classical silver-promoted Minisci-type reaction to aryl systems, by replacing carboxylic acids with arylboronic acids (Scheme 11-16) [37]. The direct coupling between azines and boronic acids occurs under very mild conditions to furnish the corresponding arylated heterocycles in good yields. While the reaction of unsubstituted pyridines is often plagued by poor regioselectivity, the mild reaction conditions can allow the late-stage direct functionalization of complex natural products without endangering competitive racemization of chiral centers.

While not heteroaromatic, benzene derivatives also take part in direct arylation reactions, with the aid of directing groups such as carboxylic acid [38] and ketone [39] functional groups (Scheme 11-17). In the latter case, pinacolone works as a hydride acceptor to facilitate catalyst turnover.

There are significant benefits associated with the use of organosilicon compounds in arylation reactions, including tractability, low toxicity, and environmental benignity. To

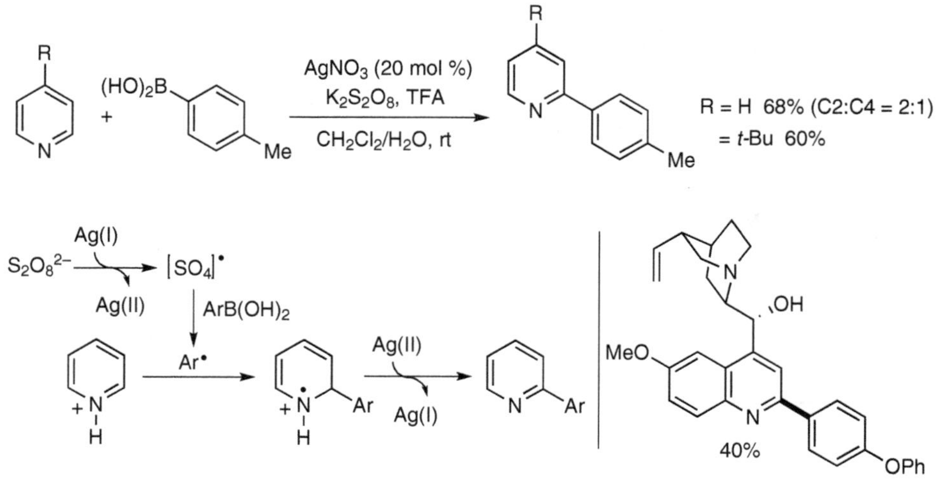

SCHEME 11-16. Direct arylation of pyridines and quinolines with an arylboronic acid by Ag-promoted Minisci-type reaction.

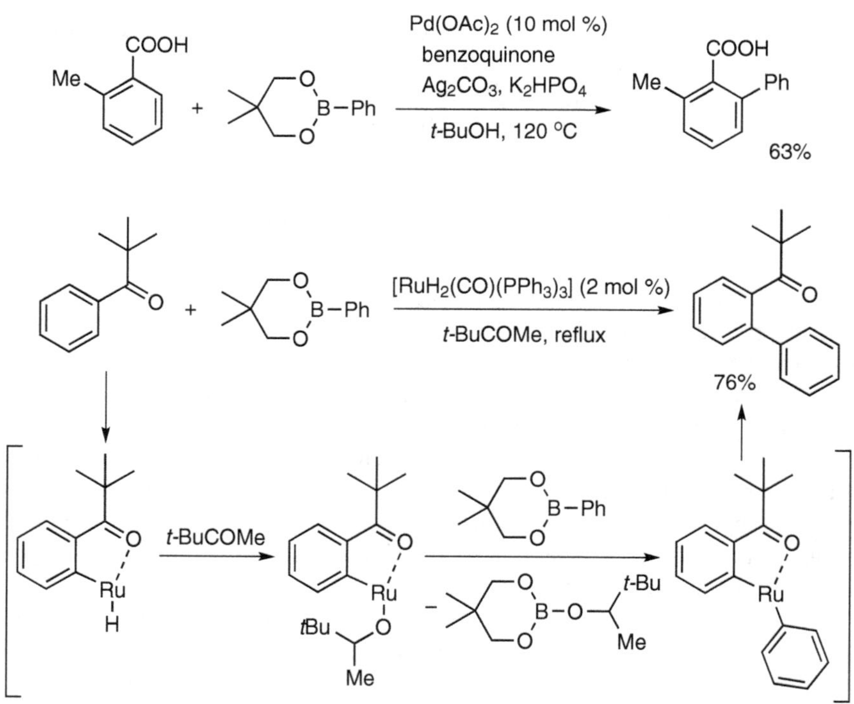

SCHEME 11-17. Pd- and Ru-catalyzed and coordination-assisted direct arylation of aromatic ketones and carboxylic acids with arylboronate esters.

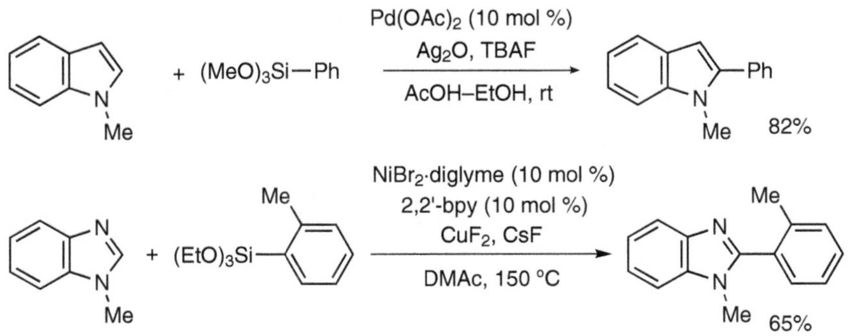

SCHEME 11-18. Pd- and Ni-catalyzed direct arylation of heteroarenes with arylsiloxanes.

date, palladium [40] and nickel [41] complexes have been reported to promote the direct coupling of heteroarenes with arylsiloxanes in the presence of fluoride activators (Scheme 11-18).

Other organometallic reagents can also be deployed. Aryl stannanes undergo C−C bond formation in the presence of an Rh- or a bimetallic Pd/Cu-catalyst system [42]. Under iron [43] and cobalt catalysis [44], benzoquinoline derivatives couple directly with diphenylzinc and phenyl Grignard reagents, respectively, at lower temperature (Scheme 11-19). In these systems, the presence of 1,2-dichloroethane derivatives as oxidants is critical. On the other hand, a combination of [$Ni(COD)_2$] and PCy_3 realizes the direct C-2 arylation of the simple, unsubstituted pyridines with diarylzincs (Scheme 11-20 versus Scheme 11-6) [45]. A plausible course of reaction involves the unique 1,2-addition of arylnickelate intermediate and subsequent oxidation with the arylzinc. At present, this is the only general protocol for the catalytic direct arylation of pyridines with high C-2 selectivity.

Under appropriate oxidative conditions, some aromatic carboxylic acids can couple with heteroaromatic C−H bonds to afford the heterobiaryls with the liberation of CO_2 (Scheme 11-21) [46]. Currently, the coupling components are somewhat limited to certain

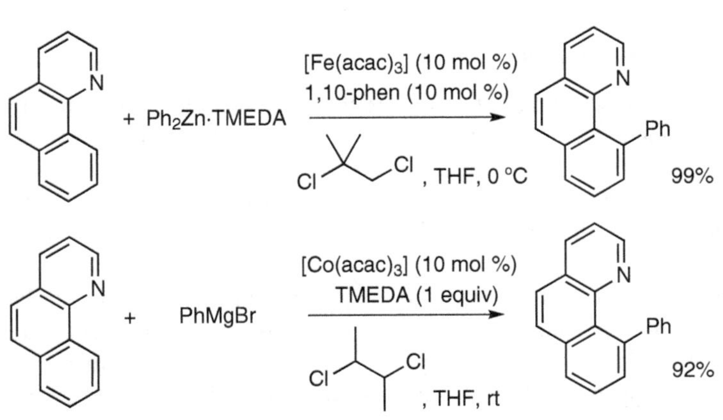

SCHEME 11-19. Fe- and Co-catalyzed direct arylation of benzoquinolines with diphenylzinc and phenyl Grignard reagent.

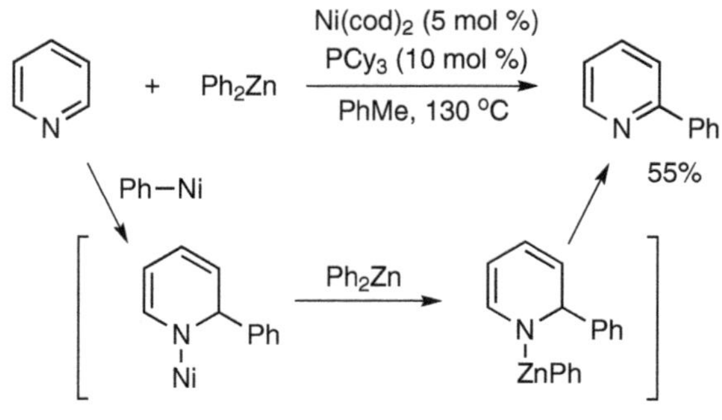

SCHEME 11-20. Ni-catalyzed direct arylation of pyridine with diphenylzinc.

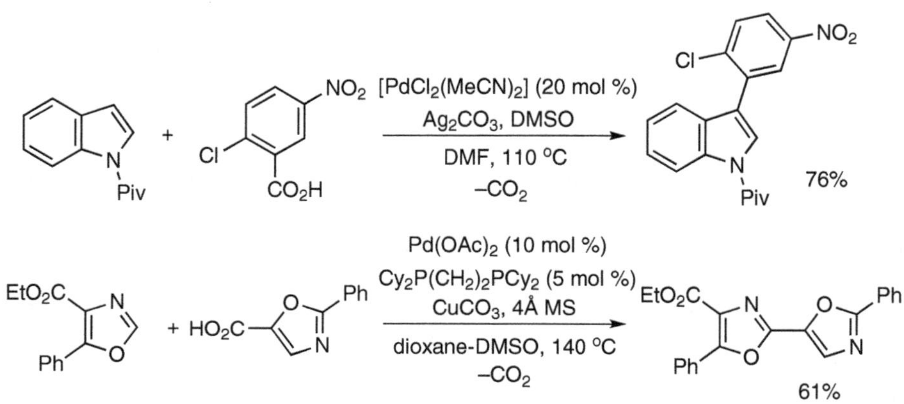

SCHEME 11-21. Pd-catalyzed arylation of heteroarenes with aromatic carboxylic acids.

heterocycles or *ortho*-substituted systems, but their synthetic utilities deserve special attention, due to the stability and accessibility of carboxylic acids.

11.2.3 C–H/C–H Coupling

The direct oxidative coupling of two different arenes via dual C–H bond cleavage is the most attractive and ideal approach to the biaryl structure, from the viewpoint of atom and step economy. Thus, the development of this type of transformation is one of the hottest topics in C–H activation chemistry. Most reported procedures employ palladium catalysts and require the presence of metal oxidants such as copper and silver salts (Scheme 11-22) [47].

On the other hand, the direct biaryl coupling between 1,3-azoles and arylazines can be effected using $Cu(OAc)_2$ alone (Scheme 11-23) [48]. Although an excess of the copper salt is essential for the full conversion, the palladium-free process shows that there is potential application of copper in C–H functionalization chemistry. Furthermore, metal-free coupling between thiophenes and pyrroles can also be achieved by using a hypervalent-iodine(III) reagent [49]. These leading examples are significant from the view that these "oxidants" can effect coupling reactions on their own.

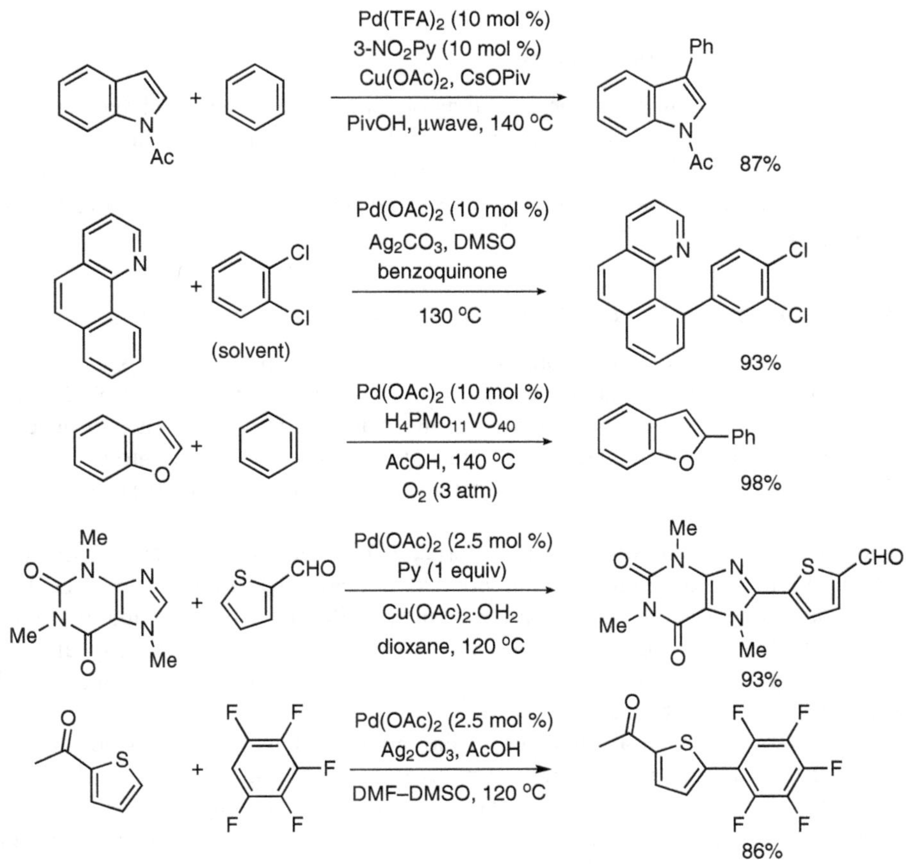

SCHEME 11-22. Pd-catalyzed oxidative direct C–H/C–H couplings.

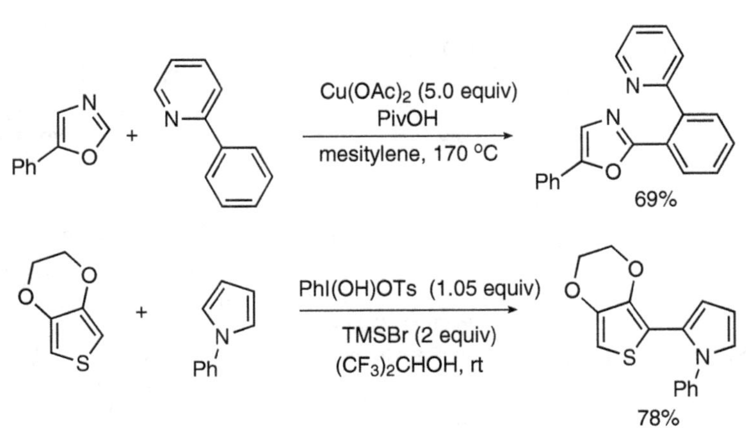

SCHEME 11-23. Pd-free oxidative direct C–H/C–H couplings.

11.3 DIRECT ALKENYLATION

11.3.1 Coupling with Alkenyl Halides

The direct alkenylation with the corresponding organic halides is less explored because of the superiority of the more atom-economical Fujiwara–Moritani coupling (see below). Nevertheless, some 1,3-azoles, *N*-iminopyridinium ylides, and imidazopyridines are known to undergo direct alkenylation reactions in the presence of copper [50] or palladium [51] catalysts (Scheme 11-24).

11.3.2 Coupling with Alkenes (Fujiwara–Moritani Reaction)

The palladium-catalyzed direct coupling of arenes with alkenes, also known as the Fujiwara–Moritani reaction, is the most powerful and attractive preparation of alkenylarenes from two simple, readily available starting materials [52]. The reaction generally proceeds through (i) C–H palladation of arene, (ii) insertion of alkene into the resultant C–Pd bond, (iii) β-H elimination, (iv) reductive elimination of HX, and (v) reoxidation of Pd(0) to Pd(II) (Scheme 11-25).

The presence of a suitable oxidant is crucial for catalytic turnover. For example, *t*-BuOOH [52c] and a combination of polyoxometalate $H_7PMo_8V_4O_{40}$ and O_2 [52e] prove to be efficient and cost-effective oxidants, while metal-based oxidants, such as $Cu(OAc)_2$ and AgOAc, also work well in the direct alkenylation of thiophenes and thiazoles, respectively, and the presence of free hydroxyl or amide groups can be tolerated (Scheme 11-26) [53].

The Fujiwara–Moritani alkenylation of the indole nucleus has been widely studied, owing to the occurrence of the structure motif in many biologically active and natural products. Herein the major challenge is the control of site-selectivity (C-2 versus C-3). In this regard, Gaunt and coworkers reported the solvent-controlled regioselective alkenylation of indoles (Scheme 11-27) [54]; the deployment of neutral conditions provides the C3-alkenylated product predominantly, whereas an acidic medium led to the preferred formation of the C2-alkenylated isomer. More reliable site-control is accomplished by

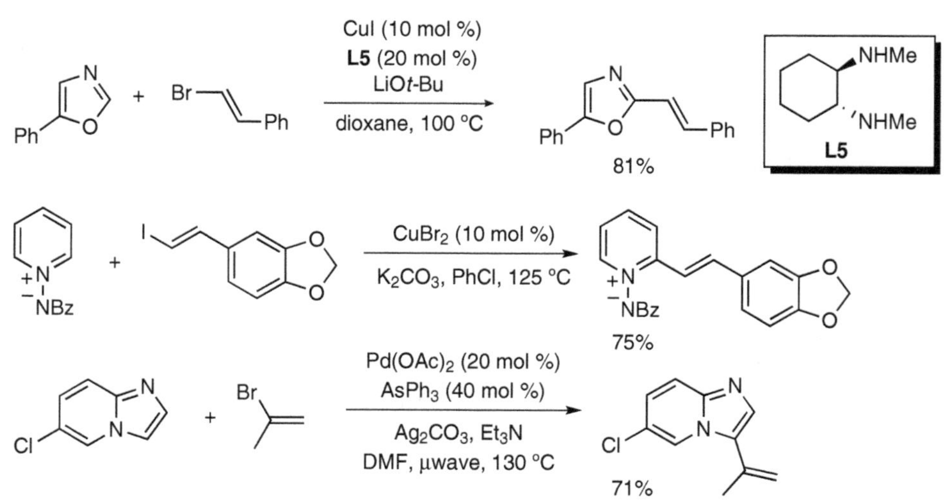

SCHEME 11-24. Cu- and Pd-catalyzed direct alkenylation of heteroarenes with alkenyl halides.

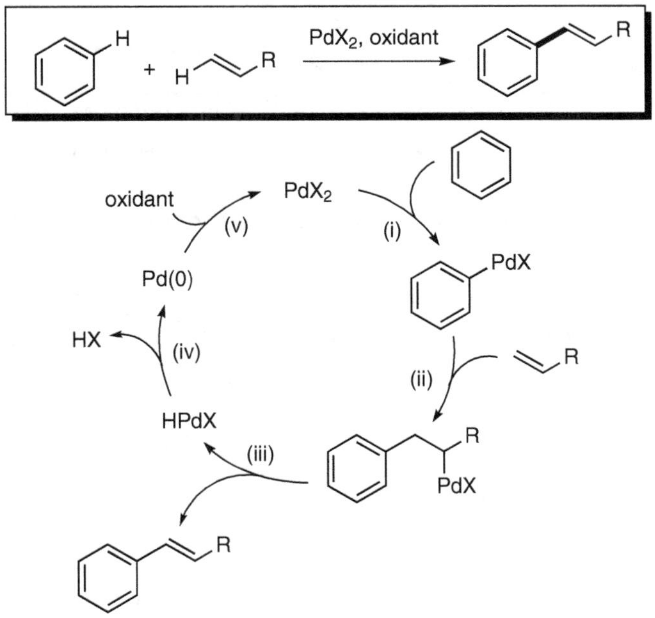

SCHEME 11-25. General mechanism for the Pd-catalyzed Fujiwara–Moritani reaction.

using the appropriate directing groups. The introduction of the 2-pyridylsulfonyl substituent to the indole nitrogen is a general protocol for obtaining exclusive C-2-selectivity (Scheme 11-28) [55]. On the other hand, the carboxyl group functions as a removable *ortho*-director (Scheme 11-29) [56]. In this case, 2- and 3-indole carboxylic acids react with the acrylate ester with concomitant decarboxylation under palladium catalysis to furnish 3- and 2-alkenylated indoles, respectively.

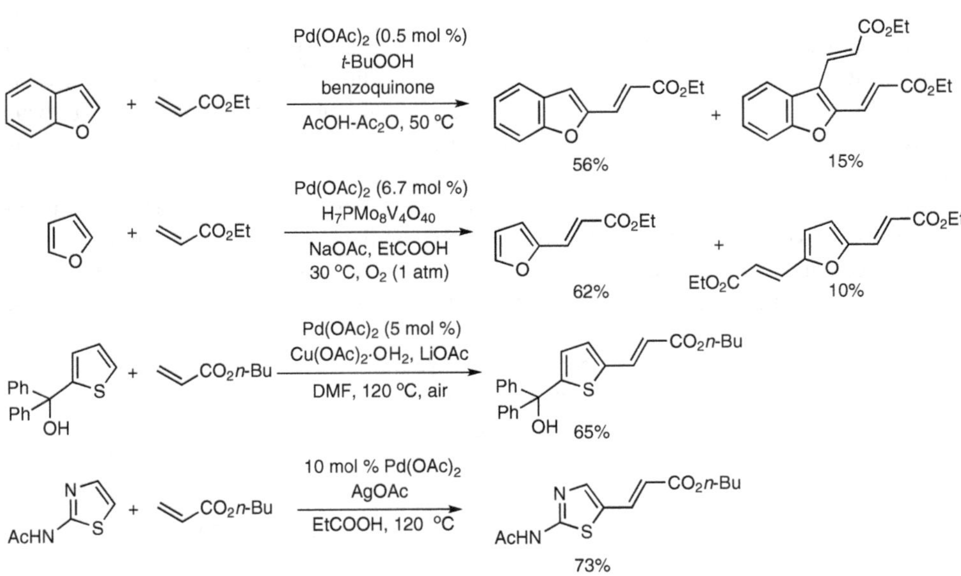

SCHEME 11-26. Pd-catalyzed Fujiwara–Moritani reaction of furans, thiophenes, and thiazoles.

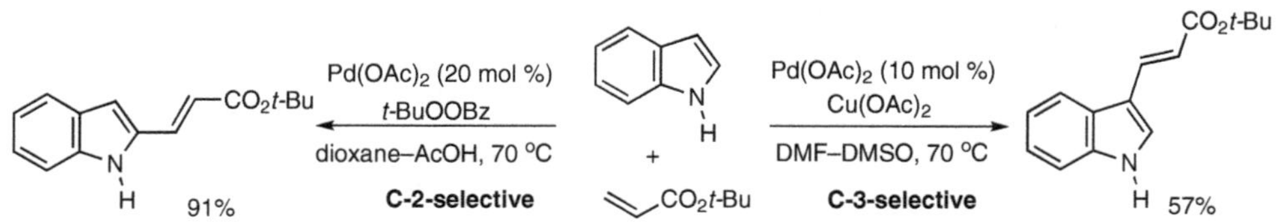

SCHEME 11-27. Solvent-controlled regioselective alkenylation of indole.

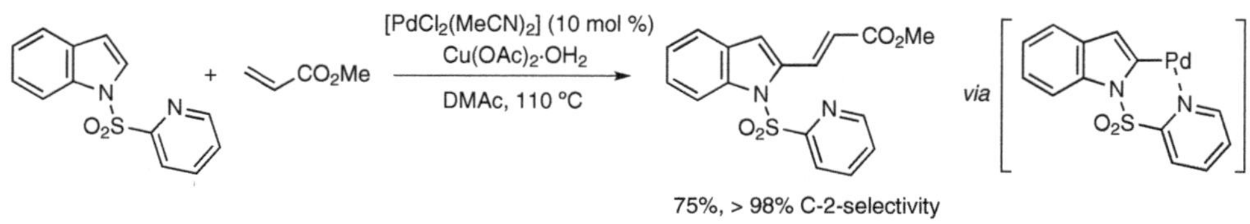

SCHEME 11-28. Highly C2-selective alkenylation of indole directed by the 2-pyridylsulfonyl group.

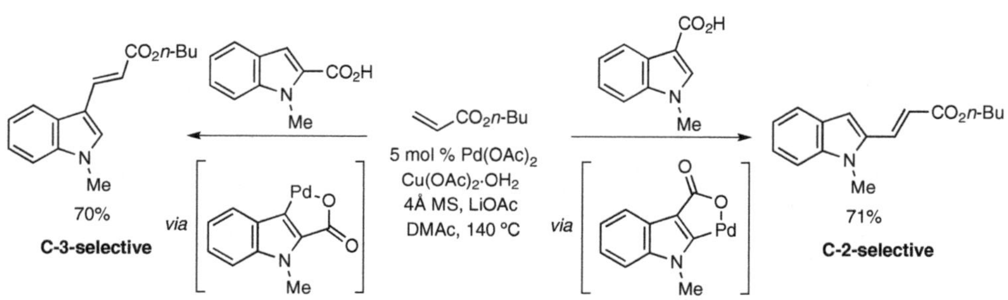

SCHEME 11-29. Regiodivergent alkenylation of indole directed by the carboxyl group.

In contrast to the electron-rich heterocycles, electron-deficient pyridines are highly resistant to alkenylation reactions under identical conditions. However, recently, Yu has reported the catalytic activity of a $Pd(OAc)_2$/1,10-phen complex for the Fujiwara–Moritani coupling of simple pyridines (Scheme 11-30) [57]. Notably, the catalysis shows a uniquely high selectivity for coupling at C-3.

Rhodium, ruthenium, and palladium complexes have been reported as catalysts for the (decarboxylative) alkenylation of heteroarene carboxylic acids. By far, rhodium gave the best performance: full conversion of indole-2-carboxylic acid is achieved using 1 mol% catalyst loading (Scheme 11-31 vs. Scheme 11-27) [58]. In comparison, 5–10 mol% of

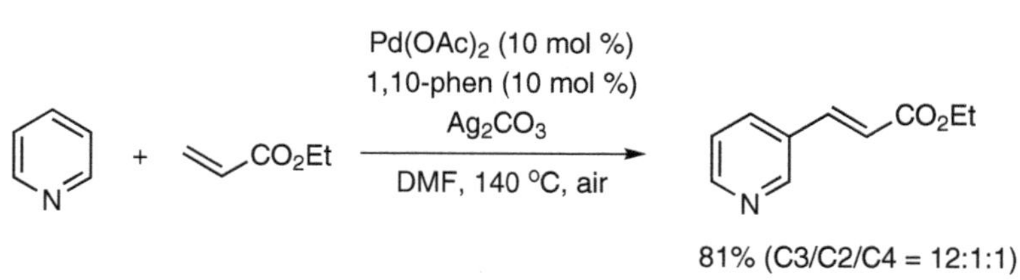

SCHEME 11-30. Pd-catalyzed direct and C-3-selective alkenylation of pyridines.

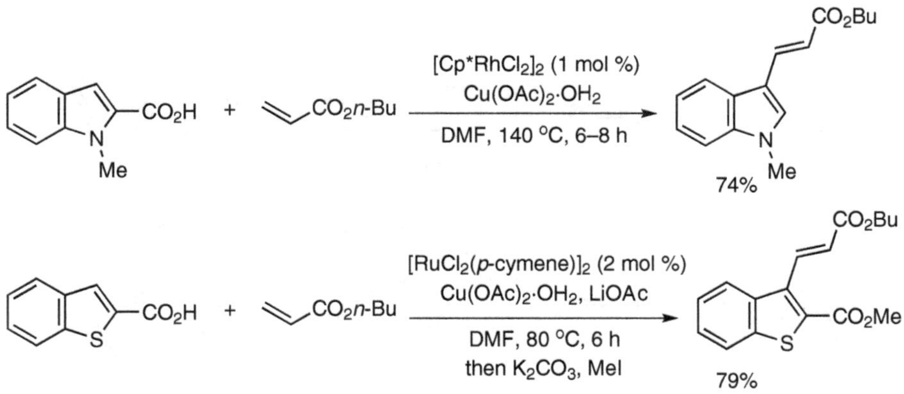

SCHEME 11-31. Rh- and Ru-catalyzed alkenylation of heteroarenes directed by the carboxyl group.

palladium is essential for the completion of the same reaction [56]. On the other hand, the competitive decarboxylation process is very slow in the ruthenium-catalyzed reaction, allowing the retention of the carboxyl group for further structural elaboration [59].

11.3.3 Coupling with Alkynes (Hydroarylation)

Catalytic alkyne hydroarylation allows for the direct synthesis of alkenylated arenes. Its synthetic utility is as high as that of the Fujiwara–Moritani reaction, since the addition reaction is overall redox neutral, and thus requires no external oxidant. This type of reaction is usually initiated by the direct metalation or oxidative addition of aromatic C–H bonds (Scheme 11-1, i, ii, or iv). Although some π-acidic metals, such as gold, catalyze similar transformations, these reactions are activated by the metal binding to the alkyne instead of the aromatic ring, and are thus excluded from the present discussion [60].

The pioneering work of Murai and coworkers, using ruthenium catalysis to effect coupling between aromatic ketones and C–C unsaturated molecules, is a significant milestone for C–H alkenylation (and alkylation) chemistry (Scheme 11-32). The catalytic

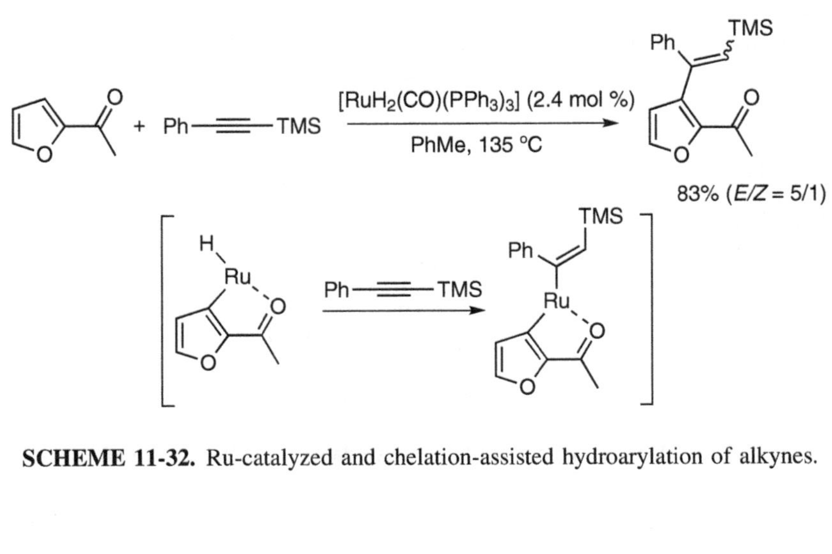

SCHEME 11-32. Ru-catalyzed and chelation-assisted hydroarylation of alkynes.

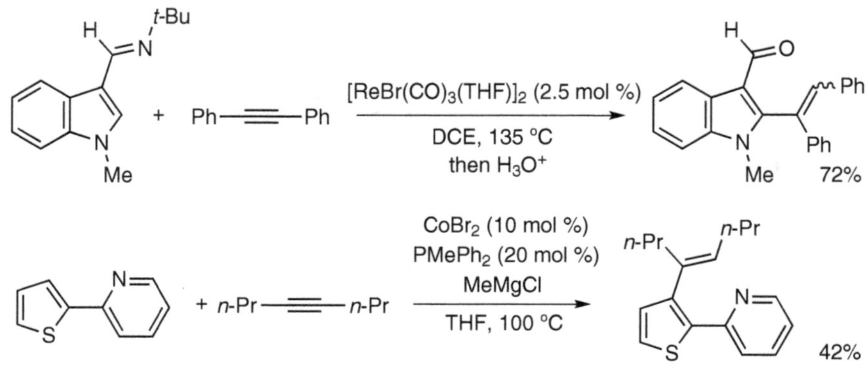

SCHEME 11-33. Re- and Co-catalyzed and chelation-assisted hydroarylation of alkynes.

cycle is triggered by the chelation-assisted oxidative addition of C–H bond at the *ortho*-position, and subsequent insertion and reductive elimination to produce the alkenylated product [61]. Since then, a rhenium catalyst has also been found effective for the reaction of heteroarenes bearing an imino directing group [62], while a low-valent cobalt species (generated *in situ*) has found application in the reaction of certain azoles, typified by Scheme 11-33 [63].

Nakao and Hiyama described a much higher activity of the low-valent nickel complex for the hydroarylation of a variety of heteroarenes [64]. As exemplified by Scheme 11-34, the direct addition of caffeine across 4-hexyne occurs even at 35 °C with high efficiency.

In the presence of appropriate Lewis acids, the scope of the reaction may be extended to less reactive heterocycles, such as imidazole and pyridone (Scheme 11-35). In this system, the nickel and Lewis acid act cooperatively in the C–H activation, as illustrated. The double insertion of alkynes is also possible by changing the Lewis acid.

A cationic rhodium species, $[Cp^*Rh(MeCN)_3][SbF_6]_2$, promotes the hydroarylation of pyrroles and indoles via a nonoxidative metalation with the aid of PivOH (Scheme 11-36) [65]. A deuterium-labeling experiment and control reaction without the acid additive suggest the mechanism to consist of: (i) reversible rhodation of arene with concomitant loss of a proton; (ii) migratory insertion of alkyne; and (iii) protonolysis of the carbon–rhodium bond.

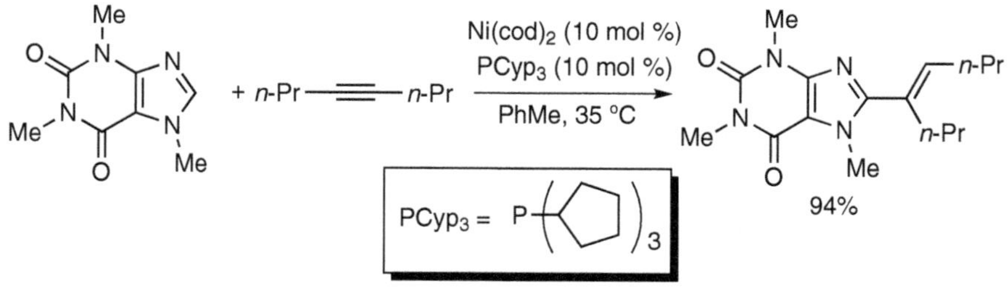

SCHEME 11-34. Ni-catalyzed direct alkynylation of caffeine with alkynes.

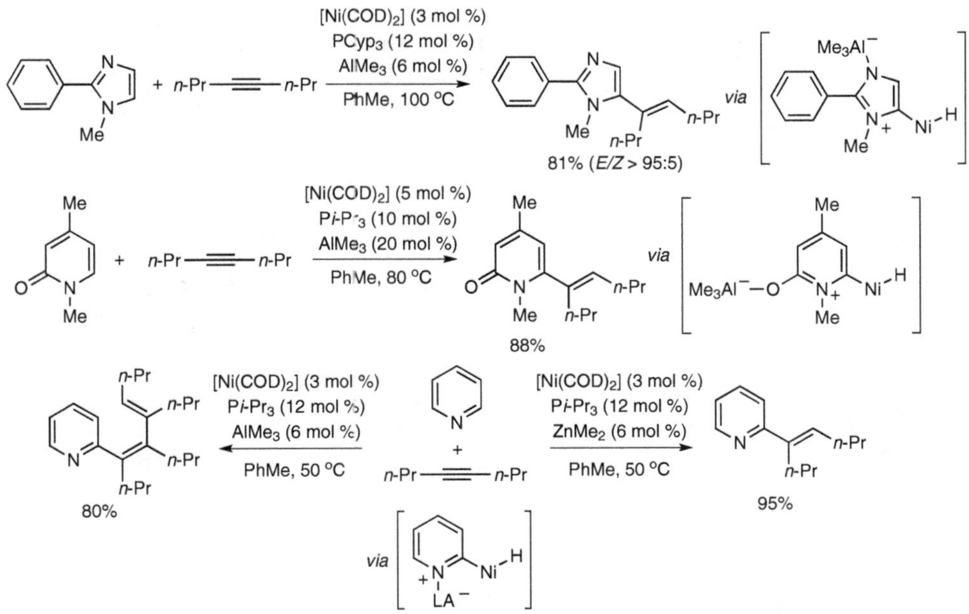

SCHEME 11-35. Nickel/Lewis acid co-catalyzed direct alkenylation of imidazoles and pyridines.

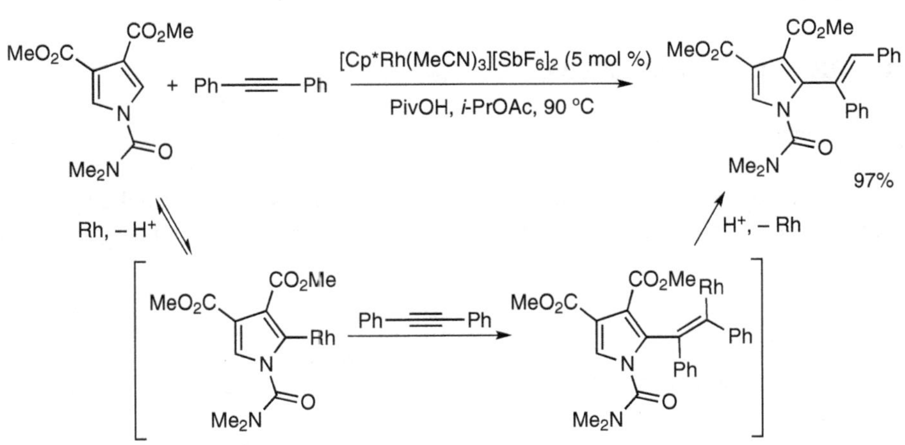

SCHEME 11-36. Suggested mechanism for cationic Rh-catalyzed hydroarylation reaction of alkynes.

By employing a stoichiometric amount of $Cu(OAc)_2$ as an oxidant, similar rhodium complexes catalyze the benzene ring homologation through C–H bond cleavage (Scheme 11-37). When anionic directing groups are employed, the annulation reactions take place to produce the corresponding fused heteroaromatics. Some representative examples are listed.

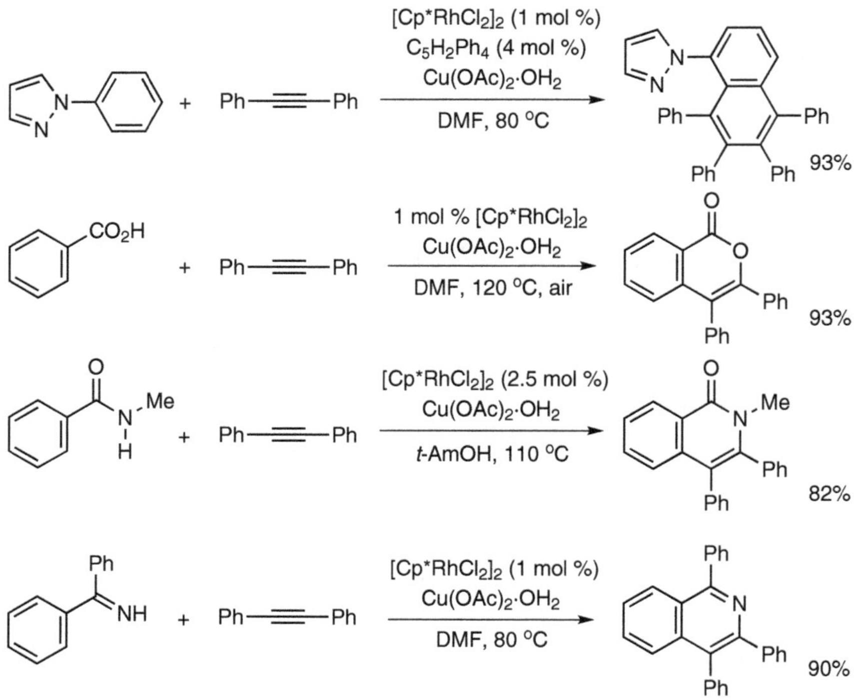

SCHEME 11-37. Rh/Cu systems for annulation reaction with alkynes via C–H bond cleavage.

11.4 DIRECT ALKYNYLATION

Arylacetylenes are among the most fundamental and important π-conjugated systems in various areas of organic chemistry. A powerful and reliable approach to these molecules is the palladium/copper-catalyzed cross-coupling of aryl halides with terminal alkynes, also known as Sonogashira coupling (Scheme 11-38, top). On the other hand, the metal-

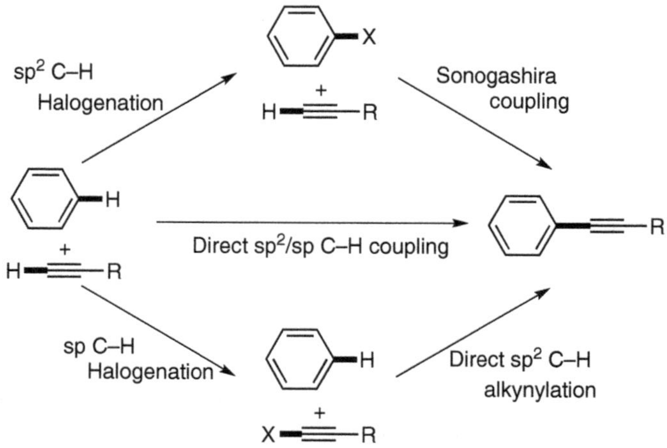

SCHEME 11-38. Disparate approaches to arylacetylenes.

catalyzed direct alkynylation of arene C–H bonds with alkynyl halides or pseudohalides is also useful as a complementary process (bottom). Ultimately, the direct coupling between arenes and terminal alkynes *via* twofold C–H bond cleavage of both substrates is an ideal goal since no preactivation step is required (middle). In this section, recent progress of the direct sp^2 C–H alkynylation with alkynyl halides and direct sp^2/sp C–H coupling is described.

11.4.1 Coupling with Alkynyl Halides or Pseudohalides

The catalytic direct alkynylation of arenes with alkynyl halides was first developed by Yamaguchi and coworkers in 2002 [67]. They employed $GaCl_3$ as a catalyst and succeeded in the *ortho*-alkynylation of phenols and anilines. The reaction is believed to proceed through the carbogallation of the triple bond, followed by β-Cl elimination (Scheme 11-39). The limitation of the process is its narrow scope, which is restricted to silyl-substituted acetylenes.

In 2007, Gevorgyan et al. made significant progress in this field by the report of a palladium-catalyzed process for the direct alkynylation of electron-rich *N*-fused heteroaromatics (Scheme 11-40) [68], which allows various bromoalkynes to be employed as substrates. The observed minor KIE value suggests the operation of an electrophilic metalation pathway in the C–H activation step (Scheme 11-1, i). A similar catalyst has subsequently been applied to the selective alkynylation (at C-3) of indoles by Gu and coworkers [69].

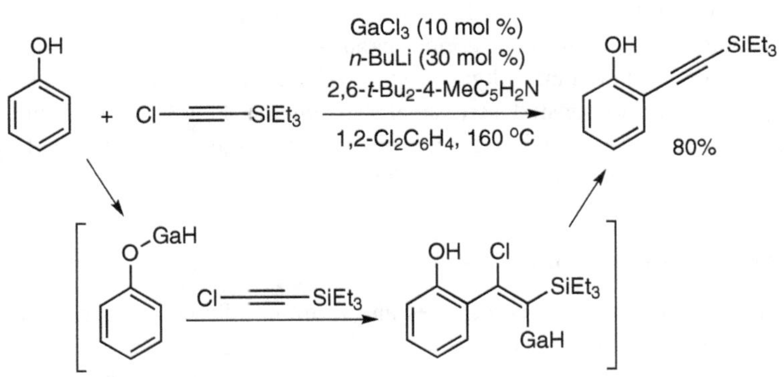

SCHEME 11-39. Ga-catalyzed direct alkynylation of phenols with alkynyl chlorides.

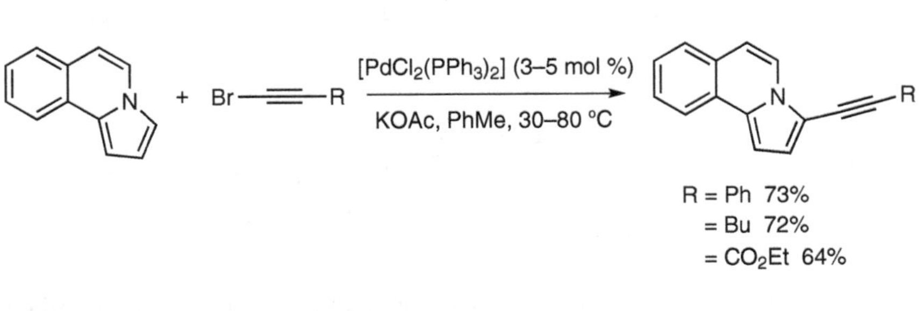

SCHEME 11-40. Pd-catalyzed direct alkynylation of *N*-fused heteroaromatics with alkynyl halides.

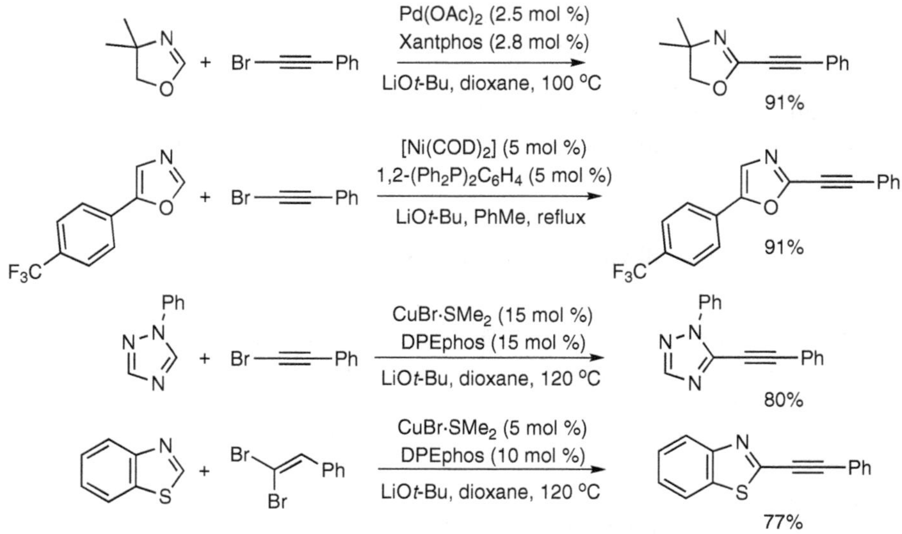

SCHEME 11-41. Pd-, Ni-, and Cu-catalyzed direct alkynylation of azoles with alkynyl halides or 1,1-dibromoethenes.

The relatively acidic 1,3-azoles also undergo the alkynylation under palladium, nickel, and copper catalysis (Scheme 11-41) [70]. All reactions require the strong base, LiO*t*-Bu, so that the C–H activation step is most likely to proceed via the base-assisted metalation pathway (Scheme 11-1, iii). More stable and convenient 1,1-dibromoethenes also work as the alkynyl halide equivalent under copper catalysis [70e].

A well-designed hypervalent alkynyl iodide (III) couples with thiophene derivatives under cooperative gold/Brønsted acid catalyst (Scheme 11-42) [71]. While the exact mechanism, as well as the role of the acid, is obscure at present, this is the single successful example of the direct alkynylation of thiophenes.

11.4.2 Coupling with Terminal Alkynes

Dehydrogenative coupling between arenes and terminal alkynes is a much more atom-efficient and attractive alternative to Sonogashira coupling, but remains a challenge in synthetic chemistry. The first successful attempt using $CuCl_2$ as a catalyst was reported by Miura in 2010 (Scheme 11-43) [72]. Although the performance and heteroarene scope are limited, the pioneering work has stimulated further discoveries of other transition metal catalysts. These include the use of nickel [73] and palladium [74] complexes for the

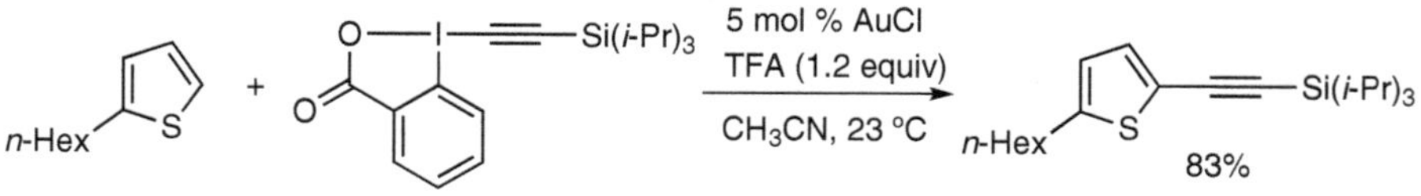

SCHEME 11-42. Gold/Brønsted acid co-catalyzed direct alkynylation of thiophenes with hypervalent alkynyl iodides.

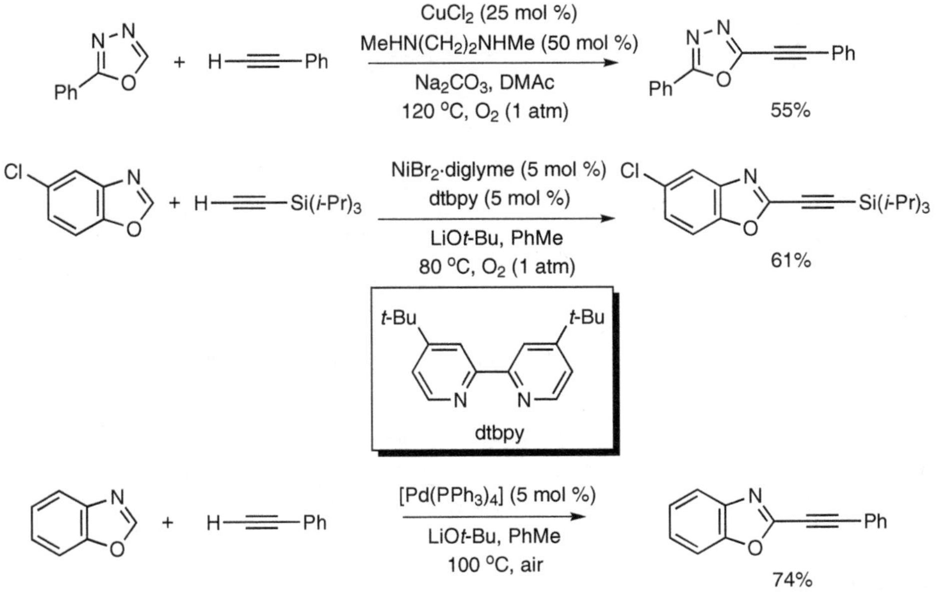

SCHEME 11-43. Cu-, Ni-, and Pd-catalyzed direct alkynylation of azoles with terminal alkynes.

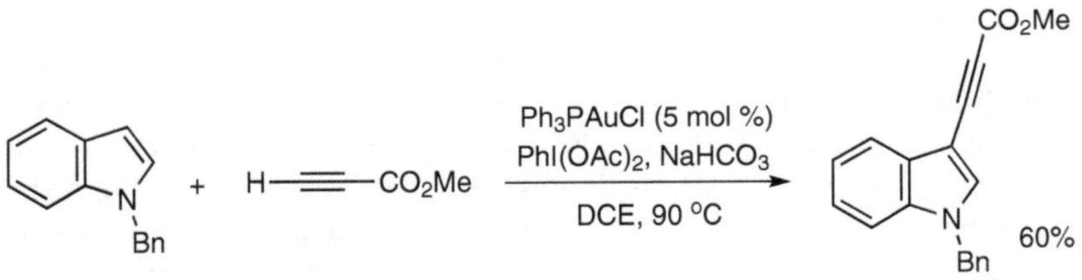

SCHEME 11-44. Au-catalyzed direct alkynylation of indoles with terminal alkynes.

alkynylation of several 1,3-azoles. It is worth noting that molecular oxygen is the sole oxidant in these transformations.

The alkynylation of electron-rich indoles and pyrroles is achieved by a Ph_3PAuCl catalyst (Scheme 11-44) [75]. However, the use of $PhI(OAc)_2$ as an oxidant is critical for success. Nevertheless, the scope of the reaction extends to anisole derivatives.

11.5 DIRECT ALKYLATION

11.5.1 Benzylation and Allylation

Diarylmethanes are an important motif in pharmaceuticals and biologically active compounds. The C–H benzylation complements the conventional Lewis-acid-promoted Friedel–Crafts benzylation, providing a more straightforward access to the desired structure. The direct benzylation of heteroarenes with benzyl chlorides is possible under

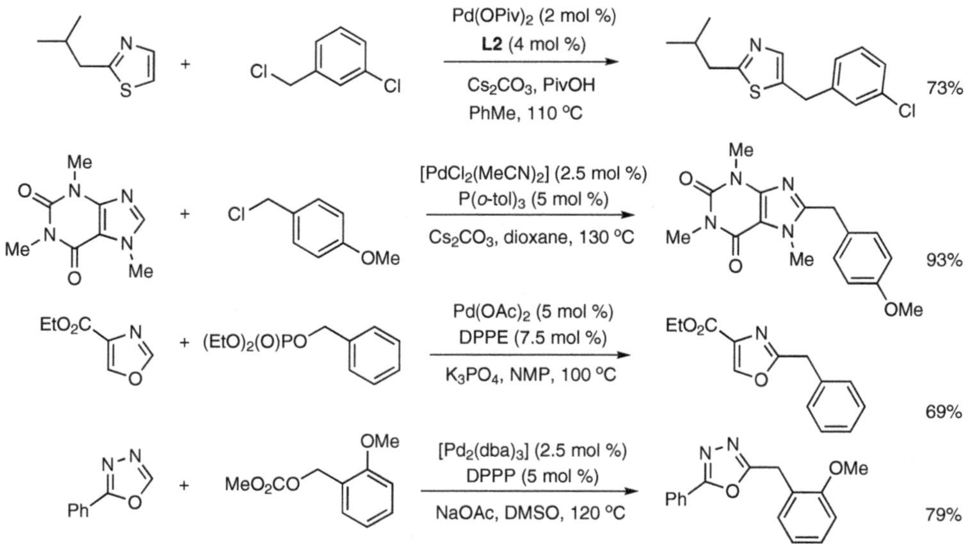

SCHEME 11-45. Pd-catalyzed direct benzylation of heteroarenes with benzyl (pseudo)halides.

some palladium-catalyzed conditions (Scheme 11-45) [76]. In addition, phosphates and carbonates, which are readily available from the parent benzylic alcohols, can also be used in these reactions [77].

The major drawback of these systems is the limited application to primary benzylic systems, probably due to the competitive β-H elimination of the alkylpalladium intermediate. Recently, Wang and coworkers addressed the problem by using *N*-tosylhydrazones (Scheme 11-46) [78]. Under basic conditions, *N*-tosylhydrazone is transformed into the diazo compound, followed by decomposition upon treatment with the heteroarylcopper species, to produce the copper carbenoid intermediate. The benzylated product then follows from subsequent 1,2-migration and protonolysis.

Catalytic insertion of styrenes into a heteroaromatic C–H bond would be a more atom-economical procedure toward the construction of diarylmethane skeletons, provided that the C–C bond formation can be regioselectively achieved at the more congested benzylic position. So far, Ni/IMes [79] and Co/PCy_3 [80] catalyst systems have been developed for

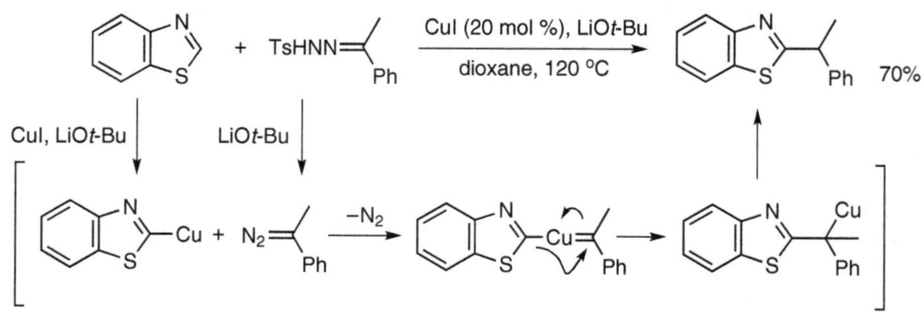

SCHEME 11-46. Cu-catalyzed direct benzylation of benzothiazole with *N*-tosylhydrazone.

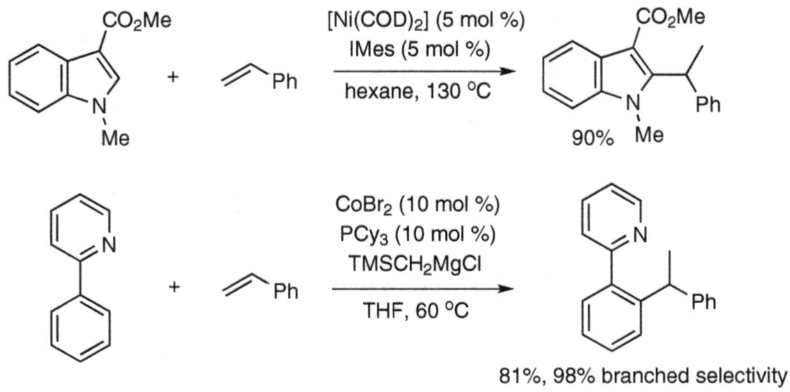

SCHEME 11-47. Ni- and Co-catalyzed direct benzylation of heteroarenes with styrenes.

this reaction (Scheme 11-47). This selectivity is attributed to the thermodynamic preference for the formation of a benzylmetal intermediate.

Direct allylation reactions still remain largely elusive. Only one practical procedure has been reported so far, which is a copper-based system effected with allyl phosphates, where very good retention of stereochemistry is observed (Scheme 11-48) [81].

11.5.2 Alkylation with Unactivated Systems

The introduction of longer alkyl side chains to heterocyclic structures is known to generally enhance their lipophilicity and solubility for greater bioavailability. Thus, this class of reactions is of considerable interest in organic synthesis.

Alkylation reactions with unactivated alkyl halides possessing β-hydrogens are challenging owing to difficulty in controlling undesired β-H elimination. The limited number of successful examples are listed in Scheme 11-49. By using palladium catalysts with appropriate ancillary phosphine ligands to suppress the side reaction [82], the coupling with unactivated alkyl bromides as well as chlorides is possible by using σ-donating $\mathrm{P}n\mathrm{Bu}_3$. On the other hand, Hu has developed a series of NNN-pincer nickel complexes for the direct alkylation of heteroarenes [83]. It was found that the pincer catalyst gradually degraded to the heterogeneous, nickel metal particle over the course of the reaction, which is responsible for the coupling, supported by the performance of a mercury test.

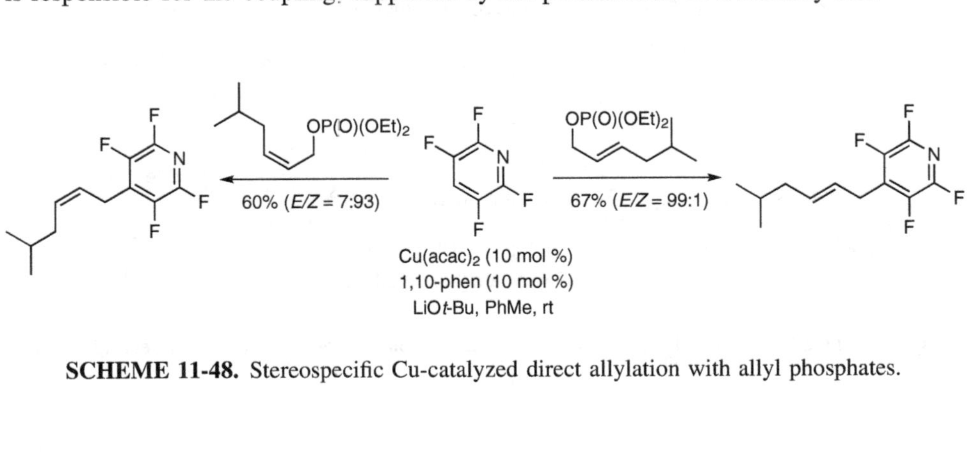

SCHEME 11-48. Stereospecific Cu-catalyzed direct allylation with allyl phosphates.

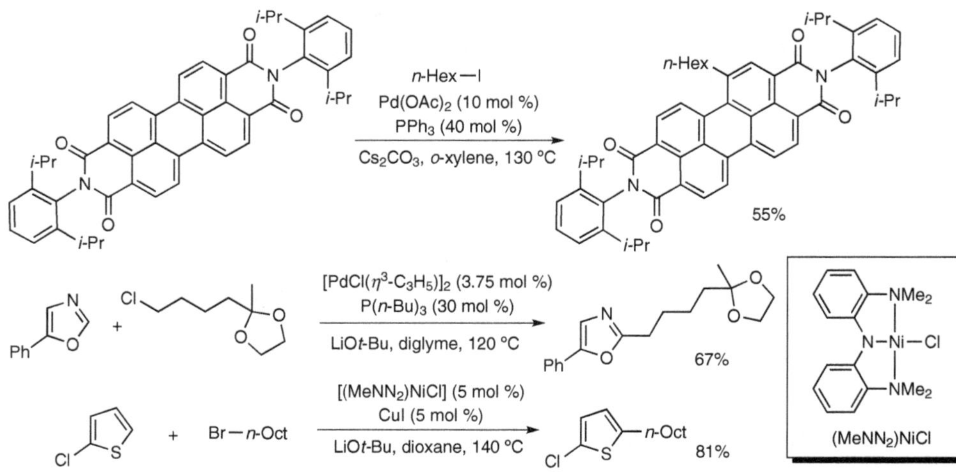

SCHEME 11-49. Pd- and Ni- catalyzed direct alkylation of heteroarenes with unactivated alkyl halides.

There are few reports of alkylation reactions with secondary alkyl halides. Promising catalytic activity was displayed by ruthenium [84] and iron [85] catalysts with cyclic systems (Scheme 11-50); further expansion of substrate scope is still to be demonstrated.

The addition of heteroaromatics to alkenes catalyzed by low-valent transition metals such as Ru, Rh, and Ni is a useful tool for the synthesis of alkyl-substituted π-systems. The reaction is proposed to proceed via oxidative addition of the C–H bond, insertion of alkene, and productive reductive elimination, as described in Section 11.3.3. Two significant results were achieved with cooperative catalysts, which are illustrated in Scheme 11-51. Benzimidazole undergoes the intermolecular coupling with a terminal alkene in the presence of $[RhCl(COE)_2]_2/PCy_3$ and lutidine·HCl to provide the alkylated product [86]; the intermediacy of *N*-heterocyclic carbene Rh(I) was demonstrated in this reaction. Recently, an unprecedented C-4-selective functionalization of pyridine has been accomplished by an Ni/Lewis acid system, similar to those previously described in Scheme 11-35. The unique catalytic activity arises from a combination of bulky ligand,

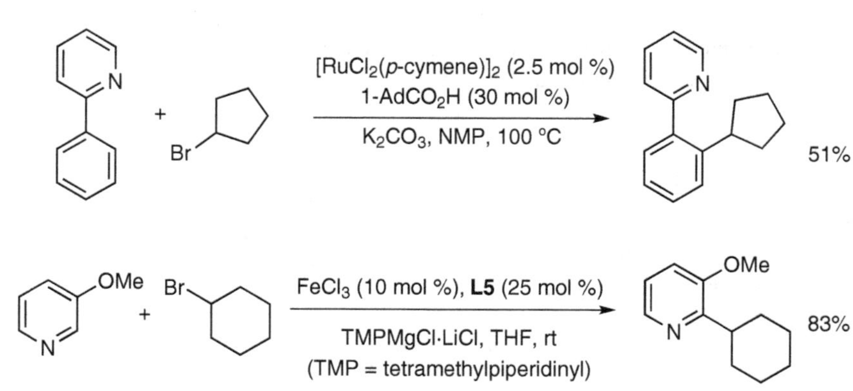

SCHEME 11-50. Ru- and Fe-catalyzed direct alkylation of hereroarenes with secondary alkyl halides.

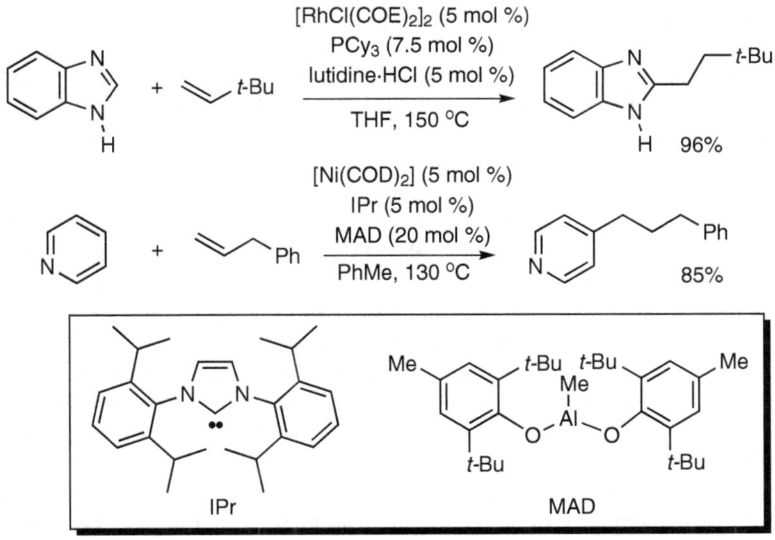

SCHEME 11-51. Cooperative catalysts for direct alkylation of imidazoles and pyridines with terminal alkynes.

IPr and a Lewis acid, MAD. A deuterium-labeling experiment indicates that C–H activation occurs irreversibly at the C-4 position, but reversibly at the C-2 and C-3 positions. Therefore, the product distribution results from the exclusive reductive elimination at the C-4 position [87].

11.6 CONCLUSION

Over the last decade, transition-metal-mediated or -catalyzed reactions involving C–H bond cleavage have developed very rapidly because of their ability to functionalize ubiquitous C–H bonds in one synthetic operation, thus providing versatile synthetic routes for the concise, direct transformation of various heterocycles, complex natural product synthesis, and late-stage functionalization of biologically active compounds [88]. Further development of new and sustainable catalytic conditions, including asymmetric catalysis [89], will increase the C–H activation chemistry to truly useful and practical levels, thus enabling the discovery of new biological and physical properties of heteroarene molecules.

REFERENCES

1. (a) de Meijere A, Diederich F *Metal-Catalyzed Cross-Coupling Reactions*. 2nd edition, Wiley-VCH, Weinheim; 2004. (b) Tsuji J *Palladium Reagents and Catalysts*. 2nd edition, Wiley, Chichester; 2004. (c) Miyaura N, *Cross-Coupling Reactions*. Topics in Current Chemistry, Vol. 219. Springer, Berlin; 2002.
2. (a) Recent reviews: Satoh T, Miura M (2007). Catalytic direct arylation of heteroaromatic compounds. *Chem. Lett.* 36, 200–205. (b) Daugulis O, Do HQ, Shabashov D (2009). Palladium- and copper-catalyzed arylation of carbon–hydrogen bonds. *Acc. Chem. Res.* 42, 1074–1086.

(c) Kulkarni AA, Daugulis O (2009). Direct conversion of carbon–hydrogen into carbon–carbon bonds by first-row transition-metal catalysis. *Synthesis* 4087–4109. (d) Chen X, Engle KM, Wang DH, Yu JQ (2009). Palladium(II)-catalyzed C–H activation/C–C cross-coupling reactions: versatility and practicality. *Angew. Chem., Int. Ed.* 48, 5094–5115. (e) Ackermann L, Vicente R, Kapdi AR (2009). Transition-metal-catalyzed direct arylation of (hetero)arenes by C–H bond cleavage. *Angew. Chem., Int. Ed.* 48, 9792–9826. (f) Sun CL, Li BJ, Shi ZJ (2010). Pd-catalyzed oxidative coupling with organometallic reagents via C–H activation. *Chem. Commun.* 677–685. (g) Lyons TW, Sanford MS (2010). Palladium-catalyzed ligand-directed C–H functionalization reactions. *Chem. Rev.* 110, 1147–1169. (h) Dudnik AS, Gevorgyan V (2010). Formal inverse Sonogashira reaction: direct alkynylation of arenes and heterocycles with alkynyl halides. *Angew. Chem., Int. Ed.* 49, 2096–2098. (i) Satoh T, Miura M (2010). Transition-metal-catalyzed regioselective arylation and vinylation of carboxylic acids. *Synthesis* 3395–3409. (j) Ackermann L (2010). Metal-catalyzed direct alkylations of (hetero) arenes via C–H bond cleavages with unactivated alkyl halides. *Chem. Commun.* 4866–4877. (k) Hirano K, Miura M (2011). Direct carbon-hydrogen bond functionalization of heterocyclic compounds. *Synlett* 294–307.

3. (a) Pivsa-Art S, Satoh T, Kawamura Y, Miura M, Nomura M (1998). Palladium-catalyzed arylation of azole compounds with aryl halides in the presence of alkali metal carbonates and the use of copper iodide in the reaction. *Bull. Chem. Soc. Jpn.* 71, 467–473. (b) Park CH, Ryabova V, Seregin IV, Sromek AW, Gevorgyan V (2004). Palladium-catalyzed arylation and heteroarylation of indolizines. *Org. Lett.* 6, 1159–1162.
4. (a) Mota AJ, Dedieu A, Bour C, Suffer J (2005). Cyclocarbopalladation involving an unusual 1,5-palladium vinyl to aryl shift as termination step: theoretical study of the mechanism. *J. Am. Chem. Soc.* 127, 7171–7182. (b) Garcia-Cuadrado D, Braga AAC, Maseras F, Echavarren AM (2006). Proton abstraction mechanism for the palladium-catalyzed intramolecular arylation. *J. Am. Chem. Soc.* 128, 1066–1067. (c) Lafrance M, Rowley CN, Woo TK, Fagnou K (2006). Catalytic intermolecular direct arylation of perfluorobenzenes. *J. Am. Chem. Soc.* 128, 8754–8756. (d) Lapointe D, Fagnou K (2010). Overview of the mechanistic work on the concerted metallation–deprotonation pathway. *Chem. Lett.* 39, 1119–1126.
5. (a) Do HQ, Daugulis O (2007). Copper-catalyzed arylation of heterocycle C–H bonds. *J. Am. Chem. Soc.* 129, 12404–12405. (b) Do HQ, Khan RMK, Daugulis O (2008). A general method for copper-catalyzed arylation of arene C–H bonds. *J. Am. Chem. Soc.* 130, 15185–15192.
6. (a) Wiedemann SH, Lewis JC, Ellman JA, Bergman RG (2006). Experimental and computational studies on the mechanism of *N*-heterocycle C–H activation by Rh(I). *J. Am. Chem. Soc.* 128, 2452–2462. (b) Kakiuchi F, Kouchi T, Mizushima E, Murai S (2010). Room-temperature regioselective C–H/olefin coupling of aromatic ketones using an activated ruthenium catalyst with a carbonyl ligand and structural elucidation of key intermediates. *J. Am. Chem. Soc.* 132, 17741–17750.
7. (a) Davies DL, Al-Duaij O, Fawcett J, Giardiello M, Hilton ST, Russell DR (2003). Room-temperature cyclometallation of amines, imines and oxazolines with $[MCl_2Cp^*]_2$ (M = Rh, Ir) and $[RuCl_2(p\text{-cymene})]_2$. *J. Chem. Soc., Dalton Trans.* 4132–4138. (b) Li L, Brennessel WW, Jones WD (2008). An efficient low-temperature route to polycyclic isoquinoline salt synthesis via C–H activation with $[Cp^*MCl_2]_2$ (M = Rh, Ir). *J. Am. Chem. Soc.* 130, 12414–12419.
8. (a) Davies DL, Donald SMA, Macgregor SA (2005). Computational study of the mechanism of cyclometalation by palladium acetate. *J. Am. Chem. Soc.* 127, 13754–13755. (b) Hull KL, Sanford MS (2009). Mechanism of benzoquinone-promoted palladium-catalyzed oxidative cross-coupling reactions. *J. Am. Chem. Soc.* 131, 9651–9653.
9. (a) Hassan J, Sévignon M, Gozzi C, Schulz E, Lemaire M (2002). Aryl–aryl bond formation one century after the discovery of the Ullmann reaction. *Chem. Rev.* 102, 1359–1469. (b) Corbet JP,

Mignani G (2006). Selected patented cross-coupling reaction technologies. *Chem. Rev.* 106, 2651–2710.

10. (a) Nakamura N, Tajima Y, Sakai K (1982). Direct phenylation of isoxazoles using palladium catalysts. Synthesis of 4-phenylmuscimol. *Heterocycles* 17, 235–245. (b) Akita Y, Itagaki Y, Takizawa S, Ohta A (1989). Cross-coupling reactions of chloropyrazines with 1-substituted indoles. *Chem. Pharm. Bull.* 37, 1477–1480. (c) Aoyagi Y, Inoue A, Koizumi I, Hashimoto R, Tokunaga K, Gohma K, Komatsu J, Sekine K, Miyafuji A, Kunoh J, Homma R, Akita Y, Ohta A (1992). Palladium-catalyzed cross-coupling reactions of chloropyrazines with aromatic heterocycles. *Heterocycles* 33, 257–272.
11. (a) Chiong HA, Daugulis O (2007). Palladium-catalyzed arylation of electron-rich heterocycles with aryl chlorides. *Org. Lett.* 9, 1449–1451. (b) Nadres ET, Lazareva A, Daugulis O (2011). Palladium-catalyzed indole, pyrrole, and furan arylation by aryl chlorides. *J. Org. Chem.* 76, 471–483.
12. Ackermann L, Althammer A, Fenner S (2009). Palladium-catalyzed direct arylations of heteroarenes with tosylates and mesylates. *Angew. Chem., Int. Ed.* 48, 201–204.
13. Lewis JC, Berman AM, Bergman RG, Ellman JA (2008). Rh(I)-catalyzed arylation of heterocycles via C–H bond activation: expanded scope through mechanistic insight. *J. Am. Chem. Soc.* 130, 2493–2500.
14. Join B, Yamamoto T, Itami K (2009). Iridium catalysis for C–H bond arylation of heteroarenes with iodoarenes. *Angew. Chem., Int. Ed.* 48, 3644–3647.
15. (a) Yoshizumi T, Tsurugi H, Satoh T, Miura M (2008). Copper-mediated direct arylation of benzoazoles with aryl iodides. *Tetrahedron Lett.* 49, 1598–1600. (b) Yoshizumi T, Satoh T, Hirano K, Matsuo D, Orita A, Otera J, Miura M (2009). Synthesis of 2,5-diaryloxazoles through van Leusen reaction and copper-mediated direct arylation. *Tetrahedron Lett.* 50, 3273–3276. (c) Kawano T, Yoshizumi T, Hirano K, Satoh T, Miura M (2009). Copper-mediated direct arylation of 1,3,4-oxadiazoles and 1,2,4-triazoles with aryl iodides. *Org. Lett.* 11, 3072–3075.
16. Canivet J, Yamaguchi J, Ban I, Itami K (2009). Nickel-catalyzed biaryl coupling of heteroarenes and aryl halides/triflates. *Org. Lett.* 11, 1733–1736.
17. Hachiya H, Hirano K, Satoh T, Miura M (2009). Nickel-catalyzed direct arylation of azoles with aryl bromides. *Org. Lett.* 11, 1737–1740.
18. (a) Campeau LC, Rousseaux S, Fagnou K (2005). A solution to the 2-pyridyl organometallic cross-coupling problem: regioselective catalytic direct arylation of pyridine *N*-oxides. *J. Am. Chem. Soc.* 127, 18020–18021. (b) Larivée A, Mousseau JJ, Charette AB (2008). Palladium-catalyzed direct C–H arylation of *N*-iminopyridinium ylides: application to the synthesis of (±)-anabasine. *J. Am. Chem. Soc.* 130, 52–54.
19. Berman AM, Lewis JC, Bergman RG, Ellman JA (2008). Rh(I)-catalyzed direct arylation of pyridines and quinolines. *J. Am. Chem. Soc.* 130, 14926–14927.
20. Okazawa T, Satoh T, Miura M, Nomura M (2002). Palladium-catalyzed multiple arylation of thiophenes. *J. Am. Chem. Soc.* 124, 5286–5287.
21. Nakano M, Tsurugi H, Satoh T, Miura M (2008). Palladium-catalyzed perarylation of 3-thiophene- and 3-furancarboxylic acids accompanied by C–H bond cleavage and decarboxylation. *Org. Lett.* 10, 1851–1854.
22. Campeau LC, Bertrand-Laperle M, Leclerc JP, Villemure E, Gorelsky S, Fagnou K (2008). C2, C5, and C4 azole *N*-oxide direct arylation including room-temperature reactions. *J. Am. Chem. Soc.* 130, 3276–3277.
23. (a) Yanagisawa S, Sudo T, Noyori R, Itami K (2006). Direct C–H arylation of (hetero)arenes with aryl iodides via rhodium catalysis. *J. Am. Chem. Soc.* 128, 11748–11749. (b) Yanagisawa S, Ueda K, Sekizawa H, Itami K (2009). Programmed synthesis of tetraarylthiophenes through sequential C–H arylation. *J. Am. Chem. Soc.* 131, 14622–14623.

24. Lu J, Tan X, Chen C (2007). Palladium-catalyzed direct functionalization of imidazolinone: synthesis of dibromophakellstatin. *J. Am. Chem. Soc.* 129, 7768–7769.

25. Deprez NR, Kalyani D, Krause A, Sanford MS (2006). Room temperature palladium-catalyzed 2-arylation of indoles. *J. Am. Chem. Soc.* 128, 4972–4973.

26. Lebrasseur N, Larrosa I (2008). Room temperature and phosphine free palladium catalyzed direct C-2 arylation of indoles. *J. Am. Chem. Soc.* 130, 2926–2927.

27. Roy D, Mom S, Beaupérin M, Doucet H, Hierso JC (2010). A versatile palladium/triphosphane system for direct arylation of heteroarenes with chloroarenes at low catalyst loading. *Angew. Chem., Int. Ed.* 49, 6650–6654.

28. Iwasaki M, Yorimitsu H, Oshima K (2007). Microwave-assisted palladium-catalyzed direct arylation of 1,4-disubstituted 1,2,3-triazoles with aryl chlorides. *Chem. Asian J.* 2, 1430–1435.

29. Wang Q, Takita R, Kikuzaki Y, Ozawa F (2010). Palladium-catalyzed dehydrohalogenative polycondensation of 2-bromo-3-hexylthiophene: an efficient approach to head-to-tail poly(3-hexylthiophene). *J. Am. Chem. Soc.* 132, 11420–11421.

30. Yanagisawa S, Ueda K, Taniguchi T, Itami K (2008). Potassium *t*-butoxide alone can promote the biaryl coupling of electron-deficient nitrogen heterocycles and haloarenes. *Org. Lett.* 10, 4673–4676.

31. (a) Sun CL, Li H, Yu DG, Yu M, Zhou X, Lu XY, Huang K, Zheng SF, Li BJ, Shi ZJ (2010). An efficient organocatalytic method for constructing biaryls through aromatic C–H activation. *Nat. Chem.* 2, 1044–1049. (b) Shirakawa E, Itoh K, Higashino T, Hayashi T (2010). *tert*-Butoxide-mediated arylation of benzene with aryl halides in the presence of a catalytic 1,10-phenanthroline derivative. *J. Am. Chem. Soc.* 132, 15537–15539. (c) Liu W, Cao H, Zhang H, Zhang H, Chung KH, He C, Wang H, Kwong FY, Lei A (2010). Organocatalysis in cross-coupling: DMEDA-catalyzed direct C–H arylation of unactivated benzene. *J. Am. Chem. Soc.* 132, 16737–16740.

32. Yang SD, Sun CL, Fang Z, Li BJ, Li YZ, Shi ZJ (2008). Palladium-catalyzed direct arylation of (hetero)arenes with aryl boronic acids. *Angew. Chem., Int. Ed.* 47, 1473–1476.

33. Kirchberg S, Tani S, Ueda K, Yamaguchi J, Studer A, Itami K (2011). Oxidative biaryl coupling of thiophenes and thiazoles with arylboronic acids through palladium catalysis: otherwise difficult C4-selective C–H arylation enabled by boronic acids. *Angew. Chem., Int. Ed.* 50, 2387–2391.

34. Ban I, Sudo T, Taniguchi T, Itami K (2008). Copper-mediated C–H bond arylation of arenes with arylboronic acids. *Org. Lett.* 10, 3607–3609.

35. Wen J, Qin S, Ma LF, Dong L, Zhang J, Liu SS, Duan YS, Chen SY, Hu CW, Yu XQ (2010). Iron-mediated direct Suzuki–Miyaura reaction: a new method for the *ortho*-arylation of pyrrole and pyridine. *Org. Lett.* 12, 2694–2697.

36. Hachiya H, Hirano K, Satoh T, Miura M (2010). Oxidative nickel–air catalysis in C–H arylation: direct cross-coupling of azoles with arylboronic acids using air as sole oxidant. *ChemCatChem* 2, 1403–1406.

37. Seiple IB, Su S, Rodriguez RA, Gianatassio R, Fujiwara Y, Sobel AL, Baran PS (2010). Direct C–H arylation of electron-deficient heterocycles with arylboronic acids. *J. Am. Chem. Soc.* 132, 13194–13196.

38. Giri R, Maugel N, Li JJ, Wang DH, Breazzano SP, Saunders LB, Yu JQ (2007). Palladium-catalyzed methylation and arylation of sp^2 and sp^3 C–H bonds in simple carboxylic acids. *J. Am. Chem. Soc.* 129, 3510–3511.

39. Kakiuchi F, Matsuura Y, Kan S, Chatani N (2005). A $RuH_2(CO)(PPh_3)_3$-catalyzed regioselective arylation of aromatic ketones with arylboronates via carbon–hydrogen bond cleavage. *J. Am. Chem. Soc.* 127, 5936–5945.

40. Liang Z, Yao B, Zhang Y (2010). $Pd(OAc)_2$-catalyzed regioselective arylation of indoles with arylsiloxane in acidic medium. *Org. Lett.* 12, 3185–3187.

41. Hachiya H, Hirano K, Satoh T, Miura M (2010). Nickel-catalyzed direct C−H arylation and alkenylation of heteroarenes with organosilicon reagents. *Angew. Chem., Int. Ed.* 49, 2202–2205.

42. (a) Oi S, Fukita S, Inoue Y (1998). Rhodium-catalyzed direct ortho arylation of 2-arylpyridines with arylstannanes via C−H activation. *Chem. Commun.* 2439–2440. (b) Kawai H, Kobayashi Y, Oi S, Inoue Y (2008). Direct C–H bond arylation of arenes with aryltin reagents catalyzed by palladium complexes. *Chem. Commun.* 1464–1466.

43. Norinder J, Matsumoto A, Yoshikai N, Nakamura E (2008). Iron-catalyzed direct arylation through directed C−H bond activation. *J. Am. Chem. Soc.* 130, 5858–5859.

44. Li B, Wu ZH, Gu YF, Sun CL, Wang BQ, Shi ZJ (2011). Direct cross-coupling of C−H bonds with Grignard reagents through cobalt catalysis. *Angew. Chem., Int. Ed.* 50, 1109–1113.

45. Tobisu M, Hyodo I, Chatani N (2009). Nickel-catalyzed reaction of arylzinc reagents with *N*-aromatic heterocycles: a straightforward approach to C−H bond arylation of electron-deficient heteroaromatic compounds. *J. Am. Chem. Soc.* 131, 12070–12071.

46. (a) Wang C, Piel I, Glorius F (2009). Palladium-catalyzed intramolecular direct arylation of benzoic acids by tandem decarboxylation/C−H activation. *J. Am. Chem. Soc.* 131, 4194–4195. (b) Cornella J, Lu P, Larrosa I (2009). Intermolecular decarboxylative direct C-3 arylation of indoles with benzoic acids. *Org. Lett.* 11, 5506–5509. (c) Xie K, Yang Z, Zhou X, Li X, Wang S, Tan Z, An X, Guo CC (2010). Pd-catalyzed decarboxylative arylation of thiazole, benzoxazole, and polyfluorobenzene with substituted benzoic acids. *Org. Lett.* 12, 1564–1567. (d) Zhang F, Greaney MF (2010). Decarboxylative C−H cross-coupling of azoles. *Angew. Chem., Int. Ed.* 49, 2768–2771.

47. (a) Stuart DR, Fagnou K (2007). The catalytic cross-coupling of unactivated arenes. *Science* 316, 1172–1175. (b) Hull KL, Sanford MS (2007). Catalytic and highly regioselective cross-coupling of aromatic C−H substrates. *J. Am. Chem. Soc.* 129, 11904–11905. (c) Dwight TA, Rue NR, Charyk D, Josselyn R, DeBoef B (2007). C−C bond formation via double C−H functionalization: aerobic oxidative coupling as a method for synthesizing heterocoupled biaryls. *Org. Lett.* 9, 3137–3139. (d) Xi P, Yang F, Qin S, Zhao D, Lan J, Gao G, Hu C, You J (2010). Palladium(II)-catalyzed oxidative C−H/C−H cross-coupling of heteroarenes. *J. Am. Chem. Soc.* 132, 1822–1824. (e) He CY, Fan S, Zhang X (2010). Pd-catalyzed oxidative cross-coupling of perfluoroarenes with aromatic heterocycles. *J. Am. Chem. Soc.* 132, 12850–12852.

48. Kitahara M, Umeda N, Hirano K, Satoh T, Miura M (2011). Copper-mediated intermolecular direct biaryl coupling. *J. Am. Chem. Soc.* 133, 2160–2162.

49. Kita Y, Morimoto K, Ito M, Ogawa C, Goto A, Dohi T (2009). Metal-free oxidative cross-coupling of unfunctionalized aromatic compounds. *J. Am. Chem. Soc.* 131, 1668–1669.

50. (a) Besselièvre F, Piguel S, Mahuteau-Betzer F, Grierson DS (2008). Stereoselective direct copper-catalyzed alkenylation of oxazoles with bromoalkenes. *Org. Lett.* 10, 4029–4032. (b) Mousseau JJ, Bull JA, Charette AB (2010). Copper-catalyzed direct alkenylation of *N*-iminopyridinium ylides. *Angew. Chem., Int. Ed.* 49, 1115–1118.

51. (a) Gottumukkala AL, Derridj F, Djebbar S, Doucet H (2008). Alkenyl bromides: useful coupling partners for the palladium-catalyzed coupling with heteroaromatics via a C−H bond activation. *Tetrahedron Lett.* 49, 2926–2930. (b) Koubachi J, Kazzouli SE, Berteina-Raboin S, Mouaddib A, Guillaumet G (2008). New and efficient palladium(0)-mediated microwave-assisted direct C3 alkenylation of imidazo[1,2-a]pyridines. *Synthesis* 2537–2542.

52. (a) Fujiwara Y, Moritani I, Danno S, Asano R, Teranishi S (1969). Aromatic substitution of olefins. VI. Arylation of olefins with palladium(II) acetate. *J. Am. Chem. Soc.* 91, 7166–7169. (b) Fujiwara Y, Maruyama O, Yoshidomi M, Taniguchi H (1981). Palladium-catalyzed alkenylation of aromatic heterocycles with olefins. Synthesis of functionalized aromatic heterocycles. *J. Org.*

Chem. 46, 851–855. (c) Jia C, Lu W, Kitamura T, Fujiwara Y (1999). Highly efficient Pd-catalyzed coupling of arenes with olefins in the presence of *tert*-butyl hydroperoxide as oxidant. *Org. Lett.* 1, 2097–2100. (d) Jia C, Kitamura T, Fujiwara Y (2001). Catalytic functionalization of arenes and alkanes via C–H bond activation. *Acc. Chem. Res.* 34, 633–639. (e) Yokota T, Tani M, Sakaguchi S, Ishii Y (2003). Direct coupling of benzene with olefin catalyzed by $Pd(OAc)_2$ combined with heteropolyoxometalate under dioxygen. *J. Am. Chem. Soc.* 125, 1476–1477.

53. (a) Maehara A, Satoh T, Miura M (2008). Palladium-catalyzed direct oxidative vinylation of thiophenes and furans under weakly basic conditions. *Tetrahedron* 64, 5982–5986. (b) Miyasaka M, Hirano K, Satoh T, Miura M (2010). Palladium-catalyzed direct oxidative alkenylation of azoles. *J. Org. Chem.* 75, 5421–5424.
54. Grimster NP, Gauntlett C, Godfrey CRA, Gaunt MJ (2005). Palladium-catalyzed intermolecular alkenylation of indoles by solvent-controlled regioselective C–H functionalization. *Angew. Chem., Int. Ed.* 44, 3125–3129.
55. García-Rubia A, Arrayás RG, Carretero JC (2009). Palladium(II)-catalyzed regioselective direct C2 alkenylation of indoles and pyrroles assisted by the *N*-(2-pyridyl)sulfonyl protecting group. *Angew. Chem., Int. Ed.* 48, 6511–6515.
56. Maehara A, Tsurugi H, Satoh T, Miura M (2008). Regioselective C–H functionalization directed by a removable carboxyl group: palladium-catalyzed vinylation at the unusual position of indole and related heteroaromatic rings. *Org. Lett.* 10, 1159–1162.
57. Ye M, Gao GL, Yu JQ (2011). Ligand-promoted C-3 selective C–H olefination of pyridines with Pd catalysts. *J. Am. Chem. Soc.* 132, 6964–6967.
58. Mochida S, Hirano K, Satoh T, Miura M (2011). Rhodium-catalyzed regioselective olefination directed by a carboxylic group. *J. Org. Chem.* 76, 3024–3033.
59. Ueyama T, Mochida S, Hukutani T, Hirano K, Satoh T, Miura M (2011). Ruthenium-catalyzed oxidative vinylation of heteroarene carboxylic acids with alkenes via regioselective C–H bond cleavage. *Org. Lett.* 12, 706–708.
60. (a) Selected examples: Tunge JA, Foresee LN (2005). Mechanistic studies of Fujiwara hydroarylation. C–H activation versus electrophilic aromatic substitution. *Organometallics* 24, 6440–6444. (b) Soriano E, Marco-Contelles J (2006). Mechanisms of the transition metal-mediated hydroarylation of alkynes and allenes. *Organometallics* 25, 4542–4553. (c) Zhang X, Teo WT, Chan PWH (2009). Ytterbium(III) triflate catalyzed tandem Friedel–Crafts alkylation/hydroarylation of propargylic alcohols with phenols as an expedient route to indenols. *Org. Lett.* 11, 4990–4993. (d) Suarez-Pantiga S, Palomas D, Rubio E, Gonzalez JM (2009). Consecutive C–H functionalization reactions of arenes: synthesis of carbo- and heteropolycyclic skeletons. *Angew. Chem., Int. Ed.* 48, 7857–7861. (e) Menon RS, Findlay AD, Bissember AC, Banwell MG (2009). The Au(I)-catalyzed intramolecular hydroarylation of terminal alkynes under mild conditions: application to the synthesis of 2*H*-chromenes, coumarins, benzofurans, and dihydroquinolines. *J. Org. Chem.* 74, 8901–8903. (f) Hashimoto T, Izumi T, Kutubi MS, Kitamura T (2010). Iron(III)-catalyzed hydroarylation of propiolic acid with activated arenes. *Tetrahedron Lett.* 51, 761–763.
61. (a) Murai S, Kakiuchi F, Sekine S, Tanaka Y, Kamatani A, Sonoda M, Chatani N (1993). Efficient catalytic addition of aromatic carbon–hydrogen bonds to olefins. *Nature* 366, 529–531. (b) Kakiuchi F, Yamamoto Y, Chatani N, Murai S (1995). Catalytic addition of aromatic C–H bonds to acetylenes. *Chem. Lett.* 8, 681–682. (c) Kakiuchi F, Uetsuhara T, Tanaka Y, Chatani N, Murai S (2002). Ruthenium-catalyzed addition of olefinic C–H bonds in conjugate enones to acetylenes to give conjugate dienones. *J. Mol. Catal. A: Chem.* 182, 511–514.
62. (a) Kuninobu Y, Tokunaga Y, Kawata A, Takai K (2006). Insertion of polar and nonpolar unsaturated molecules into carbon–rhenium bonds generated by C–H bond activation: synthesis of phthalimidine and indene derivatives. *J. Am. Chem. Soc.* 128, 202–209. (b) Kuninobu Y, Kikuchi K, Tokunaga Y, Nishina Y, Takai K (2008). Hydroarylation of acetylenes, acrylates, and

isocyanates with heteroaromatic compounds under rhenium catalysis. *Tetrahedron* 64, 5974–5981.

63. (a) Gao K, Lee PS, Fujita T, Yoshikai N (2010). Cobalt-catalyzed hydroarylation of alkynes through chelation-assisted C–H bond activation. *J. Am. Chem. Soc.* 132, 12249–12251. (b) Ding Z, Yoshikai N (2010). Cobalt-catalyzed addition of azoles to alkynes. *Org. Lett.* 12, 4180–4183.
64. (a) Nakao Y, Kanyiva KS, Oda S, Hiyama T (2006). Hydroheteroarylation of alkynes under mild nickel catalysis. *J. Am. Chem. Soc.* 128, 8146–8147. (b) Nakao Y, Kanyiva KS, Hiyama T (2008). A strategy for C–H activation of pyridines: direct C-2 selective alkenylation of pyridines by nickel/Lewis acid catalysis. *J. Am. Chem. Soc.* 130, 2448–2449. (c) Kanyiva KS, Löbermann F, Nakao Y, Hiyama T (2009). Regioselective alkenylation of imidazoles by nickel/Lewis acid catalysis. *Tetrahedron Lett.* 50, 3463–3466. (d) Nakao Y, Idei H, Kanyiva KS, Hiyama T (2009). Direct alkenylation and alkylation of pyridone derivatives by $Ni/AlMe_3$ catalysis. *J. Am. Chem. Soc.* 131, 15996–15997.
65. Schipper DJ, Hutchinson M, Fagnou K (2010). Rhodium(III)-catalyzed intermolecular hydroarylation of alkynes. *J. Am. Chem. Soc.* 132, 6910–6911.
66. (a) Selected examples: Ueura K, Satoh T, Miura M (2007). An efficient waste-free oxidative coupling via regioselective C–H bond cleavage: Rh/Cu-catalyzed reaction of benzoic acids with alkynes and acrylates under air. *Org. Lett.* 9, 1407–1409. (b) Umeda N, Tsurugi H, Satoh T, Miura M (2008). Fluorescent naphthyl- and anthrylazoles from the catalytic coupling of phenylazoles with internal alkynes through the cleavage of multiple C–H bonds. *Angew. Chem., Int. Ed.* 47, 4019–4022. (c) Fukutani T, Umeda N, Hirano K, Satoh T, Miura M (2009). Rhodium-catalyzed oxidative coupling of aromatic imines with internal alkynes via regioselective C–H bond cleavage. *Chem. Commun.* 5141–5143. (d) Hyster TK, Rovis T (2010). Rhodium-catalyzed oxidative cycloaddition of benzamides and alkynes via C–H/N–H activation. *J. Am. Chem. Soc.* 132, 10565–10569. (e) Stuart DR, Alsabeh P, Kuhn M, Fagnou K (2010). Rhodium(III)-catalyzed arene and alkene C–H bond functionalization leading to indoles and pyrroles. *J. Am. Chem. Soc.* 132, 18326–18339. (f) Mochida S, Umeda N, Hirano K, Satoh T, Miura M (2010). Rhodium-catalyzed oxidative coupling/cyclization of benzamides with alkynes via C–H bond cleavage. *Chem. Lett.* 39, 744–746.
67. (a) Kobayashi K, Arisawa M, Yamaguchi M (2002). $GaCl_3$-catalyzed *ortho*-ethynylation of phenols. *J. Am. Chem. Soc.* 124, 8528–8529. (b) Amemiya R, Fujii A, Yamaguchi M (2004). $GaCl_3$-catalyzed *ortho*-ethynylation reaction of *N*-benzylanilines. *Tetrahedron Lett.* 45, 4333–4335.
68. Seregin IV, Ryabova V, Gevorgyan V (2007). Direct palladium-catalyzed alkynylation of *N*-fused heterocycles. *J. Am. Chem. Soc.* 129, 7742–7743.
69. Gu Y, Wang XM (2009). Direct palladium-catalyzed C-3 alkynylation of indoles. *Tetrahedron Lett.* 50, 763–766.
70. (a) Pd: Kim SH, Chang S (2010). Highly efficient and versatile Pd-catalyzed direct alkynylation of both azoles and azolines. *Org. Lett.* 12, 1868–1871. Ni: (b) Matsuyama N, Hirano K, Satoh T, Miura M (2009). Nickel-catalyzed direct alkynylation of azoles with alkynyl bromides. *Org. Lett.* 11, 4156–4159. Cu: (c) Besselièvre F, Piguel S (2009). Copper as a powerful catalyst in the direct alkynylation of azoles. *Angew. Chem., Int. Ed.* 48, 9553–9556. (d) Kawano T, Matsuyama N, Hirano K, Satoh T, Miura M (2010). Room temperature direct alkynylation of 1,3,4-oxadiazoles with alkynyl bromides under copper catalysis. *J. Org. Chem.* 75, 1764–1766. (e) Berciano BP, Lebrequier S, Besselièvre F, Piguel S (2010). 1,1-Dibromo-1-alkenes as valuable partners in the copper-catalyzed direct alkynylation of azoles. *Org. Lett.* 12, 4038–4041.
71. (a) Brand JP, Charpentier J, Waser J (2009). Direct alkynylation of indole and pyrrole heterocycles. *Angew. Chem., Int. Ed.* 48, 9346–9349. (b) Brand JP, Waser J (2010). Direct alkynylation of thiophenes: cooperative activation of TIPS–EBX with gold and Brønsted acids. *Angew. Chem., Int. Ed.* 49, 7304–7307.

72. (a) Kitahara M, Hirano K, Tsurugi H, Satoh T, Miura M (2010). Copper-mediated direct cross-coupling of 1,3,4-oxadiazoles and oxazoles with terminal alkynes. *Chem. Eur. J.* 16, 1772–1775. (b) Wei Y, Zhao H, Kan J, Su W, Hong M (2010). Copper-catalyzed direct alkynylation of electron-deficient polyfluoroarenes with terminal alkynes using O_2 as an oxidant. *J. Am. Chem. Soc.* 132, 2522–2523.

73. Matsuyama N, Kitahara M, Hirano K, Satoh T, Miura M (2010). Nickel- and copper-catalyzed direct alkynylation of azoles and polyfluoroarenes with terminal alkynes under O_2 or atmospheric conditions. *Org. Lett.* 12, 2358–2361.

74. Kim SH, Yoon J, Chang S (2011). Palladium-catalyzed oxidative alkynylation of heterocycles with terminal alkynes under air conditions. *Org. Lett.* 13, 1474–1477.

75. (a) de Haro T, Nevado C (2010). Gold-catalyzed ethynylation of arenes. *J. Am. Chem. Soc.* 132, 1512–1513. Stoichiometric version: (b) Fuchita Y, Utsonomiya Y, Yasutake M (2001). Synthesis and reactivity of arylgold(III) complexes from aromatic hydrocarbons via C–H bond activation. *J. Chem. Soc., Dalton Trans.* 2330–2334.

76. (a) Verrier C, Hoarau C, Marsais F (2009). Direct palladium-catalyzed alkenylation, benzylation and alkylation of ethyl oxazole-4-carboxylate with alkenyl-, benzyl- and alkyl halides. *Org. Biomol. Chem.* 7, 647–650. (b) Lapointe D, Fagnou K (2009). Palladium-catalyzed benzylation of heterocyclic aromatic compounds. *Org. Lett.* 11, 4160–4163. (c) Sahnoun S, Messaoudi S, Brion JD, Alami M (2011). Palladium-catalyzed direct benzylation of xanthines. *ChemCatChem* 3, 893–897.

77. (a) Ackermann L, Barfüssser S, Pospech J (2010). Palladium-catalyzed direct arylations, alkenylations, and benzylations through C–H bond cleavages with sulfamates or phosphates as electrophiles. *Org. Lett.* 12, 724–726. (b) Mukai T, Hirano K, Satoh T, Miura M (2010). Palladium-catalyzed direct benzylation of azoles with benzyl carbonates. *Org. Lett.* 12, 1360–1363.

78. Zhao X, Wu G, Zhang Y, Wang J (2011). Copper-catalyzed direct benzylation or allylation of 1,3-azoles with *N*-tosylhydrazones. *J. Am. Chem. Soc.* 133, 3296–3299.

79. (a) Nakao Y, Kashihara N, Kanyiva KS, Hiyama T (2010). Nickel-catalyzed hydroheteroarylation of vinylarenes. *Angew. Chem., Int. Ed.* 49, 4451–4454. (b) Mukai T, Hirano K, Satoh T, Miura M (2009). Nickel-catalyzed C–H alkenylation and alkylation of 1,3,4-oxadiazoles with alkynes and styrenes. *J. Org. Chem.* 74, 6410–6413.

80. Gao K, Yoshikai N (2011). Regioselectivity-switchable hydroarylation of styrenes. *J. Am. Chem. Soc.* 133, 400–402.

81. Yao T, Hirano K, Satoh T, Miura M (2011). Stereospecific copper-catalyzed C–H allylation of electron-deficient arenes with allyl phosphates. *Angew. Chem., Int. Ed.* 50, 2990–2994.

82. (a) Yue W, Li Y, Jiang W, Zhen Y, Wang Z (2009). Direct meta-selective alkylation of perylene bisimides via palladium-catalyzed C–H functionalization. *Org. Lett.* 11, 5430–5433. (b) Yao T. Hirano K, Satoh T, Miura M (2010). Palladium- and nickel-catalyzed direct alkylation of azoles with unactivated alkyl bromides and chlorides. *Chem. Eur. J.* 16, 12307–12311.

83. Vechorkin O, Proust V, Hu X (2010). The nickel/copper-catalyzed direct alkylation of heterocyclic C–H bonds. *Angew. Chem., Int. Ed.* 49, 3061–3064.

84. (a) Ackermann L, Novák P, Vicente R, Hofmann N (2009). Ruthenium-catalyzed regioselective direct alkylation of arenes with unactivated alkyl halides through C–H bond cleavage. *Angew. Chem., Int. Ed.* 48, 6045–6048. (b) Ackermann L, Hofmann N, Vicente R (2011). Carboxylate-assisted ruthenium-catalyzed direct alkylations of ketimines. *Org. Lett.* 13, 1875–1877.

85. Tran LD, Daugulis O (2010). Iron-catalyzed heterocycle and arene deprotonative alkylation. *Org. Lett.* 12, 4277–4279.

86. (a) Tan KL, Bergman RG, Ellman JA (2002). Intermediacy of an *N*-heterocyclic carbene complex in the catalytic C–H activation of a substituted benzimidazole. *J. Am. Chem. Soc.* 124,

3202–3203. (b) Tan KL, Bergman RG, Ellman JA (2002). Intermolecular coupling of isomerizable alkenes to heterocycles via rhodium-catalyzed C–H bond activation. *J. Am. Chem. Soc.* 124, 13964–13965. (c) Rech JC, Yato M, Duckett D, Ember B, LoGrasso PV, Bergman RG, Ellman JA (2007). Synthesis of potent bicyclic bisarylimidazole c-jun *N*-terminal kinase inhibitors by catalytic C–H bond activation. *J. Am. Chem. Soc.* 129, 490–491.

87. (a) Tsai CC, Shih WC, Fang CH, Ong TG, Yap GPA (2010). Bimetallic nickel aluminun mediated *para*-selective alkenylation of pyridine: direct observation of η^2,η^1-pyridine Ni(0)–Al (III) intermediates prior to C–H bond activation. *J. Am. Chem. Soc.* 132, 11887–11889. (b) Nakao Y, Yamada Y, Kashihara N, Hiyama T (2010). Selective C-4 alkylation of pyridine by nickel/Lewis acid catalysis. *J. Am. Chem. Soc.* 132, 13666–13668.

88. Wang DH, Yu JQ (2011). Highly convergent total synthesis of (+)-lithospermic acid via a late-stage intermolecular C–H olefination. *J. Am. Chem. Soc.* 133, 5767–5769.

89. (a) Thalji RK, Ellman JA, Bergman RG (2004). Highly efficient and enantioselective cyclization of aromatic imines via directed C–H bond activation. *J. Am. Chem. Soc.* 126, 7192–7193. (b) Shi BF, Maugel N, Zhang YH, Yu JQ (2008). Pd^{II}-catalyzed enantioselective activation of C(sp^2)–H and C(sp^3)–H bonds using monoprotected amino acids as chiral ligands. *Angew. Chem., Int. Ed.* 47, 4882–4886. (c) Albicker MR, Cramer N (2009). Enantioselective palladium-catalyzed direct arylations at ambient temperature: access to indanes with quaternary stereocenters. *Angew. Chem., Int. Ed.* 48, 9139–9142. (d) Shi BF, Zhang YH, Lam JK, Wang DH, Yu JQ (2010). Pd(II)-catalyzed enantioselective C–H olefination of diphenylacetic acids. *J. Am. Chem. Soc.* 132, 460–461.

12

THE DISCOVERY OF A NEW Pd/Cu CATALYTIC SYSTEM FOR C−H ARYLATION AND ITS APPLICATIONS IN A PHARMACEUTICAL PROCESS

JINKUN HUANG,[1,2] XIANG WANG,[1,3] AND JOHANN CHAN[1,3]

[1]*Chemical Process R&D, Amgen Inc., Thousand Oaks, CA, USA*
[2]*Chengdu Suncadia Pharmaceutical Co., Ltd., A Subsidiary of Hengrui Medicine Co., Ltd., China*
[3]*Chemical Development, Gilead Sciences, Foster City, CA, USA*

12.1 INTRODUCTION

Transition metal-catalyzed coupling reactions of aryl halides (or their equivalents) with a variety of nucleophilic reagents, especially organo-boron compounds (Suzuki–Miyaura reaction), have been well documented and frequently applied in pharmaceutical manufacturing. Conversely, the direct C−H bond arylation of arenes such as azoles, despite many recent publications in related fields, is scarcely utilized in large-scale pharmaceutical syntheses. This is because existing C−H bond arylation methodologies, when applied to process chemistry, often face significant challenges such as high catalyst loading and long reaction times, as well as high reaction temperature that often exceed the boiling point of the reaction solvent at ambient pressure. From a process chemistry standpoint, this lack of practical direct C−H bond arylation methodologies is unfortunate, as these reactions are not only more atom-economical than the corresponding Suzuki reaction, but can often also provide a shorter and more cost-effective synthetic route to the desired products from commercially available materials.

This chapter presents a case study whereby Amgen's process group overcame the above challenges by developing a new catalyst system that allowed for a scale-up of direct C−H arylation of azoles under milder conditions to deliver kilogram quantities of a pharmaceutical intermediate. A further discussion of catalyst design and mechanistic insight is also provided.

Sustainable Catalysis: Challenges and Practices for the Pharmaceutical and Fine Chemical Industries, First Edition. Edited by Peter J. Dunn, K. K. (Mimi) Hii, Michael J. Krische, and Michael T. Williams.

12.2 DEVELOPMENT OF INITIAL PROCESS FOR THE AGONIST OF S1P1

The lysophospholipid sphingosine-1-phosphate receptor group, including S1P1, S1P2, S1P3, S1P4, and S1P5, is a family of high-affinity G protein-coupled receptors that plays an important role in the regulation of the passage of cells, pathogens, fluids, nutrients, ions, and signaling molecules between anatomical compartments during homeostasis and disease [1]. Different subsets of S1P receptors have different and sometimes opposing effects on the regulation of cellular functions. In general, activation of the different S1P receptor subtypes at the surface of endothelial cells favors barrier maintenance under basal conditions, but this equilibrium seems to change in favor of barrier disruption in pathological states. Synthetic S1P receptor modulators have shown therapeutic efficacy in clinical trials for multiple sclerosis, as well as in a range of animal disease models, including acute respiratory distress syndrome (ARDS), disseminated intravascular coagulation, ischaemia–reperfusion injury, solid graft transplantation, and experimental autoimmune encephalomyelitis [2]. For example, Novartis' discovery and development of fingolimod (marketed as Gilenya®) has garnered significant impact as a treatment for relapsing forms of multiple sclerosis [3].

As part of a collaborative effort toward a selective S1P1 agonist, a number of aryl-substituted benzothiazoles (Figure 12-1) [4] were submitted for early toxicological studies. Among these candidates, AMG 369 (**3**) was ultimately selected by the project team to be advanced into the early clinical pipeline. Originally, a great deal of medicinal chemistry effort was put into the development of the biaryl cores, and their first syntheses relied on a strategy that reacted a fluoroarylbromide with isothiocyanates followed by cyclization, as shown in Scheme 12-1. For example, candidates **1** and **2** were synthesized from the commercially available 4-bromo-3-fluorobenzaldehyde, protected in one-step as the corresponding 1,3-dioxane, which was lithiated with *n*-butyllithium at −78 °C in THF. To this mixture 5-bromo-2-fluorophenyl isothiocyanate was added to generate the corresponding thioamide, which was cyclized to form the benzothiazole **4** with a sodium carbonate suspension in DMF. Exposure of **4** to HCl in warm THF resulted in cleavage of the acetal protecting group, and reductive amination of the aldehyde **5** subsequently provided methyl azetidine carboxylate **6**. Negishi coupling of **6** with benzylzinc (or 2-fluorobenzylzinc) chloride, followed by saponification and pH adjustment of the reaction mixture, accomplished the formation of drug candidates **1** and **2**.

In the original synthesis of **1** and **2**, the key aryl benzothiazole core structure was constructed via the cyclization of isothiocyanates. This strategy performs well from a medicinal chemistry standpoint. However, when applied to the large-scale process synthesis of AMG 369, it faces significant hurdles. Specifically, (1) the whole synthetic route is linear, which is lengthy and inefficient (Scheme 12-2) [1b]; (2) the aryl lithium species derived from 2-(4-bromo-3-fluorophenyl)-1,3-dioxane (**7**) has a very limited

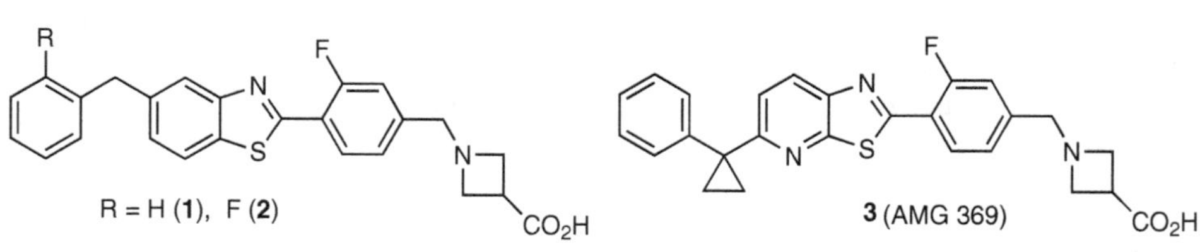

FIGURE 12-1. Selected S1P1 agonist candidates.

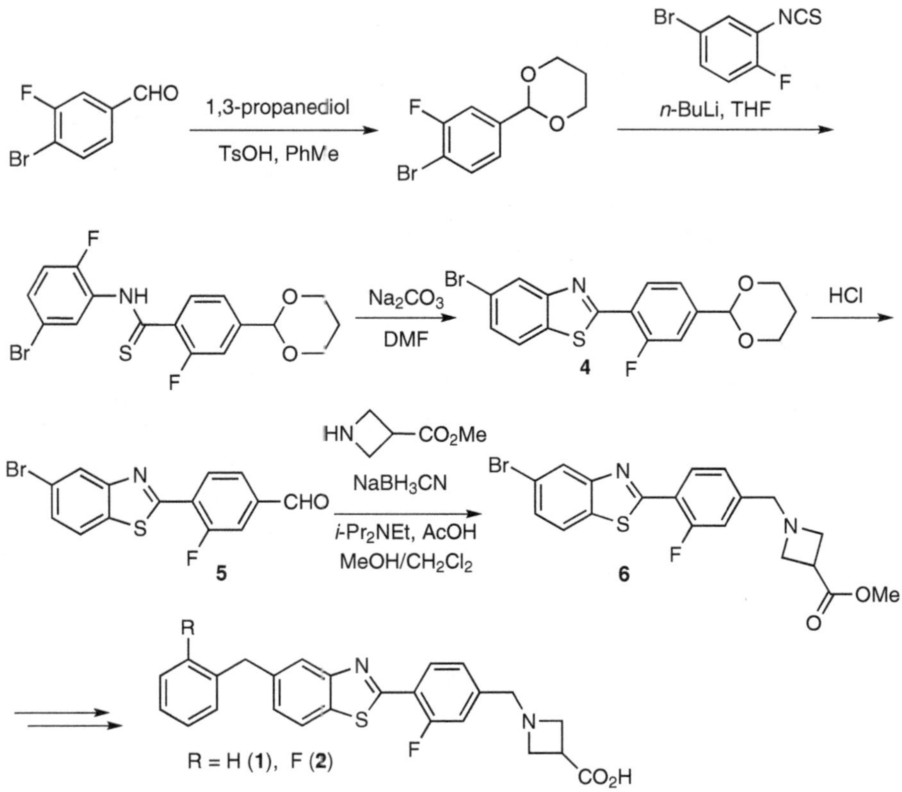

SCHEME 12-1. Overview of medicinal chemistry route to **1** and **2**.

stability profile, requiring the use of cryogenic conditions (−78 °C) and related special equipment, which severely hinders its application for the large-scale synthesis of AMG 369; (3) the protection and deprotection of the aldehyde bring an additional operational cost to the synthesis; and (4) toxic reagents (thiophosgene) and environmentally disfavored (dichloromethane) solvents used in the process also raise long-term concerns. Although this discovery route (Scheme 12-2) was successfully utilized to generate the first gram quantities of AMG 369 for early toxicology investigations, it became clear to the process team that an alternative and more efficient methodology for the construction of aryl azoles needed to be developed in order to support late-stage development of AMG 369.

When considering alternative approaches to AMG 369 to avoid the issues mentioned above, the project team evaluated two convergent strategies to assemble the aryl azabenzothiazole core (Scheme 12-3). Strategy A would form the azabenzothiazole ring via a condensation of an aminothiol with the corresponding carboxylic acid or aldehyde. Alternatively, strategy B would rely on a bold, but direct C-2 arylation of the azabenzothiazole intermediates via activation of the C−H bond, which represents a shorter and more atom-economical route to AMG 369.

Both strategies were evaluated and demonstrated to be feasible for potential scaling up. Strategy A commenced with the Pd-catalyzed carbonylation of the unprotected 4-bromo-3-

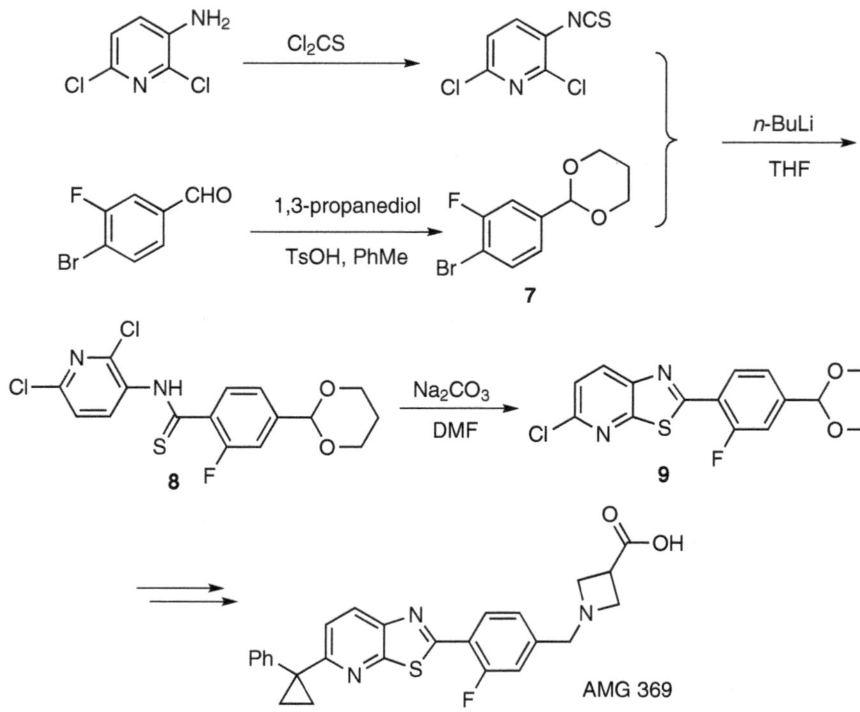

SCHEME 12-2. Medicinal chemistry route to AMG 369.

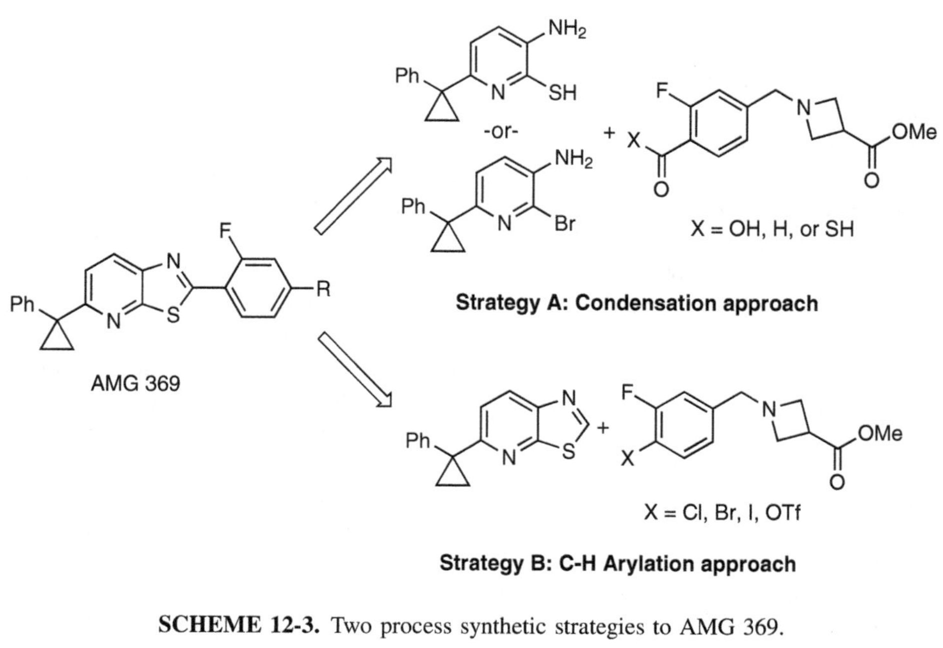

SCHEME 12-3. Two process synthetic strategies to AMG 369.

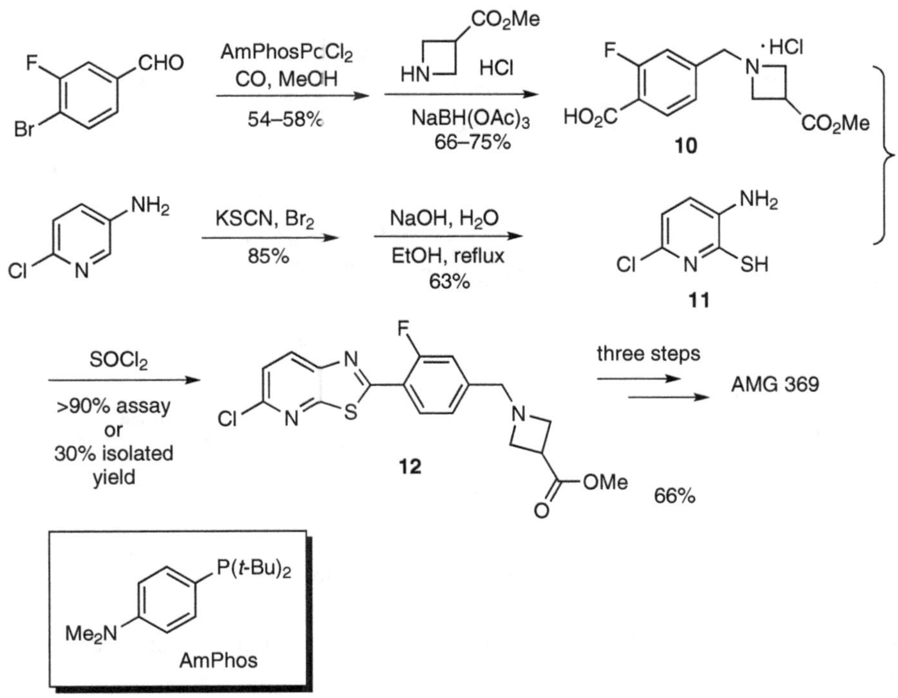

SCHEME 12-4. Synthetic strategy A for AMG 369.

fluorobenzaldehyde (Scheme 12-4).[1] Reductive amination of the resulting aldehyde with methyl azetidine carboxylate provided the condensation partner **10**. Synthesis of partner **11** was achieved by bromination/isocyanation of 5-amino-2-chloropyridine, followed by hydrolysis of the azobenzothiazole. The condensation of **10** via its intermediate acid chloride with **11** was low yielding, and the subsequent assembly of 1-styrene moiety by Suzuki coupling followed by Corey–Chaykovsky cyclopropanation and deprotection led to the isolation of AMG 369. Overall, strategy A is an eight-step synthesis with six linear steps starting with commercially available 4-bromo-3-fluorobenzaldehyde. This process was demonstrated on a small scale, with major issues associated with the Pd-catalyzed carbonylation and a challenging condensation step due to the inherent instability of **11**.

Strategy B also utilized an overall eight-step and linear six-step synthesis from commercial materials, by taking advantage of a palladium-catalyzed direct C–H bond functionalization of azabenzothiazole **13** with an aryl bromide **14** (Scheme 12-5). The process team needed to overcome two key challenges in order to implement this C–H bond arylation reaction on large scale. Namely, (1) the functionality in the reaction partners (a disubstituted olefin, a tertiary amine, and a carboxylic ester that bears an acidic alpha proton), and (2) high temperature (>110 °C) and high catalyst loading (5%) used in the arylation step. Despite these two significant challenges, strategy B was more promising and favored over strategy A, because the preparations of the coupling partners **13** and **14**, as well as the conversion of the coupling product **15** to AMG 369, were straightforward and

[1] AmPhos = di-*tert*-butyl(4-dimethylaminophenyl)phosphine.

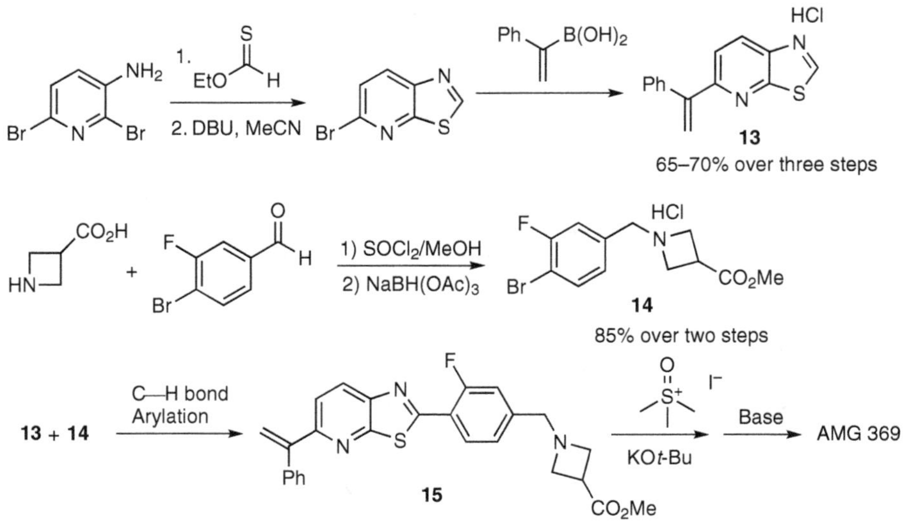

SCHEME 12-5. Synthetic strategy B for AMG 369.

relatively high yielding. In Section 12.3, we will discuss in detail how the process team at Amgen tackled this problem.

12.3 DEVELOPMENT OF C–H ARYLATION FOR THE SYNTHESIS OF AMG 369

12.3.1 Initial Results

A survey of the literature identified several methodologies that could be employed for the C–H bond arylation of azoles [5,6]. These methodologies, however, were only demonstrated on simple and mostly unfunctionalized substrates, and their applicability to our complicated substrate was unknown [7]. Indeed, when they were applied to the synthesis of AMG 369, it was quickly found that the literature conditions failed to afford significant amounts of the product, due to the lack of tolerance to functional groups. In this specific case (Scheme 12-5), the instability of the aryl bromide (**14**) with respect to azetidine ring opening, hydrolysis of the methyl ester, and the potential for Pd-catalyzed arylation of the carboxylate enolate, together with the undesired Heck reaction of the styrene of **13**, all prevented use of current literature conditions. Therefore, a mild catalytic system for C–H coupling was required in order to develop a reliable and robust process.

With the challenging task of high functional group tolerance under consideration, the process team set out to explore reaction conditions for the direct C–H arylation to the synthesis of the simpler product **1** from benzothiazole **16** and aryl bromide **17**. By doing so, the potential side reaction from Heck reaction of the olefin can be avoided and simplified the screening procedure. Based on observations from initial studies of the reaction, catalyst, base, solvent, and temperature were initially screened. The selected results are summarized in Table 12-1. As described earlier, the results obtained using literature conditions were very disappointing. For example, the latest Daugulis conditions [7b], which have been demonstrated to be very successful for coupling with simple substrates,

TABLE 12-1. Selected Initial Screening Results

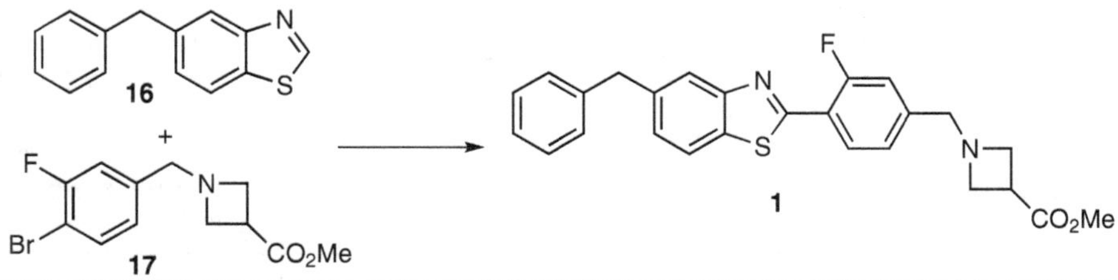

Entry	Ligand	Solvent	t (h)	Additive	Assay (%)
1	Ad_2Pn-Bu	*m*-Xylene	20	None	<1
2	XPhos	DMF	24	None	11
3	XPhos	*m*-Xylene	20	None	4
4	Xantphos	DMF	24	None	7.5
5	Xantphos	*m*-Xylene	20	None	21.6
6	XPhos	Toluene	20/44	CuI (1 equiv.)	55/72
7	Xantphos	*m*-Xylene	20	CuI (1 equiv.)	83
8	**Xantphos**	**Toluene**	**44**	**CuI (1 equiv.)**	**89**

XPhos = 2-dicyclohexylphosphino-2′,4′,6′-triisopropylbiphenyl; Xantphos = 9,9-dimethyl-4,5-bis(diphenylphosphine)xanthene.

gave essentially no desired product (entry 1). Polar solvents, such as DMF, DMAc, DMSO, and NMP, often used in this type of coupling, were found to be detrimental for the reaction, leading to fragmentation and hydrolysis of the azetidine moiety of the aryl bromide. Conversely, non-polar, high boiling point solvents such as *m*-xylene and toluene stood out as the solvents of choice. A catalyst screening performed with more than 20 different palladium/ligand[2] combinations identified Xantphos, XPhos, and RuPhos as potential ligands for the coupling, which gave appreciable conversions to the desired product.[3] Importantly, there was no significant decomposition of either one of the coupling partners. Furthermore, use of CuI as an additive was found to significantly promote the C–H arylation reaction.[4] The assay yield of the desired product was increased from 22 to 83% when 50 mol% of CuI was used in the catalytic system of $[Pd_2(dba)_3]$ (1 mol%)/Xantphos (3 mol%) (Table 12-1). The reaction was further optimized for bases, solvents, stoichiometry, and temperature. Using the catalytic system of $[Pd_2(dba)_3]$/Xantphos/CuI, screening of bases (Na_2CO_3, K_2CO_3, $CsHCO_3$, Cs_2CO_3, K_3PO_4, KOAc, CsOAc, CsOPiv, KO^tBu, Et_3N) showed Cs_2CO_3 to be optimal.[5] A solvent screen with Cs_2CO_3 as the base revealed that toluene was the best solvent for this reaction. A reaction temperature of 100–110 °C was optimal, as lower temperatures led to a slower reaction rate, while higher temperatures led to complicated reaction mixtures. Eventually, $[Pd_2(dba)_3]$ (1 mol%)/Xantphos (3 mol%)/CuI (100 mol%)/Cs_2CO_3 (3.0 equiv.)/ArBr (1.2 equiv.) in toluene (10 volumes) at 100–110 °C were identified as the best conditions. Depending on the isolation methods, the product yields ranged from 59 to 79% with >95% purity as established by HPLC analysis.

[2] Bulky ligands were chosen for screening. Either palladium complexes (2 mol% of Pd) or Pd ($[Pd_2(dba)_3]$, 1 mol%)/ligand (monodentate, 6 mol%; bidentate, 3 mol%) were used.

[3] The assay yields for Xantphos, XPhos, and Ru-phos were 22, 15, and 8%, respectively, after 20 h.

[4] Although CuI has been used as an additive, the function of CuI has never been clarified in the literature.

[5] KOt-Bu led to a messy reaction mixture, while organic base such as Et_3N gave no product.

TABLE 12-2. Optimization of Direct Arylation of Benzothiazole

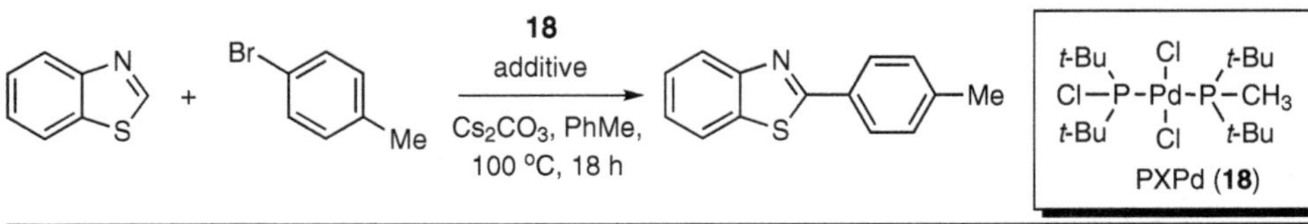

Entry	PXPd[a] (mol%)	Additive	Cu(I) (mol%)	Yield[b] (%)
1	2.5	None	0	30
2	1	None	0	16
3	1	CuI	4	32
4	0.25	None	0	<5
5	0.25	CuX[c]	1	<5
6	0.25	CuTc[d]	1	<5
7	0.25	CuI/Phen[e]	1	31
8	0.25	[Cu(PPh_3)(Phen)Br][e]	1	63
9	0.25	[Cu(PPh_3)$_3$Br]	1	52
10	0.25	[Cu(DPPF)I]	1	65
11	**0.25**	**[Cu(Xantphos)I] (19)**	**1**	**97**
12	0	**19**	1	0
13	0.25	[Cu(IPr)I]	1	<5

[a]PXPd = dichlorobis(chlorodi-*tert*-butylphosphine) palladium(II).
[b]Yield determined by HPLC based on a purified standard.
[c]X = I, Br, Cl, OTf.
[d]CuTc = copper(I)-thiophene-2-carboxylate.
[e]Phen = 1,10-phenanthroline.

12.3.2 Discovery of a New Cocatalyst

This reaction protocol, when applied to the structurally similar compound AMG 369, was found to be poorly reproducible. By analyzing the reaction components, it appears that a ligand competition between Pd and Cu might be the culprit for these poor results. This hypothesis prompted new studies to identify a better cocatalyst. The follow-up study for the C–H arylation used a model system of benzothiazole and 4-bromotoluene as coupling partners (Table 12-2). In most cases, low yields of the heteroarene coupling product were observed, except for conditions that utilized phosphorus or nitrogen ligands bound to copper(I). After extensive screening, it was discovered that the commercially available, air-stable PXPd complex **18** [8] in combination with [Cu(Xantphos)I] (**19**) [9] forms an excellent cocatalyst system. Specifically, the combination of **18** (0.25 mol%) and **19** (1 mol%) gave a 97% yield of the desired arylation product in 18 h (entry 11) at 100 °C. Notably, neither **18** nor **19**, at its respective loading, can individually catalyze the C–H arylation reaction (entries 4 and 12). Interestingly, the similar bidentate [Cu(DPPF)I] is less effective (entry 10) and the bulky monodentate [Cu(IPr)I][6] shows no catalytic activity.

This new catalytic system was found to be general for the coupling of a broad range of heteroarenes with a variety of aryl halides (Table 12-3). Electron-poor or -rich aryl bromides containing functionalities such as nitrile, trifluoromethyl, vinyl, ester, or aldehyde as well as pyridinyl rings are well tolerated. Activated aryl chlorides (entries 3 and 9)

[6] IPr = 1,3-bis(2,6-diisopropylphenyl)imidazol-2-ylidene. Cu(IPr)I was obtained as a free gift from Prof. Steven P. Nolan.

TABLE 12-3. Scope of 18/19-Catalyzed C–H Arylation

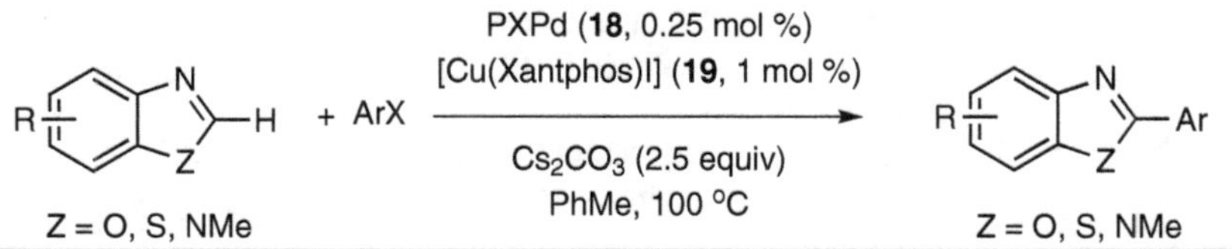

Entry	Z, R	Ar	Product	Yield[a] (%)
1	S, H	Ph	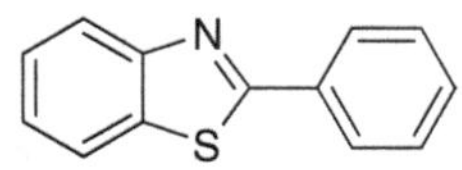	X = Cl, 6 X = Br, 97 X = OTf, 88 X = I, 68
2	S, H	4-Me-C_6H_4	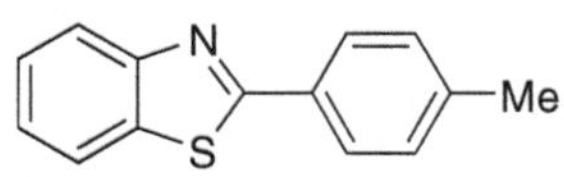	95
3	S, H	4-CF_3-C_6H_4	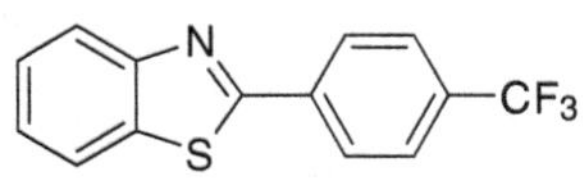	X = Br, 98 X = Cl, 90
4[b]	S, H	4-OHC-C_6H_4	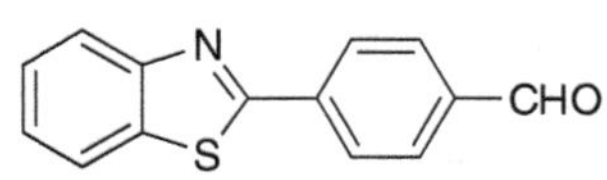	68
5[b]	S, H	4-MeO_2C-C_6H_4	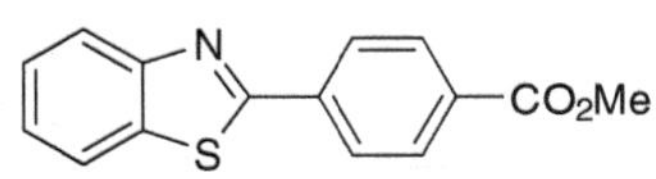	84
6	S, H	4-MeO-C_6H_4	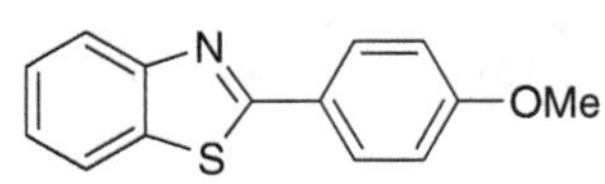	X = Br, 95 X = Cl, 0
7	S, H	2-Me-C_6H_4	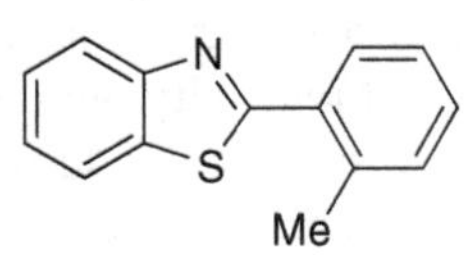	73
8	S, H	3-Me-C_6H_4	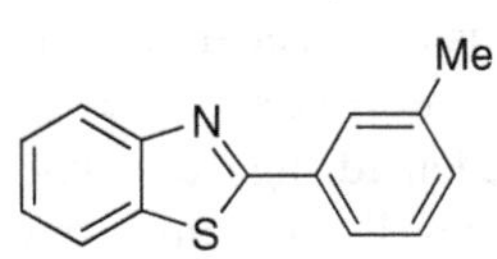	91
9	S, H	2-X-Py	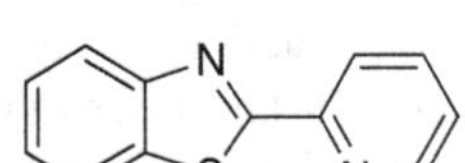	X = Cl, 88 X = Br, 95
10	S, H	3-Br-Py	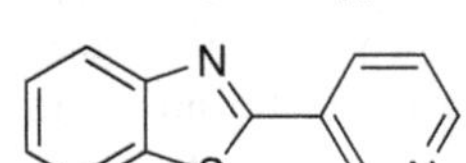	88
11	S, Bn	Ph	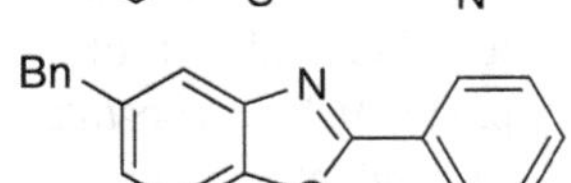	98
12	S, CF_3	Ph	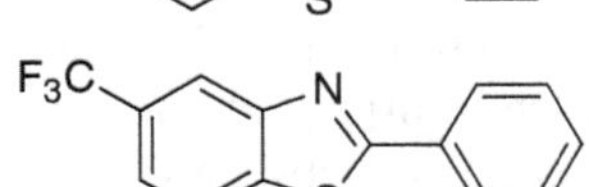	85

(continued)

TABLE 12-3. (*Continued*)

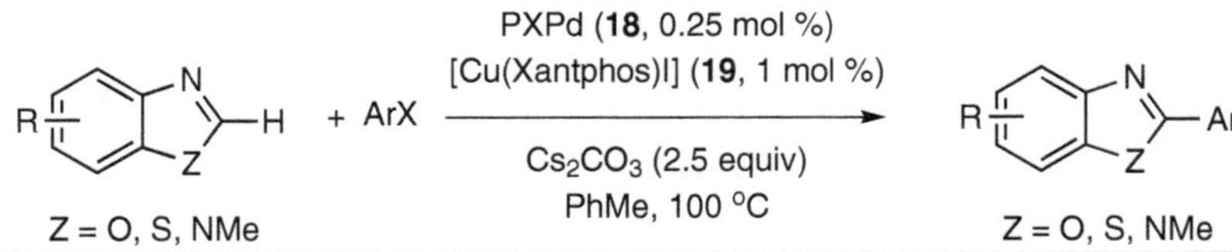

Entry	Z, R	Ar	Product	Yield[a] (%)
13	S, piperidine	Ph	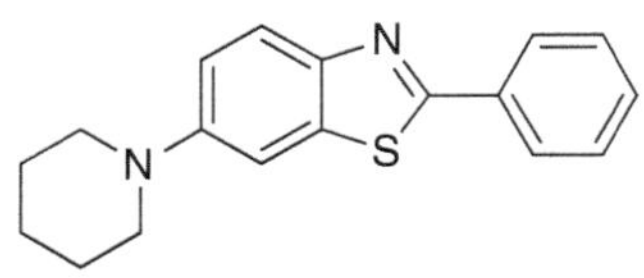	94
14	S, $CH(Ph){=}CH_2$	Ph	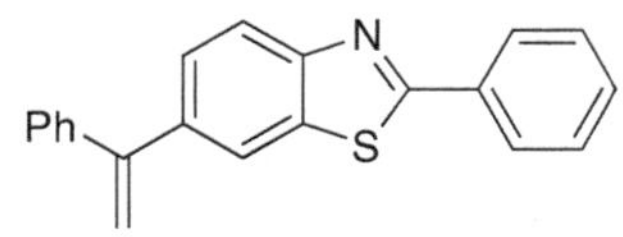	96
15	O, H	$R'C_6H_4$	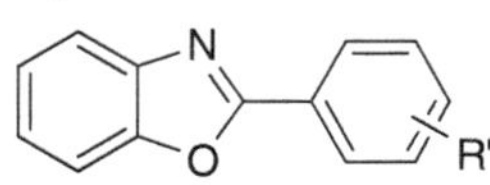	R′ = H, 99 R′ = 4-MeO, 95 R′ = = 4-CN, 88
16[b]	NMe, H	$R'C_6H_4$	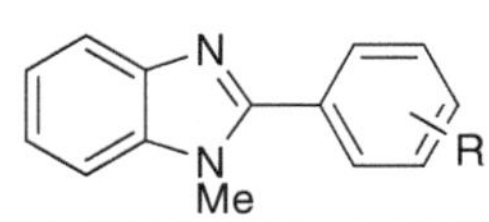	R′ = H, 62 R′ = 3-MeO, 48 R′ = 4-CF_3, 56

[a]Isolated yield. The reactions were monitored by LC–MS.
[b]Reaction conducted with 0.5 mol% of **18** and 2 mol% of **19**.

are also competent substrates, providing excellent conversion to the desired products. Phenyl triflate shows good reactivity but the corresponding reaction with iodobenzene is sluggish affording only a moderate yield (entry 1) [10]. Electron-rich benzothiazoles are more reactive (entries 11 and 13) than their electron-poor counterparts (entry 12, see below). Benzoxazole is an excellent coupling partner (entry 15) while *N*-methyl benzoimidazole is much less reactive (entry 16).

While the unprecedented high efficiency of the current catalytic system is not well understood, it is believed to stem from the introduction of [Cu(Xantphos)I] (**19**). Accordingly, its role in the reaction was elucidated. To probe its function, an equal molar mixture of tri-coordinated **19** and benzothiazole was dissolved in methylene chloride at room temperature, from which a 1:1 complex (**20**) was cleanly formed to afford crystals of sufficient quality for X-ray crystallography (Figure 12-2).

The lone pair of electrons on the nitrogen atom in benzothiazole binds to the copper to form a tetrahedral complex **20**. This solid-state structure correlates well with ^{1}H NMR data (Scheme 12-6), whereby the proton chemical shift of the 2-position shifts downfield from 9.00 ppm in benzothiazole to 9.19 ppm in **20**. It was postulated that the effect of this coordination lowers the pK_a of the adjacent C–H bond, allowing for facile deprotonation [11]. Therefore, the eletron-rich benzothiazole is more reactive than the electron-poor benzothiazole, which may not coordinate to [Cu(Xantphos)I] as indicated by the chemical shift (R = CN).

Competition experiments between this direct C–H arylation and the well-established Heck and Sonogashira reactions were conducted to gain insight into the reaction

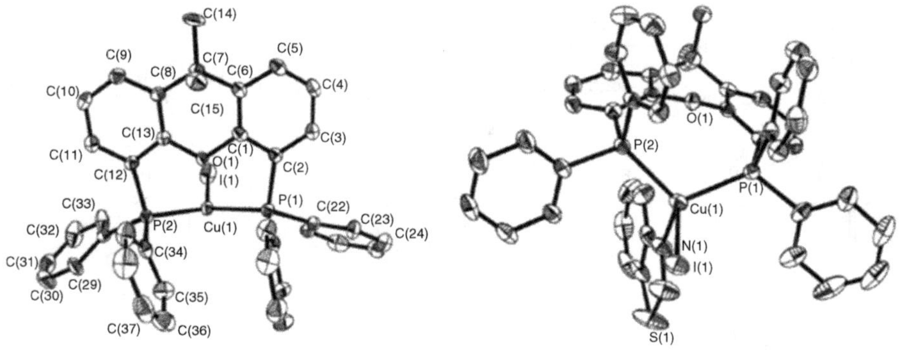

19 20

FIGURE 12-2. ORTEP view of crystal structures of [Cu(Xantphos)I] (**19**) and 3 [Cu(Xantphos)(benzothiazole)I] (**20**, 50% probability). Reprinted with permission from Ref. 9.

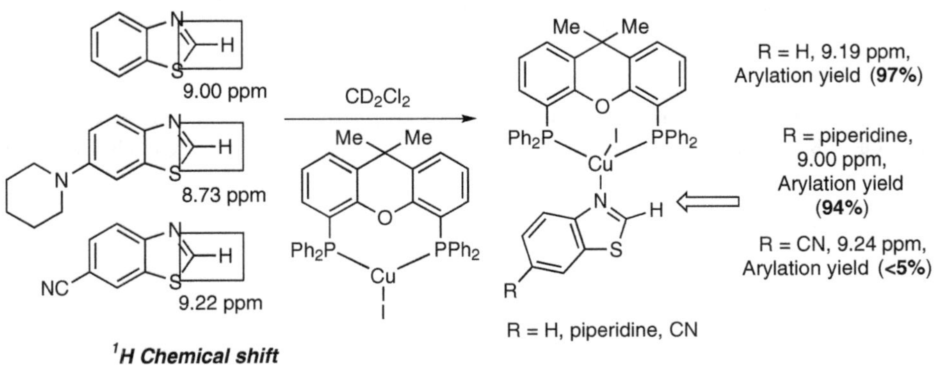

SCHEME 12-6. Relationship between key ^{1}H NMR shifts with reaction yields.

mechanism (Scheme 12-7). Under optimized reaction conditions, the coupling of bromobenzene with a mixture of styrene and benzothiazole (eq 1), or with vinyl-substituted benzothiazole (eq 2), yielded the C–H arylation product of benzothiazoles at the 2-position exclusively; no product resulting from the potential Heck reaction was detected. In contrast, using phenylacetylene instead of styrene, Sonogashira coupling was the dominating reaction (eq 3). Mechanistically, these results suggest that the C–H arylation of azoles under these conditions is different from the Heck reaction but more closely related to the Sonogashira reaction.

Based on the experimental observations so far, we propose a mechanism (Scheme 12-8) in which the copper complex takes part in the critical transmetalation step in the Pd/Cu dual catalytic cycles. The formation of **20** and its subsequent deprotonation[7]/rearrangement would afford **21** [12]. Transmetalation of the 2-benzothiazole-copper species **21** to the palladium (II) complex via oxidative addition with arylbromide followed by reductive elimination would give the desired product and release both the copper and palladium catalysts [13]. The presumed presence of such organocopper intermediates is in agreement

[7] The low isotopic effect ($K_H/K_D = 1.3$) indicates that the deprotonation is not the rate-determining step.

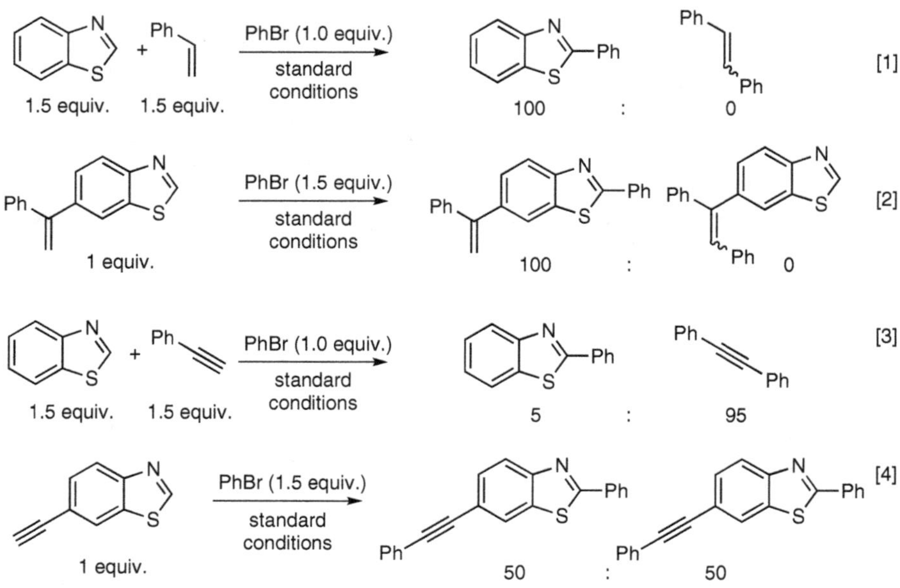

SCHEME 12-7. Competition of C–H arylation with Heck and Sonogashira reactions. Standard conditions: **18** (0.25 mol%)/**19** (1 mol%)/Cs_2CO_3 (2.5 equiv.)/substrates in toluene (10×) at 100–110 °C, 24 h.

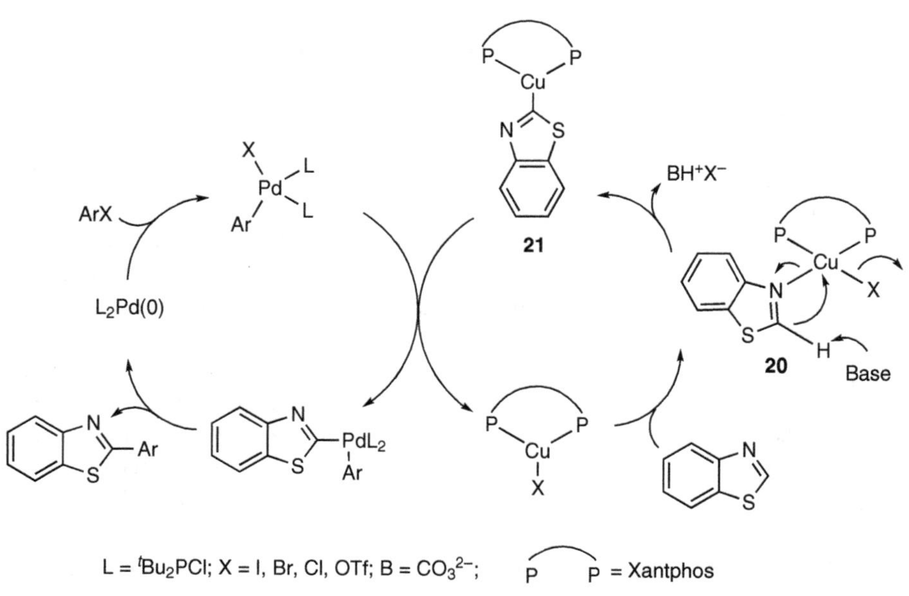

SCHEME 12-8. Proposed mechanism for the direct C–H arylation.

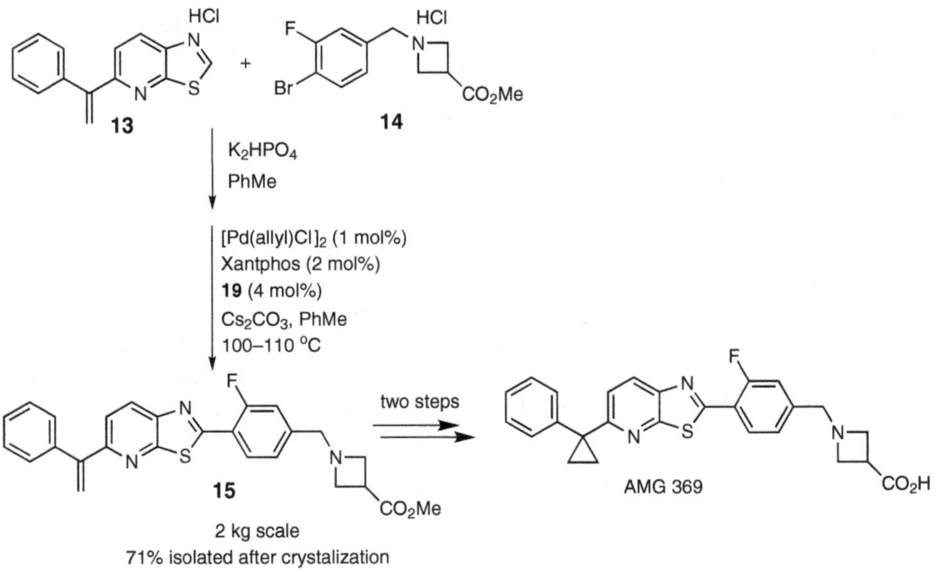

SCHEME 12-9. Application of direct C–H bond functionalization on kilogram scale synthesis of AMG 369.

with the role of copper(I) reported recently in a Suzuki coupling [14] and the well-known Sonogashira reaction [15]. However, other mechanisms [7b] including aromatic electrophilic substitution proposed by others [7c,7g], cannot be excluded at this point (see also Chapter 11).

12.3.3 Applications to the Synthesis of AMG 369

These conditions originally developed for the benzothiazole series were subsequently applied to the synthesis of AMG 369. Further investigation of reaction conditions using the phenylvinyl azabenzothiazole (**13**) led to the development of the following catalytic system: $[Pd(allyl)Cl]_2$ (1 mol%)/Xantphos (2 mol%)/[Cu(XantPhos)I] (4 mol%)/Cs_2CO_3 (3.0 equiv.)/toluene (8 volumes). To circumvent an intellectual issue, PXPd was replaced by $[Pd(allyl)Cl]_2$/Xantphos. This new system provided a robust process at 2 kg scale with full conversion of the limiting reactant (Scheme 12-9). The reaction assay yield consistently ranged from 75 to 88% after aqueous work-up, depending on the ratio of the two coupling partners.[8]

Both of the coupling partners, **13** and **14**, were used as HCl salts because their free base forms are oils. However, in the coupling reaction, free bases are preferable because the salts and Cs_2CO_3 are not soluble in toluene. In practice, the salts were neutralized to the free base using aqueous K_2HPO_4 (6 equiv., end point pH = 7)[9] and extracted into toluene. After azeotropic distillation of the organic phase to remove water (KF, <100 ppm H_2O), the toluene solution was directly used in the coupling reaction.

Work-up of the C–H coupling reaction proved to be challenging due to complications associated with copper residues. Although only 4 mol% Cu(I) was used, a rag layer formed,

[8] The use of tailpiece **14** as limiting reagents is preferable due to its lower stability.

[9] Aqueous $NaHCO_3$ was used earlier, and avoided due to gas evolution.

which sometimes led to an emulsion resulting in an inseparable mixture. Using aqueous NH_4Cl to remove the copper salts also resulted in a rag layer. Filtration of the salts followed by an aqueous wash of the organic layer did not prove to be advantageous. Eventually, the addition of brine (4 equiv.) to the reaction mixture, together with aqueous NH_4Cl, resulted in an improved separation of the phases.[10] Back extraction of the aqueous layer with toluene minimized the loss of product.

Following aqueous work-up, the crude dark brown toluene solution contains the product and some notable by-products and impurities. Treatment of the toluene solution with SiliCycle thiourea (30% w/w) at 60 °C eliminated most of the palladium and copper residues,[11] also facilitated the subsequent crystallization process. Based on the solubility of **15**, and taking advantage of the fact that most impurities are soluble in MeOH, a distillative crystallization using MeOH as the antisolvent was employed to provide 65–80% yield of crystalline **15** of 98.7% purity (Scheme 12-9), and under 400 ppm of residual Pd and copper.

12.3.4 Latest Developments and Future Perspective

Using emerging technology as well as our internal innovation, we realized a direct C–H bond functionalization process to prepare kilogram quantities of drug substance AMG 369 for clinical trials. However, a few issues remained that could negatively impact large-scale manufacturing: (1) cyclopropanation added to the linear steps in the synthesis and also complicated the impurity profile; (2) catalyst loading is relatively high which may lead to higher manufacturing cost and difficult removal of metal residues; (3) Cs_2CO_3 is costly, which limits its use on large scale. To address these challenges, further optimization of the process was undertaken.

As mentioned above, the difficult cyclopropanation of the vinylazabenzothiazole **15** at the end of the route compromised the synthesis. However, it was found that the Ni-catalyzed cyclopropanation was applicable to compound **13** (Scheme 12-10). Surprisingly, the unoptimized process provided up to 80% isolated yield of the desired cyclopropane product **22**, while earlier efforts using classical methods gave essentially no desired product. Although the results are preliminary, this proof of concept again demonstrates the potential power of catalysis.

With the new heteroarene **22** in hand, the C–H bond arylation process was further improved. The expensive Cs_2CO_3 was replaced by K_3PO_4 and catalytic amount potassium pivalate that was formed *in situ* from K_3PO_4 and pivalic acid. More importantly, the

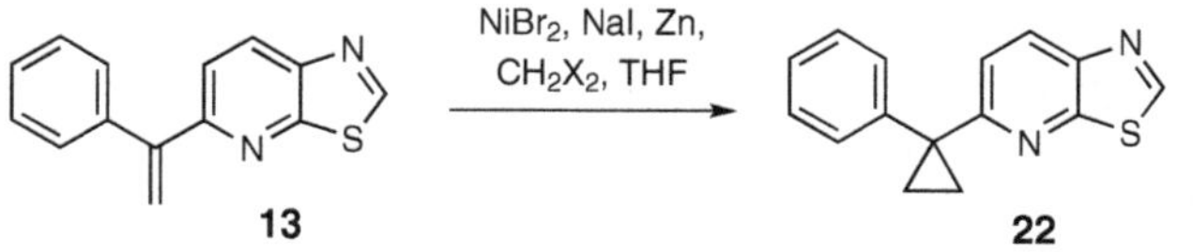

SCHEME 12-10. Cyclopropanation on early intermediate **13**.

[10] Slow agitation is critical to avoid emulsion formation.

[11] A species related to [Cu(Xantphos)X] was detected by LC–MS in the crude reaction mixture. It was difficult to remove from the product if this treatment with SiliCycle thiourea was not employed. The remaining Pd and Cu were efficiently removed in subsequent steps.

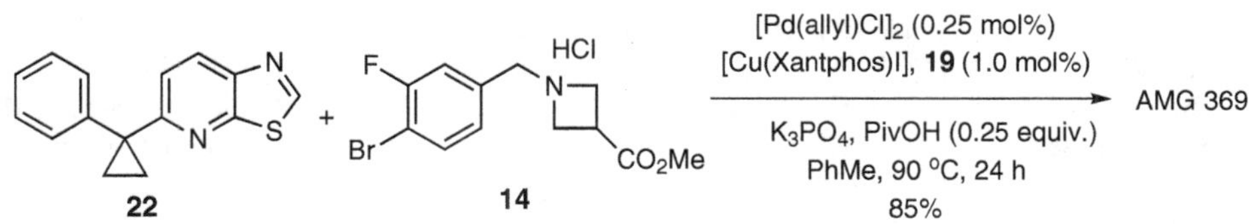

SCHEME 12-11. Improved C–H bond arylation synthesis for AMG 369.

catalyst loadings were reduced to 0.5 mol% for Pd and 1.0 mol% for Cu, which renders the process more practical for large scale manufacturing (Scheme 12-11).

12.4 CONCLUSION

In summary, a challenging direct sp^2 C–H bond arylation reaction was a roadblock in a large-scale synthesis of drug candidate AMG 369. Previous literature conditions failed to be applicable in this case due to high catalyst loading, as well as the harsh reaction conditions, which resulted in functional group intolerance. To solve this problem, a mild and efficient Pd/L and [Cu(Xantphos)I] cocatalytic system was discovered for the direct arylation of heteroarenes. This new catalytic system was demonstrated to be suitable for the couplings of a broad range of substrates under mild conditions. Notable features of this system included the unprecedented high efficiency achieved by very low metal loadings attributed to the unique role of [Cu(Xantphos)I]. As a result, an efficient process route to AMG 369 was developed with high atom economy. We are confident that as the field of C–H bond activation/functionalization continues to evolve, more large-scale applications in the pharmaceutical industry will soon follow.

REFERENCES

1. (a) Rosen H, Gonzalez-Cabrera PJ, Sanna MG, Brown S (2009). Sphingosine-1-phosphate receptor signaling. *Annu. Rev. Biochem.* 78, 743–768. (b) Cyster JG (2005). Chemokines, sphingosine-1-phosphate, and cell migration in secondary lymphoid organs. *Annu. Rev. Immunol.* 23, 127–159. (c) Takabe K, Paugh SW, Milstien S, Spiegel S (2008). Inside-out signaling of sphingosine-1-phosphate: therapeutic targets. *Pharmacol. Rev.* 60, 181–195. (d) Rivera J, Proi RL, Olivera A (2008). The alliance of sphingosine-1-phosphate and its receptors in immunity. *Nat. Rev. Immunol.* 8, 753–763.
2. (a) Buzard DJ, Thatte J, Lerner M, Edwards J, Jones RM (2008). Recent progress in the development of selective S1P1 receptor agonists for the treatment of inflammatory and autoimmune disorders. *Expert Opin. Ther. Pat.* 18, 1141–1159. (b) Cooke N, Zecri F (2007). Sphingosine 1-phosphate type 1 receptor modulators: recent advances and therapeutic potential. *Annu. Rep. Med. Chem.* 42, 245–263.
3. Cohen JA, Barkhof F, Comi G, Hartung HP, Khatri BO, Montalban X, Pelletier J, Capra R, Gallo P, Izquierdo G, Tiel-Wilck K, de Vera A, Jin J, Stites T, Wu S, Aradhye S, Kappos L (2010). A placebo-controlled trial of oral fingolimod in relapsing multiple sclerosis. *N. Engl. J. Med.* 36, 387–401.
4. (a) Pennington LD, Sham KKC, Pickrell AJ, Harrington PE, Frohn MJ, Lanman BA, Reed AB, Croghan MD, Lee MR, Xu H, McElvain M, Xu Y, Zhang X, Fiorino M, Horner M, Morrison HG, Arnett HA, Fotsch C, Wong M, Cee VJ (2011). 4-Methoxy-*N*-[2-(trifluoromethyl) biphenyl-4-

ylcarbamoyl]nicotinamide: a potent and selective agonist of S1P1. *ACS Med. Chem. Lett.* 2, 752–757. (b) Cee VJ, Frohn M, Lanman BA, Golden J, Muller K, Neira S, Pickrell A, Arnett H, Buys J, Gore A, Fiorino M, Horner M, Itano A, Lee MR, McElvain M, Middleton S, Schrag M, Rivenzon-Segal D, Vargas HM, Xu H, Xu Y, Zhang X, Siu J, Wong M, Burli RW (2011). Discovery of AMG 369, a thiazolo[5,4-*b*]pyridine agonist of S1P1 and S1P5. *ACS Med. Chem. Lett.* 2, 107–112. (c) Pennington LD, Sham KKC, Pickrell AJ, Harrington PE, Frohn MJ, Lanman BA, Reed AB, Croghan MD, Lee MR, Xu H, McElvain M, Xu Y, Zhang X, Fiorino M, Horner M, Morrison HG, Arnett HA, Fotsch C, Wong M, Cee VJ (2011). 4-Methoxy-*N*-[2-(trifluoromethyl)biphenyl-4-ylcarbamoyl]nicotinamide: a potent and selective agonist of S1P1. *ACS Med. Chem. Lett.* 2, 752–757.

5. (a) Dieterich F, Stang PJ, editors. *Metal-Catalyzed Cross-Coupling Reactions*, Wiley-VCH, New York, 1998. (b) Ackermann L, editor. *Modern Arylation Methods*, Wiley-VCH, Weinheim, 2009.
6. (a) For reviews on direct C−H arylation: Alberico D, Scott ME, Lautens M (2007). Aryl–aryl bond formation by transition-metal-catalyzed direct arylation. *Chem. Rev.* 107, 174–238. (b) Seregin IV, Gevorgyan V (2007). Direct transition metal-catalyzed functionalization of heteroaromatic compounds. *Chem. Soc. Rev.* 36, 1173–1193.
7. (a) Ackermann L, Althammer A, Fenner S (2009). Palladium-catalyzed direct arylations of heteroarenes with tosylates and mesylates. *Angew. Chem., Int. Ed.* 48, 201–204. (b) Do HQ, Khan RMK, Daugulis O (2008). A general method for copper-catalyzed arylation of arene C−H bonds. *J. Am. Chem. Soc.* 130, 15185–15192. (c) Chiong HA, Daugulis O (2007). Palladium-catalyzed arylation of electron-rich heterocycles with aryl chlorides. *Org. Lett.* 9, 1449–1451. (d) Nandurkar NS, Bhanushali MJ, Bhor MD, Bhanage BM (2008). Palladium bis(2,2,6,6-tetramethyl-3,5-heptanedionate): an efficient catalyst for regioselective C-2 arylation of heterocycles. *Tetrahedron Lett.* 49, 1045–1048. (e) Bellina F, Calandri C, Cauteruccio S, Rossi R (2007). Efficient and highly regioselective direct C-2 arylation of azoles, including free (NH)-imidazole, -benzimidazole and -indole, with aryl halides. *Tetrahedron* 63, 1970–1980. (f) Yokooji A, Okazawa T, Satoh T, Miura M, Nomura M (2003). Palladium-catalyzed direct arylation of thiazoles with aryl bromides. *Tetrahedron* 59, 5685–5689. (g) Pivsa-Art S, Satoh T, Kawamura Y, Miura M, Nomura M (1998). Palladium-catalyzed arylation of azole compounds with aryl halides in the presence of alkali metal carbonates and the use of copper iodide in the reaction. *Bull. Chem. Soc. Jpn.* 71, 467–473. (h) Wang X, Lane BS, Sames D (2005). Direct C-arylation of free (NH)-indoles and pyrroles catalyzed by Ar-Rh(III) complexes assembled *in situ*. *J. Am. Chem. Soc.* 127, 4996–4997. (i) Lewis JC, Berman AM, Bergman RG, Ellman JA (2008). Rh(I)-catalyzed arylation of heterocycles via C−H bond activation: expanded scope through mechanistic insight. *J. Am. Chem. Soc.* 130, 2493–2500. (j) Zhao D, Wang W, Yang F, Lan J, Yang L, Gao G, You J (2009). Copper-catalyzed direct C-arylation of heterocycles with aryl bromides: discovery of fluorescent core frameworks. *Angew. Chem., Int. Ed.* 48, 3296–3300. (k) Do HQ, Daugulis O (2007). Copper-catalyzed arylation of heterocycle C−H bonds. *J. Am. Chem. Soc.* 129, 12404–12405. (l) Ackermann L, Potukuchi HK, Landsberg D, Vicente R (2008). Copper-catalyzed "Click" reaction/direct arylation sequence: modular syntheses of 1,2,3-triazoles. *Org. Lett.* 10, 3081–3084.
8. **18** is commercially available from Combiphos Catalysts, Inc. For examples in coupling reactions, see Li GY (2001). The first phosphine oxide ligand precursors for transition metal catalyzed cross-coupling reactions: C−C, C−N, and C−S bond formation on unactivated aryl chlorides. *Angew. Chem., Int. Ed.* 40, 1513–1516.
9. Huang J, Chan J, Chen Y, Borths CJ, Baucom KD, Larsen RD, Faul MM (2010). A highly efficient palladium/copper cocatalytic system for direct arylation of heteroarenes: an unexpected effect of Cu(Xantphos)I. *J. Am. Chem. Soc.* 132, 3674–3675.
10. Similar unexpectedly poor reactivity of aryl iodides in C–H arylation has been previously observed: Campeau LC, Parisien M, Jean A, Fagnou K (2006). Catalytic direct arylation with

aryl chlorides, bromides, and iodides: intramolecular studies leading to new intermolecular reactions. *J. Am. Chem. Soc.* 128, 581–590.

11. The electron rich 6-piperidinyl benzothiazole (Table 12-2, entry 12) is a superior substrate to its 6-nitrile analog (<5% conversion). ^{1}H NMR studies show strong binding of **2** with the former to form the corresponding complex, but a similar interaction was not observed with the latter. A similar explanation is proposed in *N*-oxide C–H arylations; see Gorelsky SI, Lapointe D, Fagnou K (2008). Analysis of the concerted metalation-deprotonation mechanism in palladium-catalyzed direct arylation across a broad range of aromatic substrates. *J. Am. Chem. Soc.* 130, 10848–10849.
12. The wide bite angle of Xantphos may play a crucial rule: Kamer PCJ, van Leeuwen PWNM, Reek JNH (2001). Wide bite angle diphosphines: Xantphos ligands in transition metal complexes and catalysis. *Acc. Chem. Res.* 34, 895–904. Attempts to isolate this intermediate have not yet been successful.
13. A similar mechanism was proposed in Pd/Cu catalyzed decarboxylative coupling: Gooβen LJ, Deng G, Levy LM (2006). Synthesis of biaryls via catalytic decarboxylative coupling. *Science* 313, 662–664.
14. Deng JZ, Paone DV, Ginnetti AT, Kurihara H, Dreher SD, Weissman SA, Stauffer SR, Burgey CS (2009). Copper-facilitated Suzuki reactions: application to 2-heterocyclic boronates. *Org. Lett.* 11, 345–347.
15. Osakada K, Sakata R, Yamamoto T (1997). Preparation and properties of *trans*-[Pd(Ar)(C≡CPh)(PEt_3)$_2$]. Intermolecular alkynyl ligand transfer between copper(I) and palladium(II) complexes relevant to palladium complex catalyzed cross-coupling of terminal alkyne with haloarene in the presence of CuI cocatalyst. *Organometallics* 16, 5354–5364.

13

DIARYLPROLINOL SILYL ETHERS: DEVELOPMENT AND APPLICATION AS ORGANOCATALYSTS

HIROAKI GOTOH[1] AND YUJIRO HAYASHI[2]

[1]*Department of Applied Chemistry, Graduate School of Engineering, Yokohama National University, Hodogaya-ku, Yokohama, Japan*

[2]*Department Chemistry, Graduate School of Science, Tohoku University, Aoba-ku, Sendai, Japan*

13.1 INTRODUCTION AND BACKGROUND

Improving the efficiency of the synthesis of biologically active and complex organic molecules constitutes one of the most exciting challenges to the synthetic chemist. In recent years, the field of synthetic organic chemistry has shifted from merely synthesizing molecules to achieving the synthesis in a greener and more effective fashion. These syntheses often include novel concepts, innovative strategies, and improved methods.

Catalysis with small organic molecules, where an inorganic element does not form part of the active catalyst, has become a highly dynamic area in chemical research. Also known as "organocatalysis" [1–14], this type of reactions boast a wide range of practical advantages compared to metal catalysts, including air and water stability, low cost, availability, operational simplicity, mild reaction conditions, and lower toxicity. In this area, chiral amines are widely employed in various asymmetric reactions, which commonly involve enamine or iminium intermediates.

The first asymmetric aldol reaction that proceeds via an enamine intermediate was developed by the Hajos and Wiechert groups in the early 1970s, involving intramolecular, asymmetric, direct aldol reactions of triketones catalyzed by proline [1,2]. Three decades later, in 2000, List, Barbas, and Lerner discovered the intermolecular version of this proline-catalyzed aldol reaction [3]. The reaction proceeds by the condensation of the amine catalyst with a carbonyl compound to form a nucleophilic enamine intermediate,

Sustainable Catalysis: Challenges and Practices for the Pharmaceutical and Fine Chemical Industries, First Edition. Edited by Peter J. Dunn, K. K. (Mimi) Hii, Michael J. Krische, and Michael T. Williams.

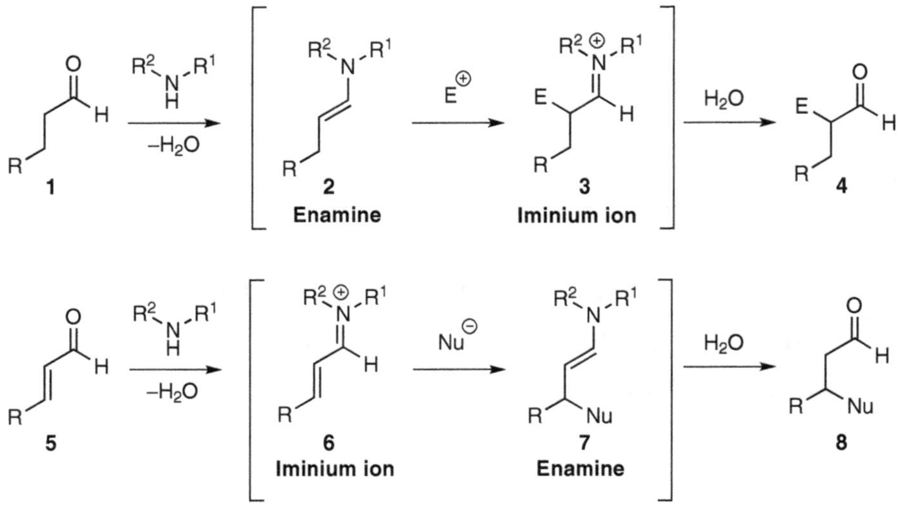

SCHEME 13-1. Enamine and iminium ion intermediates.

which then reacts with various electrophiles. Subsequent hydrolysis gives the α-substituted products and regenerates the amine catalyst (Scheme 13-1, top). On the other hand, an iminium intermediate may be generated *in situ* from an α,β-unsaturated carbonyl compound and amine catalyst. This lowers the LUMO energy of the conjugated π-system, consequently facilitating 1,4-nucleophilic additions and cycloadditions (Scheme 13-1, bottom). The first highly enantioselective catalytic Diels–Alder reaction mediated by a phenylalanine derivative was reported by the MacMillan group in 2000 [4]. Since then, a wide variety of highly enantioselective 1,4-additions and cycloadditions has been developed by using chiral amine catalysts.

Following the original discovery by List, Barbas, and Lerner, proline has been widely used in a number of asymmetric reactions. However, success was limited to electrophiles that are also hydrogen bond-acceptors. Furthermore, the practicality of proline is also limited by its low solubility in most organic solvents, thus requiring polar organic solvents such as DMF or DMSO. At this time, no catalyst is capable of promoting *both* enamine and iminium ion-mediated reactions with good stereoselectivities. In 2005, the Jørgensen group [5] and our group [6] independently reported the use of diarylprolinol derivatives for asymmetric reactions. Since then, it has been used in a number of asymmetric transformations. From a library of synthesized diarylprolinol silyl ethers, two of them, **9** and **10** (TMS = trimethylsilyl), proved to have the broadest generality across a number of different reactions (Figure 13-1, left).

Excellent enantiocontrol is provided by the bulky side-chain of the catalyst, forcing the reacting electrophile/nucleophile to approach from the less sterically hindered side of the reactive functionality (Figure 13-1, **11** and **12**). X-ray analysis of these intermediates was reported by the Seebach group [7], where the crystal structures of the enamine and the iminium ions showed that the OTMS group is positioned over the π-system. The stability of these structures was also confirmed by DFT calculations [8].

To date, there are two general reviews for the catalytic applications of diarylprolinol silyl ether catalysis [9,10]. In this chapter, the development of this class of organocatalysts

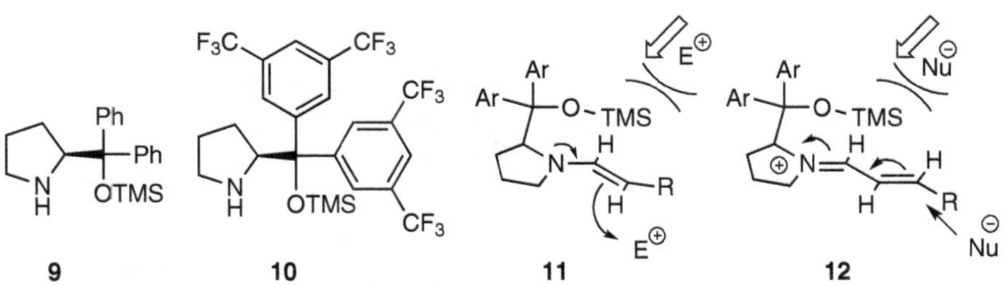

FIGURE 13-1. Most popular prolinol silyl ethers (left) and enamine and iminium ion models for these catalysts.

will be described, including work conducted in our laboratory, as well as recent applications of these catalysts in one-pot reactions.

13.2 ENAMINE INTERMEDIATE

13.2.1 Michael Reaction

At the outset, the Michael reaction was selected as a model for catalyst screening, as it is one of the most important C–C bond-forming reactions. In 2001, the Barbas group reported the reaction between aldehydes and nitro-olefins catalyzed by chiral diamines [11]. Since then, similar Michael reactions of nitro-olefins with aldehydes and ketones have been reported by several groups, using a wide range of organocatalysts, including chiral amines [12], pyrrolidinyltetrazole [13], aminomethylpyrrolidine [14], 2,2′-bipyrrolidine [15], and pyrrolidine (with pyridine as the conjugate base) [16]. However, high enantioselectivities are often difficult to achieve, especially when an aldehyde was used as nucleophile.

Using the Michael reaction of propanal **13** and nitrostyrene **14** as the model reaction (Scheme 13-2), the use of proline **16** as catalyst led to a low yield and low enantioselectivity. In contrast, excellent enantioselectivity was observed when diphenylprolinol **17**

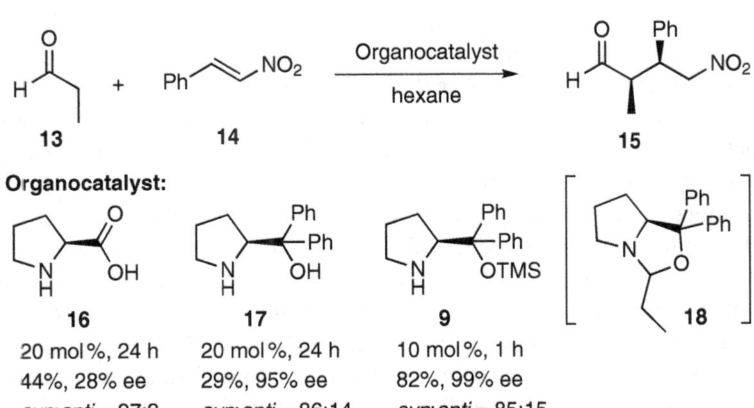

SCHEME 13-2. Catalyst screening of the Michael reaction between propanal **13** and nitrostyrene **14**.

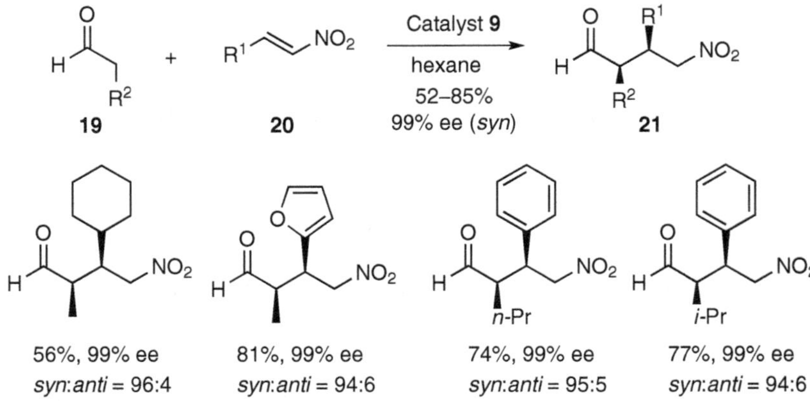

SCHEME 13-3. Michael reaction between nitro-olefins **20** and aldehydes **19**.

was employed; however, the progress of the reaction was slow, affording a low yield even after 24 h. This was attributed to the formation of oxazolidine species **18**, which would inhibit the catalyst. Indeed, when the hydroxyl group was exchanged for a siloxy group, the reactivity of the catalyst increased dramatically, along with the associated enantioselectivity (1 h, 82%, 99% ee) [6]. It should be noted that diphenylprolinol silyl ether **9** is made in one step from commercially available diphenylprolinol **17**. During our investigations, the synthesis of the 3,5-bis(trifluoromethyl)phenyl-containing diarylprolinol silyl ether **10** was reported by the Jørgensen group for the α-sulfenylation of aldehydes [5].

The reaction has broad applicability with respect to both the Michael acceptor and the donor (Scheme 13-3). With regard to the nitro-olefin, electron-rich and electron-poor aryl, and heteroaryl, as well as alkyl substituents can be accommodated. Conversely, both linear and branched alkyl aldehydes can be employed successfully as Michael donors.

The effect of acid additives in the Michael addition of aldehydes and nitroalkenes was investigated [17]. Further understanding of this transformation can be attained by combining a systematic screening of acid additives with NMR analysis (Scheme 13-4). The study revealed a direct correlation between the pK_a value of the acid and reaction efficiency. Of the 13 acids tested, 4-nitrophenol ($pK_a = 7.15$ in H_2O) turned out to be the most effective additive. ^{1}H-NMR analysis of the reaction mixture allowed us to observe the formation of the cyclobutane intermediate **24**. The formation of the formal [2+2] cycloadduct was also discovered independently by the group of Blackmond using ^{1}H NMR spectroscopy and reaction calorimetry [18].

The cyclobutane could be prepared quantitatively by addition of nitroalkene to preformed enamine **22** in the presence of molecular sieves, and each step of the reaction was investigated separately under noncatalytic conditions. Based on our studies of the reactivity of the cyclobutane intermediate toward water and acid, the mechanism of the Michael addition of aldehydes and nitroalkenes catalyzed by **9** was subsequently revised (Scheme 13-4). The stereo-defining step is the C–C bond formation between the β-C-atoms of enamine **22** and nitroalkene **20** to give the zwitterionic intermediate **23**, which is in equilibrium with the cyclobutane **24** and its precursors (enamine and nitroalkene). The protonation of the zwitterion's nitronate anion gives rise to the iminium ion **25**, which is

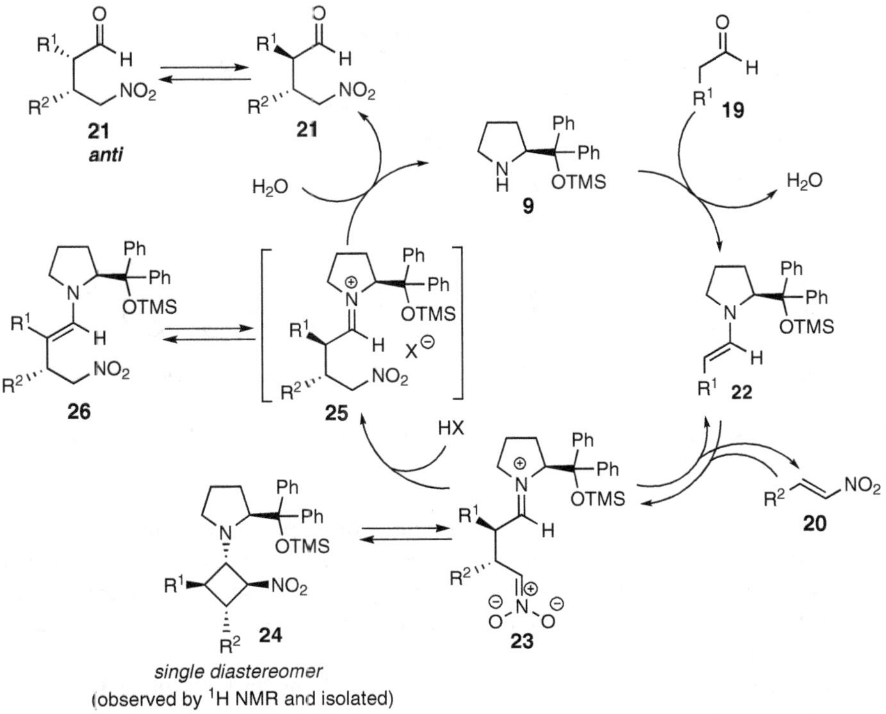

SCHEME 13-4. Proposed mechanism of the amine-catalyzed Michael addition of aldehydes **19** to nitro-olefins **20**.

hydrolyzed to the desired aldehyde **21**. In terms of relative rates, formation of the enamine **22** and cyclobutane **24** are fast, compared to the liberation of the Michael product **21**. Thus, the cyclobutane **24** is an off-cycle reservoir of the key intermediate **23**. In this mechanism, the acid additive 4-nitrophenol plays multiple roles: (1) it accelerates the enamine formation; (2) it assists the [2+2] cycloaddition, probably by interaction with the *O*-atom of the developing nitronate, lowering the transition state energy; and (3) it accelerates the cleavage of the cyclobutane ring to the Michael adduct **21**. At the same time, the acid additive is also expected to suppress the deprotonation of the iminium ion **25** to the enamine **26**, therefore maintaining an excellent level of diastereoselectivity.

13.2.2 Acetaldehyde as Nucleophile

Prior to 2008, there were many enantioselective reactions of aldehydes involving the enamine-type mechanism, but no reports of the successful use of acetaldehyde **27** as a substrate. Up to this point, controlling the reaction of acetaldehyde was considered to be difficult because of its high reactivity, both as a nucleophile and also as an electrophile.

Having developed highly enantioselective Michael reactions of aldehydes and nitroalkenes catalyzed by **9** that can afford α-substituted γ-nitro aldehydes in nearly enantiopure form (Section 13.2.1), it was hoped that the methodology can be extended to include the use of acetaldehyde, for the preparation of chiral α-unsubstituted γ-nitroaldehydes. After screening of several reaction conditions, **9** was found to be effective in this reaction

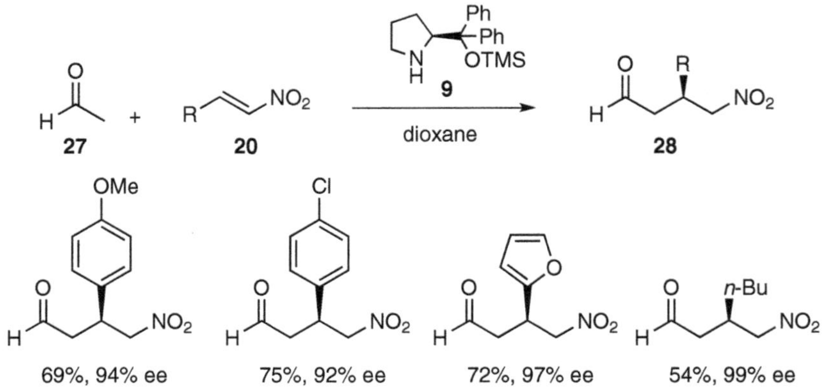

SCHEME 13-5. Catalytic asymmetric Michael reaction of acetaldehyde **27** with various nitroalkenes **20**.

when 1,4-dioxane was used as the solvent, providing the expected Michael product in good yield with excellent enantioselectivity [19]. The reaction proceeded efficiently with aryl-substituted nitroalkenes, having both electron-rich, and electron-deficient aryl groups, as well as heteroaryl and alkyl substituted groups, in high yield and excellent enantioselectivity (Scheme 13-5).

The Mannich reaction is one of the most useful reactions for the synthesis of nitrogen-containing molecules. In 2000, the List group was the first to report a three-component direct Mannich reaction mediated by proline [20]. Since then, many asymmetric variants of the reaction involving organocatalysis have been developed [21–25]. In 2008, the List group also reported a proline-catalyzed Mannich reaction between acetaldehyde (**27**) and *N*-Boc-imine [26].

In our work [27–33], reactions were conducted using acetaldehyde and *N*-benzoyl-*N*-benzylidene amine (**29**) as model substrates, and a diarylprolinol silyl ether as the key catalyst; the resulting β-amino aldehyde product **30** was isolated by its conversion into the corresponding alcohol by reduction with $LiAlH_4$. When proline **16** was employed as the catalyst, excellent enantioselectivity was obtained in CH_3CN with moderate yield, due to competitive hydrolysis of the imine under the acidic conditions. After optimization of reaction conditions, the desired product can be obtained with good yield (87%) and excellent enantioselectivity (97% ee) by using a combination of 3,5-bis(trifluoromethyl) phenyl-substituted diarylprolinol silyl ether **10** with *p*-NO_2PhCO_2H in THF. Under these conditions, the reaction with acetaldehyde also proceeded efficiently for *N*-benzoylimine, *N*-Boc-imine, and *N*-Ts-imines, to afford the requisite Mannich products (Scheme 13-6) [34]. In some cases, the Mannich adduct can be isolated as the β-amino aldehyde derivatives by rapid column chromatography.

Using diarylprolinol silyl ethers as catalysts, acetaldehyde can be employed as a nucleophile for the synthesis of β-aminoaldehydes and β-substituted-γ-nitroaldehydes. These building blocks are pivotal for the preparation of products of chemical and biological relevance. For example, the conversion of β-aminoaldehyde intermediates for the synthesis of natural products, pharmaceuticals, β-amino acids, and other building blocks had been demonstrated by List et al. [26], while we used the methodology to access a β-substituted-γ-nitroaldehyde intermediate for the synthesis of ABT-341 (**127**) (Section 13.6.3) [35].

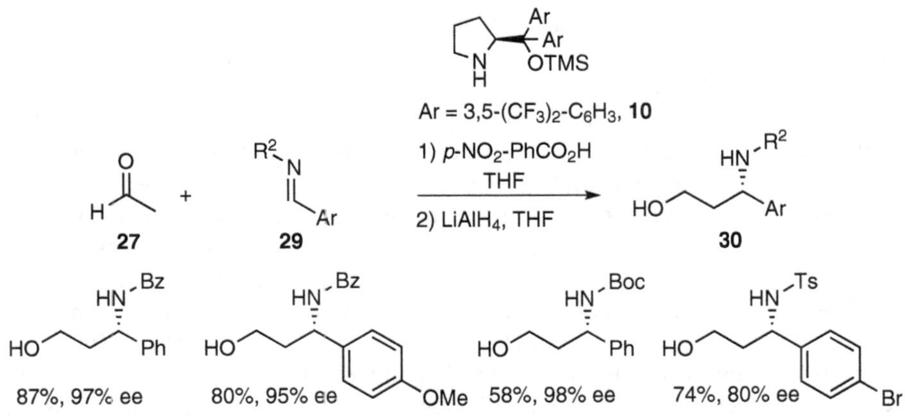

SCHEME 13-6. Catalytic asymmetric Mannich reaction of acetaldehyde with various imines.

13.2.3 α-Oxidation Using Benzoyl Peroxide

Optically active α-acyloxy- or α-hydroxyaldehydes are important intermediates in organic synthesis, and α-oxygenation of aldehydes is one of the most direct and straightforward methods for their synthesis. In spite of several enantioselective and catalytic methods available from the literature [36], there were no reports of a direct, catalytic, α-acyloxylation of aldehydes. Our interest in this area led us to examine the application of **9** for α-oxidation reactions, using benzoyl peroxide (**31**) as an oxidant (Scheme 13-7). Benzoyl peroxide is widely employed as a radical initiator in organic synthesis. Prior to our work in this area, there was only one report where it has been used in combination with a chiral

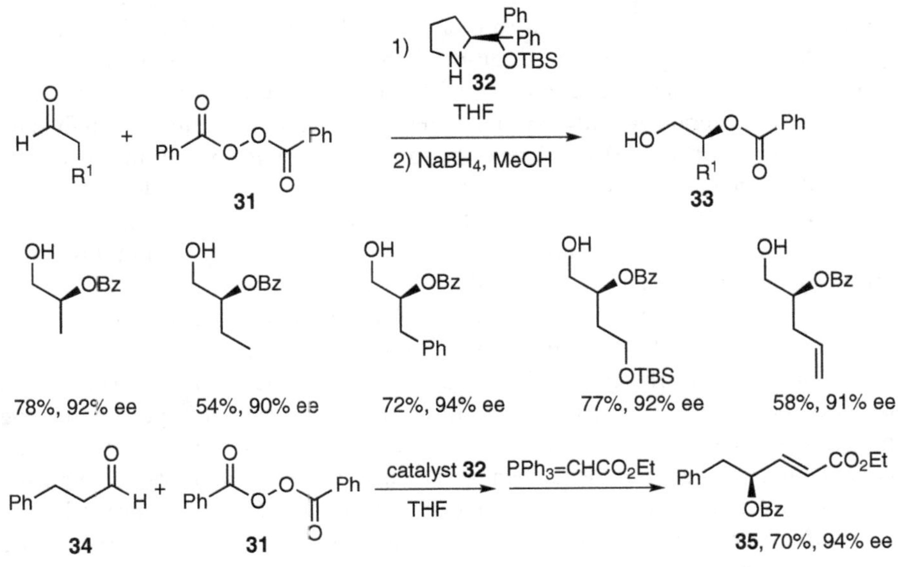

SCHEME 13-7. α-Oxidation of aldehydes using benzoyl peroxide **31**.

metalloenamine, for the asymmetric α-benzoxylation of β-ketoesters [37]. During our investigations, a similar transformation was also reported by the Maruoka group using a 2-tritylpyrrolidine catalyst [38].

The reaction of propanal and benzoyl peroxide was selected as a model, which was subjected to catalyst screening. As the generated 2-benzoyloxypropanal product is easily racemized, isolation and characterization were performed after converting it to the corresponding 2-benzoyloxypropanol, by treatment of the reaction mixture with $NaBH_4$. Proline was found to be a poor catalyst, affording very low enantioselectivity for the reaction, while diphenylprolinol trimethylsilyl ether **9** gave good enantioselectivity, but with low yield. This is because catalyst **9** deactivates by its reaction with benzoyl peroxide (**31**), forming the *N*-benzoyloxy derivative as a side product. This can be suppressed by employing the bulkier *tert*-butyldimethylsilyl ether **32**, which gave a better yield and excellent enantioselectivities not only for propanal, but also for a number of linear chain aldehydes, containing aromatic, silyl ether moieties, and alkenes (Scheme 13-7) [39].

By the addition of a Wittig reagent to the reaction mixture of 3-phenylpropanal **34**, benzoyl peroxide, and catalyst **32**, the unstable α-benzoyloxyaldehydes intermediate can be directly transformed into γ-benzoyloxy α,β-unsaturated ester **35** in a good yield without compromising the enantioselectivity. It should be noted that this transformation was performed in a one-pot operation without isolation of the racemizable α-benzoyloxyaldehyde.

The method is presently one of the most useful enantioselective α-oxidation methods of aldehydes as it can afford excellent enantioselectivities by using an inexpensive oxidant. The synthesis (−)-6-acetoxy-5-hexadecanolide, a mosquito oviposition attractant pheromone, was achieved by the Tae group, using this methodology as the key step [40].

13.2.4 Tandem Reaction Between Nitro-olefin and Pentane-1,5-dial

The catalysts were subsequently tested for their ability to promote tandem reactions, for example, the reaction between nitro-olefin **20** and pentane-1,5-dial **37**, which could be generated under equilibrium conditions from 2,6-dihydroxytetrahydropyran **36** (Scheme 13-8). Commercially available as an aqueous solution, the reagent could be used directly to generate enamine **38** in the presence of catalyst **9**; this would then react with a nitro-olefin **20** in a Michael reaction to form the zwitterionic intermediate **39**, which would then react with the aldehyde moiety in an intramolecular Henry reaction to provide **40**, hydrolyzing to provide substituted nitrocyclohexane carbaldehyde **41**. The order of these last two reactions might be reversed (that is, hydrolysis of iminium ion **39** followed by the Henry reaction to provide **40**).

After some optimization, compound **41** was obtained in good yield with high diastereoselectivity and excellent enantioselectivity when THF was used as solvent (Scheme 13-9). The reaction proceeded efficiently with aryl-substituted nitro-olefins with both electron-rich and electron-deficient aryl groups, successfully afforded the substituted cyclohexane ring in high yield with excellent enantioselectivity. Successive isomerization under two different sets of conditions can convert the tandem product into two different diastereomers without compromising the enantioselectivity. First, the formyl group in **42** can be isomerized from the axial to the equatorial position under the weakly acidic conditions of silica gel. Second, when **43** was treated with a catalytic amount of DBU in MeOH, retro-Henry, and Henry reactions can proceed to afford isomer **44** [41].

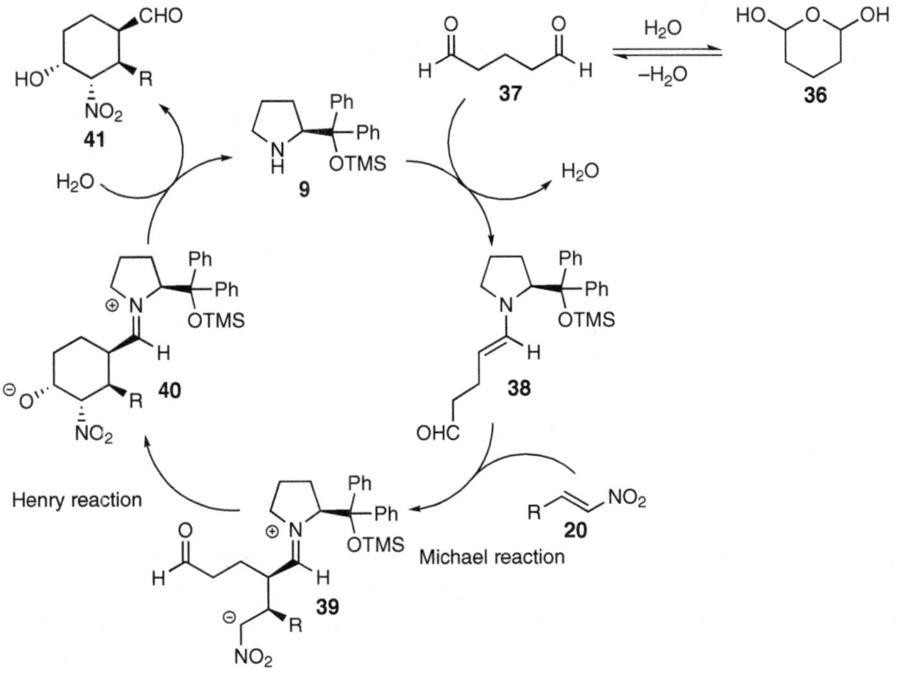

SCHEME 13-8. Proposed mechanism for the tandem reaction between pentane-1,5-dial **37** and nitro-olefin **20**.

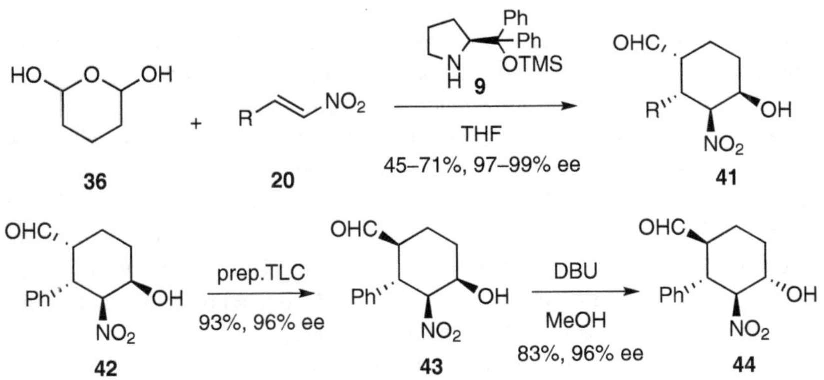

SCHEME 13-9. Tandem reaction of a pentane-1,5-dial equivalent with nitro-olefins.

13.2.5 Multicomponent Reactions

Following the demonstration of intramolecular tandem reactions, intermolecular multicomponent reactions were also investigated. These reactions can furnish heterocyclic six-membered rings such as tetrahydropyrans and piperidines, which are synthetically important structural units in many natural products. We hypothesized that substituted tetrahydropyran and piperidine derivatives **48** could be assembled by a one-pot reaction

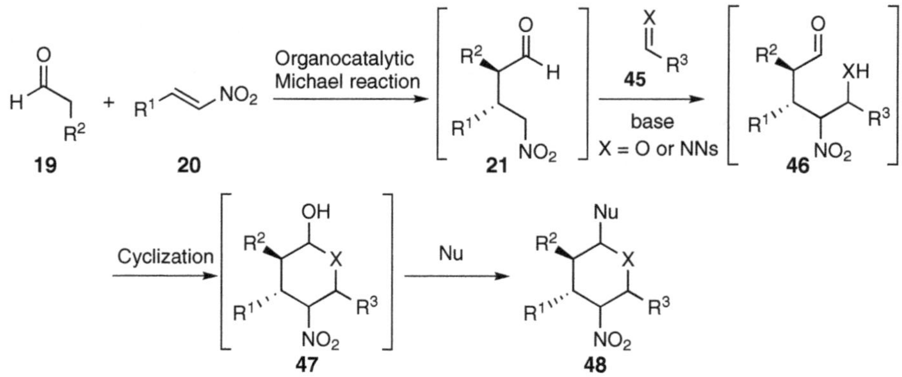

SCHEME 13-10. Synthetic scheme for the formation of a heterocyclic six-membered ring using multicomponent reactions.

of four components (aldehyde **19**, nitro-olefin **20** and imine or aldehyde **45**, allylsilane, and a nucleophile). Synthetic plans for heterocyclic six-membered ring formation were devised as follows (Scheme 13-10); tetrahydropyran or piperidine derivatives **48** could be produced if an initial Michael adduct **21** were to react with aldehyde or imine **45** through a Henry or aza-Henry reaction to produce cyclized product **46**. The hydroxy or amine group of **46** would then react intramolecularly with the aldehyde moiety to form the acetal or aminal, and thus generate the heterocyclic six-membered ring **47**. Finally, heterocycle **47** would react with a nucleophile to afford substituted heterocyclic compound **48**.

After several trials, a one-pot synthesis of substituted tetrahydropyran **47** was achieved. The initial Michael reaction of nitrostyrene and propanal proceeded in the presence of 5 mol% of **9** with *p*-nitrophenol in toluene. After removing excess propanal and solvent under reduced pressure, MeCN, benzaldehyde, and DBU (20 mol%) were added. After the formation of the 2-hydroxytetrahydropyran, the solvent was exchanged from MeCN to CH_2Cl_2, and allylsilane and $TiCl_4$ (2 equiv.) were added at −78 °C, to furnish allylated tetrahydropyran **50** in 72% yield as a single isomer with excellent enantioselectivity—all of these reactions were performed in a single flask. The reaction proceeds efficiently with various aldehydes **19** and nitroethenes **20** in the initial Michael reaction, and aldehydes **49** in the subsequent Henry reaction (Scheme 13-11) [42].

A similar methodology was devised for the synthesis of substituted piperidines using imine **51**. The *p*-nitrophenylsulfonyl (Ns) substituent was selected as the protecting group on nitrogen, because it can be easily removed under mildly basic conditions and is a good electron-withdrawing group. After optimization, a one-pot reaction sequence for the synthesis of substituted piperidine derivatives was achieved via a Michael reaction of aldehyde **19** and nitroalkene **20**, followed by an aza-Henry reaction mediated by K_2CO_3, a hemiaminalization reaction, and a Lewis acid-mediated allylation (Scheme 13-11).

The methodology provides highly substituted tetrahydropyran and piperidine rings, that contain different groups on each position of the heterocyclic ring, with complete control of the relative and absolute stereochemistry of the five contiguous carbon centers. As the allyl moiety can be easily transformed into other functional groups, the 2-allyl tetrahydropyran **50** and piperidine **52** are useful chiral building blocks.

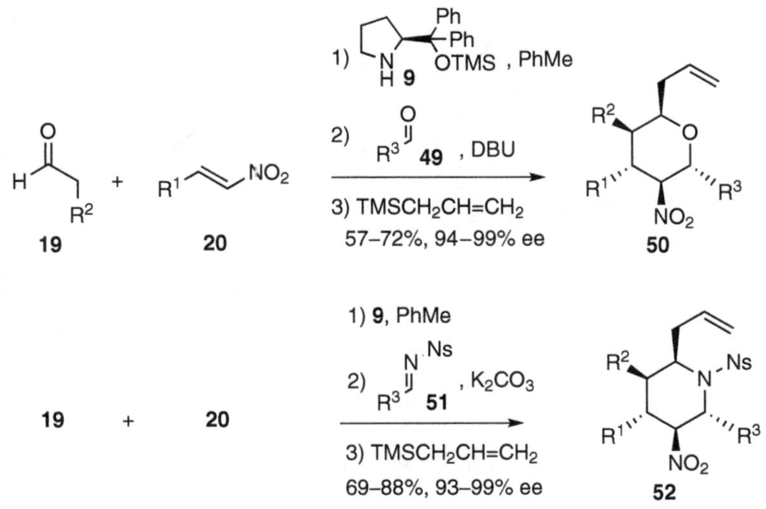

SCHEME 13-11. Multicomponent tandem reactions, mediated by **9**, for the synthesis of tetrahydropyrans or piperidines.

13.2.6 [6+2] Cycloaddition

Domino reactions are also important for the construction of useful molecules in one-pot reactions. Cycloaddition reactions are among the most powerful methods for the construction of cyclic skeletons. Compared to the well-established [4+2] cycloaddition reactions, very few [6+2] cycloaddition reactions are known [43]. Although a racemic version of this intramolecular [6+2] cycloaddition reaction (and its reaction mechanism) has been demonstrated by the Houk group [44], an asymmetric catalytic version of this chemistry had not been previously reported.

When a linear substrate **53** was exposed to a catalytic amount of **9**, a [6+2] cycloaddition reaction proceeded to afford the tricyclic compound **54** [45]. Two new C–C bonds are formed in this reaction, via domino enamine/iminium ion intermediates. The reaction proceeds well with several substrates, and *cis* isomers were obtained in every example (Scheme 13-12).

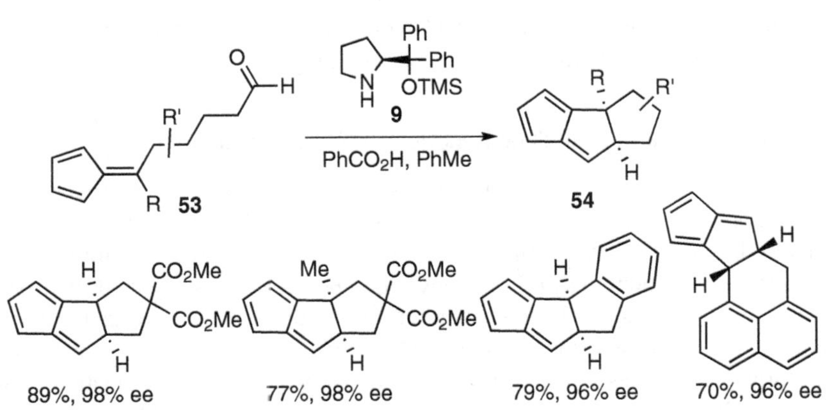

SCHEME 13-12. Formal [6+2] cycloaddition mediated by **9**.

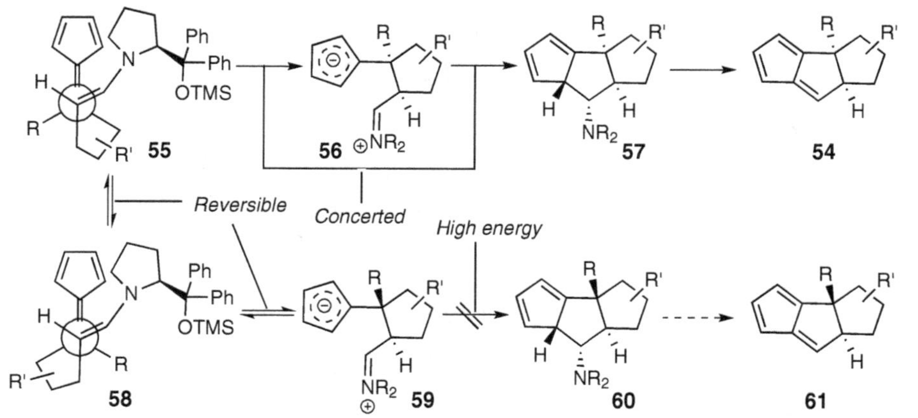

SCHEME 13-13. Proposed mechanism for the intramolecular [6+2] reaction mediated by **9**.

The proposed reaction mechanism was supported by computational analysis (Scheme 13-13), which showed that the model intermolecular [6+2] cycloaddition reaction proceeds in a stepwise mechanism via a zwitterionic intermediate [45]. In contrast, the intramolecular [6+2] cycloaddition leading to the *cis*-fused linear triquinane skeleton **57** proceeds in a concerted mechanism via a highly asynchronous transition state. The energy profiles suggest that formation of the *trans*-zwitterionic intermediate **59** is reversible. Thus, the experimental observation of exclusive formation of the *cis*-fused triquinane derivatives **54** can be rationalized as a consequence of reversible formation of *trans*-zwitterionic intermediate **59**, and energetically favorable concerted formation of the *cis*-fused tricyclic product **57**.

13.3 IMINIUM ION INTERMEDIATE

So far, the discussion has been limited to the reactivity of optically active enamine intermediates, formed from aldehyde and amine catalyst, toward several electrophiles. In the following sections, LUMO activation of α,β-unsaturated aldehydes via the formation of iminium ion intermediates (formed by a reaction between the amino catalyst and a α,β-unsaturated aldehyde) will be described, which is also an important activation pathway commonly utilized in the field of amine catalysis [46].

13.3.1 Diels–Alder and Ene-Type Reactions of Cyclopentadiene

The Diels–Alder reaction is a synthetically powerful method for the construction of regio- and stereo-chemically defined cyclohexane frameworks. Several asymmetric reactions promoted by Lewis acid have been reported [47], but the selective formation of the *exo*-isomer remains a challenging problem since the *endo*-isomer is often formed predominantly. The first asymmetric Diels–Alder reaction catalyzed by an organocatalyst was reported by the MacMillan group, where the cycloaddition between cyclopentadiene and α,β-unsaturated aldehydes proceeded with excellent enantioselectivity via an iminium ion intermediate. However, diastereoselectivity of the process was low [4].

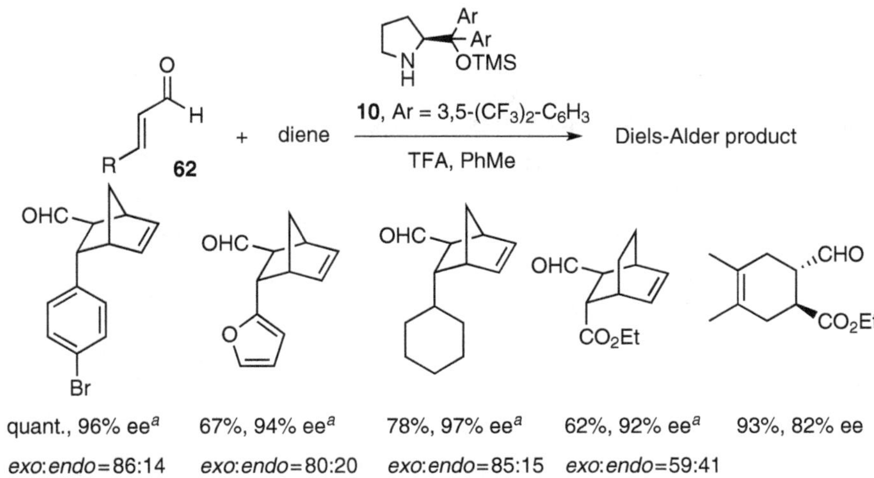

SCHEME 13-14. Diels–Alder reactions mediated by diarylprolinol silyl ether **10**. [a]ee of exo isomer.

To explore the reactivity of the iminium ion derived from a diarylprolinol silyl ether, the reaction between cyclopentadiene (**63**) and cinnamaldehyde (**65**) was examined as a model reaction. It was found that the additive, solvent, and the choice of catalyst are important for the Diels–Alder reaction to proceed. When the 3,5-bis(trifluoromethyl)phenyl-containing diarylprolinol silyl ether **10** was employed with CF_3CO_2H in toluene, the Diels–Alder adduct was obtained in excellent optical purity, and in good yield, whereas the reaction did not proceed in the presence of a weak acid such as CH_3CO_2H [42]. The presence of CF_3 groups on the catalyst's aromatic ring is essential for the Diels–Alder reaction, as the reactivity is lower without these groups. Under the optimized conditions, the reaction proceeded efficiently with not only cyclopentadiene, but also with other dienes such as 2,3-dimethyl-1,3-butadiene, 1,3-cyclohexadiene, and isoprene, with excellent enantioselectivities (Scheme 13-14). It should be noted that the *exo*-isomer was obtained predominantly in most cases.

A unique reactivity of cyclopentadiene **63** was uncovered during the course of these investigations. When the TBS-protected diphenylprolinol silyl ether **32** was employed in combination with *p*-nitrophenol in MeOH, the reaction of **63** and α,β-unsaturated aldehydes **62** did not afford the Diels–Alder product, but the ene-type products **64a** and **64b** [48]. Interestingly, the 3,5-bis(trifluoromethyl)phenyl-containing catalyst **10**, previously shown to be the best catalyst for Diels–Alder reactions, was completely inactive in this reaction. These ene-type reactions proceeded efficiently not only with cinnamaldehyde (R = Ph), but also enals containing other aromatic rings (Scheme 13-15).

This reaction is interesting, in terms of the novel reactivity of cyclopentadiene, as well as the product, since the cyclopentadiene moiety is synthetically versatile. An intramolecular Diels–Alder reaction and intramolecular cyclization were investigated as synthetic applications of these ene-type products. After treatment of cinnamaldehyde **65** and cyclopentadiene **63** in the presence of diphenylprolinol silyl ether **32** and *p*-nitrophenol for 20 h, *i*-Bu_2NH and *p*-nitrophenol were added to the reaction mixture. This afforded the bicyclo[3.3.0]octatriene **66** in good yield without compromising the enantioselectivity (63%, 95% ee) [49]. Subsequently, tricyclic derivatives **67** were also

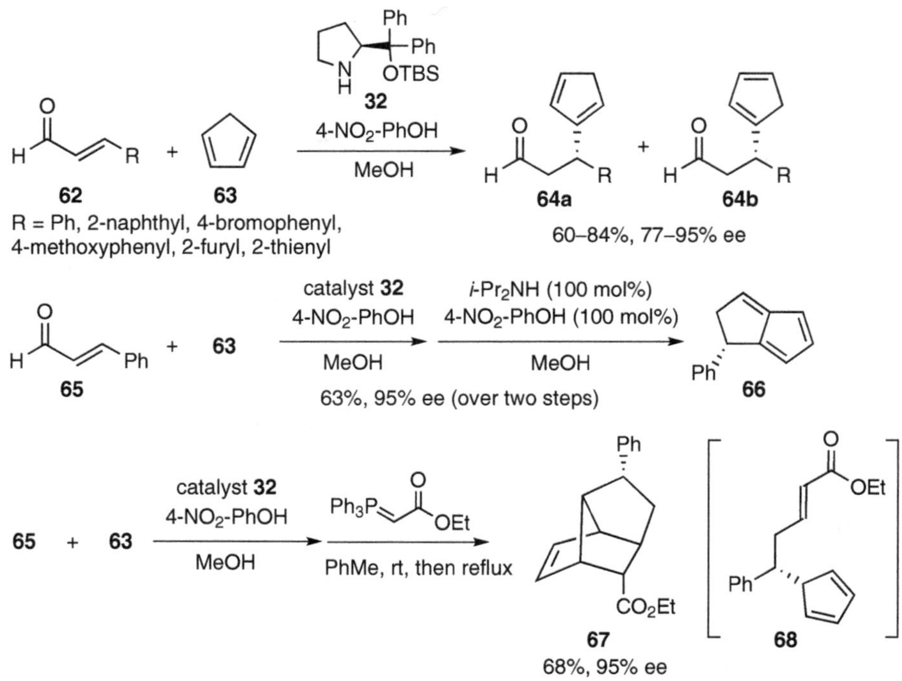

SCHEME 13-15. Ene-type of reaction mediated by diphenylprolinol silyl ether **32** and application of ene-type products **64a** and **64b**.

obtained via successive Wittig and Diels–Alder reactions in a one-pot reaction (68%, 95% ee); the Diels–Alder product resulted from the reactive ene-adduct **68**, to afford **67** as a single diastereomer. It is noteworthy that isolation of regioisomers is not necessary for these products.

13.3.2 Nitroalkane as a Nucleophile

Michael addition products of nitroalkanes to α,β-unsaturated carbonyl compounds can provide synthetically versatile intermediates such as γ-amino acid derivatives. Several organocatalysts have been applied to the reaction of nitroalkanes via an iminium activation strategy using α,β-unsaturated ketones (enones) as Michael acceptors [50–57]. However, the Michael reaction between α,β-unsaturated aldehydes **62** and nitromethane **69** does not often proceed with high enantioselectivity and yield, because nitromethane **69** is a small nucleophile and enantiofacial control is difficult. Control of its reactivity is also difficult due to the propensity of nitromethane to undergo 1, 2-addition. The Maruoka group has developed Michael reactions mediated by a chiral phase-transfer catalyst, in which silyl nitronates, prepared from nitroethane and nitropropane, were employed [58–60]; notably, the use of nitromethane was not reported.

After screening several reaction conditions, it was found that excellent yields and enantioselectivities were achieved when **9** was combined with $PhCO_2H$ as the catalyst, and MeOH was used as solvent [61]. The reaction proceeded efficiently not only for phenyl, but also for acrolein derivatives bearing electron-rich, electron-deficient, and heteroaromatic groups. In the case of alkyl-substituted acroleins, good yields and excellent

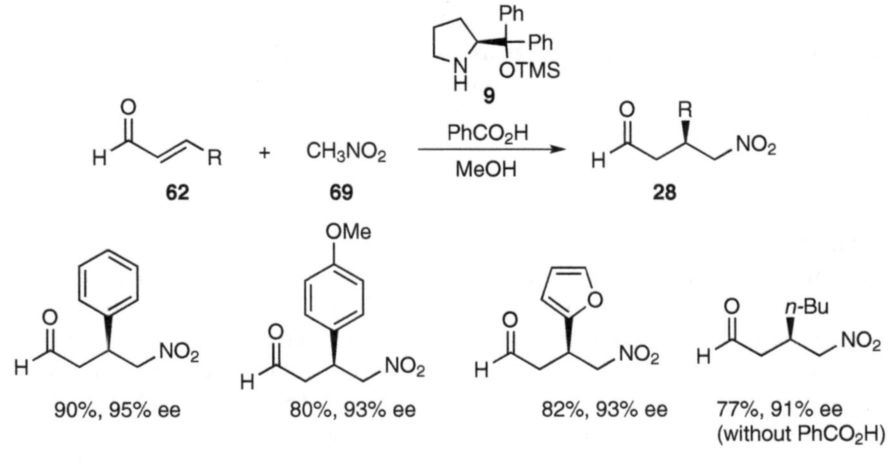

SCHEME 13-16. The Michael reaction of nitromethane **69** mediated by **9**.

enantioselectivities were obtained without addition of $PhCO_2H$ (Scheme 13-16). Moreover, other nitroalkane nucleophiles can be used successfully. On the other hand, cinnamaldehyde reacts with nitroethane in MeOH to afford the corresponding adduct in 95% yield without any diastereoselectivity (1:1), although excellent enantioselectivity was achieved. Finally, the reactivity of a β,β-disubstituted nitroalkane was low, for example, 2-nitropropane reacts with cinnamaldehyde in the absence solvent to afford the desired Michael adduct only in moderate yield, but with high enantioselectivity. During this investigation, intermolecular Michael reactions of nitromethane **69** were also reported by Palomo [62] and the Wang [63] groups, independently.

The synthetic versatility of the nitro and aldehyde groups allowed for the short syntheses of a broad range of chiral pharmaceuticals by employing the present Michael reaction as a key step (Scheme 13-17). Pregabalin **72**, an important anticonvulsant drug, was

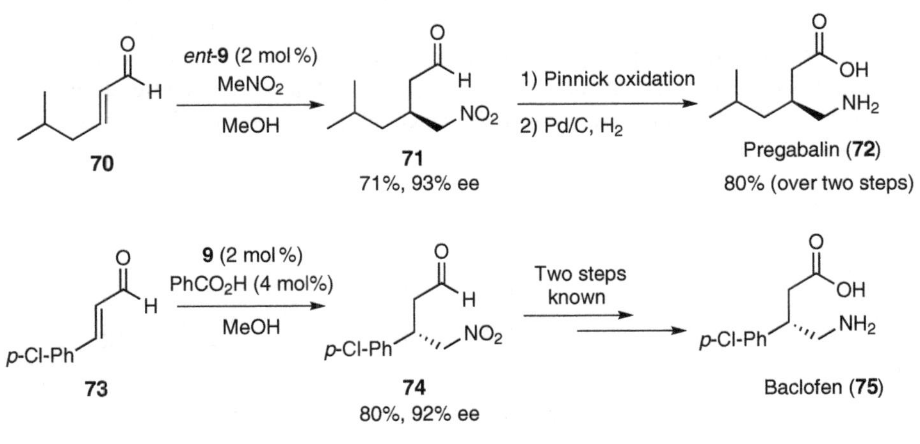

SCHEME 13-17. Synthesis of therapeutically useful compounds using an asymmetric Michael reaction of nitromethane **69**.

synthesized via an oxidation and a reduction, from Michael adducts **71**, which had been prepared using *ent*-**9** [64]. On the other hand, a therapeutically useful $GABAA_B$ receptor agonist, baclofen **75** [63], was also synthesized via a similar two-step sequence from Michael adduct **74.**

13.3.3 Nitroethanol as Nucleophile

Following the discovery of the Michael reaction of nitroalkanes mediated by diphenylprolinol silyl ethers, nitroethanol **76** was next examined as a nucleophile. Since nitroethanol possesses hydroxyl functionality, it was anticipated that a tandem reaction of the hydroxyl group with the aldehyde after the Michael reaction, as well as several other transformations, is possible. For example, following the Michael addition of nitroethanol (**76**) to α,β-unsaturated aldehyde **62** in the presence of diphenylprolinol silyl ether, the resultant Michael adduct **77** would cyclize to afford substituted tetrahydropyrans **78** and **79**. In view of the previous results for nitroethane, the diastereoselectivity of the process was expected to be low. Nevertheless, it was envisaged that the *cis*-isomer **79** could convert into the more thermodynamically stable *trans*-isomer **78**.

To realize this transformation, the reaction of cinnamaldehyde and nitroethanol **76** was examined using **9** as a catalyst in the presence of benzoic acid in MeOH. As expected, the cyclized product was obtained with excellent enantioselectivity and yield, but with low diastereoselectivity. By treatment with $NaHCO_3$, the *trans* isomer was obtained predominantly. The Michael reaction, acetalization, and isomerization reactions were conducted in one pot. The reaction proceeded efficiently not only for cinnamaldehyde, but also for acrolein derivatives with electron-rich, electron-deficient, and heteroaromatic groups (Scheme 13-18) [65].

Tetrahydropyran **78** would be a useful synthetic intermediate, as it contains several functional groups that can be elaborated further for the synthesis of nitrogen-containing molecules. Thus, further transformations of **78** were subsequently investigated (Scheme 13-19). The hydroxypyran **80** can be converted into acetoxytetrahydropyran **81**, which was reduced with Et_3SiH in the presence of $BF_3{\cdot}OEt_2$ to afford *trans*-3,4-disubstituted tetrahydropyran **82** in 93% yield. Conversely, **80** may be treated with H_2 in the presence of $Pd(OH)_2$. The reduction of the nitro group triggered an intramolecular reductive amination to provide the prolinol derivative **83**. The *cis*-3-phenylproline derivative **84** was obtained following a two-step oxidation with *o*-iodoxybenzoic acid (IBX) and Pinnick oxidation. Finally, starting with acetoxytetrahydropyran **81**, an α-amino ester derivative **86** was synthesized stereoselectively by a sequence of reduction, Wittig, oxidation, and esterification reactions.

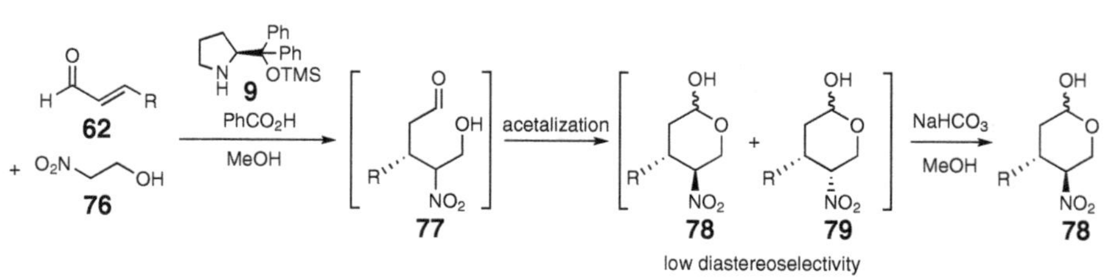

SCHEME 13-18. Catalytic asymmetric Michael reaction of nitroethanol **76** mediated by **9**.

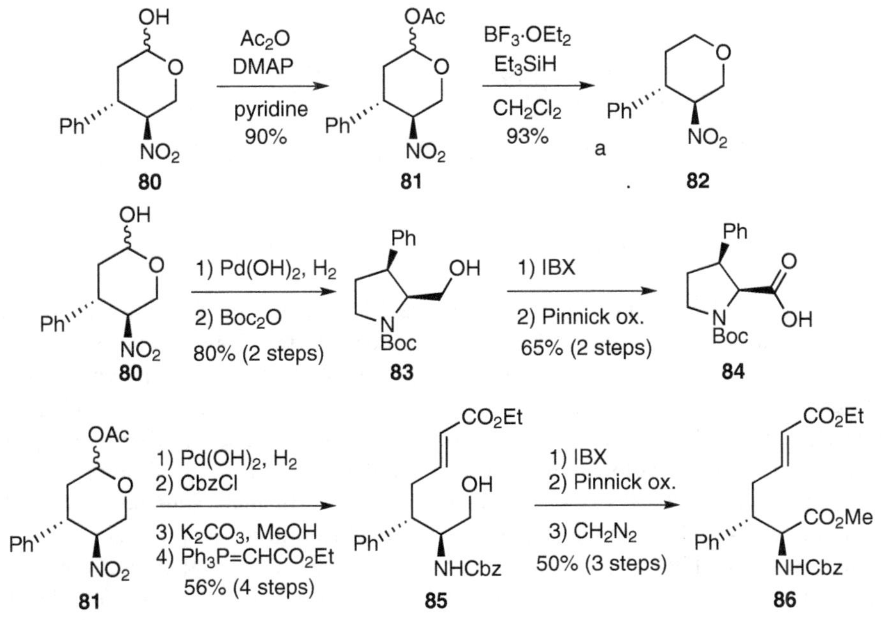

SCHEME 13-19. Transformation of the Michael adduct afforded in the previous scheme.

13.3.4 Formal Aza- and Carbo-[3+3] Cycloaddition

Using enecarbamates or enamide **87** as nucleophile, another type of tandem reaction can be realized using diarylprolinol silyl ether catalysts. The reaction of **87** with α,β-unsaturated aldehydes **62** proceeded via a sequence of ene/isomerization/hydrolysis/aminoacetalization reactions in tandem, representing a formal aza-[3+3] cycloaddition reaction (Scheme 13-20). This is a very rare example where a piperidine derivative can be assembled using a [3+3] cycloaddition reaction, in an enantioselective and catalytic fashion [66].

Realizing the synthetic importance of this reaction, substantial optimization of the reaction conditions was performed in detail, whereby the key parameter was found to be

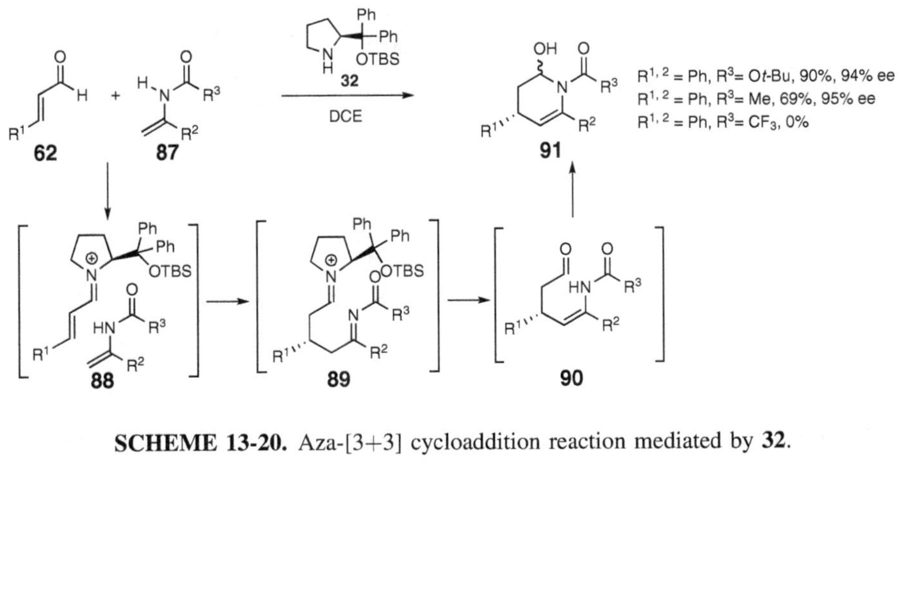

SCHEME 13-20. Aza-[3+3] cycloaddition reaction mediated by **32**.

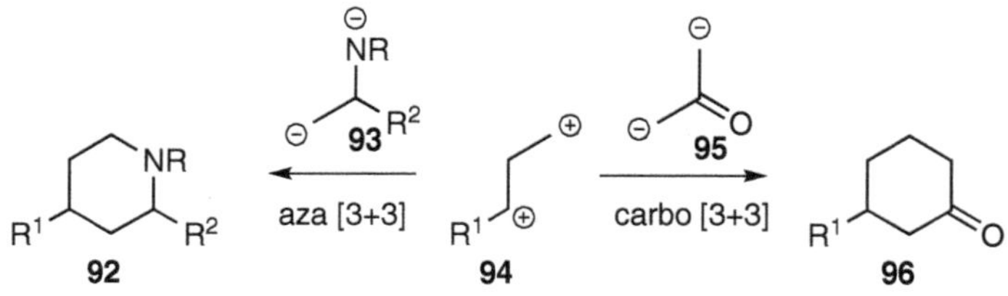

SCHEME 13-21. Aza- and carbo-[3+3] cycloadditions.

the nature of R^3 substituent of the ene carbamate. This dictates the formation of **91**, which depends on a subtle balance of several competing reactions such as isomerization/hydrolysis/aminoacetalization. The best R^3 substrate was O*t*-Bu, which allowed the reaction to proceed efficiently with several different substituents at R^1 and R^2, such as phenyl, electron-deficient aromatic, electron-rich aromatic, and heteroaromatic groups.

Encouraged by the development of the successful aza-[3+3] cycloaddition reaction, the corresponding carbo-[3+3] cycloaddition was also investigated (Scheme 13-21). Compared to the well-established Diels–Alder [4+2] cycloaddition, enantioselective catalytic carbo-[3+3] cycloaddition reactions are very rare. It was expected that the formal carbo-[3+3] cycloaddition reaction would proceed between an α,β-unsaturated aldehyde **62** and 3-oxopentanedioate **97**, via tandem Michael/enamine hydration/Knoevenagel condensation reactions. The reaction was first reported by the Jørgensen group using the diphenylprolinol silyl ether **9** to catalyze the reaction between **62** and **97**, affording the 2:1 addition product **99** in high diastereo- and enantioselectivities [67].

In our investigation, the reaction between cinnamaldehyde (**65**) and dimethyl 3-oxopentanedioate (**97**) was facilitated by using the diphenylprolinol silyl ether **32** and $PhCO_2H$ in CH_2Cl_2. The desired cyclohexenone was isolated, following its reduction to the corresponding alcohol **98**, as a single isomer. The reactions proceed equally efficiently with electron deficient aromatic and electron-rich aromatic and heteroaromatic substituents on the α,β-unsaturated aldehyde (Scheme 13-22) [68]. In this reaction, the

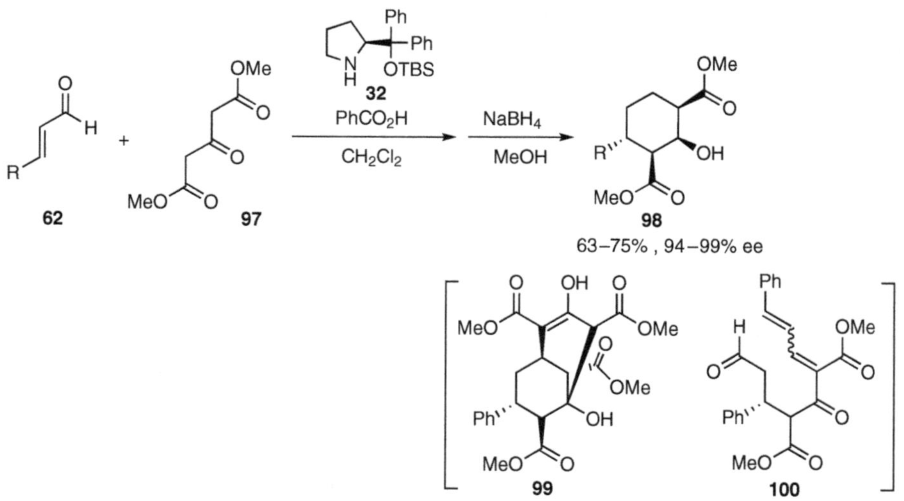

SCHEME 13-22. Carbo-[3+3] cycloaddition mediated by diphenylprolinol silyl ether **32**.

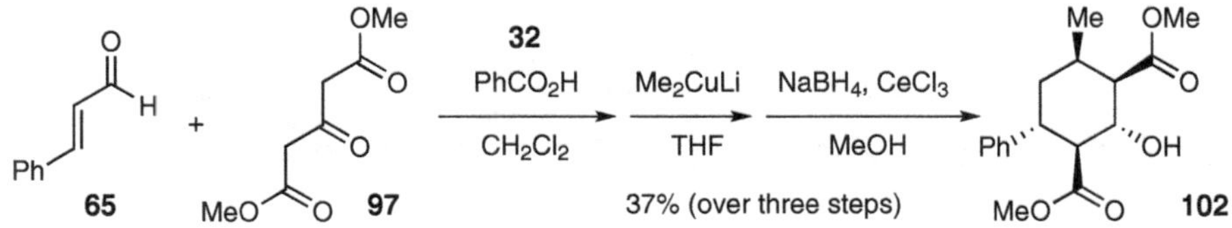

SCHEME 13-23. Synthesis of functionalized cyclohexane derivatives using one-pot reaction.

molar ratio of the reagents was crucial: when 1.5 equiv. of tricarbonyl compound **97** was used, the 2:1 adduct **99** was formed. However, when 0.5 equiv. of **97** was employed, the 1:2 adduct **100** was formed instead of the desired product.

The formal [3+3] cycloaddition product is synthetically useful, because it possesses several functional groups, including an electron-deficient double bond activated by keto and ester moieties. Further tandem one-pot transformations were investigated, whereby a solution of Me_2CuLi was added to the reaction mixture containing the cycloadduct, to generate the methyl addition product in a mixture of keto and enol forms (Scheme 13-23). Reduction under Luche's conditions afforded the highly substituted cyclohexanol **102**. It should be noted that the reaction is both high yielding and clean, affording a method to construct complex cyclohexanones in one-pot reactions.

13.4 FORMAL C–H INSERTION

One of the greatest modern synthetic challenges is to transform the unreactive C–H bond into a new C–C bond, and to control the absolute configuration of the newly generated stereogenic center at the same time. Recently, great progress has been achieved in the field of transition-metal catalysis in the activation of unreactive C–H bonds. To investigate the potential of organocatalysts in this type of reaction, the enamine-iminium ion manifold was accessed via the generation of the enamine **2** from an aldehyde **1** and an amine, which reacts with a hydride-abstracting reagent to afford an iminium ion **6**. The enamine **2** is an electron-rich alkene, and DDQ can act as a hydride-abstracting reagent from allylic hydrogen. The iminium ion **6** formed by using **9** is known to be a reactive intermediate toward different nucleophiles, to afford a β-substituted aldehyde **8** after hydrolysis (Scheme 13-24).

3-Phenylpropanal, DDQ and nitromethane were selected as model aldehyde, hydride-abstracting reagent, and nucleophile, respectively, as the asymmetric Michael reaction of nitromethane with α,β-unsaturated aldehydes catalyzed by **9** was known. Although the domino reaction was not realized, a two-step, one-pot formal C–H insertion was achieved with high yield and enantioselectivity [69]. The initial oxidation reaction was completed

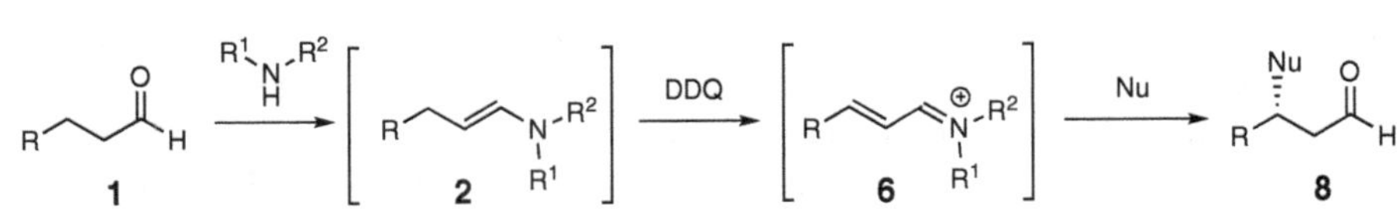

SCHEME 13-24. Synthetic planning for a formal C–H insertion mediated by enamine and iminium ion.

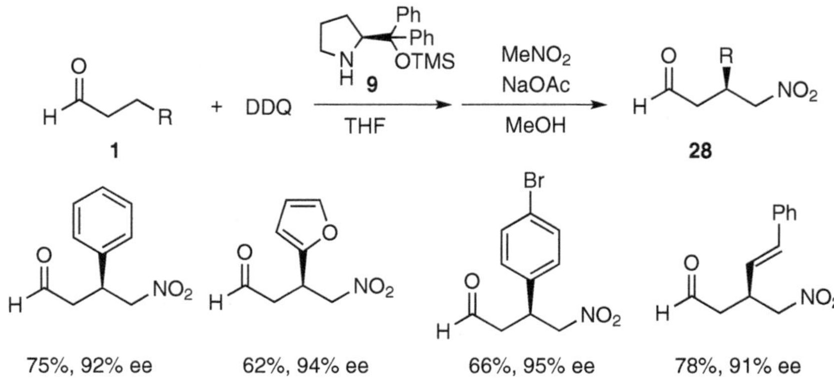

SCHEME 13-25. Formal C–H insertion reaction mediated by **9**.

within 1 h at room temperature in quantitative yield when 20 mol% of **9** was used in THF. This reaction did not proceed in the absence of the secondary amine catalyst—its replacement by a tertiary amine (such as Et_3N) did not afford any product. In the second step, the addition of MeOH and NaOAc is key to smooth reaction. Upon completion of the oxidation reaction, NaOAc was added, followed shortly after by MeOH and nitromethane, to give the C–H insertion product in high yield and enantioselectivity. The addition of NaOAc caused an immediate precipitation of the sodium salt of hydroquinone, effectively removing the acidic hydroquinone. The reaction proceeded efficiently with β-substituted aldehydes containing electron-rich, electron-deficient, heteroaromatic groups and olefins (Scheme 13-25).

13.5 REACTIONS IN THE PRESENCE OF WATER

Water is considered an environmentally benign medium and exhibits distinct reactivity and properties compared to organic solvents. As a result, processes that utilize water as the reaction medium have also attracted considerable interest in recent years as they can also confer unusual reactivity [70–80]. In the Diels–Alder reaction, for instance, the reaction can be accelerated "in water" and "on water". "In water" effect was first observed by Rideout and Breslow, in which all the reagents dissolved homogeneously [71,72]. In contrast, "on water" reactions, first reported by Sharpless and coworkers, vigorous stirring was essential for the reaction to proceed at the interface of two phases [73–75]. With this background, we investigated the organocatalyzed aldol reaction in a concentrated organic phase, with water present as a second phase and influencing the reaction in the organic phase [76–78], which we termed "in the presence of water" [79]. At the same time, the group of Barbas also reported the same reaction by using a diamine catalyst [80].

Our interest in this area includes the application of diphenylprolinol silyl ether catalyst to the Diels–Alder reaction with water as the reaction medium [70]. As already described earlier, the combination of diarylprolinol silyl ether **10** and CF_3CO_2H is an effective catalyst system. The optimization of the reaction revealed that the combination of $HClO_4$ and 3,5-bis(trifluorometthy)-phenyl-containing diarylprolinol silyl ether allowed the reaction to proceed in the presence of water. Notably, it was much faster than the

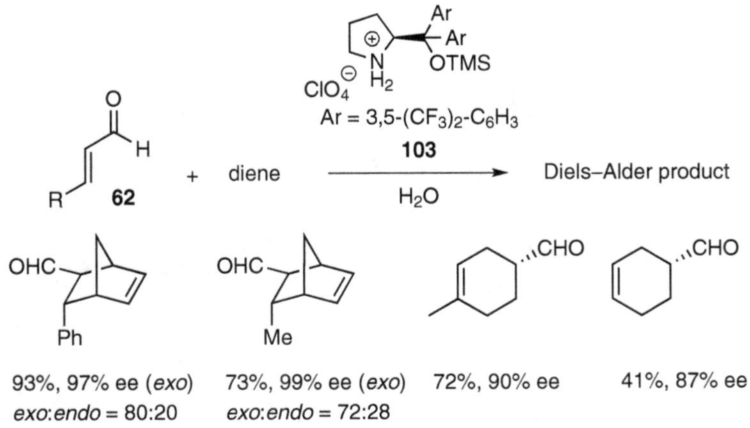

SCHEME 13-26. Diels–Alder reaction mediated by diarylprolinol silyl ether salt **103** in the presence of water.

corresponding reaction performed with the catalyst and CF_3CO_2H in toluene. Importantly, the system includes some novel combinations of diene and dienophile for enantioselective Diels–Alder reactions (Scheme 13-26); while butadiene showed low reactivity, it reacted with acrolein to provide the adduct with high enantioselectivity (41%, 87% ee).

The effect of water on the reaction was examined for the reaction between cinnamaldehyde **65** and cyclopentadiene **63**. Even in the presence of a large amount of water, the reaction proceeded efficiently with the hydrophobic cinnamaldehyde between the two phases. In contrast, too much water reduced the reactivity of acrolein (R = H) that dissolves in water. Moreover, stirring is not essential; for a reaction in which there was initial stirring for 1 min, the reaction efficiency was unchanged without additional stirring. These results suggest that: (1) the reaction proceeds in the organic phase and does not proceed at the interface between the water and organic phases; and (2) a small amount of water dissolved in the organic phase accelerates the reaction and affects the transition state, resulting in increased enantioselectivity. This water effect is completely different from "in water reaction" and it is also different from the effect of the "on water reaction".

13.6 SYNTHESIS OF BIOLOGICALLY ACTIVE MOLECULES

As described above, diphenylprolinol derivatives are some of the most useful catalysts for the synthesis of optically active synthetic intermediates. In this section, their applications to the synthesis of natural products and non-natural biologically active molecules are described.

13.6.1 Synthesis of (−)-β-Santalol

The enantioselective synthesis of (−)-β-santalol was reported by the Fehr group, using diarylprolinol silyl ether as a catalyst in the first step of the reaction sequence (Scheme 13-27) [81]. Reaction conditions were optimized by using equimolar amounts

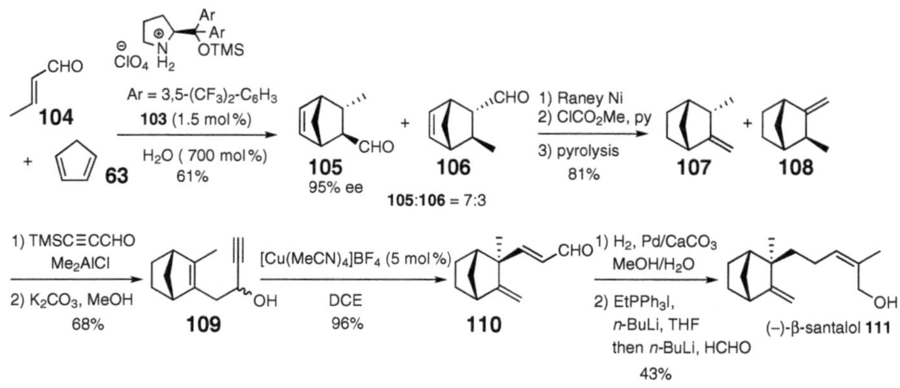

SCHEME 13-27. Synthesis of (−)-β-santalol **111**.

of cyclopentadiene **63** and crotonaldehyde **104** in the presence of the diarylprolinol silyl ether salt **103**, and in the presence of water. The cycloadducts **105** and **106** were then hydrogenated as a mixture to give the corresponding saturated alcohols, which were converted into methyl carbonates and pyrolyzed to afford hydrocarbons **107** and **108**. The ene reaction with trimethylsilylpropynal follows, affording **109** after desilylation. Exposure to $[Cu(CH_3CN)_4]BF_4$ (5 mol%) promoted cyclization–fragmentation reaction to the desired aldehyde **110**. Finally, site-selective hydrogenation with $Pd/CaCO_3$, followed by a Wittig reaction afforded the target-**111**.

Because water was employed as a solvent and excellent enantioselectivity and reactivity has been obtained, this Diels–Alder reaction is among the greener and more practical procedures for enantioselective Diels–Alder reaction of α,β-unsaturated aldehydes. Actually, the multigram-scale preparation of the Diels−Alder adduct of crotonaldehyde and cyclopentadiene under optimal conditions was achieved. After stirring a reaction mixture contacting crotonaldehyde (**104**, 2.1 g), cyclopentadiene (**63**, 2.0 g) and water 3.8 ml in the presence of 1.5 mol% catalyst **103** (314 mg), the two-phase system was separated. Distillation of the crude mixture gave the Diels–Alder product (2.1 g, 61%, *exo*:*endo* = 7:3) [81].

13.6.2 Synthesis of Oseltamivir

(−)-Oseltamivir (**119**) phostate (Tamiflu) is a neuraminidase inhibitor, and is one of the most promising therapeutics used in the treatment of both type A and type B human influenza [82]. The first step of the synthesis is the addition of aldehyde **112** to nitroalkene **113** (Scheme 13-28). Catalyzed by the diphenylprolinol silyl ether ***ent* -9,** the Michael addition proceeded in an almost quantitative yield and excellent enantioselectivity. This intermediate then enters into a tandem reaction by first reacting with vinylphosphonate **115** via a domino/Henry-type Michael addition, followed by an intramolecular Horner–Wadsworth–Emmons olefination, which assembles the cyclohexene core **116**. Unfortunately, a mixture of (5*R*)-**116** and (5*S*)-**116** was obtained. However, by adding *p*-thiocresol (tolSH) at the end of the one-pot reaction, the resulting sulfa-Michael product **117** can be generated predominantly as the desired 5*S* isomer by the Michael reaction of thiol and isomerization. At this stage, the only chromatographic separation of the synthesis

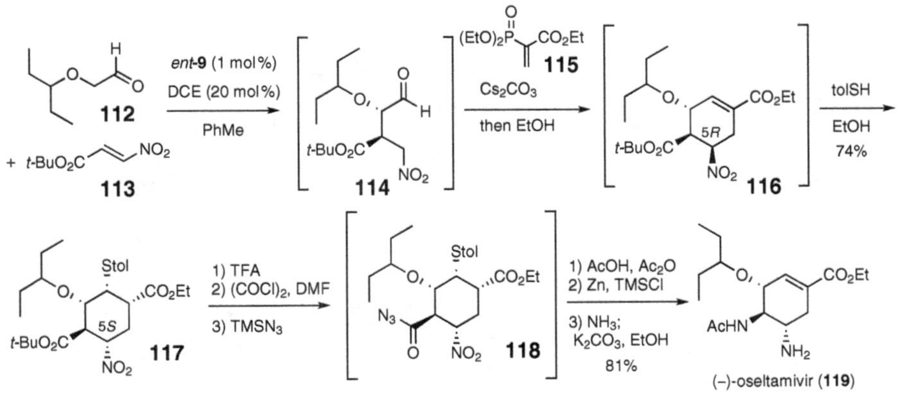

SCHEME 13-28. Synthesis of oseltamivir **119**.

was performed. The second "one-pot" process was initiated by cleavage of the *tert*-butyl ester and subsequent conversion of the product into an acyl azide **118**. A Curtius rearrangement at room temperature, reduction of the nitro group, and addition of potassium carbonate induced a retro-Michael reaction during the second one-pot operation. Finally, acid/base extraction provided (−)-oseltamivir (**119**). The synthesis was completed by employing two one-pot reaction sequences with a total yield of 60% [83–86].

Since oseltamivir **119** is an important drug for the treatment of influenza, its syntheses have been achieved by several groups [87–91]. The key advantages of this highly efficient and practical synthesis are provided by the use of the diarylprolinol silyl ether catalyst and one-pot synthesis. First, the use of diphenylprolinol silyl ether affords Michael adduct **114**, which possesses nitro and aldehyde groups, in high yield. Moreover, only 1 mol% loading of the catalyst is needed to promote the asymmetric Michael reaction, which can be implemented for the one-pot reaction and also large-scale synthesis. Because the Michael adduct contains two highly reactive positions, it can be used in the subsequent domino Michael/Horner–Wadsworth–Emmons reactions in one-pot, and the catalyst does not disturb this domino sequence. As a result, this tandem reaction affords the key substrate **116** which has the entire carbon skeleton of oseltamivir **119**. This synthetically distinct route should enable the novel preparation of tamiflu derivatives.

13.6.3 Synthesis of ABT-341 (127)

As the organocatalytic Michael reaction mediated by **9** has broad applicability with respect to both the Michael acceptor and the donor, this method could be applied to the preparation of pharmaceutical products that possess a six-membered ring. Dipeptidyl peptidase IV (DPP4, also known as CD26), a 110 kDa serine protease that is ubiquitously distributed in the body, deactivates glucose-regulating hormones, such as GLP-1 and GIP. Thus, DPP4 inhibition has become a useful therapy for type 2 diabetes. In this regard, ABT-341 (**127**) is a highly potent, selective, and orally bioavailable DPP4 inhibitor developed by Abbott Laboratories [92].

The synthesis of ABT-341 (**127**) began with a Michael addition of acetaldehyde (**24**) to the nitroalkene **120** using ***ent*-9**, followed by the addition of the resulting Michael product **121** to the vinylphosphonate **122**, followed by a Horner–Wadsworth–Emmons reaction

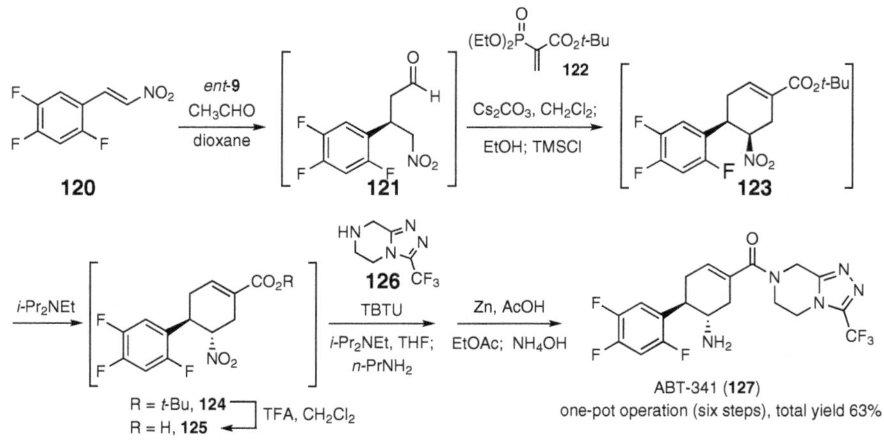

SCHEME 13-29. Synthesis of ABT-341 (**127**).

(Scheme 13-29) [35]. After a base-promoted epimerization, the *tert*-butyl ester was cleaved, and the resulting carboxylic acid **125** was subjected to a TBTU-mediated amide formation with amine **126**. Reduction of the nitro group with Zn/AcOH, and subsequent purification by chromatography, afforded the target compound **127** in a one-pot operation. Compound **121** was prepared not only by the Michael reaction of acetaldehyde and nitroalkene **120**, but also by the Michael reaction of nitromethane and the α,β-unsaturated aldehyde [61]. In this alternative reaction sequence, four components could be assembled in one-pot. Remarkably, the catalyst **9** used in the first step does not interfere with the subsequent five reactions, and the final product was obtained in high yield, as each reaction is nearly quantitative.

13.7 CONCLUSION

Jørgensen [5] and our group [6] independently found that diarylprolinol derivatives promote asymmetric reactions. Since then, these catalysts have been used in a number of asymmetric transformations. These reactions can either proceed via enamine or iminium ion intermediates, or both at the same time. Notably, diarylprolinol silyl ethers could be used in many different solvent systems, including water.

One of the problems for the diarylprolinol silyl ether is its lower reactivities compared to other metal or amino catalysts. Consequently, the substrate scope is limited to aldehydes or α,β-unsaturated aldehydes; ketones (which can be employed as substrates in certain reactions using MacMillan's catalyst or proline) or esters are currently not compatible with this catalyst system. However, the lower reactivity of catalyst offers cleaner reactions, as side reactions, such as self-condensation of aldehyde, are suppressed. Conversely, more reactive aldehyde species, including acetaldehyde, can be used as substrates in diarylprolinol silyl ether-catalyzed reactions. Additionally, diarylprolinol silyl ether has a remarkable and general reactivity for aldehydes and α,β-unsaturated aldehydes. Catalyst loading could be reduced to 1 mol% in some reactions. Since aldehydes have higher reactivity than ketones, it is ideally suited for one-pot reactions. Moreover, reactions mediated by small catalytic amount of diarylprolinol silyl ether provide products in high yield and excellent enantioselectivity, resulting in minimal side product formation. The unique reactivity of the reactive catalyst also ensures that it does not interfere in subsequent reactions, making it compatible for applications in multistep one-pot reactions.

The ultimate aim of a "one-pot reaction" is not to carry out as many reactions as possible, but to prepare useful compounds in a practical manner. It is particularly powerful for the construction of complex molecules with minimal operational steps [93]. Asymmetric catalytic reactions often require extensive tuning of sensitive reaction conditions, which can be mitigated by employing more powerful screening tools. In this regard, we believe that diarylprolinol silyl ether has the potential to mediate many more one-pot reactions for practical organic synthesis.

REFERENCES

1. Hajos ZG, Parrish DR (1974). Asymmetric synthesis of bicyclic intermediates of natural product chemistry. *J. Org. Chem.* 39, 1615–1621.
2. Eder U, Sauer G, Wiechert R (1971). New type of asymmetric cyclization to optically active steroid CD partial structures. *Angew. Chem., Int. Ed.* 10, 496–497.
3. List B, Lerner RA, Barbas, III, CF (2000). Proline-catalyzed direct asymmetric aldol reactions. *J. Am. Chem. Soc.* 122, 2395–2396.
4. Ahrendt KA, Borths CJ, MacMillan DWC (2000). New strategies for organic catalysis: the first highly enantioselective organocatalytic Diels–Alder reaction. *J. Am. Chem. Soc.* 122, 4243–4244.
5. Marigo M, Wabnitz TC, Fielenbach D, Jørgensen KA (2005). Enantioselective organo-catalyzed α−sulfenylation of aldehydes. *Angew. Chem., Int. Ed.* 44, 794–797.
6. Hayashi Y, Gotoh H, Hayashi T, Shoji M (2005). Diphenylprolinol silyl ethers as efficient organocatalysts for the asymmetric Michael reaction of aldehydes and nitroalkenes. *Angew. Chem., Int. Ed.* 44, 4212–4215.
7. Seebach D, Groselj U, Badine DM, Schweizer WB, Beck AK (2008). Isolation and X-ray structures of reactive intermediates of organocatalysis with diphenylprolinol ethers and with imidazolidinones. A survey and comparison with computed structures and with 1-acyl-imidazolidinones: the 1,5-repulsion and the geminal-diaryl effect at work. *Helv. Chim. Acta* 91, 1999–2034.
8. Groselj U, Seebach D, Badine DM, Schweizer WB, Beck AK, Krossing I, Klose P, Hayashi Y, Uchimaru T (2009). Structures of the reactive intermediates in organocatalysis with diarylprolinol ethers. *Helv. Chim. Acta* 92, 1225–1259.
9. Mielgo A, Palomo C (2008). α,α-Diarylprolinol ethers: new tools for functionalization of carbonyl compounds. *Chem. Asian J.* 3, 922–948.
10. Xu LW, Li L, Shi ZH (2010). Asymmetric synthesis with silicon-based bulky amino organocatalysts. *Adv. Synth. Catal.* 352, 243–279.
11. Betancort JM, Barbas, III, CF (2001). Catalytic direct asymmetric Michael reactions: taming naked aldehyde donors. *Org. Lett.* 3, 3737–3740.
12. Sulzer-Mosse S, Alexakis A (2007). Chiral amines as organocatalysts for asymmetric conjugate addition to nitroolefins and vinyl sulfones via enamine activation. *Chem. Commun.* 3123–3135.
13. Cobb AJA, Longbottom DA, Shaw DM, Ley SV (2004). 5-Pyrrolidin-2-yltetrazole as an asymmetric organocatalyst for the addition of ketones to nitro-olefins. *Chem. Commun.* 1808–1809.

14. Mase N, Thayumanavan R, Tanaka F, Barbas, III, CF (2004). Direct asymmetric organocatalytic Michael reactions of α,α-disubstituted aldehydes with β-nitrostyrenes for the synthesis of quaternary carbon-containing products. *Org. Lett.* 6, 2527–2530.
15. Alexakis A, Andrey O (2002). Diamine-catalyzed asymmetric Michael additions of aldehydes and ketones to nitrostyrene. *Org. Lett.* 4, 3611–3614.
16. Ishii T, Fujioka S, Sekiguchi Y, Kotsuki H (2004). A new class of chiral pyrrolidine–pyridine conjugate base catalysts for use in asymmetric Michael addition reactions. *J. Am. Chem. Soc.* 126, 9558–9559.
17. Patora-Komisarska K, Benohoud M, Ishikawa H, Seebach D, Hayashi Y (2011). Organocatalyzed Michael addition of aldehydes to nitro alkenes – generally accepted mechanism revisited and revised. *Helv. Chim. Acta* 94, 719–745.
18. Bures J, Armstrong A, Blackmond DG (2011). Mechanistic rationalization of organocatalyzed conjugate addition of linear aldehydes to nitro-olefins. *J. Am. Chem. Soc.* 133, 8822–8825.
19. Hayashi Y, Itoh T, Ohkubo M, Ishikawa H (2008). Asymmetric Michael reaction of acetaldehyde catalyzed by diphenylprolinol silyl ether. *Angew. Chem., Int. Ed.* 47, 4722–4724.
20. List B (2000). The direct catalytic asymmetric three-component Mannich reaction. *J. Am. Chem. Soc.* 122, 9336–9337.
21. Cordova A (2004). The direct catalytic asymmetric Mannich reaction. *Acc. Chem. Res.* 37, 102–112.
22. Notz W, Tanaka F, Barbas, III, CF (2004). Enamine-based organocatalysis with proline and diamines: the development of direct catalytic asymmetric aldol, Mannich, Michael, and Diels–Alder reactions. *Acc. Chem. Res.* 37, 580–591.
23. Marques MM (2006). Catalytic enantioselective cross-Mannich reaction of aldehydes. *Angew. Chem., Int. Ed.* 45, 348–352.
24. Ting A, Schaus SE (2007). Organocatalytic asymmetric Mannich reactions. New methodology, catalyst design, and synthetic applications. *Eur. J. Org. Chem.* 5797–5815.
25. Bhadury PS, Song BA (2010). Chemistry of organocatalytic asymmetric Mannich reactions. *Curr. Org. Chem.* 14, 1989–2006.
26. Yang JW, Chandler C, Stadler M, Kampen D, List B (2008). Proline-catalyzed Mannich reactions of acetaldehyde. *Nature* 452, 453–455.
27. Hayashi Y, Tsuboi W, Shoji M, Suzuki N (2003). Application of high pressure induced by water-freezing to the direct catalytic asymmetric three-component List–Barbas–Mannich reaction. *J. Am. Chem. Soc.* 125, 11208–11209.
28. Hayashi Y, Tsuboi W, Ashimine I, Urushima T, Shoji M, Sakai K (2003). The direct and enantioselective, one-pot, three-component, cross-Mannich reaction of aldehydes. *Angew. Chem., Int. Ed.* 42, 3677–3680.
29. Hayashi Y, Urushima T, Shin M, Shoji M (2005). The stereoselective synthesis of α-substituted β-amino secondary alcohols based on the proline-mediated, asymmetric, three-component Mannich reaction and its application to the formal total synthesis of nikkomycins B and Bx. *Tetrahedron* 61, 11393–11404.
30. Hayashi Y, Urushima T, Shoji M, Uchimaru T, Shiina I (2005). The direct, enantioselective, one-pot, three-component, cross-Mannich reaction of aldehydes: the reason for the higher reactivity of aldimine versus aldehyde in proline-mediated Mannich and aldol reactions. *Adv. Synth. Catal.* 347, 1595–1604.
31. Asahi K, Fu HY, Hayashi Y, Eguchi H, Murata H, Tsujii M, Tsuji S, Tanimura H, Kawano S (2007). Helicobacter pylori infection affects toll-like receptor 4 expression in human gastric mucosa. *Hepatogastroenterology* 54, 1941–1944.

32. Hayashi Y, Urushima T, Aratake S, Okano T, Obi K (2008). Organic solvent-free, enantio- and diastereoselective, direct Mannich reaction in the presence of water. *Org. Lett.* 10, 21–24.

33. Urushima T, Ishikawa H, Hayashi Y (2011). Asymmetric Mannich reaction of imines derived from aliphatic and aromatic aldehydes catalyzed by diarylprolinol silyl ether. *Chem. Eur. J.* 17, 8273–8276.

34. Hayashi, Y, Okano T, Itoh T, Urushima T, Ishikawa H, Uchimaru T (2008). Direct organocatalytic Mannich reaction of acetaldehyde: an improved catalyst and mechanistic insight from a computational study. *Angew. Chem., Int. Ed.* 47, 9053–9058.

35. Ishikawa H, Honma M, Hayashi Y (2011). One-pot high-yielding synthesis of the DPP4-selective inhibitor ABT-341 by a four-component coupling mediated by a diphenylprolinol silyl ether. *Angew. Chem., Int. Ed.* 50, 2824–2827.

36. Vilaivan T, Bhanthumnavin W (2010). Organocatalyzed asymmetric α-oxidation, α-aminoxylation and α-amination of carbonyl compounds. *Molecules* 15, 917–958.

37. Lee J, Oya S, Snyder JK (1991). Asymmetric oxidation of β-ketoesters with benzoyl peroxide; enantioselective formation of protected tertiary alcohols. *Tetrahedron Lett.* 32, 5899–5902.

38. Kano T, Mii H, Maruoka K (2009). Direct asymmetric benzoyloxylation of aldehydes catalyzed by 2-tritylpyrrolidine. *J. Am. Chem. Soc.* 131, 3450–3451.

39. Gotoh H, Hayashi Y (2009). Diphenylprolinol silyl ether as a catalyst in an asymmetric, catalytic and direct α-benzoyloxylation of aldehydes. *Chem. Commun.* 3083–3085.

40. Park Y, Tae J (2010). Facile synthesis of (−)-6-acetoxy-5-hexadecanolide by organocatalytic α-oxygenation-allylation-RCM strategy. *Synthesis* 3627–3630.

41. Hayashi Y, Okano T, Aratake S, Hazelard D (2007). Diphenylprolinol silyl ether as a catalyst in an enantioselective, catalytic, tandem Michael/Henry reaction for the control of four stereocenters. *Angew. Chem., Int. Ed.* 46, 4922–4925.

42. Gotoh H, Hayashi Y (2007). Diarylprolinol silyl ether as catalyst of an exo-selective, enantioselective Diels–Alder reaction. *Org. Lett.* 9, 2859–2862.

43. Toselli N, Martin D, Achard M, Tenaglia A, Bürgi T, Buono G (2008). Enantioselective cobalt-catalyzed [6+2]-cycloadditions of cycloheptatriene with alkynes. *Adv. Synth. Catal.* 350, 280–286.

44. Wu TC, Houk KN (1985). Construction of linear-fused tricyclopentanoids by intramolecular [6 + 2] cycloadditions of fulvenes with enamines. *J. Am. Chem. Soc.* 107, 5308–5309.

45. Hayashi Y, Gotoh H, Honma M, Sankar K, Kumar I, Ishikawa H, Konno K, Yui H, Tsuzuki S, Uchimaru T (2011). Organocatalytic, enantioselective intramolecular [6+2] cycloaddition reaction for the formation of tricyclopentanoids and insight on its mechanism from a computational study. *J. Am. Chem. Soc.* 133, 20175–20185.

46. Erkkila A, Majander I, Pihko PM (2007). Iminium catalysis. *Chem. Rev.* 107, 5416–5470.

47. Hayashi Y, *Catalytic Asymmetric Diels–Alder Reactions*. Wiley-VCH, Weinheim, 2002, pp 5–55.

48. Gotoh H, Masui R, Ogino H, Shoji M, Hayashi Y (2006). Enantioselective ene reaction of cyclopentadiene and α,β-enals catalyzed by a diphenylprolinol silyl ether. *Angew. Chem., Int. Ed.* 45, 6853–6856.

49. Gotoh H, Ogino H, Ishikawa H, Hayashi Y (2010). One-pot synthesis of chiral bicyclo[3.3.0] octatrienes using diphenylprolinol silyl ether-mediated ene-type reaction. *Tetrahedron* 66, 4894–4899.

50. Yamaguchi M, Shiraishi T, Igarashi Y, Hirama M (1994). Catalytic asymmetric Michael addition of nitroalkane to enone and enal. *Tetrahedron Lett.* 35, 8233–8236.

51. Yamaguchi M, Igarashi Y, Reddy RS, Shiraishi T, Hirama M (1997). Asymmetric Michael addition of nitroalkanes to prochiral acceptors catalyzed by proline rubidium salts. *Tetrahedron* 53, 11223–11236.

52. Hanessian S, Pham V (2000). Catalytic asymmetric conjugate addition of nitroalkanes to cycloalkenones. *Org. Lett.* 2, 2975–2978.
53. Halland N, Hazell RG, Jørgensen KA (2002). Organocatalytic asymmetric conjugate addition of nitroalkanes to α,β-unsaturated enones using novel imidazoline catalysts. *J. Org. Chem.* 67, 8331–8338.
54. Prieto A, Halland N, Jørgensen KA (2005). Novel imidazolidine-tetrazole organocatalyst for asymmetric conjugate addition of nitroalkanes. *Org. Lett.* 7, 3897–3900.
55. Mitchell CE, Brenner SE, Ley SV (2005). A versatile organocatalyst for the asymmetric conjugate addition of nitroalkanes to enones. *Chem. Commun.* 5346–5348.
56. Mitchell CE, Brenner SE, Garcia-Fortanet J, Ley SV (2006). An efficient, asymmetric organocatalyst-mediated conjugate addition of nitroalkanes to unsaturated cyclic and acyclic ketones. *Org. Biomol. Chem.* 4, 2039–2049.
57. Hojabri L, Hartikka A, Moghaddam FM, Arvidsson PI (2007). A new imidazole-containing imidazolidinone catalyst for organocatalyzed asymmetric conjugate addition of nitroalkanes to aldehydes. *Adv. Synth. Catal.* 349, 740–748.
58. Ooi T, Doda K, Maruoka K (2003). Highly enantioselective Michael addition of silyl nitronates to α,β-unsaturated aldehydes catalyzed by designer chiral ammonium bifluorides: efficient access to optically active γ-nitro aldehydes and their enol silyl ethers. *J. Am. Chem. Soc.* 125, 9022–9023.
59. Ooi T, Morimoto K, Doda K, Maruoka K (2004). Evaluation of the relationship between the catalyst structure and regio- as well as stereoselectivity in the chiral ammonium bifluoride-catalyzed asymmetric addition of silyl nitronates to α,β -unsaturated aldehydes. *Chem. Lett.* 33, 824–825.
60. Ooi T, Doda K, Takada S, Maruoka K (2006). Asymmetric Michael addition of silyl nitronates to cyclic α,β-unsaturated ketones catalyzed by chiral quaternary ammonium bifluorides: isolation and selective functionalization of enol silyl ethers of optically active γ-nitro ketones. *Tetrahedron Lett.* 47, 145–148.
61. Gotoh H, Ishikawa H, Hayashi Y (2007). Diphenylprolinol silyl ether as catalyst of an asymmetric, catalytic, and direct Michael reaction of nitroalkanes with α,β-unsaturated aldehydes. *Org. Lett.* 9, 5307–5309.
62. Palomo C, Landa A, Mielgo A, Oiarbide M, Puente A, Vera S (2007). Water-compatible iminium activation: organocatalytic Michael reactions of carbon-centered nucleophiles with enals. *Angew. Chem., Int. Ed.* 46, 8431–8435.
63. Zu L, Xie H, Li H, Wang J, Wang W (2007). Highly enantioselective organocatalytic conjugate addition of nitromethane to α,β-unsaturated aldehydes: three-step synthesis of optically active baclofen. *Adv. Synth. Catal.* 349, 2660–2664.
64. Gotoh H, Ishikawa H, Hayashi Y (2007). Diphenylprolinol silyl ether as catalyst of an asymmetric, catalytic, and direct Michael reaction of nitroalkanes with α,β-unsaturated aldehydes. *Org. Lett.* 9, 5307–5309.
65. Gotoh H, Okamura D, Ishikawa H, Hayashi Y (2009). Diphenylprolinol silyl ether as a catalyst in an asymmetric, catalytic, and direct Michael reaction of nitroethanol with α,β-unsaturated aldehydes. *Org. Lett.* 11, 4056–4059.
66. Hayashi Y, Gotoh H, Masui R, Ishikawa H (2008). Diphenylprolinol silyl ether as a catalyst in an enantioselective, catalytic, formal aza [3 + 3] cycloaddition reaction for the formation of enantioenriched piperidines. *Angew. Chem., Int. Ed.* 47, 4012–4015.
67. Bertelsen S, Johansen RL, Jørgensen KA (2008). Controlling the formation of 1 out of 64 stereoisomers using organocatalysis. *Chem. Commun.* 3016–3018.
68. Hayashi Y, Toyoshima M, Gotoh H, Ishikawa H (2009). Diphenylprolinol silyl ether catalysis in an asymmetric formal carbo [3 + 3] cycloaddition reaction via a domino Michael/Knoevenagel condensation. *Org. Lett.* 11, 45–48.

69. Hayashi Y, Itoh T, Ishikawa H (2011). Oxidative and enantioselective cross-coupling of aldehydes and nitromethane catalyzed by diphenylprolinol silyl ether. *Angew. Chem., Int. Ed.* 50, 3920–3924.

70. Hayashi Y, Samanta S, Gotoh H, Ishikawa H (2008). Asymmetric Diels–Alder reactions of α,β-unsaturated aldehydes catalyzed by a diarylprolinol silyl ether salt in the presence of water. *Angew. Chem., Int. Ed.* 47, 6634–6637.

71. Rideout DC, Breslow R (1980). Hydrophobic acceleration of Diels–Alder reactions. *J. Am. Chem. Soc.* 102, 7816–7817.

72. Breslow R (1991). Hydrophobic effects on simple organic reactions in water. *Acc. Chem. Res.* 24, 159–164.

73. Narayan S, Muldoon J, Finn MG, Fokin VV, Kolb HC, Sharpless KB (2005). "On water": unique reactivity of organic compounds in aqueous suspension. *Angew. Chem., Int. Ed.* 44, 3275–3279.

74. Jung Y, Marcus RA (2007). On the theory of organic catalysis "on water". *J. Am. Chem. Soc.* 129, 5492–5502.

75. Chanda A, Fokin VV (2009). Organic synthesis "on water". *Chem. Rev.* 109, 725–748.

76. Hayashi Y, Sumiya T, Takahashi J, Gotoh H, Urushima T, Shoji M (2006). Highly diastereo- and enantioselective direct aldol reactions in water. *Angew. Chem., Int. Ed.* 45, 958–961.

77. Hayashi Y, Aratake S, Okano T, Takahashi J, Sumiya T, Shoji M (2006). Combined proline-surfactant organocatalyst for the highly diastereo- and enantioselective aqueous direct cross-aldol reaction of aldehydes. *Angew. Chem., Int. Ed.* 45, 5527–5529.

78. Aratake S, Itoh T, Okano T, Nagae N, Sumiya T, Shoji M, Hayashi Y (2007). Highly diastereo- and enantioselective direct aldol reactions of aldehydes and ketones catalyzed by siloxyproline in the presence of water. *Chem. Eur. J.* 13, 10246–10256.

79. Hayashi Y (2006). In water or in the presence of water? *Angew. Chem., Int. Ed.* 45, 8103–8104.

80. Mase N, Nakai Y, Ohara N, Yoda H, Takabe K, Tanaka F, Barbas, III, CF (2006). Organocatalytic direct asymmetric aldol reactions in water. *J. Am. Chem. Soc.* 128, 734–735.

81. Fehr C, Magpantay I, Arpagaus J, Marquet X, Vuagnoux M (2009). Enantioselective synthesis of (−)-β-santalol by a copper-catalyzed enynol cyclization–fragmentation reaction. *Angew. Chem., Int. Ed.* 48, 7221–7223.

82. Shibasaki M, Kanai M (2008). Synthetic strategies for oseltamivir phosphate. *Eur. J. Org. Chem.* 1839–1850.

83. Ishikawa H, Suzuki T, Hayashi Y (2009). High-yielding synthesis of the anti-influenza neuramidase inhibitor (–)-oseltamivir by three "one-pot" operations. *Angew. Chem., Int. Ed.* 48, 1304–1307.

84. Ishikawa H, Hayashi Y (2010). Three-step synthesis of antiinfluenza drug oseltamivir. Total synthesis using diphenylprolinol silyl ether as catalyst. *Kagaku to Seibutsu* 48, 156–157.

85. Ishikawa H, Suzuki T, Orita H, Uchimaru T, Hayashi Y (2010). High-yielding synthesis of the anti-influenza neuraminidase inhibitor (–)-oseltamivir by two "one-pot" sequences. *Chem. Eur. J.* 16, 12616–12626.

86. Ishikawa H, Bondzic BP, Hayashi Y (2011). Synthesis of (−)-oseltamivir by using a microreactor in the Curtius rearrangement, *Eur. J. Org. Chem.* 6020–6031.

87. Yeung YY, Hong S, Corey EJ (2006). A short enantioselective pathway for the synthesis of the anti-influenza neuramidase inhibitor oseltamivir from 1,3-butadiene and acrylic acid. *J. Am. Chem. Soc.* 128, 6310–6311.

88. Fukuta Y, Mita T, Fukuda N, Kanai M, Shibasaki M (2006). De novo synthesis of tamiflu via a catalytic asymmetric ring-opening of meso-aziridines with $TMSN_3$. *J. Am. Chem. Soc.* 128, 6312–6313.

89. Cong X, Yao ZJ (2006). Ring-closing metathesis-based synthesis of (3*R*,4*R*,5*S*)-4-acetylamino-5-amino-3-hydroxy-cyclohex-1-ene-carboxylic acid ethyl ester: a functionalized cycloalkene skeleton of GS4104. *J. Org. Chem.* 71, 5365–5368.

90. Satoh N, Akiba T, Yokoshima S, Fukuyama T (2007). A practical synthesis of (−)-oseltamivir. *Angew. Chem., Int. Ed.* 46, 5734–5736.

91. Bromfield KM, Graden H, Hagberg DP, Olsson T, Kann N (2007). An iron carbonyl approach to the influenza neuraminidase inhibitor oseltamivir. *Chem. Commun.* 3183–3185.

92. Pei Z, Li X, von Geldern TW, Madar DJ, Longenecker K, Yong H, Lubben TH, Stewart KD, Zinker BA, Backes BJ, Judd AS, Mulhern M, Ballaron SJ, Stashko MA, Mika AK, Beno DWA, Reinhart GA, Fryer RM, Preusser LC, Kempf-Grote AJ, Sham HL, Trevillyan JM (2006). Discovery of ((4*R*,5*S*)-5-amino-4-(2,4,5-trifluorophenyl)cyclohex-1-enyl)-(3-(trifluoromethyl)-5,6-dihydro-[1,2,4]triazolo[4,3-*a*]pyrazin-7(8*H*)-yl)methanone (ABT-341), a highly potent, selective, orally efficacious, and safe dipeptidyl peptidase IV inhibitor for the treatment of type 2 diabetes. *J. Med. Chem.* 49, 6439–6442.

93. Albrecht L, Jiang H, Jørgensen KA (2011). A simple recipe for sophisticated cocktails: organocatalytic one-pot reactions—concept, nomenclature, and future perspectives. *Angew. Chem., Int. Ed.* 50, 8492–8509.

14

ORGANOCATALYSIS FOR ASYMMETRIC SYNTHESIS: FROM LAB TO FACTORY

FENG XU

Department of Process Research, Merck Research Laboratories, Rahway, NJ, USA

14.1 INTRODUCTION

The terminology of "organocatalyst" discussed in this chapter refers to small *chiral* organic molecules that catalyze and facilitate asymmetric transformations. Since the coining of the name "organocatalysis" by MacMillan in 2000 [1], the field of asymmetric organocatalysis has blossomed dramatically [2]. The impact of organocatalysis on synthetic chemistry over the past decade has been revolutionary, though the general concept of using small chiral organic molecules to catalyze asymmetric transformations is not new. In 1954, for example, Prelog reported the use of cinchona alkaloids for the asymmetric preparation of cyanohydrins [3]. However, prior to the recent "gold rush" of organocatalysis, which was catalyzed by the work of List [4] and MacMillan [1] at the turn of the millennium, two organocatalysis processes suitable for large-scale synthesis are noteworthy [5].

In the early 1970s, the Hajos–Parrish–Eder–Sauer–Wiechert reaction, an asymmetric aldol reaction catalyzed by proline through an enamine species for the preparation of synthetically useful progesterone intermediates (Scheme 14-1), was discovered independently by two groups [5a,b,c,d] at Schering AG and Hoffmann–La Roche. However, detailed mechanistic understanding of this reaction was not realized at the time until List published his results in 2003 [6]. For nearly three decades, the significance of this discovery was therefore not recognized, leaving the field of enamine-mode organocatalysis largely undeveloped.

In 1984, Dolling and coworkers at Merck reported the first example of an asymmetric alkylation by applying a chiral phase-transfer catalyst, a cinchonine-derived quaternary ammonium salt [5e,f]. Thorough mechanistic studies led to the discovery of the fine-tuned catalyst **5** for substrate **4** in terms of enantioselectivity. The high enantioselectivity

Sustainable Catalysis: Challenges and Practices for the Pharmaceutical and Fine Chemical Industries, First Edition. Edited by Peter J. Dunn, K. K. (Mimi) Hii, Michael J. Krische, and Michael T. Williams.

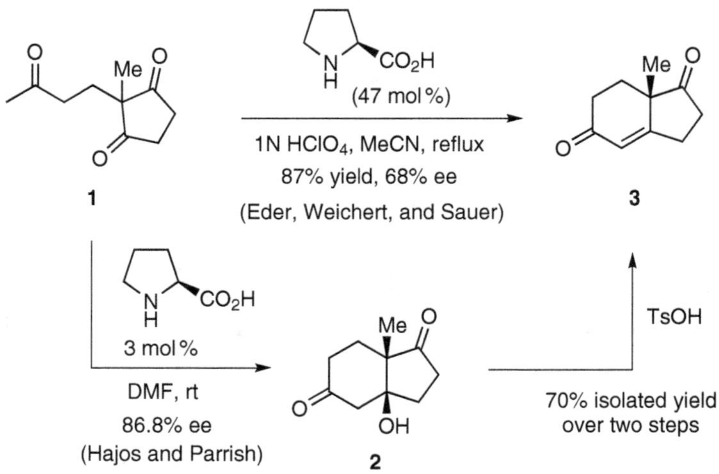

SCHEME 14-1. Organocatalysis via an enamine intermediate.

achieved was rationalized by the preferential formation of ion-pair complex **6** through a combination of ionic, hydrogen-bonding, and $\pi-\pi$ stacking interactions (Scheme 14-2). More importantly, a practical process based on this discovery was further developed and applied at large scale to the synthesis of (+)-indacrinone **8**, a potential uricosuric-diuretic agent [5e,f].

Since the "gold rush" of organocatalysis, numerous novel reactions and transformations, including some that were difficult to realize previously, can now effectively be achieved by applying small organic molecules tailored to mimic biological processes [2]. To date, the applications of organocatalysis in the pharmaceutical industry have been

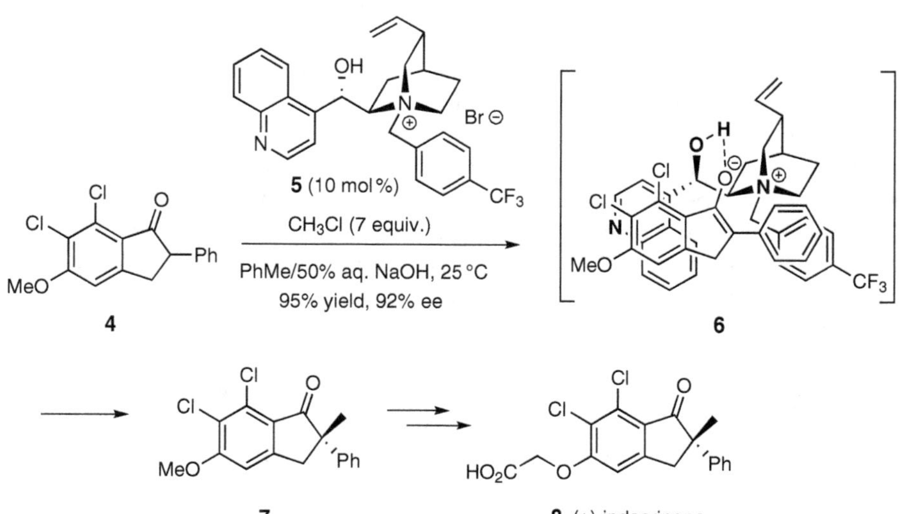

SCHEME 14-2. Enantioselective phase-transfer-catalyzed alkylation.

focused primarily in the area of medicinal chemistry to discover drug candidates on a small laboratory scale [2f,h]. Nevertheless, implementation of practical organocatalytic processes on industrial scale, which is recognized as one of the expected important advances in this field, has been emerging [7–11]. In addition, the development and demonstration of scaleable processes on a laboratory scale by academic researchers also reveal the wide range of possible applications of recently discovered novel organocatalysts [12].

It is important to recall that the use of small achiral organic molecules to catalyze classical organic reactions on a large scale has been well established. For example, *N,N*-dimethylaminopyridine (DMAP), one of the most popular achiral catalysts, has been widely used on an industrial scale to catalyze ester or amide synthesis by formation of an acyl pyridinium species in the presence of a carboxylic acid halide or anhydride. In principle, application of these recently developed organocatalysts for asymmetric synthesis on large scale can be achieved, when those typical development issues including the availability and cost of organocatalysts, catalyst loading, and stability under reaction conditions are managed effectively.

Whereas the interests of academia are usually to explore the general scope and novelty of a new reaction, the typical development issues encountered in industry are often substrate specific, as the synthetic goal for a large-scale industrial application is specific and explicit. Hence, process development in industry usually focuses on optimizing a substrate-specific transformation, in lieu of exploring the scope of the application, the general knowledge of which gained during the development is considered as a "nice-to-have bonus" to the chemistry community.

This chapter describes two recent applications [7,8] of organocatalysis on an industrial scale at Merck and also addresses solutions to development issues encountered, from the lab scale to the industrial scale. The idea of what constitutes "large scale" is subjective, but in this context, the industrial scale refers to reactions that have been demonstrated on pilot plant scale and above.

14.2 PREPARATION OF TELCAGEPANT, AN APPLICATION OF IMINIUM ORGANOCATALYSIS

14.2.1 Background and Synthetic Strategy

Migraine is a neurovascular disorder characterized by severe, debilitating, and throbbing unilateral headache [13]. Though a leading cause of disability, migraine is a highly prevalent disease with a clear unmet medical need [13]. Telcagepant (**9**) [14], a selective and potent neuropeptide calcitonin gene-related peptide (CGRP) receptor antagonist, is a late-stage clinical candidate for the acute treatment of migraine [15].

Retrosynthetic analysis of the manufacturing route of telcagepant (**9**) is shown in Scheme 14-3. The key challenge in synthesizing **9** is how to prepare caprolactam **11** efficiently [16]. As precursor **12** could be obtained via a Doebner–Knoevenagel-type coupling, Xu *et al.* at Merck [7] envisioned that an enantioselective addition of $MeNO_2$ to an α,β-unsaturated iminium intermediate formed by activating cinnamaldehyde **15** with an organocatalyst would provide the precursor **13** in one operation. As such, the desired C-6 (*S*) stereochemistry in caprolactam **11** could be established effectively.

The advent of organocatalysis has had an impact on synthetic chemistry at an extraordinary level. Rather than discovered, asymmetric iminium catalysis was the first

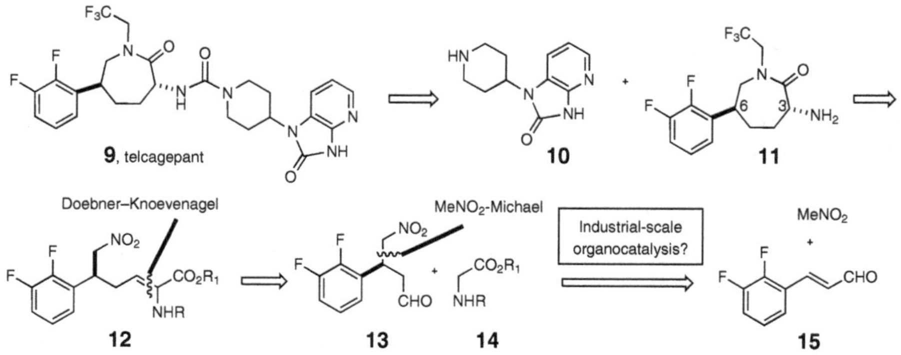

SCHEME 14-3. Retrosynthetic analysis of telcagepant **9**.

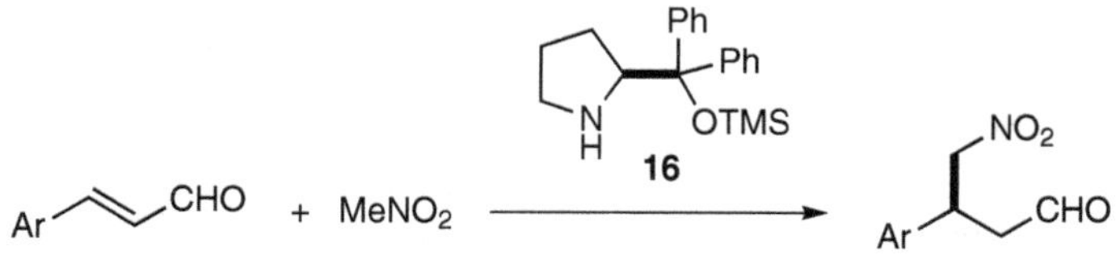

SCHEME 14-4. Asymmetric nitromethane Michael addition.

organocatalytic activation mode *designed* for organic synthesis [2f]. However, application of iminium organocatalysis on a large scale had not been realized at the time of this work [2]. With emerging interest in this relatively new field, in 2007 Hayashi first achieved asymmetric 1,4-addition of nitromethane to α,β-unsaturated aldehydes via a chiral iminium-ion species in the presence of a chiral amine catalyst (Scheme 14-4, see also Chapter 13) [17a]. Interestingly, around the same time four other groups [17b–e] independently confirmed that the use of Jørgensen–Hayashi catalyst **16** was the organocatalyst of choice for this transformation in terms of enantioselectivity.

14.2.2 Preliminary Results, Identification of By-products, and Reaction Pathway Consideration

The first attempt at an asymmetric addition of $MeNO_2$ to substrate **15** under Hayashi's conditions (**16**, 20 mol% $PhCO_2H$, MeOH) [17a] did give a promising result (50–75% yield) and, more importantly, confirmed that high enantioselectivity (95% ee) could be achieved by using the Jørgensen–Hayashi catalyst **16** [18]. However, after further scrutinizing the reaction conditions and impurity profile, it was quickly realized that the use of alcohol solvents led to the formation of several undesired by-products in the presence of the acidic cocatalyst $PhCO_2H$ (Scheme 14-5).[1] Specifically, formed by reacting alcohol with starting aldehyde **15** as well as product aldehyde **13**, the unwanted acetal by-products **17** and **18** could be present at a level of up to 20%, varying from batch to batch during the development work.

[1] $PhCO_2H$ was the optimal acid of choice in Hayashi's system, see ref. 17a.

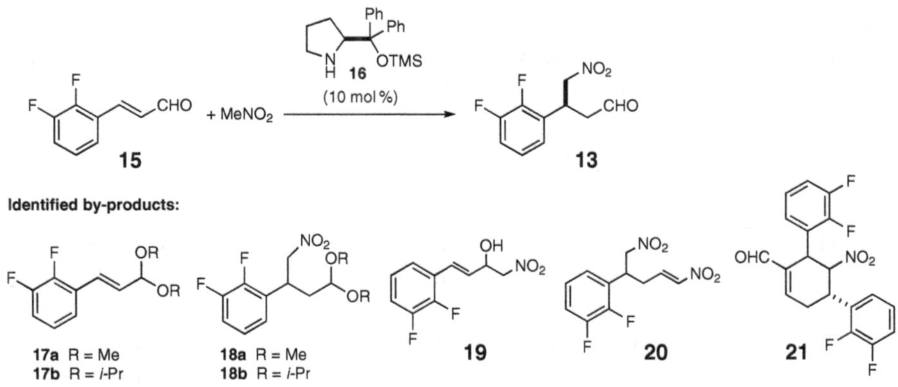

SCHEME 14-5. Asymmetric addition of $MeNO_2$ to **15** and some identified by-products.

To further understand the reaction pathways and therefore to suppress impurity formation, several additional by-products **19**, **20**, and **21** were isolated and characterized. The plausible pathways for the formation of these by-products are shown in Scheme 14-6. The formation of alcohol **19** could be via 1,2-addition of $MeNO_2$ to **15**. Not surprisingly, the product **13** could further interact with catalyst **16** to form an iminium species, which could react with $MeNO_2$ to give **20**, or be captured by alcohol solvents to form acetal **18**. On the other hand, benzoic acid could also be responsible for the formation of acetal **17** and **18** without involving an iminium species in the presence of aldehyde **15** and **13** in an alcohol solvent. It was also interesting to observe that the desired aldehyde product **13**

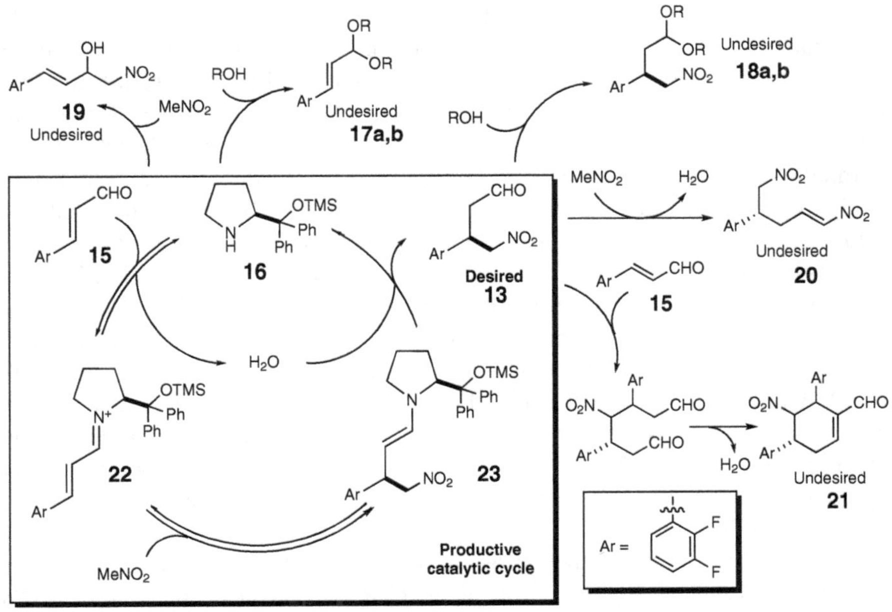

SCHEME 14-6. Plausible pathways for the formation of aldehyde **13** and related impurities.

could also further react with starting aldehyde **15** or intermediate **22** followed by an intramolecular condensation to give **21**. With this understanding in mind, further optimization for the specific substrate **15** was initiated.

14.2.3 "Cocktail" Cocatalysts in Non-alcohol Solvents

The solvent compatibility issue had to be addressed first. A quick solvent survey, however, confirmed that in Hayashi's system the use of dry alcohol solvents in combination with **16** and $PhCO_2H$ was crucial and optimal to obtain relatively high yields and reasonable reaction rate. Although the reaction rate to form acetals was slower in *i*-PrOH than in MeOH, the acetal formation was inevitable and varied from batch to batch during development work. The lack of control of the formation of these side reactions was a serious issue, which would lead to inability to predict the reaction performance upon scale-up. Attempts to suppress the formation of acetal by-products by varying the temperature were not successful. At this point, it became clear that this approach would not be suitable for further development, unless a new system in non-alcoholic solvents was developed.

Further studies in dry aprotic solvents revealed that the reaction proceeded in very low conversion (<30%) or gave a complicated reaction profile (Table 14-1, entry 3). However, high conversion could be achieved in aqueous aprotic solvents. Among the various solvents screened, aqueous THF was identified as the most promising solvent system.

Finally, a unique, effective "cocktail" catalyst system (PivOH/$B(OH)_3$/5 mol% **16**) was discovered. The use of pivalic acid (PivOH) was essential to achieve a higher reaction rate and conversion; however, overcharging PivOH also had an adverse effect on the stability of product **13**, giving rise to a higher level of impurities including **21** (Table 14-1, entry 6). In contrast, the use of $B(OH)_3$ as a cocatalyst led to fewer impurities, albeit at the expense of a much slower reaction rate (Table 14-1, entry 7). Naturally, it was concluded at this point that the use of a combination of the weak acids PivOH/$B(OH)_3$ would balance reaction efficiency and product stability. As a result, the PivOH charge was further reduced from 15 mol% to 5 mol% in the presence of 50 mol% of $B(OH)_3$ without sacrificing the reaction rate or conversion (Table 14-1, entry 8), while the formation of impurities, especially **21**, was successfully suppressed to improve the yield significantly. The use of the combination of PivOH/$B(OH)_3$ significantly suppressed the formation of the previously observed by-products and improved the reaction rate. It was believed that both the formation of the iminium ion intermediate **22** (from the condensation of **15** and **16**, Scheme 14-6) and hydrolysis of the subsequent enamine intermediate **23** (from the addition of $MeNO_2$ to the iminium ion intermediate) to liberate the catalyst **16** were accelerated with the help of the acidic cocatalyst PivOH/$B(OH)_3$ [2], while weak acids such as PivOH and $B(OH)_3$ did not interfere with the proper function of amine catalyst **16** [19]. As a result, the use of a combination of weak acids PivOH (5 mol%)/$B(OH)_3$ (50 mol%) allowed for the balance of the reaction efficiency and product stability. The formation of the by-products, such as **20** and **21**, was thus suppressed significantly.

In addition, with a reduced amount of nitromethane (4 equiv.), the reaction yield was decreased to 66%. Increasing the nitromethane charge to 8 equiv. improved the yield to 74%; however, the reaction became sluggish.

TABLE 14-1. Selected Results of Asymmetric 1,4-Addition of $MeNO_2$ in the Presence of 5 mol% of 16

Entry	Cocatalysts	$MeNO_2$ (equiv.)	Solvents	Conditions	Conversion (%)[a]	Yield (%)[a]	ee (%)[a]	By-products (%)
1	20 mol% $PhCO_2H$	6	MeOH	10 °C, 17 h	99	~50–75	95	**17a** (2–20), **19** (<0.5), **20** (1), **21** (2)
2	10 mol% $PhCO_2H$	6	*i*-PrOH	20 °C, 25 h	98	~55–70	94	**17b** (1–15), **19** (<1), **20** (1), **21** (7.9)
3	10 mol% $PhCO_2H$	6	MeCN	20 °C, 14 h	99	~20		**19** (<1), **20** (3.9), **21** (72)
4	–	3	10% H_2O–THF	20 °C, 63 h	96	58	95	**19** (<1), **21** (2.9)
5	10 mol% $PhCO_2H$	6	25% H_2O–THF	20 °C, 14 h	31	N/A	N/A	
6	15 mol% PivOH	6	25% H_2O–THF	20 °C, 14 h	94	65	96	**19** (<1), **20** (1.0), **21** (7.2)
7	100 mol% $B(OH)_3$	5	25% H_2O–THF	20 °C, 67 h	85	61	95	**19** (<1), **20** (1), **21** (0.7)
8	5 mol% PivOH, 50 mol% $B(OH)_3$	6	17% H_2O–THF	20 °C, 29 h	94	73	95	**19** (<0.1), **20** (1.0), **21** (2.8)
9	5 mol% PivOH, 25 mol% $B(OH)_3$	6	25% H_2O–THF	20 °C, 23 h	88	72	95	**19** (<0.1), **20** (0.8), **21** (4.4)
10	5 mol% PivOH, 100 mol% $B(OH)_3$	6	25% H_2O–THF	20 °C, 23 h	92	71	95	**19** (<0.1), **20** (0.9), **21** (2.9)
11	5 mol% PivOH, 50 mol% $B(OH)_3$	4	25% H_2O–THF	20 °C, 22 h	97	66	95	**19** (<0.1), **20** (0.9), **21** (3.5)
12	5 mol% PivOH, 50 mol% $B(OH)_3$	8	25% H_2O–THF	20 °C, 39 h	96	74	95	**19** (<0.1), **20** (1.6), **21** (2.9)

[a]The enantiomeric excess (ee), yield, and *Z*:*E* ratio were determined by high performance liquid chromatography (HPLC) analysis.

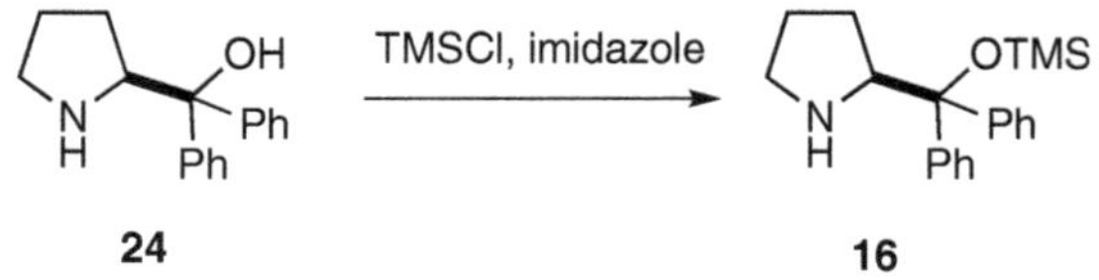

SCHEME 14-7. Preparation of Jørgensen–Hayashi catalyst **16**.

14.2.4 The Use of Crude Jørgensen–Hayashi Catalyst

Various catalysts were screened to identify an optimal catalyst for this transformation. For example, modifications of the structure of **16** were explored, including replacing the TMS group with TBS or TES and replacing the phenyl group with various aromatic rings to study the electronic as well as the steric effects. Other types of organocatalysts outside the pyrrolidine skeleton were also screened. To date, thorough screening confirmed that the use of catalyst **16** was deemed optimal for the preparation of **13**.

The availability of an organocatalyst, as well as its purity impact on the reaction profile, is also a factor that needs to be addressed for a large-scale industrial application. Although the Jørgensen–Hayashi catalyst **16** is commercially available, it was relatively expensive at the time. In addition, the catalyst **16** is glue-like thick oil, which is not easy to handle for a large-scale application. Purification of thick oil-like, noncrystalline catalyst **16** was not ideal to carry out on scale and should be avoided. Thus, studies to develop a robust process to prepare **16** were initiated.

Initial studies revealed that prolinol **24**, which is the precursor to prepare **16**, had a much lower reactivity than **16** to catalyze the asymmetric nitromethane Michael addition. This observation made it possible for the Merck team to utilize a crude catalyst **16** for the preparation of **13** without sacrificing the enantioselectivity and yield, as a small amount of the unconsumed prolinol **24** remaining in the reaction system was not active enough to be competitive with the reaction pathway catalyzed by **16**. Thus, after aqueous workup, the crude catalyst **16**, prepared by reacting prolinol **24** with TMSCl (1.3 equiv.) in the presence of imidazole (1.7 equiv.) in THF at 50 °C (Scheme 14-7) in 98% yield, was used directly to catalyze the asymmetric nitromethane Michael addition.

In terms of stability, the solution of **16** could be stored for several months without significant decomposition, which made the large-scale application of **16** more user-friendly. In addition, the crude reaction mixture of **16** without any aqueous workup could also be directly used for the nitromethane addition; albeit, the chemical yield of Michael addition could drop by 2–3%.

14.2.5 Evolution to a Streamlined Through-process

Distinct from a typical metal-catalyzed reaction, an organocatalytic process could be much less sensitive or less easily poisoned by process impurities. Indeed, as the reaction pathways were harnessed under the cocatalyst system, the iminium organocatalysis performed extremely robustly even with the *crude* **16** and *unpurified* substrate **15**. This result made it possible to develop a more effective manufacturing through-process to prepare **13** without isolation of any intermediates.

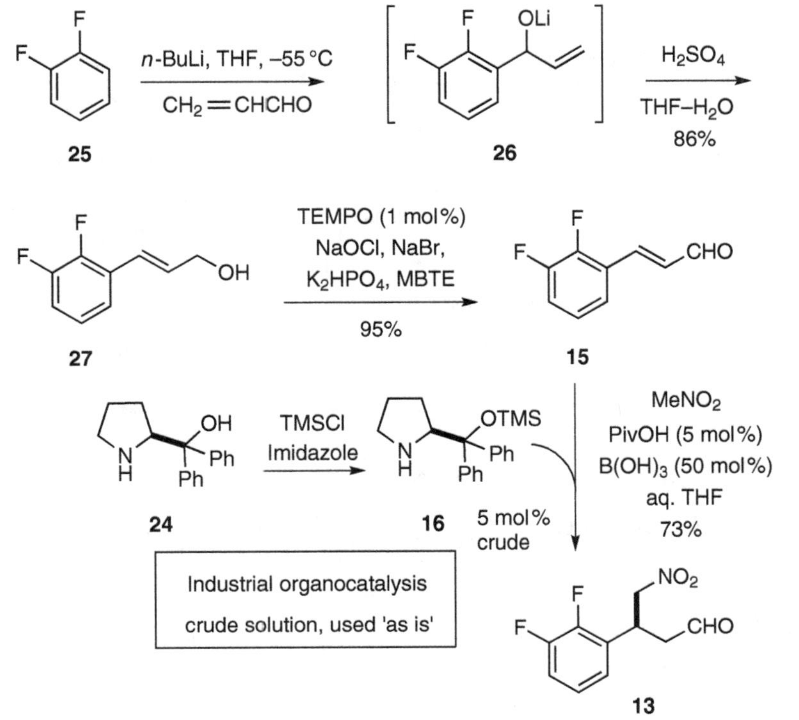

SCHEME 14-8. Through-process to prepare aldehyde **13** without isolation of intermediates[2].

Starting from 1,2-difluorobenzene **25**,[2] **15** was efficiently prepared in a two-pot process (Scheme 14-8).[3] The crude stream **15** was directly used, without further purification, for the subsequent organocatalytic nitromethane Michael addition under optimal conditions, which involved the use of nitromethane (6 equiv.) in 17% H_2O/THF in the presence of **16** (5 mol%), PivOH (5 mol%), and $B(OH)_3$ (50 mol%). Typically, after 30 h at ambient temperature, **13** could be reproducibly prepared in 73% assay yield and 95% ee. The robustness and reproducibility were successfully achieved at >250 kg scale. To the best of our knowledge, this represents the first application of iminium organocatalysis on an industrial scale.

To minimize the degradation of the relatively unstable nitroaldehyde **13**, in practice, the crude solution of **13** was directly used for the next step without purification. Through an amine-catalyzed Doebner–Knoevenagel condensation, enamine acid **29a** was prepared. The unprecedented Doebner–Knoevenagel condensation developed by the Merck team is also via an iminium species. Interestingly, bulky amine **16**, the organocatalyst used for the asymmetric Michael addition of nitromethane, was not able to promote the Doebner–Knoevenagel condensation, possibly due to the steric hindrance.[4] After screening,

[2] Unless otherwise mentioned, the processes described here were successfully carried out on >250 kg scale.

[3] Aldehyde **15** is a volatile solid and hence was not suitable for isolation via drying by suction or vacuum oven; however, its loss through concentration of solvents under reduced pressure was negligible.

[4] It was not surprising that the Doebner-Knoevenagel reaction could not be catalyzed by the bulky amine **16**. In addition, the residue of **16** in the crude stream also did not affect the performance of the subsequent reaction.

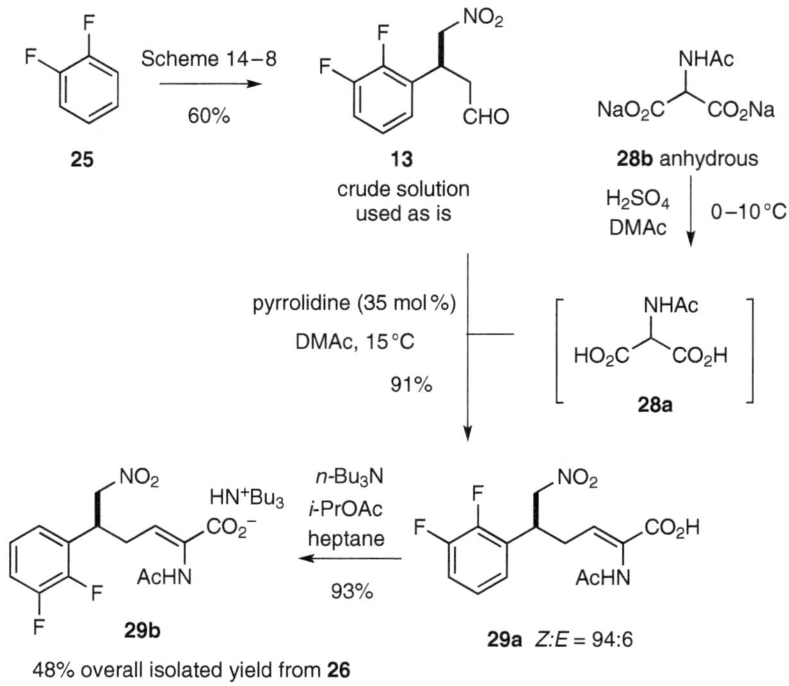

SCHEME 14-9. Five-step through-process to prepare **29b** without isolation of intermediates.

pyrrolidine, which is much smaller in size in comparison with **16**, was determined as an optimal catalyst for the condensation of aldehyde **13** and acid **28a**. Acid **28a** was generated from the treatment of the corresponding sodium salt **28b** with sulfuric acid *in situ*. Thus, the first isolated crystalline solid in the synthesis of telcagepant, tri-*n*-butylamine salt **29b**, was obtained in >99% purity and >99% ee in 48% yield over a five-step through-process (Scheme 14-9). A significant ee upgrade was also achieved during the isolation of **29b**.

14.2.6 Completion of Telcagepant Synthesis

Hydrogenation of **29b** in the presence of Pd–C, LiCl, and H_2SO_4 in *i*-PrOH afforded the desired ester **30a**. The use of LiCl suppressed the formation of corresponding *meta*-des F by-product to <0.2%, which was important in preparing pure telcagepant for pharmaceutical use. Without aqueous workup, the crude hydrogenated ester **30a** was selectively monoalkylated, followed by a basic aqueous workup to give acid **30b**. *t*-BuCOCl promoted lactamization, followed by an NaOH-catalyzed, dynamic crystallization gave the desired *trans* caprolactam acetamide **31** in >99% purity and >99% ee in 73% isolated yield over a five-step through-process without isolation of any intermediates (Scheme 14-10). Deprotection of **31** in aq *i*-PrOH–HCl gave the desired penultimate intermediate **11**, which was isolated as its HCl salt MTBE solvate in >99% purity with 83% isolated yield.

Finally, treatment of **11** with 1,1′-carbonyldiimidazole (CDI) in the presence of Et_3N followed by addition of **10** afforded telcagepant (**9**) [16], which was isolated as its potassium salt EtOH solvate in 92% yield with >99.8% purity and >99.9% ee. Isolating only three intermediates, the overall yield of this cost-effective synthesis is 27%.

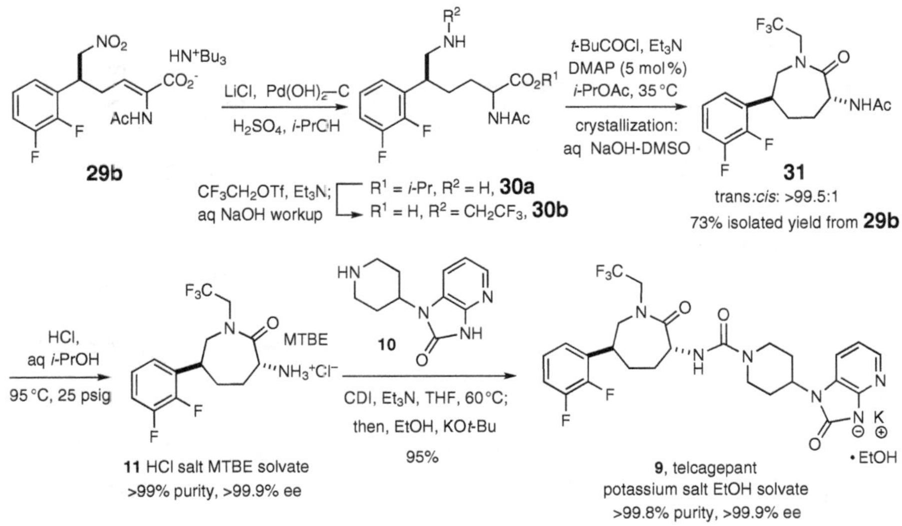

SCHEME 14-10. Synthesis of telcagepant.

The key to the success in the development of the organocatalytic Michael addition of nitromethane was to develop a unique, practical organocatalyst "cocktail" system (5 mol% PivOH/50 mol% $B(OH)_3$/5 mol% **16**) to harness the reaction pathways. The C-6 stereogenic center was thus practically and effectively established in one operation in >95% ee for the synthesis of telcagepant, via an iminium species.

14.3 PREPARATION OF MK-8613, APPLICATION OF ASYMMETRIC MICHAEL ADDITION CATALYZED BY DESMETHYL QUINIDINE

14.3.1 Synthetic Target and Strategy Analysis

MK-8613 **32** is a potent, selective protease dipeptidyl peptidase IV (DPP-4) inhibitor and a drug candidate for the treatment of type 2 diabetes mellitus (T2DM). The molecule contains a piperidine ring, which is fused in a tricyclic system and has two adjacent stereogenic centers in a *trans* relationship. The previously described synthesis of **32** was racemic and required chromatographic resolution of a late-stage racemic intermediate [20]. Asymmetric synthesis of the unique skeleton of MK-8613, as well as similar structures, was lacking at the inception of this work [20]. Xu and coworkers at Merck first developed an efficient approach to construct the piperidine-fused imidazopyridine skeleton with the desired stereochemistry and functionality through a novel three-component cascade coupling in one step [21]. As such, the retrosynthetic strategy [8] was eventually simplified to install the C-4 stereogenic center first by applying an enantioselective Michael addition on the corresponding nitrostyrene substrate **36**, which could be prepared in >80% yield through a modified Henry reaction [8] from commercially available 2,4,5-trifluorobenzaldehyde (Scheme 14-11) [22].

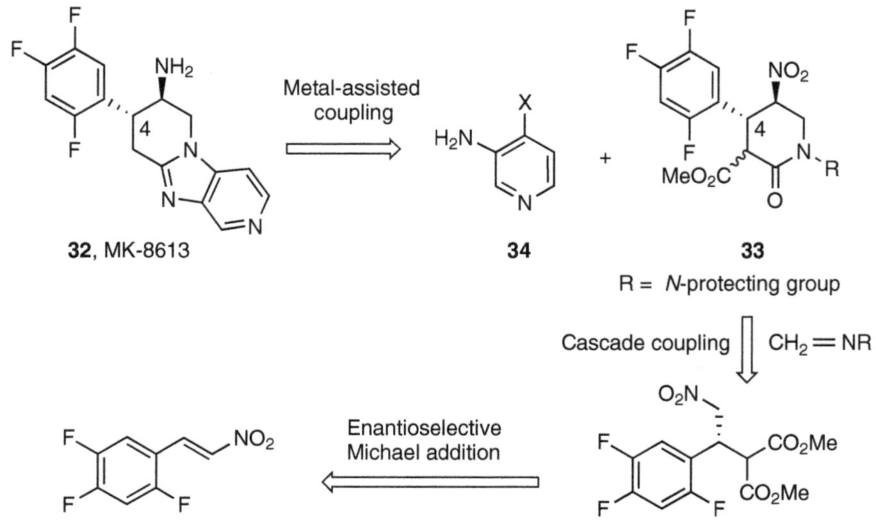

32, MK-8613 **34** **33** **36** **35**

SCHEME 14-11. Retrosynthetic analysis of MK-8613.

Hence, studies were initiated to set up the desired tertiary stereogenic center of the key intermediate **35** by applying an enantioselective Michael addition of dimethyl malonate on nitrostyrene **36** (Scheme 14-12). In fact, several catalysts had been reported to realize this type of Michael addition with high enantioselectivity at the time [23,24]. Preliminary studies showed that the use of Deng's catalyst [23d] gave a satisfactory result for the preparation of **35** in terms of ee and conversion, when 10 mol% of desmethyl quinidine (DMQ, **37**), a typical loading reported in the literature, was used. Other metal catalysts [23] also gave satisfactory results. However, the use of Deng's cinchona alkaloid derivatives [23d], which would result in a process free of heavy metal, is enticing among the methodologies reported and explored, because heavy metal residues in pharmaceutical products (typically <10 ppm) are strictly regulated by agencies, such as the FDA. In contrast, the rejection of organic reagents including organocatalysts (not limited to ppm level) can be achieved easily. To turn these initial proof-of-concept results into a process viable for the preparation of **32** on an industrial scale, several issues such as the availability

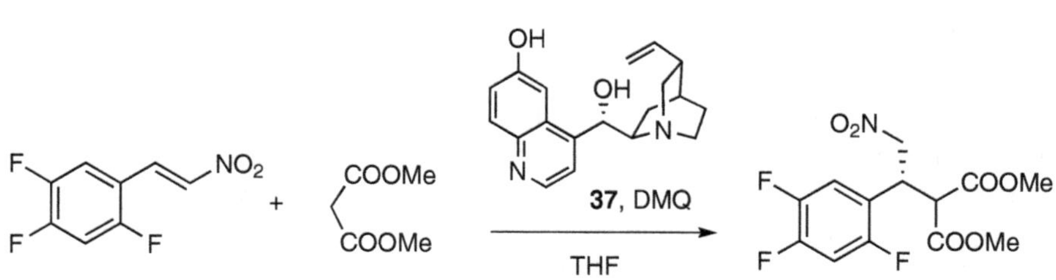

36 **35**

SCHEME 14-12. Asymmetric Michael addition of dimethyl malonate to nitrostyrene **36**.

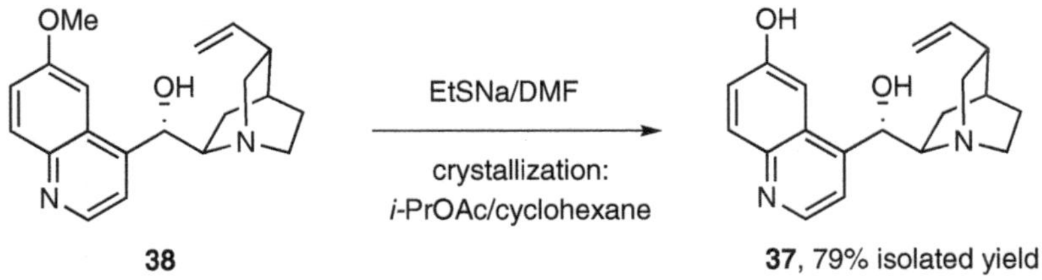

SCHEME 14-13. Preparation of catalyst **37**.

of the DMQ catalyst **37**, turnover number/organocatalyst loading, as well as the other process parameters had to be further evaluated.

14.3.2 A Practical Process to Prepare Catalyst DMQ

For large-scale applications, the availability of the reagents and raw starting materials is usually an important factor that cannot be overlooked. One of the more appealing advantages of applying Deng's methodology is the easy accessibility of desmethyl quinidine (DMQ) [23d], which can be prepared from quinidine (**38**) in one step. Adaptation of the literature procedure permitted the direct crystallization of **37** in 79% yield from *i*-PrOAc/cyclohexane upon completion of desmethylation (Scheme 14-13).

14.3.3 Substrate-Specific Process Evolution

Again, instead of seeking general application and scope, industrial large-scale process development focuses on developing a customized, substrate-specific process. Although Deng *et al.* [23d,25] had studied in-depth the application of DMQ to catalyze various types of reactions, further scrutiny of the nitromethane Michael addition chemistry with the desired specific substrate **36** was still warranted. Therefore, optimizations to lower catalyst loading, to increase volumetric productivity, and to maintain an adequate reaction rate were initiated.

Several important insights were gained during the development. The temperature effect on this transformation was investigated first. Not surprisingly, lower reaction temperatures resulted in higher ee. However, the reaction became sluggish below −25 °C. A plot of reaction enantioselectivity versus temperature is shown in Figure 14-1. The optimal temperature range was thus determined to be between −20 and −15 °C.

Interestingly, water content greatly impacted the enantioselectivity of the organocatalytic Michael addition. The presence of a small amount of water significantly decreased the enantioselectivity, while keeping the reaction system dry at −20 °C yielded **35** in 95% ee reproducibly. A plot of reaction enantioselectivity versus water content is shown in Figure 14-2.

The observed impact of water content on enantioselectivity is consistent with the postulated reaction mechanism. The plausible transition state **39** rationalizes the origin of the enantioselectivity of the Michael addition on nitrostyrenes (Figure 14-3). In this model, cinchona alkaloid DMQ **37** adopts a gauche-open conformer to simultaneously activate and orient the Michael donor (dimethyl malonate) and the acceptor (nitrostyrene **36**) by means of a network of hydrogen-bonding interactions [25]. The substituents of the two

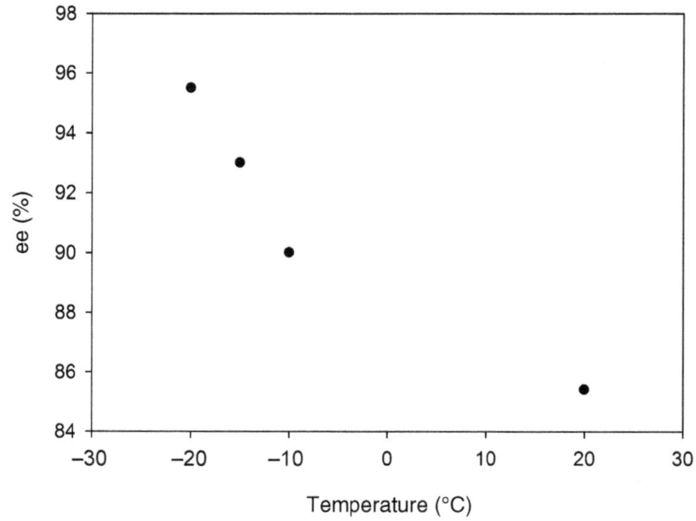

FIGURE 14-1. Temperature effect on enantioselectivity in the presence of 2 mol% DMQ **37** in dry THF.

bond-forming carbons are in a staggered rather than an eclipsed arrangement. Under an acid–base bifunctional catalysis mode, hydrogen-bonding interactions between the catalyst DMQ **37** and the two substrates play a pivotal role in aligning the reaction components perfectly for further reaction. As such, under anhydrous conditions, side reactions that bypass the acid–base bifunctional catalysis mode are minimized, and high enantioselectivity of the corresponding Michael addition could be achieved. However, in a "wet" reaction

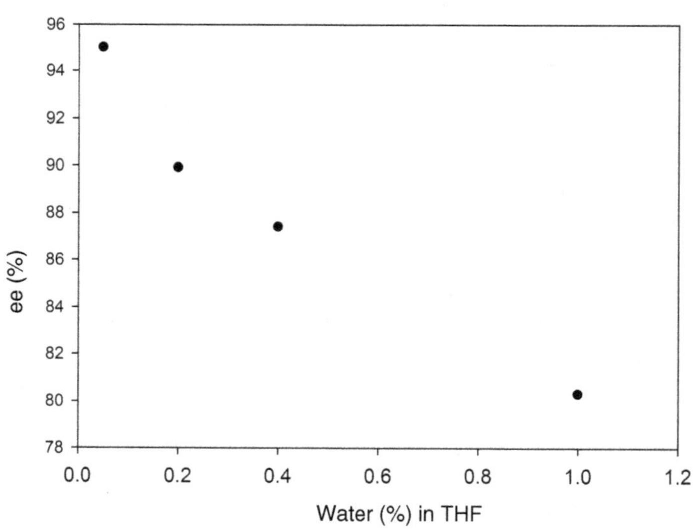

FIGURE 14-2. The effect of water content on enantioselectivity in the presence of 1 mol% DMQ **37** at −15 °C.

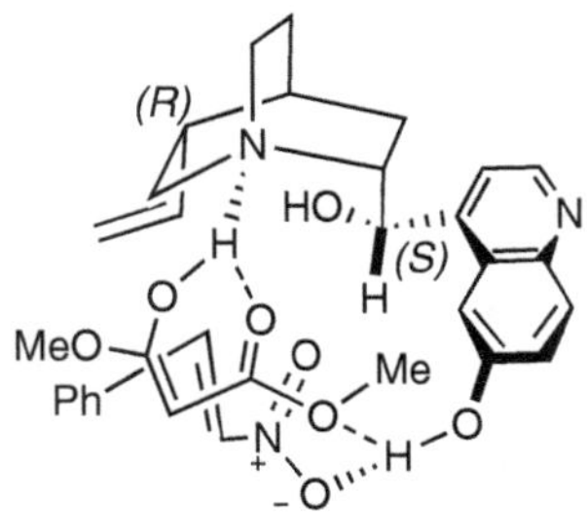

39

FIGURE 14-3. Stereochemical model for the Michael addition reaction catalyzed by DMQ **37**. (*See the color version of this figure in Color Plate section.*)

system, water molecules can competitively interfere with the hydrogen-bonding interactions and decrease the efficiency of the DMQ catalyst, resulting in an erosion of enantioselectivity.

In terms of catalyst loading for this specific electron-deficient substrate **36**, DMQ catalyzed the reaction extremely efficiently. With optimization, remarkably, an outstanding yield (>99% conversion and 95% assay yield) and enantioselectivity (95% ee) were achieved with as little as only 0.5–1 mol% of DMQ between −20 and −15 °C in dry THF (Note: 10 mol% loading of DMQ is typically adopted in the literature). Furthermore, instead of using a large excess of dimethyl malonate, which was typically up to 3 equiv. in the literature [23d], greater than 99% conversion could also be achieved with only 1.1 equiv. of dimethyl malonate at >100 kg scale runs. To the best of our knowledge, this development represents the first application of an organocatalytic Michael addition on an industrial scale.

To isolate **35**, a straightforward process was also developed. After the reaction was complete, the crude reaction solution was solvent-switched to *i*-PrOH. Water was then added to the reaction mixture and **35** directly crystallized from aq *i*-PrOH in 88% isolated yield with about 1% ee upgrade.

14.3.4 Completion of MK-8613 Synthesis

With **35** in hand, piperidinone **41b** with desired functionality could be obtained in one pot through an efficient cascade coupling [21]. Treatment of malonate **35** with 1.3 equiv. of allylamine and 1.05 equiv. of aqueous HCHO in *i*-PrOH at 50 °C gave **40** in 90% yield as a mixture of isomers (Scheme 14-14). Without workup, the crude reaction mixture **40** was treated with NaOH followed by decarboxylation of the resulting acid upon acidification to give a mixture of *cis*/*trans* **41a** and **41b**. An efficient isolation of the desired *trans* isomer **41b** was achieved through a base-catalyzed, dynamic crystallization-driven process, as the undesired **41a** was converted into **41b** under the crystallization conditions. In practice, without any aqueous workup, the one-pot cascade reaction mixture after decarboxylation was pH adjusted to pH = 7–8 by simply adding aqueous $NaHCO_3$. Upon aging at 45 °C, the desired *trans* isomer of **41b** gradually crystallized and was isolated directly from crude aqueous *i*-PrOH solution by a simple filtration in 76% overall isolated yield and >98% purity.

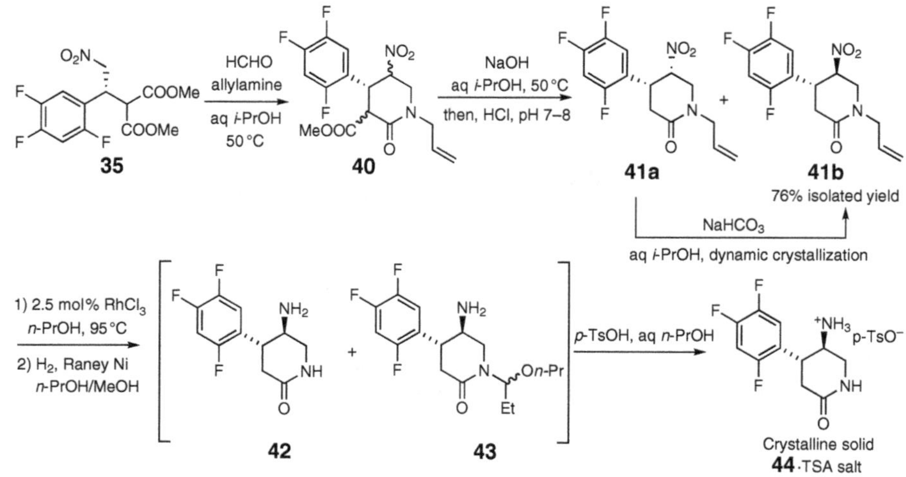

SCHEME 14-14. Preparation of **44**· *p*-TSA salt via two through-processes.

Through a rhodium-catalyzed isomerization/deallylation [26] followed by hydrogenation, a mixture of **42** and **43** was obtained. The desired amine *p*-TSA salt **44** crystallized directly from the reaction mixture by applying a controlled hydrolysis/crystallization procedure. As such, an easily operated, three-step through-process was developed to afford **44** in 62% overall yield with >99% purity and >98% ee (Scheme 14-14). A significant ee upgrade was also achieved during the isolation of the *p*-TSA salt **44**.

Copper-mediated coupling of Boc-protected pyrrolidinone **45** with aminopyridine **46** afforded **47**, which was treated with conc. HCl in EtOH at 50 °C to complete the preparation of **32** in 82% yield and >99% purity (Scheme 14-15).

Through an organocatalytic Michael addition of dimethyl malonate to nitrostyrene **36**, the first stereogenic center was efficiently established with high enantioselectivity, which paved the way for a synthetic path for the application of a subsequent novel cascade coupling [21] to assemble the core structure of MK-8613 with the desired stereochemistry. This cost-effective synthesis had an overall yield of 23% [8].

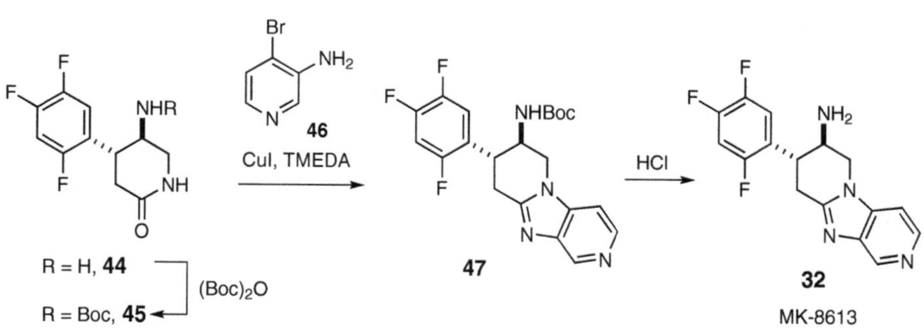

SCHEME 14-15. Preparation of **32** via Cu-assisted coupling.

14.4 CONCLUSIONS

The highlighted two examples discussed in this chapter demonstrate that organocatalysis can be practically applied on large scale to solve challenging synthetic problems. Application of organocatalysis can provide a unique approach allowing a quick access to key synthetic intermediates that are not easily prepared through other synthetic protocols. Organocatalysis represents a valuable addition to metal-catalyzed and enzymatic procedures for asymmetric synthesis on scale. Compared to classical metal-catalyzed chemistry, organocatalysts are typically air stable and are not easily poisoned by process impurities, offering robustness in developing a productive and streamlined process for industrial application. In addition, application of an organocatalytic process free of heavy metals is appealing in the pharmaceutical industry, because residues of heavy metals in active pharmaceutical ingredients (API) are strictly regulated. Other key factors, such as availability of organocatalysts, catalyst cost, and catalyst loading, could also play a significant role in influencing the application of an organocatalytic process on an industrial scale.

Overall, asymmetric organocatalysis is a useful tool in organic synthesis on large scale. Novelty and diversity resulting from the chemistry of asymmetric organocatalysis still continue to expand, and the application of organocatalysis will likely be much broader in the pharmaceutical industry in the future.

ACKNOWLEDGMENTS

This chapter represents the outstanding contributions of many scientists within Merck Research Laboratories. The author hereby expresses his sincere gratitude to the many colleagues whose hard work and dedication led to the success of the chemistry presented here. Their names appear in Refs [7,8].

REFERENCES

1. Ahrendt KA, Borths CJ, MacMillan DWC (2000). New strategies for organic catalysis: the first highly enantioselective organocatalytic Diels–Alder reaction. *J. Am. Chem. Soc.* 122, 4243–4244.
2. (a) For recent reviews, see The preceding chapter authored by Y. Hayashi in this book. (b) Marqués-López E, Herrera RP, Christmann M (2010). Asymmetric organocatalysis in total synthesis—a trial by fire. *Nat. Prod. Rep.* 27, 1138–1167. (c) Valero G, Companyo X, Bravo N, Alba ANR, Moyano A, Rios R (2010). Searching for untrodden paths in the organocatalysis territory. *Synlett* 1883–1908.(d) Bertelsen S, Jørgensen KA (2009). Organocatalysis—after the gold rush. *Chem. Soc. Rev.* 38, 2178–2189. (e) Enders D, Narine AA (2008). Lessons from nature: biomimetic organocatalytic carbon–carbon bond formations. *J. Org. Chem.* 73, 7857–7870. (f) MacMillan DWC (2008). The advent and development of organocatalysis. *Nature* 455, 304–308. (g) Walji AM, MacMillan DWC (2007). Strategies to bypass the taxol problem. Enantioselective cascade catalysis, a new approach for the efficient construction of molecular complexity. *Synlett* 1477–1489. (h) Gaunt MJ, Johansson CCC, McNally A, Vo NT (2007). Enantioselective organocatalysis. *Drug Discovery Today*, 12, 8–27. (i) Moyano A, Rios R (2011). Asymmetric organocatalytic cyclization and cycloaddition reactions. *Chem. Rev.* 111, 4703–4832.

3. Prelog V, Wihelm M (1954). Untersuchungen über asymmetrische synthesen VI. Der reaktionsmechanismus und der sterische verlauf der asymmetrischen cyanhydrin-Synthese. *Helv. Chim. Acta* 37, 1634–1660.
4. List B, Lerner RA, Barbas, III, CF (2000). Proline-catalyzed direct asymmetric aldol reactions. *J. Am. Chem. Soc.* 122, 2395–2396.
5. Few examples of asymmetric organocatalysis on large scale were reported prior to the 'gold rush' of organocatalysis. For application of enamine organocatalysis, see (a) Eder U, Sauer G, Weichert R, inventors. Optically active 1,5-indandione and 1,6-naphthalenedione derivatives. German Patent DE2014757 1971. (b) Eder U, Sauer G, Weichert R (1971). New type of asymmetric cyclization to optically active steroid CD partial structures. *Angew. Chem., Int. Ed.* 10, 496–497. (c) Hajos ZG, Parrish DR, inventors. Asymmetric synthesis of optically active polycyclic organic compounds. German Patent DE2102623 1971. (d) Hajos ZG, Parrish DR (1974). Asymmetric synthesis of bicyclic intermediates of natural product chemistry. *J. Org. Chem.* 39, 1615–1621. For a practical, asymmetric phase-transfer catalyzed alkylation, see (e) Dolling UH, Davis P, Grabowski EJJ (1984). Efficient catalytic asymmetric alkylations: 1. Enantioselective synthesis of (+)-indacrinone via chiral phase-transfer catalysis. *J. Am. Chem. Soc.* 106, 446–447. (f) Hughes DH, Dolling UH, Ryan KM, Schoenewaldt EF, Grabowski EJJ (1987). Efficient catalytic asymmetric alkylations. 3. A kinetic and mechanistic study of the enantioselective phase-transfer methylation of 6,7-dichloro-5-methoxy-2-phenyl-1-indanone. *J. Org. Chem.* 52, 4745–4752. (g) Bhattacharya A, Dolling UH, Grabowski EJJ, Karady S, Ryan KM, Weinstock LM (1986). Enantioselective Robinson annelations via phase-transfer catalysis. *Angew. Chem., Int. Ed.* 25, 476–477.
6. For detailed mechanistic studies of enamine catalysis, see (a) Hoang L, Bahmanyar S, Houk KN, List B (2003). Kinetic and stereochemical evidence for the involvement of only one proline molecule in the transition states of proline-catalyzed intra- and intermolecular aldol reactions. *J. Am. Chem. Soc.* 125, 16–17. For early examples of enamine catalysis, see (b) Woodward RB, Logusch E, Nambiar KP, Sakan K, Ward DE, Au-Yeung BW, Balaram P, Browne LJ, Card PJ, Chen CH, ChBnevert RB, Fliri A, Frobel K, Gais HJ, Garratt DG, Hayakawa K, Heggie W, Hesson DP, Hoppe D, Hoppe I, Hyatt JA, Ikeda D, Jacobi PA, Kim KS, Kobuke Y, Kojima K, Krowicki K, Lee VJ, Leutert T, Malchenko S, Martens J, Matthews RS, Ong BS, Press JB, Rajan Babu TV, Rousseau G, Sauter HM, Suzuki M, Tatsuta K, Tolbert LM, Truesdale EA, Uchida I, Ueda Y, Uyehara T, Vasella AT, Vladuchick WC, Wade PA, Williams RM, Wong HNC (1981). Asymmetric total synthesis of erythromycin: 1. Synthesis of an erythronolide A secoacid derivative via asymmetric induction. *J. Am. Chem. Soc.* 103, 3210–3213. (c) Danishefsky S, Cain P (1976). Optically specific synthesis of estrone and 19-norsteroids from 2,6-lutidine. *J. Am. Chem. Soc.* 98, 4975–4983.
7. Xu F, Zacuto M, Yoshikawa N, Desmond R, Hoerrner S, Itoh T, Journet M, Humphrey GR, Cowden C, Strotman N, Devine P (2010). Asymmetric synthesis of telcagepant, a CGRP receptor antagonist for the treatment of migraine. *J. Org. Chem.* 75, 7829–7841.
8. Xu F, Corley E, Zacuto M, Conlon DA, Pipik B, Humphrey G, Murry J, Tschaen D (2010). Asymmetric synthesis of a potent, aminopiperidine-fused imidazopyridine dipeptidyl peptidase IV inhibitor. *J. Org. Chem.* 75, 1343–1353.
9. Keen SP, Cowden CJ, Bishop BC, Brands KMJ, Davies AJ, Dolling WH, Lieberman DR, Stewart GW (2005). Palladium-catalyzed allylic acyloxylation of terminal alkenes in the presence of a base. *J. Org. Chem.* 70, 1771–1779.
10. Scott JP, Ashwood MS, Brands KMJ, Brewer SE, Cowden CJ, Dolling UH, Emerson KM, Gibb AD, Goodyear A, Oliver SF, Stewart GW, Wallace DJ (2008). Development of a phase transfer catalyzed asymmetric synthesis for an estrogen receptor beta selective agonist. *Org. Process Res. Dev.* 12, 723–730.
11. For additional recent examples of application of phase-transfer catalyzed alkylation/addition, see (a) Huffman MA, Rosen JD, Farr RN, Lynch JE (2007). Synthesis of a selective estrogen receptor

β-modulator via asymmetric phase-transfer catalysis. *Tetrahedron* 63, 4459–4463. (b) Jiang X, Gong B, Prasad K, Repič O (2008). A practical synthesis of a chiral analogue of FTY720. *Org. Process Res. Dev.* 12, 1164–1169. For recent examples of organocatalytic desymmetrization, see (c) Yue TY, McLeod DD, Albertson KB, Beck SR, Deerberg J, Fortunak JM, Nugent WA, Radesca LA, Tang L, Xiang CD (2006). Stereoselective process for a CCR3 antagonist. *Org. Process Res. Dev.* 10, 262–271. (d) Mittendorf J, Benet-Buchholz J, Fey P, Mohrs KH (2003). Efficient asymmetric synthesis of β-amino acid BAY 10-8888/PLD-118, a novel antifungal for the treatment of yeast infections. *Synthesis* 136–140. (e) Ishii Y, Fujimoto R, Mikami M, Murakami S, Miki Y, Furukawa Y (2007). Practical syntheses of chiral α-amino acids and chiral half-esters by kinetic resolution of urethane-protected α-amino acid *N*-carboxyanhydrides and desymmetrization of cyclic meso-anhydrides with new modified cinchona alkaloid catalysts. *Org. Process Res. Dev.* 11, 609–615. For examples of organocatalytic epoxidation, see (f) Ager DJ, Anderson K, Oblinger E, Shi Y, VanderRoest J (2007). An epoxidation approach to a chiral lactone: application of the Shi epoxidation. *Org. Process Res. Dev.* 11, 44–51.

12. For selected recent leading examples, see (a) Ishikawa H, Suzuki T, Hayashi Y (2009). High-yielding synthesis of the anti-influenza neuramidase inhibitor (−)-oseltamivir by three "one-pot" operations. *Angew. Chem., Int. Ed.* 48, 1304–1307. (b) Simmons B, Walji AM, MacMillan DWC (2009). Cycle-specific organocascade catalysis: application to olefin hydroamination, hydro-oxidation, and amino-oxidation, and to natural product synthesis. *Angew. Chem., Int. Ed.* 48, 4349–4353. (c) Zuend SJ, Coughlin MP, Lalonde MP, Jacobsen EN (2009). Scaleable catalytic asymmetric Strecker syntheses of unnatural α-amino acids. *Nature* 461, 968–971. (d) Jones SB, Simmons B, MacMillan DWC (2009). Nine-step enantioselective total synthesis of (+)-minfiensine. *J. Am. Chem. Soc.* 131, 13606–13607.

13. (a) Goadsby PJ, Lipton RB, Ferrari MD (2002). Migraine—current understanding and treatment. *N. Engl. J. Med.* 346, 257–270. (b) Stovner L, Hagen K, Jensen R, Katsarava Z, Lipton R, Scher A, Steiner T, Zwart JA (2007). The global burden of headache: a documentation of headache prevalence and disability worldwide. *Cephalalgia* 27, 193–210.

14. Paone DV, Shaw AW, Nguyen DN, Burgey CS, Deng JZ, Kane SA, Koblan KS, Salvatore CA, Mosser SD, Johnston VK, Wong BK, Miller-Stein CM, Hershey JC, Graham SL, Vacca JP, Williams TM (2007). Potent, orally bioavailable calcitonin gene-related peptide receptor antagonists for the treatment of migraine: discovery of *N*-[(3*R*,6*S*)-6-(2,3-difluorophenyl)-2-oxo-1-(2,2,2-trifluoroethyl)azepan-3-yl]-4- (2-oxo-2,3-dihydro-1H-imidazo[4,5-*b*]pyridin-1-yl) piperidine-1-carboxamide (MK-0974). *J. Med. Chem.* 50, 5564–5567.

15. (a) Ho TW, Ferrari MD, Dodick DW, Galet V, Kost J, Fan X, Leibensperger H, Froman S, Asaid C, Lines C, Koppen H, Winner PK (2008). Efficacy and tolerability of MK-0974 (telcagepant), a new oral antagonist of calcitonin gene-related peptide receptor, compared with zolmitriptan for acute migraine: a randomised, placebo-controlled, parallel-treatment trial. *Lancet* 372, 2115–2123. (b) Ho TW, Mannix L, Fanm X, Assaid C, Furtek C, Jones C, Lines C, Rapoport A (2008). Randomized controlled trial of an oral CGRP receptor antagonist, MK-0974, in acute treatment of migraine. *Neurology* 70, 1304–1312.

16. For early syntheses, see (a) Burgey CS, Paone DV, Shaw AW, Deng JZ, Nguyen DN, Potteiger SL, Graham SL, Vacca JP, Williams TM (2008). Synthesis of the (3*R*,6*S*)-3-amino-6-(2,3-difluorophenyl)azepan-2-one of telcagepant (MK-0974), a calcitonin gene-related peptide receptor antagonist for the treatment of migraine headache. *Org. Lett.* 10, 3235–3238. (b) Steinhuebel DP, Krska SW, Alorati A, Baxter JM, Belyk K, Bishop B, Palucki M, Sun Y, Davies IW (2010). Asymmetric hydrogenation of protected allylic amines. *Org. Lett.* 12, 4201–4203. (c) Baxter JM, Steinhuebel DP, Palucki M, Davies IW (2005). Stereoselective enol tosylation: preparation of trisubstituted α,β-unsaturated esters. *Org. Lett.* 7, 215–218. (d) Janey JM, Orella CJ, Njolito E, Baxter JM, Rosen JD, Palucki M, Sidler RR, Li W, Kowal JJ, Davies IW (2008). Raney-Co mediated reductive cyclization of an α,β-unsaturated nitrile. *J. Org. Chem.* 73, 3212–3217. (e) Palucki M, Davies I, Steinhuebel D, Rosen J, inventors; Merck & Co, Inc.,

assignee. Process for preparation of (3*R*,6*S*)-3-amino-6-(2,3-difluorophenyl)-1-(2,2,2-trifluoroethyl)azepan-2-one hydrochloride. World Patent WO/2007/120589. October 25, 2007.

17. (a) Gotoh H, Ishikawa H, Hayashi Y (2007). Diphenylprolinol silyl ether as catalyst of an asymmetric, catalytic, and direct Michael reaction of nitroalkanes with α,β -unsaturated aldehydes. *Org. Lett.* 9, 5307–5309. (b) Palomo C, Landa A, Mielgo A, Oiarbide M, Pugel A, Vera S (2007). Water-compatible iminium activation: organocatalytic Michael reactions of carbon-centered nucleophiles with enals. *Angew. Chem., Int. Ed.* 46, 8431–8435. (c) Zu L, Xie H, Li H, Wang J, Wang W (2007). Highly enantioselective organocatalytic conjugate addition of nitromethane to α,β-unsaturated aldehydes: three-step synthesis of optically active baclofen. *Adv. Synth. Catal.* 349, 2660–2664. (d) Hojabri L, Hartikka A, Moghaddam FM, Arvidsson PI (2007). A new imidazole-containing imidazolidinone catalyst for organocatalyzed asymmetric conjugate addition of nitroalkanes to aldehydes. *Adv. Synth. Catal.* 349, 740–748. (e) Wang Y, Li P, Liang X, Zhang TY, Ye J (2008). An efficient enantioselective method for asymmetric Michael addition of nitroalkanes to α,β-unsaturated aldehydes. *Chem. Commun.* 1232–1234.

18. (a) Marigo M, Wabnitz TC, Fielenbach D, Jørgensen KA (2005). Enantioselective organocatalyzed α sulfenylation of aldehydes. *Angew. Chem., Int. Ed.* 44, 794–797. (b) Hayashi Y, Gotoh H, Hayashi T, Shoji M (2005). Diphenylprolinol silyl ethers as efficient organocatalysts for the asymmetric Michael reaction of aldehydes and nitroalkenes. *Angew. Chem., Int. Ed.* 44, 4212–4215.

19. For an example of the acidity of PivOH and its interaction with amine bases, see Xu F, Armstrong, III, JD, Zhou GX, Simmons B, Hughes D, Ge Z, Grabowski EJJ (2004). Mechanistic evidence for an α-oxoketene pathway in the formation of β-ketoamides/esters via Meldrum's acid adducts. *J. Am. Chem. Soc.* 126, 13002–13009.

20. (a) Edmondson SD, Mastracchio A, Cox JM, Eiermann GJ, He H, Lyons KA, Patel RA, Patel SB, Petrov A, Scapin G, Wu JK, Xu S, Zhu B, Thornberry NA, Roy RS, Weber AE (2009). Aminopiperidine-fused imidazoles as dipeptidyl peptidase-IV inhibitors. *Bioorg. Med. Chem. Lett.* 19, 4097–4101. (b) Cox JM, Edmondson SD, Harper B, Weber AE, inventors; Merck & Co, Inc., assignee. Aminopiperidines as dipeptidyl peptidase-IV inhibitors for the treatment or prevention of diabetes. World Patent WO/2006/039325. April 13, 2006.

21. Xu F, Corley E, Murry JA, Tschaen DM (2007). Highly selective cascade couplings for the syntheses of functionalized piperidinones and bispidines. *Org. Lett.* 9, 2669–2672.

22. One-pot synthesis led to a low yield of fluoro-substituted styrenes, see Jacobo R, Cota A, Rogel E, Garcia JD, Rivero IA, Hellberg LH, Somanathan R (1994). Synthesis of some fluorinated nitro-olefins. *J. Fluorine Chem.* 67, 253–255.

23. For examples, see (a) Terada M, Ube H, Yaguchi Y (2006). Axially chiral guanidine as enantioselective base catalyst for 1,4-addition reaction of 1,3-dicarbonyl compounds with conjugated nitroalkenes. *J. Am. Chem. Soc.* 128, 1454–1455. (b) Evans DA, Seidel D (2005). Ni(II)–bis[(*R,R*)-*N,N'*-dibenzylcyclohexane-1,2-diamine]Br_2 catalyzed enantioselective Michael additions of 1,3-dicarbonyl compounds to conjugated nitroalkenes. *J. Am. Chem. Soc.* 127, 9958–9959. (c) Okino T, Hoashi Y, Furukawa T, Xu X, Takemoto Y (2005). Enantio- and diastereoselective Michael reaction of 1,3-dicarbonyl compounds to nitroolefins catalyzed by a bifunctional thiourea. *J. Am. Chem. Soc.* 127, 119–125. (d) Li H, Wang Y, Tang L, Deng L (2004). Highly enantioselective conjugate addition of malonate and β-ketoester to nitroalkenes: asymmetric C–C bond formation with new bifunctional organic catalysts based on cinchona alkaloids. *J. Am. Chem. Soc.* 127, 9906–9907. (e) Watanabe M, Ikagawa A, Wang H, Murata K, Ikariya T (2004). Catalytic enantioselective Michael addition of 1,3-dicarbonyl compounds to nitroalkenes catalyzed by well-defined chiral Ru amido complexes. *J. Am. Chem. Soc.* 126, 11148–11149. (f) Ji J, Barnes DM, Zhang J, King SA, Wittenberger SJ, Morton HE (1999). Catalytic enantioselective conjugate addition of 1,3-dicarbonyl compounds to nitroalkenes. *J. Am. Chem. Soc.* 121, 10215–10216. (g) Hynes PS, Stupple PA, Dixon DJ (2008). Organocatalytic asymmetric total synthesis of (*R*)-rolipram and

formal synthesis of (3*S*,4*R*)-paroxetine. *Org. Lett.* 10, 1389–1391. (h) Zhu Q, Huang H, Shi D, Shen Z, Xia C (2009). An efficient synthesis of chiral diamines with rigid backbones: application in enantioselective Michael addition of malonates to nitroalkenes. *Org. Lett.* 11, 4536–4539.

24. For a recent review, see Palomo C, Oiarbide M, López R (2009). Asymmetric organocatalysis by chiral Brønsted bases: implications and applications. *Chem. Soc. Rev.* 38, 632–653.
25. Li H, Wang Y, Tang L, Wu F, Liu X, Guo C, Foxman BM, Deng L (2005). Stereocontrolled creation of adjacent quaternary and tertiary stereocenters by a catalytic conjugate addition. *Angew. Chem., Int. Ed.* 44, 105–108.
26. Zacuto MJ, Xu F (2007). One-step $RhCl_3$-catalyzed deprotection of acyclic *N*-allyl amides. *J. Org. Chem.* 72, 6298–6300.

15

CATALYTIC VARIANTS OF PHOSPHINE OXIDE-MEDIATED ORGANIC TRANSFORMATIONS

Stephen P. Marsden
School of Chemistry, University of Leeds, Leeds, UK

15.1 GENERAL INTRODUCTION

The oxaphilicity of phosphorus, especially in the P(V) oxidation state, has long been exploited by organic chemists to drive reactions involving substitution or metathesis of oxygen-containing functional groups. The remarkable strength of the P=O bond (ca. 130 kcal mol^{-1}) [1] can compensate enthalpically for the conversion of relatively strong C–O and C=O bonds to weaker bonds. A further advantage of organophosphorus compounds is that the highly favorable redox transformation of phosphorus from the P(III) to P(V) oxidation state can be used to generate reactive reagents *in situ*. Despite the ubiquity of phosphorus-mediated transformations in academic and industrial discovery research programs [2], these processes manifest several undesirable characteristics when scale-up considerations are brought into play. The first is the inevitable generation of stoichiometric, high molecular weight waste. The molecular weight of triphenylphosphine oxide (278), a common stoichiometric by-product, represents a significant fraction of the waste in any transformation in which it is involved. The negative impact upon atom economy [3] or E-factor [4] of manufacturing processes is immediately apparent. The second issue is one familiar to most practising organic chemists: phosphine oxides are generally water insoluble, highly polar products that can be very difficult to separate even on a laboratory scale, with recourse frequently made to column chromatography. The practical and economic limitations of this technique mean that separation of the phosphine oxide waste is a major problem for process chemists. The third issue is that many of the activating agents utilized to generate the reactive phosphorus species have undesirable safety properties (e.g., azides and azo compounds), creating further issues on scale.

Sustainable Catalysis: Challenges and Practices for the Pharmaceutical and Fine Chemical Industries, First Edition. Edited by Peter J. Dunn, K. K. (Mimi) Hii, Michael J. Krische, and Michael T. Williams.

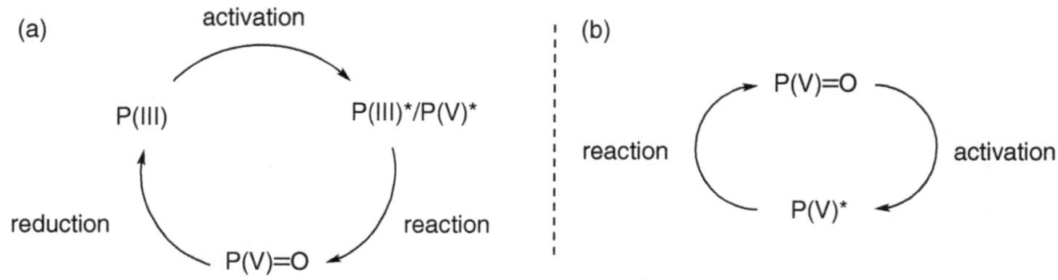

SCHEME 15-1. The two general mechanisms for organophosphorus catalysis: (a) reduction-enabled and (b) redox–neutral.

The issues associated with the use of organophosphorus reagents on manufacturing scale have been frequently highlighted by the process chemistry community [5].

Research efforts to address these shortcomings have largely focused on innovative methods to facilitate separation and reagent recycling. A more appealing and sustainable solution would be to generate variants of the reactions that are catalytic in the organophosphorus component and, where possible, avoid activating agents with negative safety properties. Reports of catalytic organophosphorus reactions have been limited and sporadic in the literature over the past 50 years, but the past 4 years has seen an intensification of interest in the area with some breakthrough processes being reported. Two distinct approaches have been adopted (Scheme 15-1). The first might be described as reduction-enabled catalysis, wherein a process that initiates with a phosphorus(III) compound is rendered catalytic by reduction of the phosphorus(V) oxide back to the phosphorus(III) state (Scheme 15-1a). The second is a redox–neutral catalysis, wherein the phosphorus is maintained at the +5 oxidation state throughout, but the phosphine oxide is converted to a reactive phosphorus(V) reagent *in situ* (Scheme 15-1b).

The current chapter aims to discuss progress and impediments in the development of catalytic organophosphorus processes for olefination, imination, halogenation, and nucleophilic substitution (Mitsunobu) reactions. In each case, we will give a perspective on current industrial applications and issues, and also where appropriate discuss the various separation/recycling approaches that have been adopted so that readers can assess the state of the art in the catalytic processes alongside the alternative protocols.

15.2 WITTIG CHEMISTRY

The generation of olefins by condensation of carbonyl compounds with phosphonium ylides was discovered in 1953 by Georg Wittig [6,7]. The phosphonium ylides (alkylidenephosphoranes) are generated by deprotonation of alkylphosphonium salts, themselves usually generated by substitution of alkyl halides (or other electrophiles) with trivalent phosphines (Scheme 15-2). Condensation with a carbonyl compound then generates the alkene product and phosphine oxide by-product. The energetically unfavorable exchange of a carbonyl C=O bond (ca. 178–179 kcal mol^{-1}) for an alkene C=C bond (ca. 146 kcal mol^{-1}) is compensated by exchange of the weak ylide C–P/C=P bond for the strong P=O bond [1].

Ylides bearing electron-withdrawing substituents on the carbon atom are relatively stable (being isolable in many cases), can be generated using relatively mild base such as

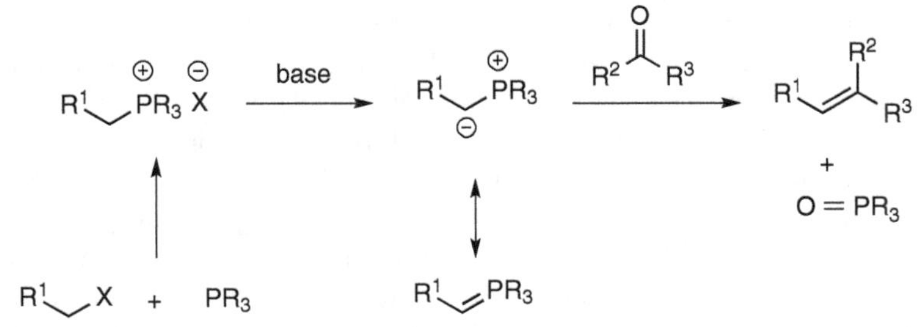

SCHEME 15-2. General scheme for the synthesis of alkylidenephosphoranes and the Wittig reaction.

metal alkoxide or hydroxide salts [8], and generally produce the *E*-olefinic isomers. On the other hand, ylides bearing solely hydrogens or alkyl substituents on the carbon atom are generally not isolable. They are typically generated using a strong base (alkyllithiums or lithium amides on a discovery scale), requiring anhydrous/aprotic conditions, although on process scale this is generally substituted for the use of metal alkoxide bases at elevated temperatures [9]. In terms of stereoselectivity, nonstabilized ylides generally produce the *Z*-olefinic isomer, although the Schlosser modification [10] gives access to the *E*-isomer, at the expense of a further equivalent of organolithium base and extended reaction at cryogenic temperatures.

The strategic advantages of this method—convergency, regiospecificity, and stereoselectivity—were quickly appreciated by industrial chemists, but the practical limitations of the reaction were apparent from the outset. The commercial synthesis of Vitamin A acetate **1** utilizing a Wittig olefination was brought online by BASF in 1971, but the exothermicity of the reaction of partially-stabilized ylide **2** with aldehyde **3** (Scheme 15-3) required a continuous process, while recycling of the triphenylphosphine oxide by-product was necessary for the synthesis to be economically viable [11].

This was achieved by selective extraction of the highly nonpolar product into hydrocarbon solvents in which the phosphine oxide is poorly soluble. The phosphine oxide was then treated with phosgene to generate triphenylphosphonium dichloride and carbon dioxide [12]. Reaction of the dichloride with elemental phosphorus regenerates the triphenylphosphine as well as phosphorus trichloride (Scheme 15-4).

SCHEME 15-3. Commercial synthesis of Vitamin A acetate **1** [11].

$$O{=}PPh_3 \;+\; COCl_2 \;\longrightarrow\; Cl_2PPh_3$$

$$3\ Cl_2PPh_3 \;+\; 2\ P \;\longrightarrow\; 3\ PPh_3 \;+\; 2\ PCl_3$$

SCHEME 15-4. Industrial recycling of triphenylphosphine oxide to triphenylphosphine [11].

Although viable for the highly nonpolar product **1**, selective separation of triphenylphosphine oxide from most organic molecules by solvent extraction is not broadly applicable owing to similar solubility profiles. Much interest has therefore focused on modified Wittig reagents that generate phosphine oxide by-products that can be readily phase-separated from the desired organic products. The use of Horner–Wadsworth–Emmons reactions of phosphonate anions (rather than phosphonium ylides) is a clear example of this, since the dialkyl phosphate by-product can be readily extracted into a mildly basic aqueous phase [13]. Additionally, reagent-driven control of stereochemistry is possible: the reactions generally return the *E*-isomers but the use of electron-withdrawing alkoxy substituents such as trifluoroethoxy [14] or diaryloxy groups [15] on the phosphonate allow access to the *Z*-isomer. This reaction is, however, generally limited to the delivery of carbon fragments that bear electron-withdrawing groups such as esters and ketones.

Generalized phase separation approaches for the Wittig reaction have included reactions generating water-soluble phosphine oxides [16], polymer-supported phosphine oxides [17], fluorous-tagged phosphine oxides [18], and PEG-supported phosphine oxides [19]. Recycling of the phosphine oxides can be achieved by the aforementioned conversion to phosphonium dichlorides using phosgene or oxalyl chloride [20] followed by treatment with phosphorus or lithium aluminum hydride. Alternatively, direct reduction of the phosphine oxides can be achieved with relatively harsh reagents including lithium aluminum hydride [21], di-iso-butylaluminum hydride [22], and trichlorosilane [23].

In terms of catalytic variants of the Wittig reaction, initial attempts focused on the replacement of phosphorus ylides by other group 15 ylides. Shi and Huang demonstrated catalytic olefination reactions using arsonium ylides, exploiting the more facile reduction of arsine oxides to arsines compared with their phosphorus analogs (Scheme 15-5) [24].

The ylides are generated *in situ* from trialkylarsines, alkyl halides and base, and following olefination the arsine oxide is recycled to the arsine by reduction with triphenyl phosphite (Scheme 15-6). Hence, although this reaction is catalytic in the ylide-bearing reagent, it does still generate stoichiometric phosphorus(V) oxide waste. Additionally, the catalyst loadings are unattractively high, while the toxicity of the arsenic reagent of course limits the broader applicability of this reaction.

In related chemistry, Tang demonstrated that telluronium ylides could be harnessed to perform catalytic Wittig reactions using triphenylphosphite or, more usefully, sodium bisulfite as the stoichiometric reducing agent (Scheme 15-7) [25]. The telluride catalyst was employed in relatively low loadings, and to further improve the utility of the process the telluride was attached to a soluble PEG-polymer chain, facilitating separation by selective precipitation. The catalyst was thus able to be recycled, although some loss of activity was seen. Again, the toxicity of tellurium-containing compounds limits likely applications on scale.

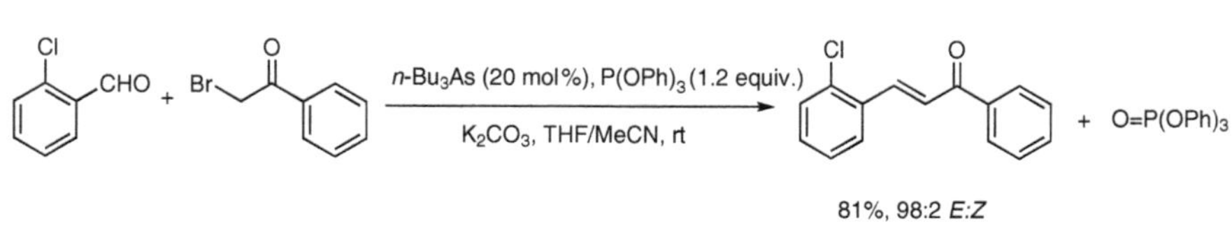

SCHEME 15-5. Arsonium ylide-mediated catalytic olefination [24].

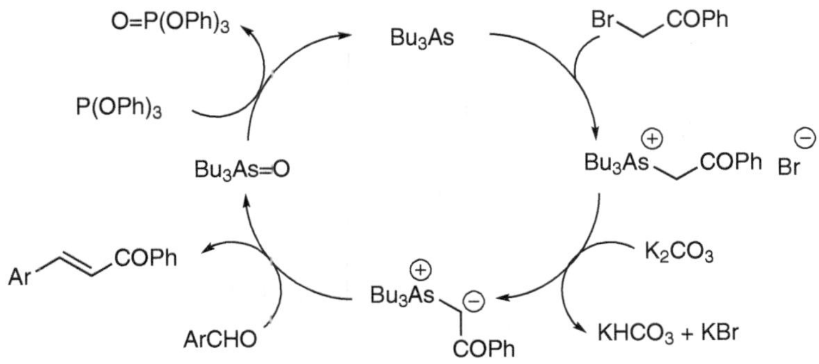

SCHEME 15-6. Catalytic cycle for arsonium ylide-mediated olefinations [24].

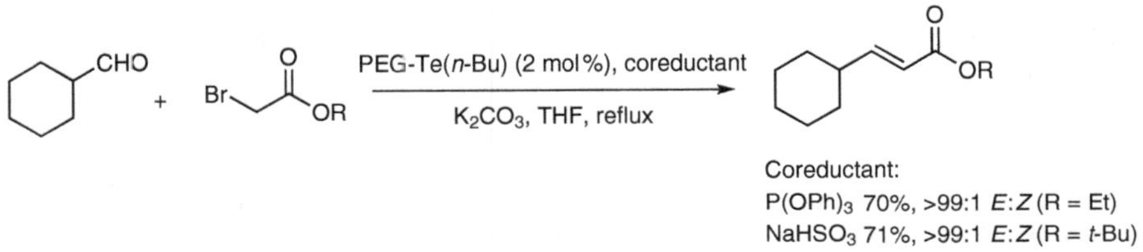

SCHEME 15-7. Telluronium ylide-mediated catalytic olefination [25].

The development of a catalytic phosphorus ylide-mediated Wittig reaction was finally reported in 2009. O'Brien and coworkers recognized that it was likely that carbon–phosphorus ylides could be generated in the presence of a carbonyl coupling partner by reaction of a phosphine with an appropriate alkyl halide, followed by deprotonation of the resulting phosphonium salt [26]. Following successful Wittig reaction of the ylide, completion of the catalytic cycle would require chemoselective reduction of the phosphine oxide to the phosphine in the presence of the stoichiometric starting materials and products—a challenging prerequisite given the relatively harsh reducing agents typically used to reduce phosphine oxides to phosphines. Ultimately, it was found that the combination of the relatively reductively labile [27] phosphine oxide **4** with diphenylsilane as a stoichiometric reducing agent was able to meet the conditions for phosphine regeneration without significant competitive reduction of educts/adducts. The complete catalytic cycle is shown below (Scheme 15-8).

The reaction works well with a diverse array of aliphatic, aromatic, and heteroaromatic-substituted aldehydes (Scheme 15-9). In terms of ylide structure, the reaction has generally been demonstrated with alkyl halides that give rise to stabilized ylides (bearing ester, ketone or cyano functions), though examples of semistabilized ylides bearing electron-poor aryl groups are given. There are no examples to date of simple alkyl-substituted ylides, which could reflect either their slower rates of phosphonium salt formation and/or the higher pK_a of the proton for removal to generate the ylide. Most reactions were carried out at 10 mol% catalyst loading. However, a preparative example on multigram scale was performed at a reduced (4 mol%) loading with little effect on yield, though an extended reaction time was required.

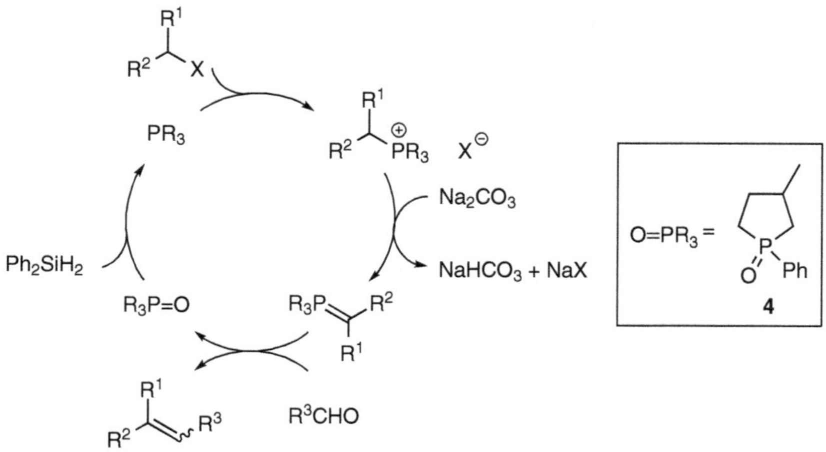

SCHEME 15-8. Catalytic cycle for the reduction-enabled phosphorus ylide-mediated olefination [26].

The reactions of ester and ketone-substituted ylides are highly stereoselective in favor of the expected *E*-isomer, though selectivities are lower with the cyano- and aryl-substituted ylides. Interestingly, detailed studies on the reactions of the ester-substituted ylide demonstrated an unexpected relationship between stereoselectivity and temperature, that is, higher selectivities were seen at higher temperatures. Control experiments demonstrated that the Wittig reactions are intrinsically only moderately *E*-selective (ca. 2:1), but that a second catalytic cycle (Scheme 15-10) mediated by the nucleophilic

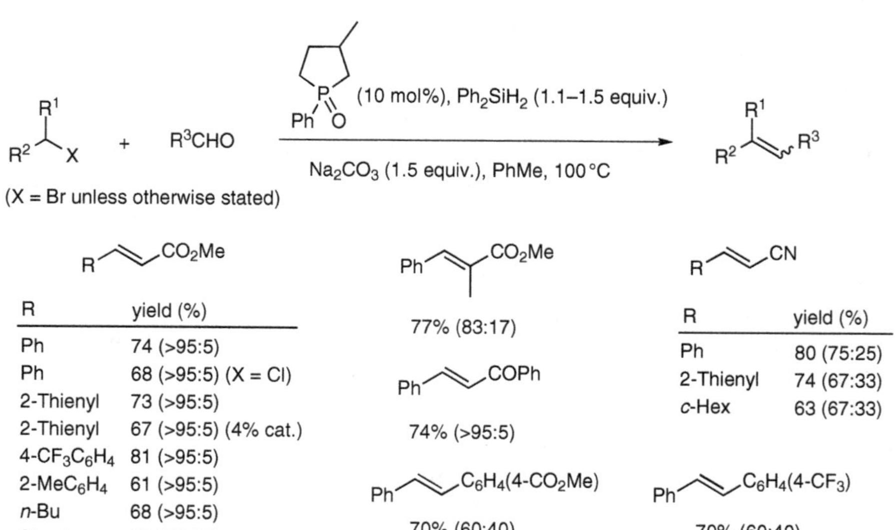

R	yield (%)
Ph	74 (>95:5)
Ph	68 (>95:5) (X = Cl)
2-Thienyl	73 (>95:5)
2-Thienyl	67 (>95:5) (4% cat.)
4-CF$_3$C$_6$H$_4$	81 (>95:5)
2-MeC$_6$H$_4$	61 (>95:5)
n-Bu	68 (>95:5)
Styryl	70 (83:17)

R	yield (%)
Ph	80 (75:25)
2-Thienyl	74 (67:33)
c-Hex	63 (67:33)

SCHEME 15-9. Scope of the catalytic Wittig olefination [26]. Values in parentheses denote *E*:*Z* ratio.

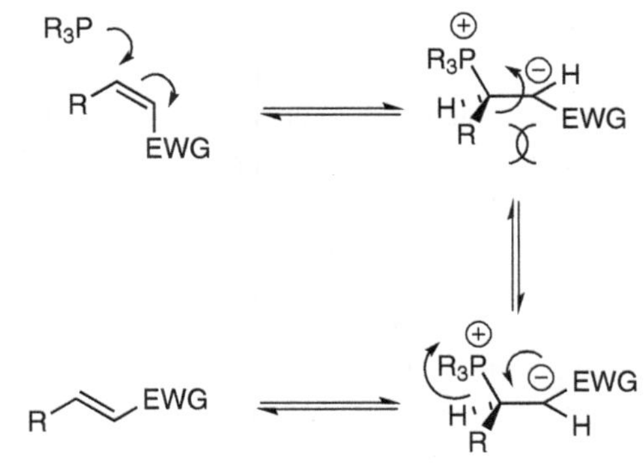

SCHEME 15-10. Phosphine-catalyzed isomerization of olefin isomers.

phosphine triggers equilibration to the thermodynamically more stable *E*-isomer (>95:5 in most cases). This isomerization pathway does not seem to operate significantly in the case of the less electrophilic cyano- and aryl-substituted alkenes. This represents a limitation of the method as it currently stands, although since it is well known that phosphine structure influences stereoselectivity in the Wittig reaction [28] it may well be that catalyst refinement will deliver synthetically useful selectivities across a broad range of substrates.

An apparently related olefination reaction had actually been reported 2 years earlier by Greaney [29]. The reaction of α-bromocarbonyl compounds with aryl aldehydes in the presence of trichlorosilane, triethylamine, and 10 mol% of triphenylphosphine oxide generates the corresponding alkenes in moderate to good yield (Scheme 15-11) and with excellent selectivity (single *E*-isomers detected in all cases). However, mechanistically the reaction is actually a silyl Reformatsky reaction, in which the trichlorosilane generates a nucleophilic silyl enol ether by displacement of halide from the α-bromocarbonyl compound, followed by nucleophilic addition to the aldehyde and elimination. It appears that the role of the phosphine oxide here is one of a Lewis base to activate the trichlorosilane, since the reaction proceeds (albeit less efficiently) in its absence.

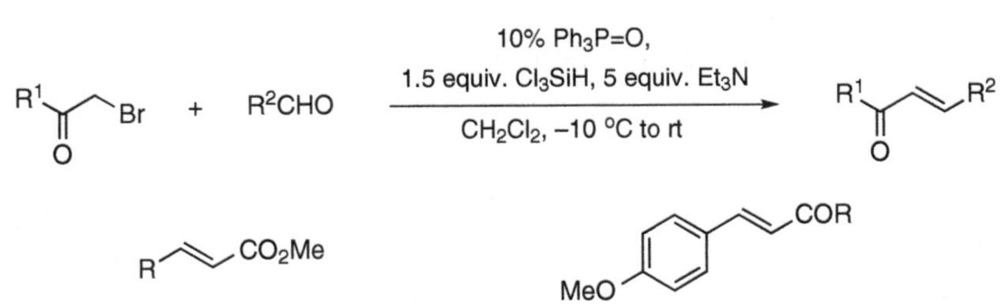

R	
4-MeOC$_6$H$_4$	69%
4-MeOC$_6$H$_4$	59% (no Ph$_3$P=O)
Ph	54%
4-FC$_6$H$_4$	50%
2-MeC$_6$H$_4$	55%

R	
Ph	73%
4-MeOC$_6$H$_4$	71%
4-ClC$_6$H$_4$	70%

SCHEME 15-11. Phosphine oxide-promoted silyl Reformatsky reaction [29].

15.3 AZA-WITTIG CHEMISTRY

The condensation of a nitrogen–phosphorus ylide (iminophosphorane) with a carbonyl compound to generate a carbon–nitrogen double bond was discovered in 1919 by Staudinger [30]. However, the synthetic potential of the method was largely unrecognized until the popularization of the Wittig olefination reaction some 40 years later, and hence the process has become known as the aza-Wittig reaction. In common with the carbon variant, the unfavorable exchange of a C=O bond (ca. 178–179 kcal mol^{-1}) for a C=N bond (ca. 147 kcal mol^{-1}) is compensated by the formation of the strong P=O bond.

Iminophosphoranes react with a wider array of carbonyl functions than the corresponding carbon-centered ylides: aldehydes, ketones, isocyanates, acid halides, carbon dioxide, esters, and (in intramolecular mode) amides and their derivatives are all capable of undergoing aza-Wittig reaction (Scheme 15-12) [31]. The condensation with carboxylate derivatives (acid halides, esters, amides) is particularly valuable for the formation of imidoyl halides, imidates, and amidines under mild, neutral conditions that contrast markedly to the strongly acidic/electrophilic conditions often used for their preparation. The reaction finds significant application in the synthesis of heterocycles by a number of strategies (direct cyclization by intramolecular aza-Wittig condensation; intermolecular aza-Wittig reaction followed by electrocyclization or cycloaddition; intermolecular aza-Wittig reaction/nucleophilic capture) [31].

Despite widespread applications in academic and industrial discovery laboratories, to our knowledge there are no examples of the scale-up of the aza-Wittig reaction. This doubtless reflects the generation of stoichiometric phosphine oxide waste, but also problems associated with the generation of the iminophosphoranes. These species are most commonly prepared by the Staudinger reaction, that is, the condensation of phosphines with organic azides, liberating nitrogen as a by-product (Scheme 15-12). This is a very clean and efficient process, taking place under mild conditions; however, the safety issues associated with the synthesis and manipulation of organic azides doubtless contribute to the lack of industrial take-up of the aza-Wittig reaction. An alternative method for the generation of iminophosphoranes is the modified Kirsanov reaction, in which (usually aromatic) amines are reacted with electrophilic phosphorus(V) species such as Appel reagents (either pre-formed or generated *in situ*) with loss of (typically) 2 equiv. of hydrohalic acid. The Kirsanov-type reactions are therefore less atom efficient than the Staudinger reaction, and the electrophilic nature of the reagents means that these processes

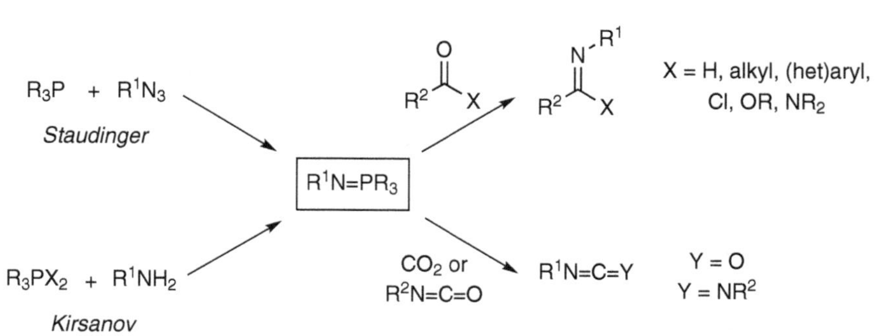

SCHEME 15-12. Generation and aza-Wittig reactions of iminophosphoranes [31].

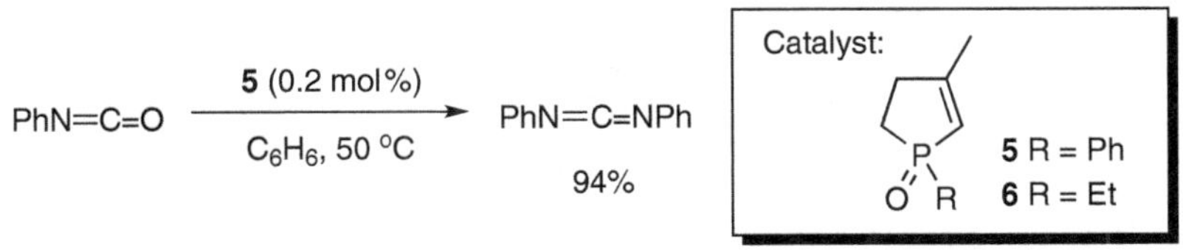

SCHEME 15-13. Catalytic self-condensation of isocyanates to carbodiimides [32,33].[1]

are also somewhat less chemoselective than the Staudinger process. To gain broader synthetic utility, it is therefore clear that the aza-Wittig requires not only a process that avoids the generation of stoichiometric phosphine oxide waste but also addresses the issues associated with iminophosphorane generation.

In 1962, workers at the DuPont company in Wilmington reported a convenient method for the generation of symmetrical carbodiimides by self-condensation of isocyanates catalyzed by phospholene oxides **5** and **6** (Scheme 15-13) [32,33].[1] The reactions were carried out with or without solvent, at ambient or elevated temperatures, and typically used just 0.1–0.8 mol% of the phospholene oxide catalyst. The stoichiometric by-product of the reaction is carbon dioxide, making this a very attractive route for the clean synthesis of carbodiimides, especially when compared with standard methods for the synthesis of this sensitive functional group (e.g., desulfurization of thioureas, dehydration of ureas).

This reaction is still widely used [34], and has been applied to the synthesis of isotopically-labeled carbodiimides for mechanistic studies [35]. Both aryl and alkyl isocyanates are substrates for the reaction, although the latter are considerably less reactive and typically require elevated temperatures to achieve high conversions. Aryl isocyanates bearing electron-withdrawing groups react faster than those bearing electron-donating and/or *ortho*-substituents. Isothiocyanates can be used as substrates but as expected are much less reactive than isocyanates. The reaction can also be used to generate polycarbodiimides by self-condensation of 1,*n*-diisocyanates [32]. Polycarbodiimides prepared in this way have been used as solid-phase coupling reagents for peptide synthesis [36] and as components of photosensitive polymers for imaging materials [37].

The mechanism of the reaction has been studied in detail [38]. The reaction was found to be reversible: in closed systems, an equilibrium was established between aryl isocyanate and carbodiimide, the position of which is determined by the aryl substituents (electron-donating groups favor isocyanate, electron-withdrawing groups favor carbodiimide). In reactions carried out at reflux in "open" systems, the reactions are effectively irreversible (due to loss of the carbon dioxide) and show clean first-order dependencies on both isocyanate and catalyst concentration. Hammett analysis revealed a positive ρ value of ca. 1.2, broadly consistent with that observed for addition of nucleophiles to isocyanates. A large negative entropy of activation indicates a highly ordered transition state. Finally, the incorporation of the oxygen atom of the phosphine oxide into the liberated carbon dioxide was verified by ^{18}O labeling studies [39]. A proposed catalytic cycle consistent with the experimental observations is shown in Scheme 15-14. The first, rate-limiting, step is the metathetical reaction of the phosphine oxide with isocyanate, proceeding via a four-membered transition-state or short-lived intermediate, generating iminophosphorane, and carbon dioxide. The second (fast) step involves the aza-Wittig reaction between the

[1] The structure of the catalysts was initially represented as the Δ-3,4-isomer, but was later shown to be the Δ-2,3-isomer.

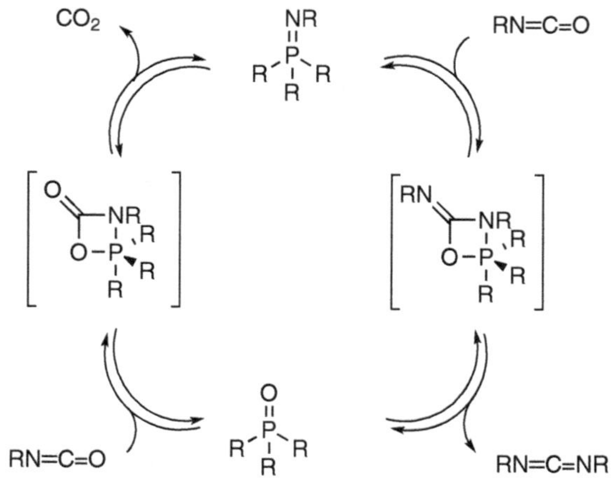

SCHEME 15-14. Catalytic cycle for the self-condensation of isocyanates.

iminophosphorane and another equivalent of isocyanate, giving the carbodiimide product and regenerating the phosphine oxide. Strong precedent exists for some of these steps: the reactions of isolated iminophosphoranes with both carbon dioxide to give isocyanates and with isocyanates to give carbodiimides are well known [31]. Further, reaction of highly hindered phosphine oxides with sulfonyl isocyanates has been shown to give isolable iminophosphoranes [40,41].

Consistent with this mechanism, a range of P=O-containing compounds (phosphine oxides, phosphinates, phosphonamides, phosphoramides, phosphoramidites, phosphates, phosphonofluoridates and aza-phosphatranes) as well as other species containing polarized X=O bonds (arsine and stibine oxides, sulfoxides, metal-oxo complexes) are also effective catalysts for the transformation [33,42–48]. As expected, simple amine-*N*-oxides, which contain a dipolar N–O bond but no multiple X–O bonding, are not reactive. The choice of **5** as catalyst was predicated at least in part upon its ready availability (by cheletropic reaction of phosphinyl dichlorides with isoprene followed by hydrolysis) but was also serendipitous, as these are highly active catalysts: **5** is ca. 20,000 times more reactive than triphenylphosphine oxide. This is attributed to the relief of strain in the phospholene ring (C–P–C bond angle ca. 90°) in the trigonal bipyramidal transition state of the metathesis reaction with isocyanate [49].

The generation of iminophosphoranes by the metathetical reaction of phosphine oxides with isocyanates obviates many of the problems associated with the classical Staudinger/Kirsanov approaches to their synthesis. The interception of this iminophosphorane by an electrophile other than a second molecule of isocyanate would therefore be of great synthetic utility, but is potentially problematic given the rapid nature of carbodiimide formation. In their initial studies, Monagle and Campbell demonstrated the potential intermediacy of the iminophosphorane in the self-condensation by carrying out a reaction in benzaldehyde as solvent, returning a modest yield of *N*-phenyl benzaldimine [33]. The reaction has been used preparatively to make imines of the highly electron-deficient hexafluoroacetone (Scheme 15-15) [50].

Froyen reported that highly active arsine oxides catalyze the crossed aza-Wittig reaction of isocyanates with reactive carbonyl compounds such as aldehydes, quinones, and α-diketones [51]. The reaction of vinyl isocyanates with 9,10-phenanthrenequinone

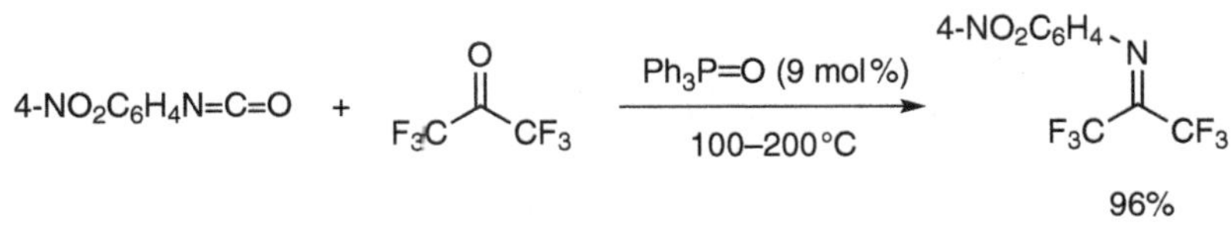

SCHEME 15-15. Intermolecular catalytic aza-Wittig reaction [50].

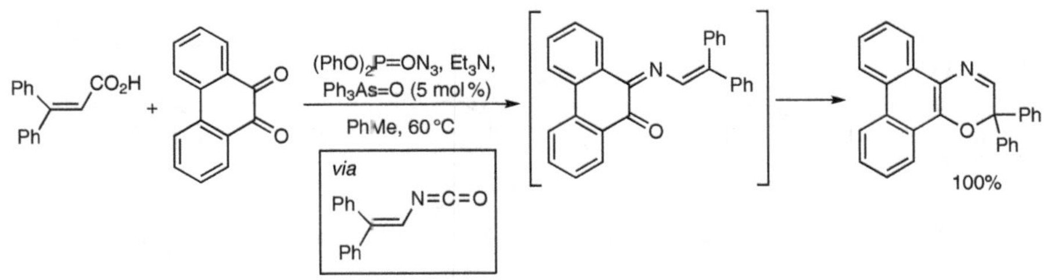

SCHEME 15-16. Intermolecular catalytic aza-Wittig reaction/electrocyclization [52].

generates intermediates that undergo electrocyclic ring closure to give photochromic oxazines (Scheme 15-16) [51,52].

In more broadly synthetically useful applications, the reaction has been harnessed by Marsden and coworkers to generate azole and azine heterocycles by intramolecular aza-Wittig reactions of aryl isocyanates bearing pendant carbonyl functionality [53]. Following the initial metathesis reaction between phosphine oxide and isocyanate, the intramolecular cyclization is entropically favored over intermolecular dimerization to form carbodiimides. The reaction works effectively for the synthesis of alkoxy/aminophenanthridines, with catalyst loadings as low as 1 mol% (Scheme 15-17).

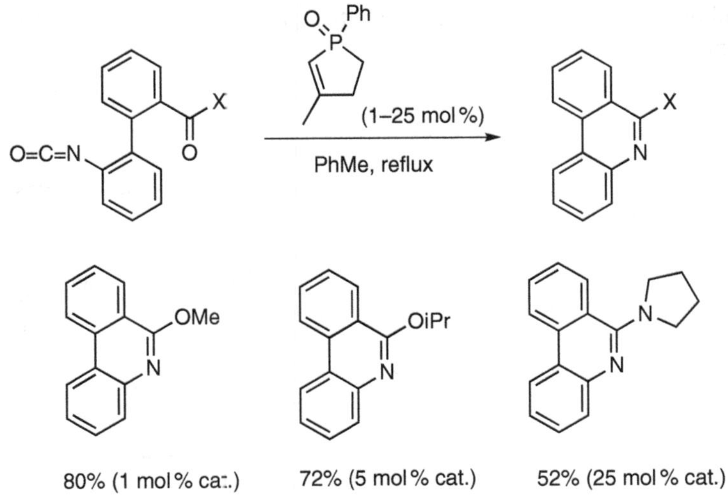

SCHEME 15-17. Catalytic intramolecular aza-Wittig reactions in azine synthesis [53].

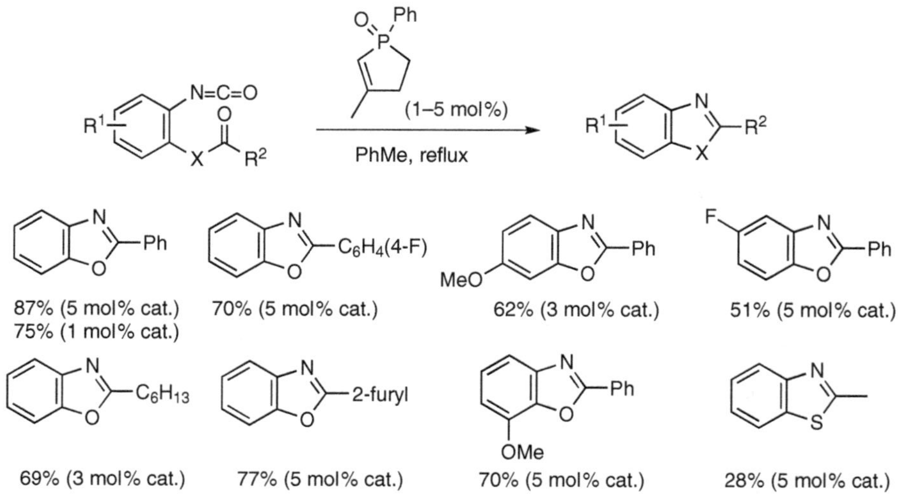

SCHEME 15-18. Catalytic intramolecular aza-Wittig reactions in azole synthesis [53].

The reaction was also used for the construction of five-membered heterocycles in the guise of benzoxazoles and benzothiazoles at catalyst loadings of 1–5 mol% (Scheme 15-18).

Monitoring of the reactions by *in situ* IR spectroscopy confirmed that no reaction occurred until addition of the catalyst, whereupon the signals for both the isocyanate and pendant carbonyl groups decayed with apparent first-order kinetics, consistent with the mechanism proposed by Monagle and Campbell, that is, isocyanate/phosphine oxide metathesis as the rate-determining step (Scheme 15-19) [38,54].

Overall, the catalytic aza-Wittig heterocyclization addresses the problems of both iminophosphorane generation and stoichiometric phosphine oxide waste and should be broadly applicable in heterocyclic synthesis.

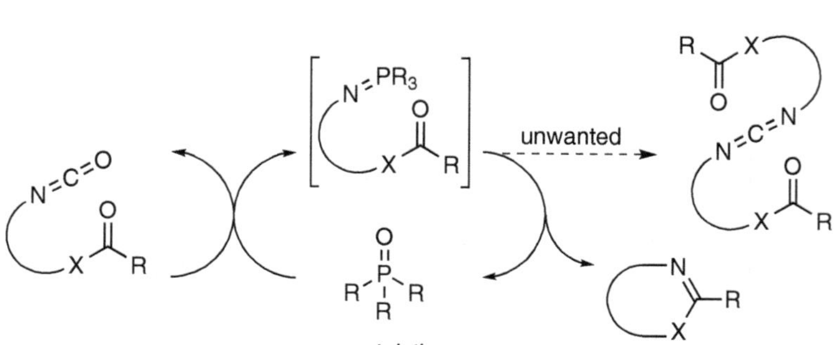

SCHEME 15-19. Catalytic cycle for the intramolecular aza-Wittig reaction.

15.4 MITSUNOBU CHEMISTRY

The Mitsunobu reaction involves the *in situ* activation and substitution of alcohols by nucleophiles with pK_as in the range of 3 to ca. 15 [55,56]. The reaction utilizes a redox reaction between a phosphine (reducing agent) and an azo compound (oxidizing agent) to generate an electrophilic phosphorus(V) species that activates the alcohol toward substitution, as well as a hydrazinyl anion that activates the nucleophilic component by deprotonation. The experimental expediency of coupling an unactivated nucleophile with an unactivated electrophile by addition of two simple reagents is clear, while the tolerance of the reaction to a wide variety of nucleophiles (carboxylic acids, phenols, thiols, phthalimides, azide, malonates) coupled with generally clean inversion of configuration in reactions of chiral secondary alcohols explains the popularity with research chemists from both industry and academia (Scheme 15-20).

The reaction nevertheless suffers from practical drawbacks. It generates two high molecular weight by-products (a phosphine oxide and a hydrazine) both of which detract from the atom efficiency of the process and create separation issues. Further, the requirement to use high-energy azo compounds as reagents is undesirable on scale (diethyl azodicarboxylate has a reported decomposition energy of $1000\,J\,g^{-1}$) [5]. As a consequence, despite its widespread use in academia and discovery chemistry, it is not as widely adopted on scale as might be expected—a survey of process chemistry reactions carried out in major pharmaceutical companies revealed that the Mitsunobu reaction was responsible for a mere 0.2% of reactions carried out [5,57].

Several solutions to the separation issue have been presented, based upon phase tagging of one or both reagents. Strategies include liquid–liquid phase separations (using aqueous soluble azodicarboxylate [58], base-labeled phosphines for extraction to aqueous acid [59] and acid-labile *tert*-butyl carboxylate-labeled azodicarboxylates [60]), tagged reagents for solid-phase extraction (e.g., fluorous azodicarboxylate [61] or MPEG-labeled azodicarboxylates and phosphines for solid-phase extraction [62]), polymer-supported azodicarboxylates (filtration) [63], or selective precipitation (phosphonium-salt labeled phosphines and dicarboxylates [64] or use of monomers for ROMP polymerization [65]). Combinations allowing ready separation of both components have also been used, employing, for example, base-labeled phosphine in conjunction with acid-labile *tert*-butyl carboxylates [66] or the use of both reagents supported on soluble polymers [67].

To date there have been no Mitsunobu reactions that are catalytic in the phosphorus component. However, Toy has demonstrated a variant that is catalytic in the potentially hazardous azodicarboxylate component (Scheme 15-21) [68]. In this process, diacetoxyiodobenzene is used as a stoichiometric cooxidant to recycle the reduced hydrazine back to the azodicarboxylate.

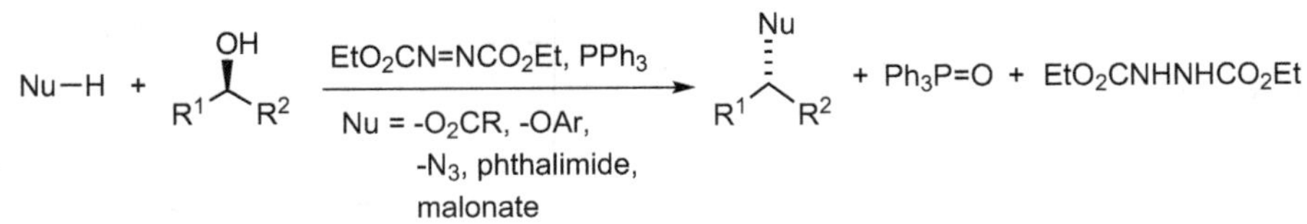

SCHEME 15-20. The Mitsunobu reaction.

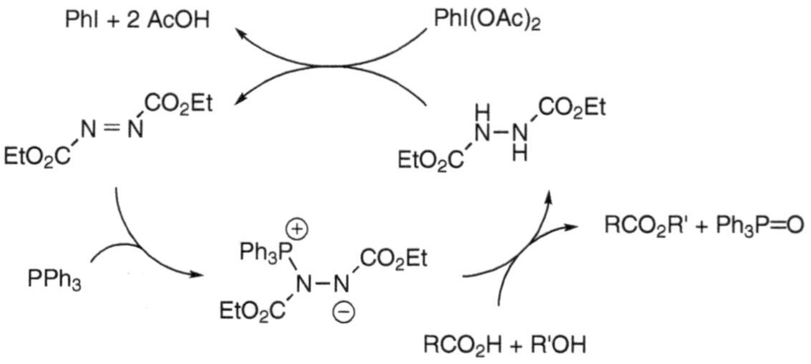

SCHEME 15-21. Catalytic cycle for the Mitsunobu reaction using sub-stoichiometric azo compound [68].

The process gives slightly lower yields than the stoichiometric process, but generally proceeds with identical stereoselectivities (Scheme 15-22). Although diacetoxyiodobenzene is less than ideal as a stoichiometric cooxidant from a process perspective (high molecular weight), the general principle of azo-recycling has been demonstrated and further developments can be anticipated.

The difficulty in developing Mitsunobu reactions catalytic in both the phosphorus and azo components is that the former requires reductive regeneration of the active species while the latter requires oxidative regeneration: issues with mutual incompatibilities of the recycling reagents will be difficult to resolve. An alternative protocol that effects Mitsunobu-like reactions using cyanomethylenephosphoranes **7** or **8** as a single reagent has been developed (Scheme 15-23) [69]. Here, protonation of the ylide generates the active nucleophile and the activated electrophilic phosphorus species for alcohol activation. The by-products are phosphine oxide and acetonitrile.

The reaction is effective for substitution by carboxylic acids, sulfonamides, trifluoroacetamides, and active methylene compounds; a range of the products prepared using this method is shown in Scheme 15-23. The reagents are less reactive than the Mitsunobu system but reactions can be forced to completion by carrying out the process at elevated temperatures. Some reactions of secondary alcohols fail or are low-yielding using this system.

Although there are no reports of catalytic variants of this reaction, the precedent for catalytic phosphine recycling/*in situ* ylide generation from the catalytic Wittig reaction [26] suggests that completion of a catalytic cycle may be feasible.

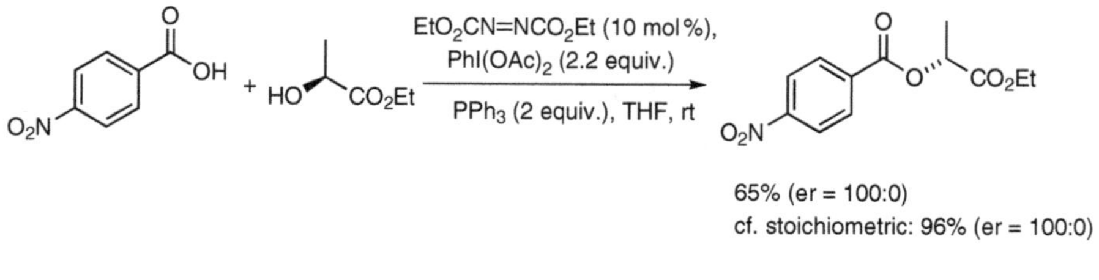

SCHEME 15-22. An exemplar of the catalytic Mitsunobu reaction [68].

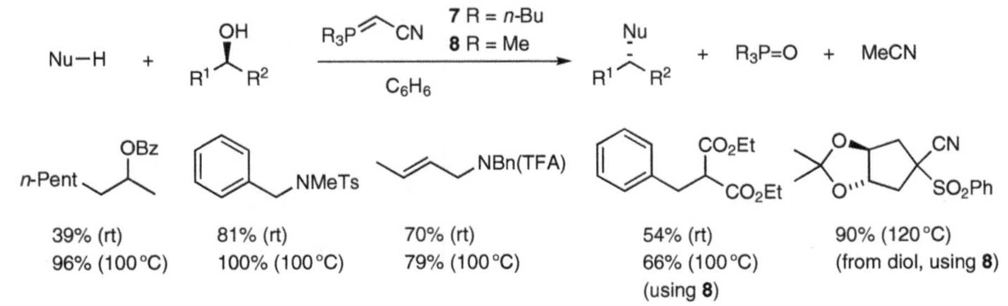

SCHEME 15-23. Cyanomethylenephosphoranes in direct substitution reactions [69].

15.5 APPEL HALOGENATIONS

Activation of alcohols by conversion to halides (or related sulfonate ester nucleofuges) is an important industrial-scale transformation: a review of the reactions carried out in process chemistry departments of three major pharmaceutical companies revealed that functional group interconversions accounted for 10% of all transformations carried out, and that 22% of these involved alcohol activation [57].

Organophosphorus-mediated halogenations are widely used on a laboratory scale, with so-called Appel halogenations [70] using triphenylphosphonium dihalides being especially popular (Scheme 15-24). These phosphorus(V) reagents can be preformed in a separate step, or generated *in situ* by reaction of triphenylphosphine with appropriate electrophilic halogen sources in a redox process. The reactions proceed by formation of an activated alkoxyphosphonium salt that then undergoes nucleophilic substitution by the liberated halide, generating the alkyl halide product and phosphine oxide waste as both a stoichiometric by-product of the reaction and also from the hydrolytic decomposition of any superstoichiometric reagent used to drive high conversions.

A patent report describes a phosphine oxide-catalyzed chlorination of alcohols by thionyl chloride (Scheme 15-25). The mechanism of the reaction is not clear but is thought not to involve the formation of phosphorus(V) dichloride.

Denton has subsequently exploited the known [20] ability of oxalyl chloride to convert triphenylphosphine oxide to the corresponding dichloride to develop a catalytic reaction for the chlorination of alcohols (Scheme 15-26) [72,73]. Although this is a rapid reaction, alcohols can react with oxalyl chloride directly to generate chlorooxalyl esters, which compromise the yield of the desired chloride. This is circumvented by slow addition of the oxalyl chloride to a solution of the alcohol and catalyst at such a rate as to ensure that significant concentrations of oxalyl chloride do not build up.

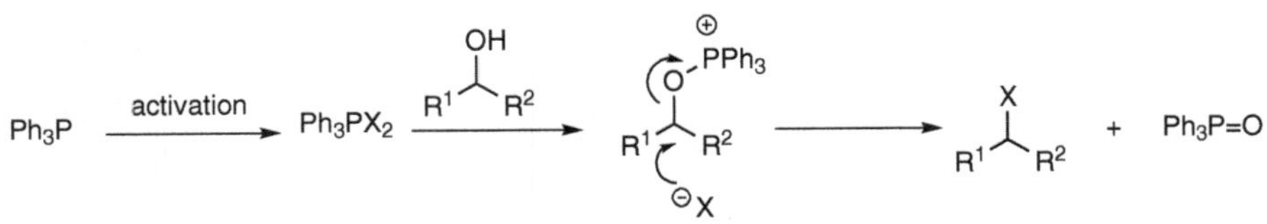

SCHEME 15-24. Appel-type halogenations.

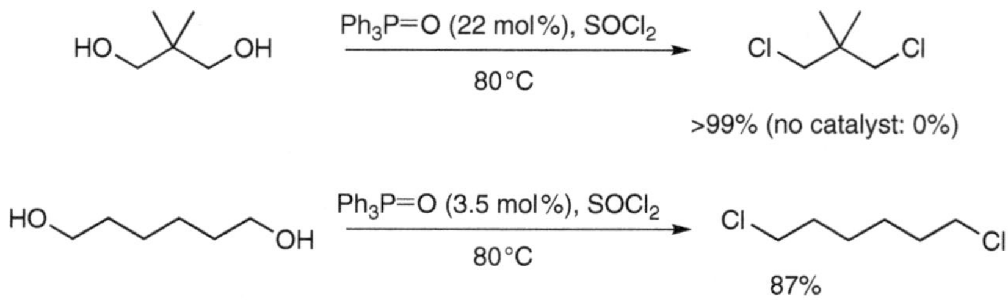

SCHEME 15-25. Phosphine oxide-catalyzed chlorination of alcohols [71].

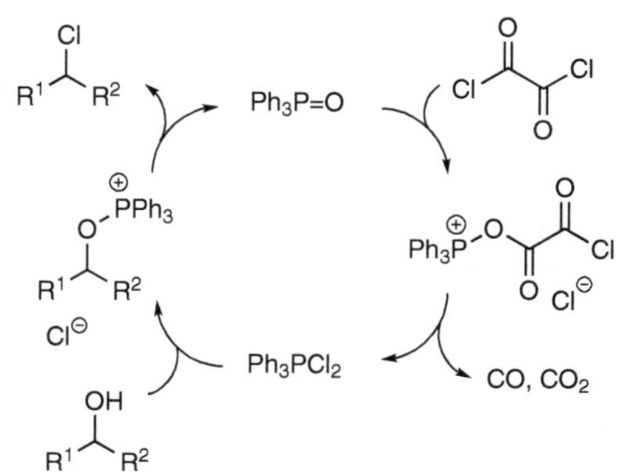

SCHEME 15-26. Catalytic cycle for the oxalyl chloride-mediated chlorination of alcohols [72,73].

The reaction proceeds well for a broad range of alcohols. Primary aliphatic, allylic, benzylic, and propargylic alcohols react cleanly in generally good yields (Scheme 15-27). The yields presented are isolated yields; in most cases the crude yield as judged by quantitative ^{1}H NMR experiments is significantly higher. Chiral secondary alcohols react with clean inversion of configuration, but the reaction fails for some hindered cyclic secondary alcohols (e.g., menthol).

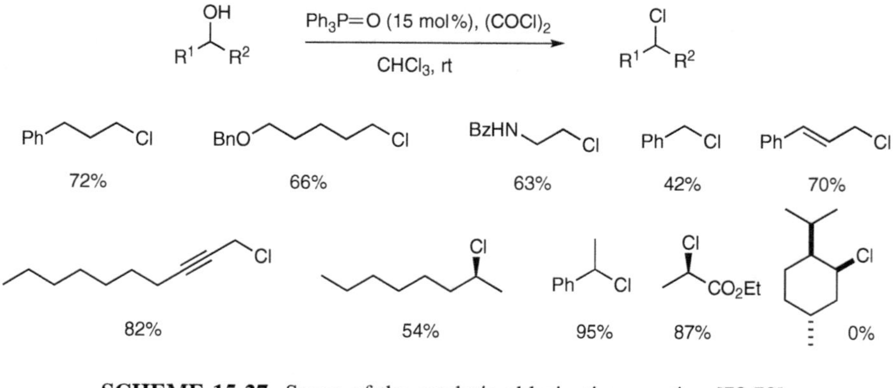

SCHEME 15-27. Scope of the catalytic chlorination reaction [72,73].

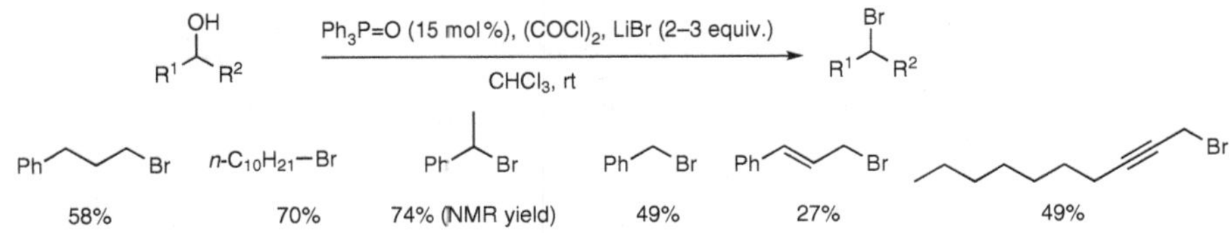

SCHEME 15-28. Catalytic bromination of alcohols [73].

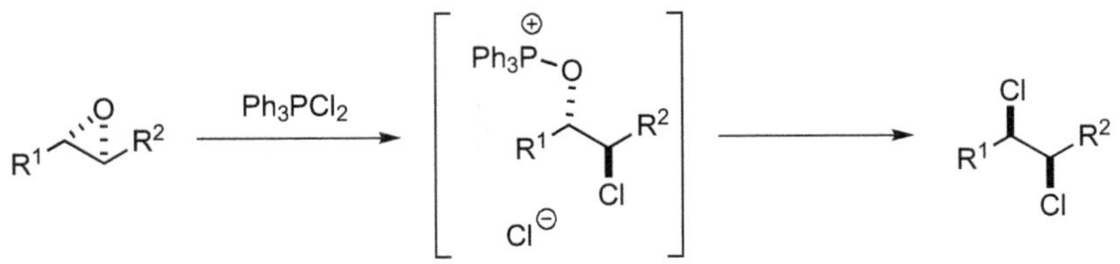

SCHEME 15-29. Dichlorination of epoxides [74].

The same group has also developed a protocol for catalytic bromination reactions [73]. Attempts to use oxalyl bromide in place of oxalyl chloride failed, largely due to competition from direct reaction of the reagent with the substrate alcohol. Recognizing that the intermediate oxyphosphonium salt generated from the phosphonyl dichloride reagent could potentially be intercepted by other nucleophiles, a protocol was developed in which the reaction was carried out in the presence of a modest excess of lithium bromide. The superior nucleophilicity toward the intermediate oxyphosphonium salt of bromide versus chloride, coupled with the higher concentration of bromide, delivers good yields of bromides directly from the alcohol (Scheme 15-28).

1,2-Dichlorides are useful reagents and can be accessed by reaction of epoxides with triphenylphosphonium dichloride (Scheme 15-29) [74]. The reaction proceeds with a double inversion of configuration, consistent with phosphonium salt-mediated nuclophilic epoxide opening followed by direct substitution of the alkoxyphosphonium salt.

Denton and coworkers have demonstrated that the same overall transformation can be effected using the catalytic triphenylphospine oxide/oxalyl chloride combination (Scheme 15-30) [75]. The stereochemical course of the transformations is as for the stoichiometric variant, supporting a similar mechanistic pathway for the process.

The Denton approach to halogenation of alcohols utilizes the "redox-neutral" catalytic pathways (Scheme 15-1b). An alternative approach based upon the "reduction-enabled" catalytic pathway (Scheme 15-1a) has been developed by van Delft [76]. In this system, a nucleophilic phosphorus(III) reagent **9** reacts with an electrophilic source of halogen (bromine in this case) to generate a halophosphonium salt. Activation of the alcohol by this salt and subsequent displacement by the liberated bromide generates the desired alkyl bromide and phosphine oxide. Reduction of the phosphine oxide back to the phosphine using diphenylsilane (as employed in the catalytic Wittig reaction [26]) completes the catalytic cycle (Scheme 15-31). The stumbling block in this process is that many electrophilic sources of bromine react competitively with the silane. Ultimately, the workers found that diethyl bromomalonate was a sufficiently mild source of bromine to coexist alongside the silane without significant cross-reactivity.

The scope of this reaction is shown in Scheme 15-32. The reaction proceeds well for primary alcohols but performs poorly for secondary alcohols, unless neighboring group

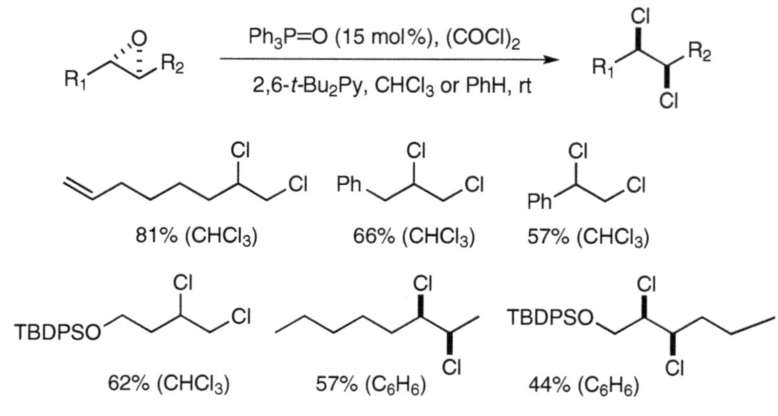

SCHEME 15-30. Catalytic dichlorination of epoxides [75].

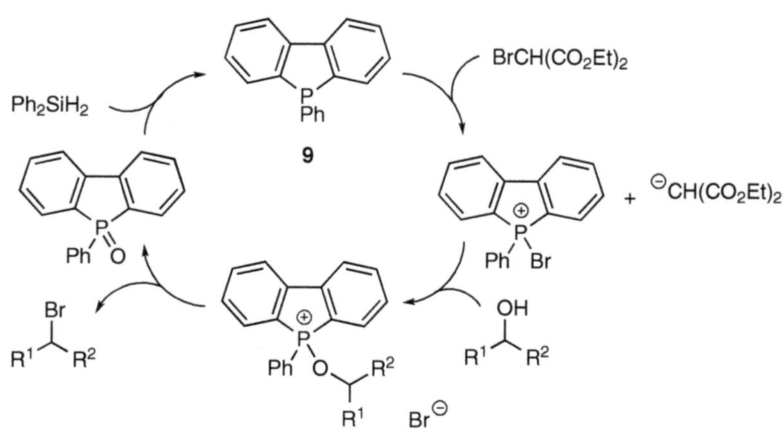

SCHEME 15-31. Catalytic cycle for the "reduction-enabled" bromination of alcohols [76].

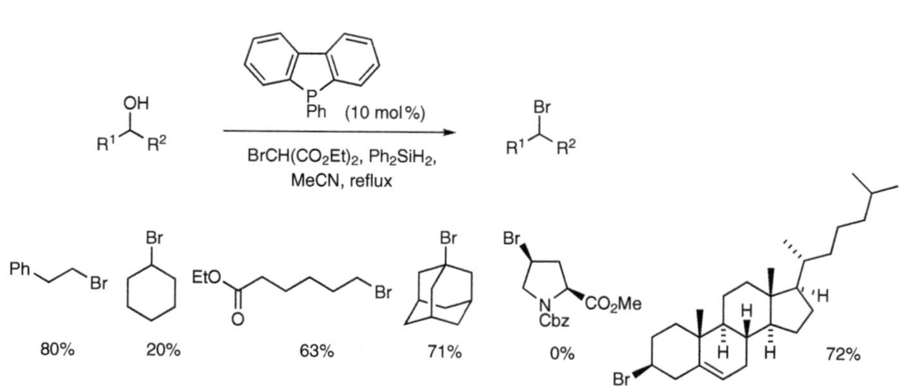

SCHEME 15-32. Scope of the "reduction-enabled" bromination of alcohols [76].

participation acts to facilitate the reaction (as in the reaction of cholest-5-enol). Substrates reacting via S_N1 pathways are also effective substrates.

15.6 CONCLUSIONS

The well-documented atom inefficiencies and waste separation issues associated with otherwise useful organophosphorus chemistry are being challenged by new developments in catalytic variants. Although still somewhat in its infancy, advances over the past 4 years have demonstrated proof of principle for phosphine oxide catalysis in several synthetically important reactions using both the "reduction-enabled" and "redox–neutral" catalytic manifolds. Optimization of these processes and further extension to other useful organophosphorus-mediated processes can be anticipated, with a truly catalytic variant of the Mitsunobu reaction a particularly tantalizing goal. The ultimate demonstration of the maturity of the catalytic methods outlined in this overview will come with their application in the development of scalable processes.

REFERENCES

1. Lide DR. editor. *CRC Handbook of Chemistry and Physics*, 84th edition. CRC Press, Boca Raton, FL, 2003–2004.
2. Quin LD. *A Guide to Organophosphorus Chemistry*. Wiley-Blackwell, New York, 2000.
3. Trost BM (1995). Atom economy—a challenge for organic synthesis: homogeneous catalysis leads the way. *Angew. Chem., Int. Ed. Engl.* 34, 259–281.
4. Sheldon RA (2007). The E factor: fifteen years on. *Green Chem.* 9, 1273–1283.
5. Constable DJC, Dunn PJ, Hayler JD, Humphrey GR, Leazer Jr., JL, Linderman RJ, Lorenz K, Manley J, Pearlman BA, Wells A, Zaks A, Zhang TY (2007). Key green chemistry research areas - a perspective from pharmaceutical manufacturers. *Green Chem.* 9, 411–420.
6. (a) Wittig G, Geissler G (1953). Zur reaktionsweise des pentaphenylphosphors und einiger derivate. *Liebigs Ann. Chem.* 580, 44–57. (b) Wittig G, Schöllkopf U (1954). Über triphenylphosphinmethylene als olefinbildende reagenzien. *Chem. Ber.* 87, 1318–1330.
7. Maryanoff BE, Reitz AB (1989). The Wittig olefination reaction and modifications involving phosphoryl-stabilized carbanions. Stereochemistry, mechanism, and selected synthetic aspects. *Chem. Rev.* 89, 863–927.
8. For a recent process example, see Stuk TL, Assink BK, Bates RC, Erdman DT, Fedij V, Jennings SM, Lassig JA, Smith RJ, Smith TL (2003). An efficient and cost-effective synthesis of pagoclone. *Org. Process Res. Dev.* 7, 851–855.
9. For recent process examples, see (a) Stivanello M, Leoni L, Bortolaso R (2002). Synthesis of 1,5-bis(triphenylphosphonium)pentan-3-ol dichloride and its application to the preparation of 1,7-di(pyridin-3-yl)heptan-4-ol. *Org. Process Res. Dev.* 6, 807–810. (b) Boulton LT, Brick D, Fox ME, Jackson M, Lennon IC, McCague R, Parkin N, Rhodes D, Ruecroft G (2002). Synthesis of the potent antiglaucoma agent, travoprost. *Org. Process Res. Dev.* 6, 138–145. (c) Ikemoto T Ito T, Hashimoto H, Kawarasaki T, Nishiguchi A, Mitsudera H, Wakimasu M, Tomimatsu K (2000). Development of a new synthetic route of a non-peptide CCR5 antagonist, TAK-779, for large-scale preparation. *Org. Process Res. Dev.* 4, 520–525.
10. Schlosser M, Christmann KF (1966). *trans*-Selective olefin syntheses. *Angew. Chem., Int. Ed.* 5, 126.

11. Pommer H (1977). The Wittig reaction in industrial practice. *Angew. Chem., Int. Ed.* 16, 423–429.
12. Relles HM, Schluenz RW (1974). Chemical transformations with regenerable, polymer-supported trisubstituted phosphine dichlorides. Efficacious incorporation of phosphorus reagents on polymer supports. *J. Am. Chem. Soc.* 96, 6469–6475.
13. Wadsworth WS, Emmons WD (1961). The utility of phosphonate carbanions in olefin synthesis. *J. Am. Chem. Soc.* 83, 1733–1738.
14. Still WC, Gennari C (1983). Direct synthesis of *Z*-unsaturated esters. A useful modification of the Horner–Emmons olefination. *Tetrahedron Lett.* 24, 4405–4408.
15. Ando K (1997). Highly selective synthesis of *Z*-unsaturated esters by using new Horner–Emmons reagents, ethyl (diarylphosphono)acetates. *J. Org. Chem.* 62, 1934–1939.
16. (a) McNulty J, Das P (2009). Highly stereoselective and general synthesis of (*E*)-stilbenes and alkenes by means of an aqueous Wittig reaction. *Eur. J. Org. Chem.* 4031–4035. (b) McNulty J, Keskar K (2008). Stereoselective Wittig olefination reactions employing a novel *ortho*-P-aryl alkoxide effect. *Tetrahedron Lett.* 49, 7054–7057. (c) Wang Q, El Khoury M, Schlosser M (2000). Bridging the final gap in stereocontrolled Wittig reactions: methoxymethoxy-armed allylic phosphorus ylides affording conjugated dienes with high *cis* selectivity. *Chem. Eur. J.* 6, 420–426. (d) Russell MG, Warren S (1998). Wittig reactions in water. Synthesis of new water-soluble phosphonium salts and their reactions with substituted benzaldehydes. *Tetrahedron Lett.* 39, 7995–7998. (e) Daniel H, Le Corre M (1987). Stereoselectivite de la reaction de Wittig. Utilisation d'une phosphine portant un groupement nucleophile. *Tetrahedron Lett.* 28, 1165–1168. (f) Trippet S, Walker DM (1961). The use, in Wittig reactions, of phosphonium salts and phosphobetaines containing a basic group. *J. Chem. Soc.* 2130–2133.
17. (a) Leung PSW, Teng Y, Toy PH (2010). Rasta resin-PPh_3 and its use in chromatography-free Wittig reactions. *Synlett* 1997–2001. (b) Westman J (2001). An efficient combination of microwave dielectric heating and the use of solid-supported triphenylphosphine for Wittig reactions. *Org. Lett.* 3, 3745–3747.
18. Galante A, Lhoste P, Sinou D (2001). Wittig reaction using perfluorinated ylides. *Tetrahedron Lett.* 42, 5425–5427.
19. Sieber F, Wentworth P, Janda KD (2000). Exploiting poly(ethylene glycol) as a matrix for liquid-phase organic synthesis. *Molecules* 5, 1018–1032.
20. Byrne PA, Rajendran KV, Muldoon J, Gilheany DG (2012). A convenient and mild chromatography-free method for the purification of the products of Wittig and Appel reactions. *Org. Biomol. Chem.* 10, 3531–3537.
21. Imamoto T, Kikuchi S, Miura T, Wada Y (2000). Stereospecific reduction of phosphine oxides to phosphines by the use of a methylation reagent and lithium aluminum hydride. *Org. Lett.* 3, 87–90.
22. Busacca CA, Raju R, Grinberg N, Haddad N, James-Jones P, Lee H, Lorenz JC, Saha A, Senanayake CH (2008). Reduction of tertiary phosphine oxides with DIBAL-H. *J. Org. Chem.* 73, 1524–1531.
23. Naumann K, Zon G, Mislow K (1969). Use of hexachlorodisilane as a reducing agent. Stereospecific deoxygenation of acyclic phosphine oxides. *J. Am. Chem. Soc.* 91, 7012–7023.
24. Shi L, Wang W, Wang Y, Huang Y (1989). The first example of a catalytic Wittig-type reaction. Tri-*n*-butylarsine-catalyzed olefination in the presence of triphenyl phosphite. *J. Org. Chem.* 54, 2027–2028.
25. Huang ZZ, Ye S, Xia W, Tang Y (2001). A practical catalytic Wittig-type reaction. *Chem. Commun.* 1384–1385.
26. O'Brien CJ, Tellez JL, Nixon ZS, Kang LJ, Carter AL, Kunkel SR, Przeworski KC, Chass GA (2009). Recycling the waste: the development of a catalytic Wittig reaction. *Angew. Chem., Int. Ed.* 48, 6836–6839.

27. Keglevich G, Gaumont AC, Denis JM (2001). Selective reductions in the sphere of organophosphorus compounds. *Heteroat. Chem.* 12, 161–167.

28. Vedejs E, Marth CF (1988). Mechanism of the Wittig reaction: the role of substituents at phosphorus. *J. Am. Chem. Soc.* 110, 3948–3958.

29. Smith JM, Greaney MF (2007). A one-pot silyl-Reformatsky olefination. *Tetrahedron Lett.* 48, 8687–8690.

30. Staudinger H, Meyer J (1919). Über neue organische phosphorverbindungen III. Phosphinmethylenderivate und phosphinimine. *Helv. Chim. Acta* 2, 635–646.

31. For recent reviews, see (a) Palacios F, Alonso C, Aparicio D, Rubiales G, de los Santos JM (2007). The aza-Wittig reaction: an efficient tool for the construction of carbon–nitrogen double bonds. *Tetrahedron* 63, 523–575. (b) Eguchi S (2005). Recent progress in the synthesis of heterocyclic natural products by the Staudinger/intramolecular aza-Wittig reaction. *ARKIVOC* 98–119. (c) Fresneda PM, Molina P (2004). Application of iminophosphorane-based methodologies for the synthesis of natural products. *Synlett* 1–17.

32. Campbell TW, Monagle JJ (1962). A new synthesis of mono- and polycarbodiimides. *J. Am. Chem. Soc.* 84, 1493.

33. Campbell TW, Monagle JJ, Foldi VS (1962). Carbodiimides. I. Conversion of isocyanates to carbodiimides with phospholine oxide catalyst. *J. Am. Chem. Soc.* 84, 3673–3677.

34. (a) Campbell TW, Monagle JJ (1963). Diphenylcarbodiimide. *Org. Synth.* 43, 31–33. (b) for a recent example, see Zhou HB, Alper H (2004). Regioselective and enantioselective synthesis of seven-membered ring cyclic arylguanidine and urea derivatives. *Tetrahedron* 60, 73–79.

35. Zhou HB, Dong C, Alper H (2004). Novel rhodium-catalyzed reaction of thiazolidine derivatives with carbodiimides. *Chem. Eur. J.* 10, 6058–6065.

36. Wolman Y, Kivity S, Frankel M (1967). The use of polyhexamethylenecarbodiimide, an insoluble condensing agent, in peptide synthesis. *Chem. Commun.* 629–630.

37. Mochizuki A, Sakamoto M, Yoshioka M, Fukuoka T, Takeshi K, Ueda M (2000). Novel photosensitive polymer based on polycarbodiimide and photoamine generator. *J. Polym. Sci. Part A: Polym. Chem.* 38, 329–336.

38. Monagle JJ, Campbell TW, McShane HF (1962). Carbodiimides. II. Mechanism of the catalytic formation from isocyanates. *J. Am. Chem. Soc.* 84, 4288–4295.

39. Monagle JJ, Mengenhauser JV (1966). Nucleophilic activity of the phosphoryl group. Oxygen-18 transfer in reactions of phenyl isocyanate with labeled phosphine oxides. *J. Org. Chem.* 31, 2321–2324.

40. Hoffmann H, Forster H, Tor-Poghossian G (1969). Über die umsetzung von *N*-fluorsulfonylisocyanat mit verschiedenen phosphorverbindungen. *Monatsh. Chem.* 100, 311–315.

41. Hall CR, Smith DJH (1974). Nucleophilic substitution at phosphoryl phosphorus. *Tetrahedron Lett.* 15, 1693–1696.

42. Ostrogovich G, Kerek F, Buzas A, Doca N (1969). Catalytic activity of some 1-aryl-3-methyl-3-phospholene-1-oxides in the disproportionation of phenyl isocyanate to diphenylcarbodiimide and carbon dioxide. *Tetrahedron* 25, 1875–1880.

43. Ulrich H, Tucker B, Sayigh AAR (1967). P–N heterocycles. Synthesis and use in the catalytic conversion of isocyanates into carbodiimides. *J. Org. Chem.* 32, 1360–1362.

44. Monagle JJ (1962). Carbodiimides. III. Conversion of isocyanates to carbodiimides. Catalyst studies. *J. Org. Chem.* 27, 3851–3855.

45. Tang J, Mohan T, Verkade JG (1994). Selective and efficient syntheses of perhydro-1,3,5-triazine-2,4,6-triones and carbodiimides from isocyanates using $ZP(MeNCH_2CH_2)_3N$ catalysts. *J. Org. Chem.* 59, 4931–4938.

46. Appleman JO, de Carlo VJ (1967). Conversion of isocyanates into carbodiimides with isopropyl methylphosphonofluoridate as catalyst. Kinetic studies. *J. Org. Chem.* 32, 1505–1507.

47. Hunger K (1966). Katalytische reaktionen von 1-oxo-1-alkoxyphospholinen. *Tetrahedron Lett.* 7, 5929–5931.

48. (a) V-oxo complexes: Birdwhistell KR, Boucher T, Ensminger M, Harris S, Johnson M, Toporek S (1993). Catalysis of phenyl isocyanate condensation to *N,N'*-diphenylcarbodiimide via vanadium oxo and imido complexes. *Organometallics* 12, 1023–1025. (b) Ti-oxo complexes: Darwish W, Seikel E, Kasmarker R, Harms K, Sundermeyer J (2011). Synthesis and X-ray crystal structures of imido and ureato derivatives of titanium(IV) phthalocyanine and their application in the catalytic formation of carbodiimides by metathesis from isocyanates. *Dalton Trans.* 40, 1787–1794. Several other metal complexes also catalyze the self-condensation of isocyanates to carbodiimides: (c) Fe/W/Mo carbonyls: Ulrich H, Tucker B, Sayigh AAR (1967). Metal carbonyls as catalysts in the conversion of isocyanates to carbodiimides. *Tetrahedron Lett.* 8, 1731–1733. (d) Fe/Mn carbonyls: Rahman AKF, Nicholas KM (2007). Catalytic conversion of isocyanates to carbodiimides by cyclopentadienyl manganese tricarbonyl and cyclopentadienyl iron dicarbonyl dimer. *Tetrahedron Lett.* 48, 6002–6004. (e) Ti/Zr/Hf alkoxides: Kumar A, Samuelson AG (2010). Room temperature metathesis of aryl isocyanates and aromatic aldehydes catalyzed by group(IV) metal alkoxides: an experimental and computational study. *J. Organomet. Chem.* 695, 338–345.

49. (a) Hudson RF, Brown C (1972). Reactivity of heterocyclic phosphorus compounds. *Acc. Chem. Res.* 5, 204–211. (b) Aksnes G, Froyen P (1969). Study of reactions between phenyl isocyanate and phosphine oxides, respectively phosphine imines. *Acta Chem. Scand.* 23, 2697–2703.

50. (a) Zeifman YV, Gambaryan NP, Knunyants IL (1963). Hexafluoroacetonimines. *Dokl. Akad. Nauk SSSR* 153, 1334–1337. (b) Roberts JD, Hall GE, Middleton WJ (1971). Nuclear magnetic resonance spectroscopy. Kinetics of isomerization of *p*-substituted hexafluoroacetone *N*-phenylimines. *J. Am. Chem. Soc.* 93, 4778–4781. (c) Petrov VA, Khasnis DD (1999). Unusual reaction of *N*-arylimines of hexafluoroacetone with *m*-chloroperoxybenzoic acid. New route to 2,2-bis(trifluoromethyl)benzoxazolidines. *Isr. J. Chem.* 39, 147–150.

51. Froyen P (1993). Synthesis and reactions of triphenylarsimines. *Phosphorus Sulfur Silicon Relat. Elem.* 81, 37–48.

52. Zhao W, Carreira EM (2011). One-pot synthesis of novel photochromic oxazine compounds. *Org. Lett.* 13, 5084–5087.

53. Marsden SP, McGonagle AE, McKeever-Abbas B (2008). Catalytic aza-Wittig cyclizations for heteroaromatic synthesis. *Org. Lett.* 10, 2589–2591.

54. An alternative mechanism for the cyclization was proposed in a review by List and Lay: List B, Lay F (2008). Catalytic aza-Wittig cyclisation. *Synfacts* 8, 872.

55. (a) Mitsunobu O (1981). The use of diethyl azodicarboxylate and triphenylphosphine in synthesis and transformation of natural products. *Synthesis* 1–28. (b) Hughes DL (1992). The Mitsunobu reaction. *Org. React.* 42, 335–656. (c) Kumara Swamy KC, Bhuvan Kumar NN, Balaraman E, Pavan Kumar KVP (2009). Mitsunobu and related reactions: advances and applications. *Chem. Rev.* 109, 2551–2651.

56. For some recent examples of Mitsunobu reactions in scale-up studies, see (a) de Koning PD, McAndrew D, Moore R, Moses IB, Boyles DC, Kissick K, Stanchina CL, Cuthbertson T, Kamatani A, Rahman L, Rodriguez R, Urbina A, Sandoval A, Rose PR (2011). Fit-for-purpose development of the enabling route to Crizotinib (PF-02341066). *Org. Process Res. Dev.* 15, 1018–1026. (b) Connolly TJ, Auguscinski W, Fung P, Galante R, Liu W, McGovern L, Sebastian A, Shen X, Shi X, Wilk B, Varsalona R, Zhong H (2010). Development of a pilot-plant-scale synthesis of an alkylated dihydrobenzothiadiazole *S,S*-dioxide: incorporation of a late-stage Mitsunobu reaction. *Org. Process Res. Dev.* 14, 868–877.

57. Carey JS, Laffan D, Thomson C, Williams MT (2006). Analysis of the reactions used for the preparation of drug candidate molecules. *Org. Biomol. Chem.* 4, 2337–2347.

58. Hagiya K, Muramoto N, Misaki T, Sugimura T (2009). DMEAD: a new dialkyl azodicarboxylate for the Mitsunobu reaction. *Tetrahedron* 65, 6109–6114.

59. (a) Camp D, Jenkins ID (1988). The use of a phosphine containing a basic group in the Mitsunobu esterification reaction. *Aust. J. Chem.* 41, 1835–1839. (b) von Itzstein M, Mocerino M (1990). (*p*-Dimethylaminophenyl)diphenylphosphine: a more practical phosphine in the Mitsunobu reaction. *Synth. Commun.* 20, 2049–2057.
60. Starkey GW, Parlow JJ, Flynn DL (1998). Chemically-tagged Mitsunobu reagents for use in solution-phase chemical library synthesis. *Bioorg. Med. Chem. Lett.* 8, 2385–2390.
61. (a) Chu Q, Henry C, Curran DP (2008). Second-generation tags for fluorous chemistry exemplified with a new fluorous Mitsunobu reagent. *Org. Lett.* 10, 2453–2456. (b) Markowicz MW, Dembinski R (2002). Fluorous, chromatography-free Mitsunobu reaction. *Org. Lett.* 4, 3785–3787.
62. (a) Figlus M, Wellaway N, Cooper AWJ, Sollis SL, Hartley RC (2011). Synthesis of arrays using low molecular weight MPEG-assisted Mitsunobu reaction. *ACS Comb. Sci.* 13, 280–285. (b) Figlus M, Tarruella AC, Messer A, Sollis SL, Hartley RC (2010). Low molecular weight MPEG-assisted organic synthesis. *Chem. Commun.* 46, 4405–4407.
63. Arnold LD, Assil HI, Vederas JC (1989). Polymer-supported alkyl azodicarboxylates for Mitsunobu reactions. *J. Am. Chem. Soc.* 111, 3973–3976.
64. Poupon JC, Boezio AA, Charette AB (2006). Tetraarylphosphonium salts as solubility-control groups: phosphonium-supported triphenylphosphine and azodicarboxylate reagents. *Angew. Chem., Int. Ed.* 45, 1415–1420.
65. Barrett AGM, Roberts RS, Schroder J (2000). Impurity annihilation: chromatography-free parallel Mitsunobu reactions. *Org. Lett.* 2, 2999–3001.
66. Kiankarimi M, Lowe R, McCarthy JR, Whitten JP (1999). Diphenyl 2-pyridylphosphine and di-*tert*-butylazodicarboxylate: convenient reagents for the Mitsunobu reaction. *Tetrahedron Lett.* 40, 4497–4500.
67. Harned AM, He HS, Toy PH, Flynn DL, Hanson PR (2005). Multipolymer solution-phase reactions: application to the Mitsunobu reaction. *J. Am. Chem. Soc.* 127, 52–53.
68. But TYS, Toy PH (2006). Organocatalytic Mitsunobu reactions. *J. Am. Chem. Soc.* 128, 9636–9637.
69. (a) Tsunoda T, Nagino C, Oguri M, Ito S (1996). Mitsunobu-type alkylation with active methine compounds. *Tetrahedron Lett.* 37, 2459–2462. (b) Tsunoda T, Nagaku M, Nagino C, Kawamura Y, Ozaki F, Hioki H, Ito S (1995). Carbon–carbon bond formation with new Mitsunobu reagents. *Tetrahedron Lett.* 36, 2531–2534. (c) Tsunoda T, Ozaki F, Ito S (1994). Novel reactivity of stabilized methylenetributylphosphorane: a new Mitsunobu reagent. *Tetrahedron Lett.* 35, 5081–5082.
70. Appel R (1975). Tertiary phosphane/tetrachloromethane, a versatile reagent for chlorination, dehydration, and P–N linkage. *Angew. Chem., Int. Ed.* 14, 801–811.
71. Rohde T, Huttenloch O, Osswald F, Wissel K, inventors; BASF SE, assignee. Method for chlorinating alcohols with thionyl chloride in the presence of a triaryl phosphinoxide. World Patent WO/2007/028761. March 15, 2007.
72. Denton RM, An J, Adeniran B (2010). Phosphine oxide-catalyzed chlorination reactions of alcohols under Appel conditions. *Chem. Commun.* 46, 3025–3027.
73. Denton RM, An J, Adeniran B, Blake AJ, Lewis W, Poulton AM (2011). Catalytic phosphorus (V)-mediated nucleophilic substitution reactions: development of a catalytic Appel reaction. *J. Org. Chem.* 76, 6749–6767.
74. Yoshimitsu T, Fukumoto N, Nakami R, Kojima N, Tanaka T (2010). Asymmetric total synthesis of (+)-hexachlorosulfolipid, a cytotoxin isolated from Adriatic mussels. *J. Org. Chem.* 75, 5425–5437.
75. Denton RM, Tang X, Przeslak A (2010). Catalysis of phosphorus(V)-mediated transformations: dichlorination reactions of epoxides under Appel conditions. *Org. Lett.* 12, 4678–4681.
76. van Kalkeren HA, Leenders SHAM, Hommerson CRA, Rutjes FPJT, van Delft FL (2011). *In situ* phosphine oxide reduction: a catalytic Appel reaction. *Chem. Eur. J.* 17, 11290–11295.

16

FORMATION OF C–C BONDS VIA CATALYTIC HYDROGENATION AND TRANSFER HYDROGENATION

JOSEPH MORAN[1,2] AND MICHAEL J. KRISCHE[1]

[1]*Department of Chemistry and Biochemistry, University of Texas at Austin, Austin, TX, USA*
[2]*ISIS, University of Strasbourg, Strasbourg, France*

16.1 INTRODUCTION: MINIMIZING PREACTIVATION FOR SYNTHETIC EFFICIENCY

Many methods for chemical synthesis exist, yet the majority of this technology is not easily brought to scale. Hence, there is an authentic need to discover and develop catalytic processes that embody the principal characteristics of "process relevance," in particular, the ability to transform abundant, renewable feedstocks to value-added products in the absence of stoichiometric by-products. This quality is embodied by both catalytic hydrogenation, which has found broad application across all segments of the chemical industry, including the manufacture of chiral pharmaceutical ingredients, and alkene hydroformylation [1], the prototypical C–C bond-forming hydrogenation and largest volume application of homogenous catalysis [2,3].

Inspired by the impact of hydrogenation and hydroformylation on industrial chemistry, systematic investigations into reductive C–C bond formation via catalytic hydrogenation were initiated in our laboratory. We have found that diverse π-unsaturated reactants are reductively coupled to polarized C=X (X = O, NR) bonds under hydrogenation conditions, offering an alternative to stoichiometric organometallic reagents in a broad range of carbonyl and imine additions. This concept is extended further by related transfer hydrogenative processes, for which hydrogen exchange between alcohols and π-unsaturated reactants triggers the generation of electrophile–nucleophile pairs that

Sustainable Catalysis: Challenges and Practices for the Pharmaceutical and Fine Chemical Industries, First Edition. Edited by Peter J. Dunn, K. K. (Mimi) Hii, Michael J. Krische, and Michael T. Williams.

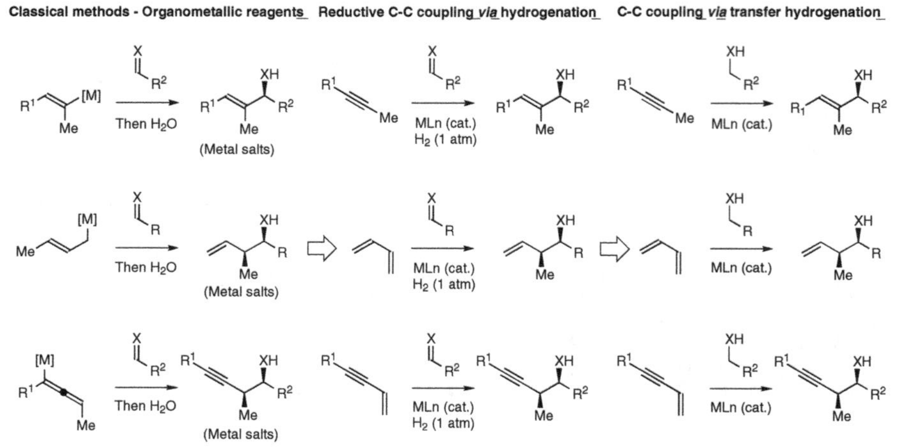

SCHEME 16-1. Hydrogen-mediated C=X (X = O, NR) addition circumvents stoichiometric by-product generation.

combine to form products of carbonyl addition directly from the alcohol oxidation level (Scheme 16-1) [4].

In terms of potential process relevance, hydrogen-mediated C–C bond formations, like the parent hydrogenations and hydroformylations, do not generate stoichiometric by-products, that is, they exhibit complete atom economy [5]. Additionally, the ability to merge redox events with C–C bond construction is step economic [6,7]. Indeed, it is noteworthy that transfer hydrogenative C–C couplings of primary alcohols bypass discrete alcohol oxidation, which is problematic on scale [8]. Finally, the use of abundant π-unsaturated reactants as synthetic equivalents to nonstabilized carbanions minimizes preactivation, that is, the degree of separation between reagent and feedstock, while avoiding discrete metallation–transmetallation sequences that often accompany the use of stoichiometric organometallic reagents. Beyond these attributes, the C–H and C–C bonds are native functional groups in all organic molecules, and so the retrosynthetic operation of C–C bond cleavage accompanied by withdrawal of hydrogen can be broadly applied.

To illustrate how these features streamline organic synthesis, it is instructive to compare classical approaches to carbonyl vinylation and crotylation with the corresponding hydrogenative processes. For example, the Oppolzer–Wipf method for enantioselective carbonyl vinylation requires alkyne hydrozirconation mediated by Cp_2ZrHCl, a highly mass-intensive reagent, with subsequent transmetallation to zinc mediated by $ZnMe_2$, a pyrophoric liquid [9,10]. Four stoichiometric organometallics are employed in advance of C–C coupling, not including reagents required for the preparation of Cp_2ZrHCl and $ZnMe_2$. Similarly, Brown's reagents for carbonyl crotylation, (*E*)- and (*Z*)-crotyldiisopinocampheylborane [11], are prepared through potassiation of butene employing Schlosser's base [12] followed by transmetallation to boron. Again, multiple manipulations and stoichiometric reagents (*n*-BuLi, KC_4H_7, Ipc_2BOMe) are required to prepare the reagent. Moreover, for each crotyl moiety that is transferred, 2 equiv. of isopinocampheol are generated, requiring separation of a secondary alcohol product from

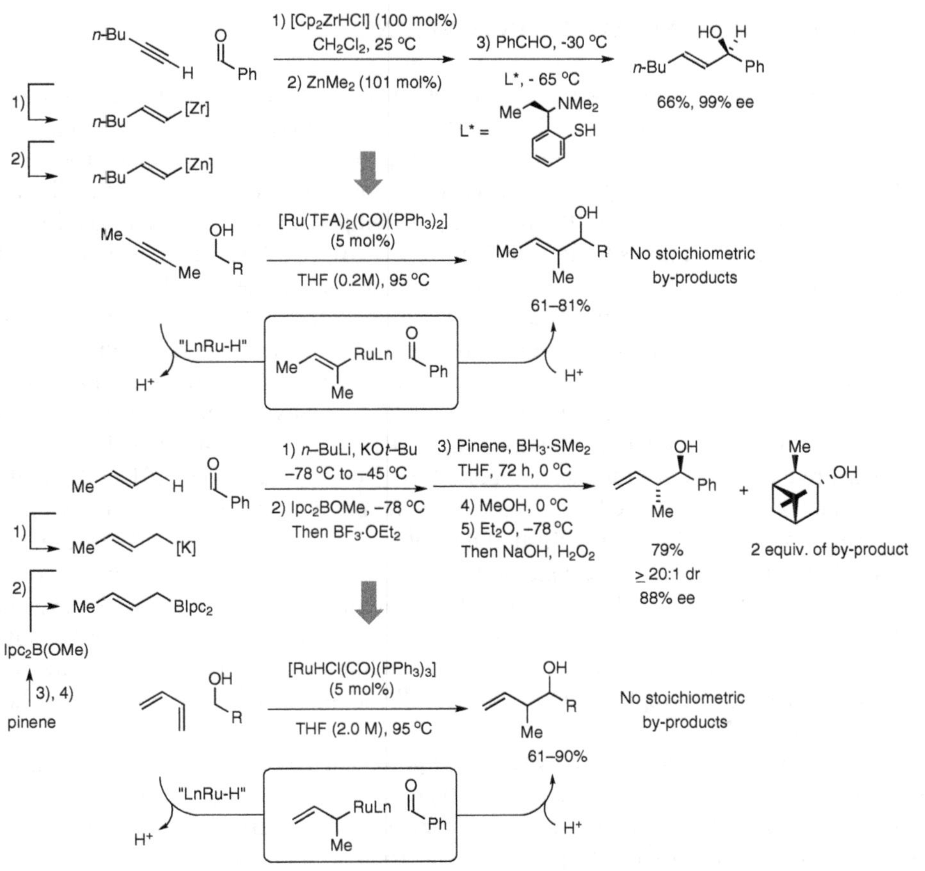

SCHEME 16-2. Minimizing preactivation for synthetic efficiency via hydrogenative C–C coupling.

a super-stoichiometric quantity of secondary alcohol by-product [13]. For the corresponding hydrogenative processes, carbonyl addition is achieved from the alcohol oxidation level through direct use of abundant π-unsaturated reactants in the absence of stoichiometric by-products, dramatically enhancing the E-factor (kg waste/kg product) [14] of each transformation (Scheme 16-2).

Catalytic hydrogenation and carbonyl addition rank among the oldest and most broadly utilized chemical transformations employed in organic chemistry. Interestingly, while Sabatier and Grignard shared the 1912 Nobel Prize for their seminal contributions to these fields, respectively, systematic efforts to merge these areas were absent until recently [4,15,16]. In this account, hydrogenative variants of major classes of carbonyl and imine additions are surveyed. Specifically, C=X (X=O, NR) vinylation, aldol and Mannich additions, carbonyl allylations, and related processes are described. This account is not exhaustive, but is meant to highlight milestones in reactivity and processes of greatest preparative utility. These results demonstrate that the union of catalytic hydrogenation and carbonyl addition chemistry has evoked broad new capabilities, and yet only a fraction of the potential of catalytic hydrogenation to serve as a method of C–C bond formation has been uncovered.

16.2 CARBONYL AND IMINE VINYLATION

Allylic alcohols and allylic amines are important building blocks in organic synthesis [17,18]. Among the methods available for their enantioselective preparation, metal-catalyzed allylic substitution employing *O*- and *N*-nucleophiles figures prominently [19]. A more convergent approach involves the catalytic enantioselective addition of vinylmetal reagents to aldehydes [9,10,20–23]. In comparison, corresponding additions of vinylmetal reagents to imines are less developed [24–26]. Notably, the requisite vinylmetal reagents employed in the aforementioned aldehyde and imine additions are prepared through alkyne hydrometallation. Thus, metal-catalyzed alkyne-C=X (X = O, NR) reductive coupling represents a more direct approach to allylic alcohols and allylic amines.

Metal-catalyzed alkyne–carbonyl reductive coupling was first reported in 1994 by Ojima and coworkers [27], who described the rhodium-catalyzed reductive cyclization of acetylenic aldehydes mediated by silane. In 1995, related silane-mediated titanocene-catalyzed reductive cyclizations were reported by Crowe and Rachita [28], and, in 1997, corresponding nickel-catalyzed cyclizations were reported by Montgomery and coworkers [29a–c,e]. Based on Montgomery's finding, intermolecular reductive alkyne–aldehyde coupling was achieved under the conditions of nickel catalysis by Jamison and coworkers in 2000 [30]. Improved nickel-based catalysts for intermolecular coupling were developed later by Takai and coworkers [31] and Montgomery and coworkers [29d], and enantioselective variants soon followed [29f,30b]. Withstanding work from the present author (see below), only a single example of intermolecular alkyne–imine reductive coupling has been reported, to date [32].

Although the aforementioned alkyne–carbonyl reductive coupling protocols bypass the use of preformed vinyl-metal reagents, they typically employ terminal reductants that are metallic, pyrophoric, or highly mass intensive (e.g., ZnR_2, BEt_3, $HSiR_3$). In contrast, hydrogenation of alkynes in the presence of carbonyl compounds and imines delivers vinylation products in the absence of stoichiometric by-products [4b]. For example, using cationic rhodium catalysts, hydrogenation of conjugated alkynes in the presence of glyoxalates and pyruvates provides the corresponding α-hydroxy esters with uniformly high levels of enantioselectivity [33a–c]. Under closely related conditions, heterocyclic aromatic aldehydes and ketones, those that are isostructural with respect to vicinal dicarbonyl compounds, couple to conjugated alkynes to provide optically enriched heteroaryl-substituted secondary and tertiary carbinols, respectively (Scheme 16-3) [33d].

Reductive couplings of conjugated enynes or diynes to activated imines in the form of (*N*-sulfinyl)iminoacetates are also possible under the conditions of rhodium-catalyzed hydrogenation [34]. Using appropriately substituted *N*-sulfinyl substituents [35], the corresponding β,γ-unsaturated α-amino acid esters are generated as single diastereomers. Exhaustive hydrogenation of the diene or enyne side chain of the C–C coupling product provides access to β-substituted α-amino acids (Scheme 16-4).

Computational and experimental data support a mechanism involving alkyne–carbonyl oxidative coupling to form metallacyclic intermediates, which are subject to turnover limiting hydrogenolysis via σ-bond metathesis with elemental hydrogen [36]. In certain processes, carboxylic acid cocatalysts promote dramatic rate enhancements. The basis for this effect appears to involve protonation of the transient metallacyclic intermediate to form a rhodium carboxylate. Computational studies from three different research groups suggest that four-centered transition structures for the hydrogenolysis of Rh–O bonds are substantially higher in energy than those occurring by way of six-centered transition

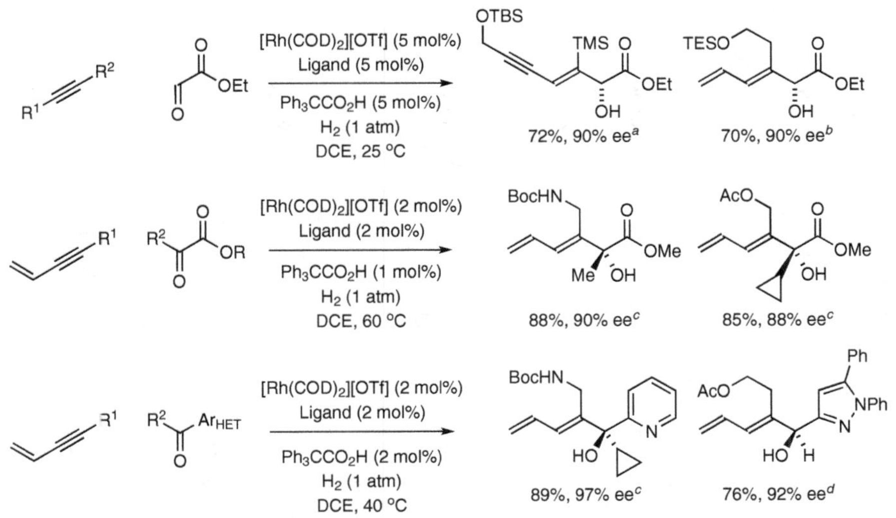

SCHEME 16-3. Rh-catalyzed hydrogenation of conjugated alkynes in the presence of activated carbonyl compounds to provide allylic alcohols. [a]Ligand = (*R*)-Cl,MeO-BIPHEP. [b]Ligand = (*R*)-3,5-*t*-Bu-4-MeO-MeO-BIPHEP. [c]Ligand = (*R*)-XylylWALPHOS. [d]Ligand = (*R*)-TolBINAP.

structures involving rhodium carboxylates [36]. The requirement of cationic rhodium catalysts also merits discussion. As revealed in mechanistic studies of the prototypical enantioselective hydrogenation of dehydroamino esters, hydrogen oxidative addition is turnover limiting when using cationic rhodium catalysts [37,38]. In C–C bond forming hydrogenations, the diminished rate of hydrogen oxidative addition combined with the availability of an additional coordination site allows alkyne-C=X (X = O, NR) oxidative coupling to become more rapid than hydrogen oxidative addition. Indeed, it is the very rapid nature of the alkyne-C=X oxidative coupling event with respect to hydrogen oxidative addition that assists in preventing overreduction of highly unsaturated diene- and enyne-containing products (Scheme 16-5).

The alkyne–C=X oxidative coupling event may be viewed as a reduction across two functional groups. For cationic rhodium(I), which is a rather weak reductant, the oxidative coupling of conjugated alkynes and activated electrophiles occurs readily by virtue of their

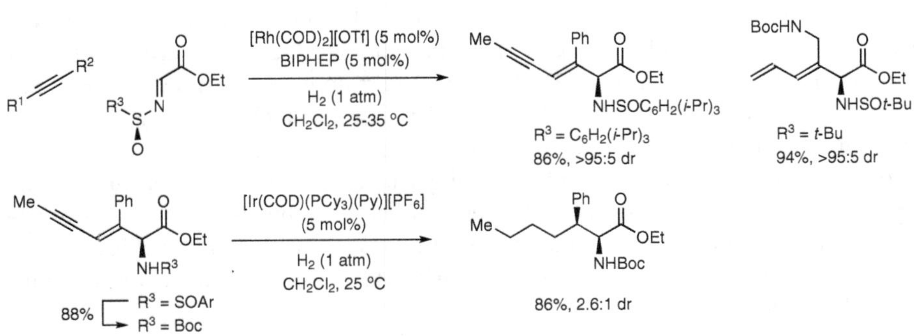

SCHEME 16-4. Rh-catalyzed hydrogenation of conjugated alkynes in the presence of activated imines to provide β-substituted α-amino acids.

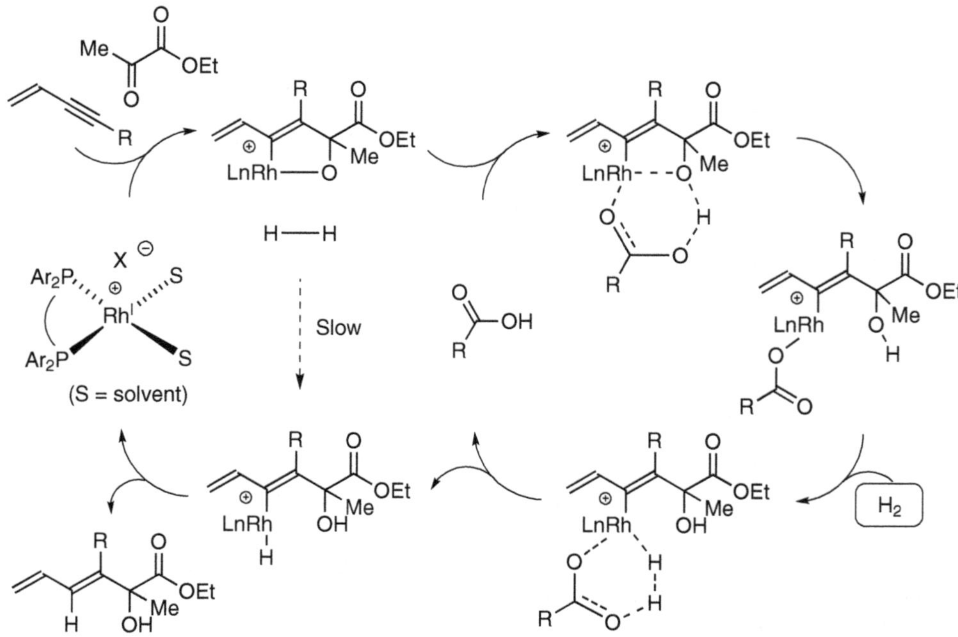

SCHEME 16-5. A general catalytic cycle for hydrogen-mediated alkyne–carbonyl reductive coupling employing cationic rhodium catalysts as supported by computational studies [36].

lower LUMO energies, meaning they are more easily reduced. To promote hydrogen-mediated reductive couplings of nonconjugated alkynes to unactivated aldehydes, which embody higher lying LUMO energies, tethering of the alkyne and aldehyde partners is required. For example, using chirally modified rhodium catalysts, hydrogenation of the indicated acetylenic aldehydes promotes highly enantioselective reductive carbocyclization (Scheme 16-6) [39].

An alternative means of activating reagents that embody higher lying LUMO energies is through the use of a more strongly reducing metal catalyst. Iridium(I) complexes are stronger π-donors than rhodium [40] due to relativistic effects associated with the lanthanide contraction [41], as supported by the relative energies of infrared radiation absorbed by isostructural carbonyl complexes of iridium and rhodium: $(Ph_3P)_2M(Cl)(CO)$, M = Ir, $\nu_{co} = 1965\ cm^{-1}$; M = Rh, $\nu_{co} = 1980\ cm^{-1}$ [40]. In accordance with the Dewar–Chatt–Duncanson model [42], and as borne out by single-crystal X-ray diffraction analysis of iridium(I) alkyne complexes [43], the increased π-donicity of iridium confers metallacyclic character to the bound alkyne, enhancing its nucleophilicity and, in turn, facilitating the alkyne–C=X oxidative coupling event. Thus, using second-generation hydrogenation catalysts based on iridium, highly enantioselective intermolecular reductive coupling of 1,2-dialkyl substituted alkynes to *N*-arylsulfonyl imines is achieved [44]. The resulting trisubstituted allylic amines are formed with complete levels of *E*:*Z* selectivity (≥95:5), and using nonsymmetric alkynes excellent regioselectivity is observed (Scheme 16-7).

As corroborated by isotopic labeling, ESI-MS, and computational studies [36c], the hydrogenative coupling of acetylene, an abundant chemical feedstock [45], occurs through a unique mechanism involving the formation of a rhodacyclopentadiene [46] followed by

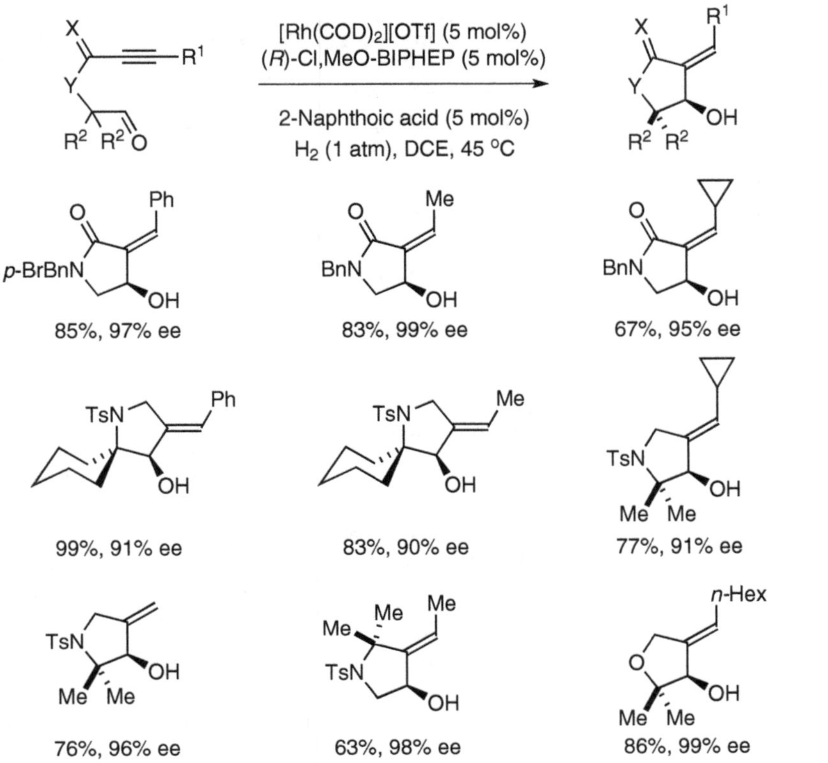

SCHEME 16-6. Enantioselective carbocyclization of acetylenic aldehydes via rhodium-catalyzed asymmetric hydrogenation.

C=X (X = O, NR) insertion and Brønsted acid cocatalyzed hydrogenolysis of the metallacycle to form products of (Z)-butadienylation [47]. Using chiral rhodium catalysts, allylic alcohols and allylic amines are generated in highly optically enriched form. Hydrogenative coupling of acetylene to α-chiral aldehydes employing enantiomeric rhodium catalysts

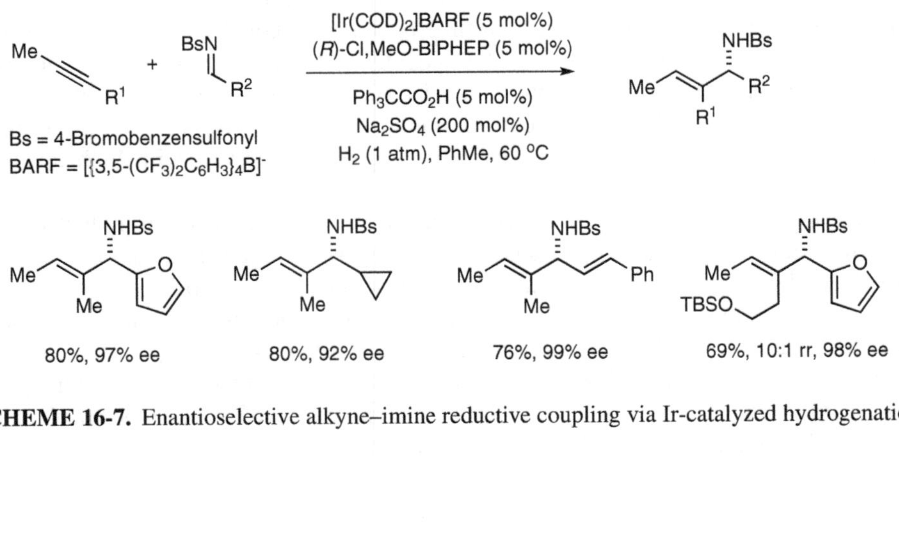

SCHEME 16-7. Enantioselective alkyne–imine reductive coupling via Ir-catalyzed hydrogenation.

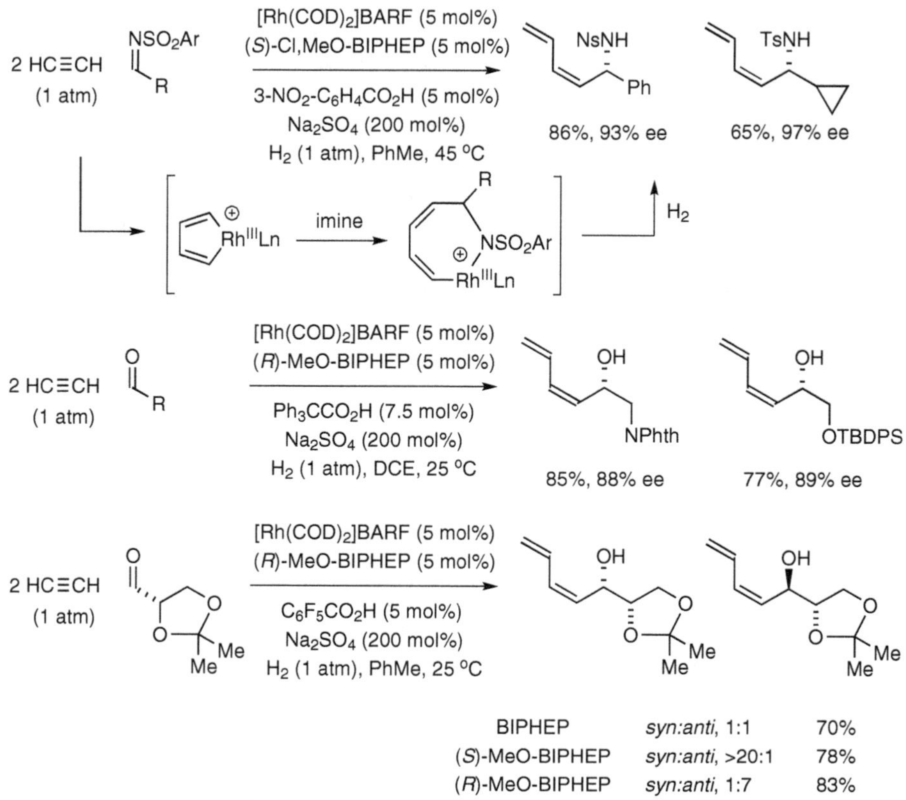

SCHEME 16-8. Carbonyl and imine (*Z*)-butadienylation via Rh-catalyzed hydrogenative coupling of acetylene.

ligated by (*S*)-MeO-BIPHEP and (*R*)-MeO-BIPHEP delivers diastereomeric products of carbonyl-(*Z*)-butadienylation with good levels of catalyst-directed diastereoselectivity (Scheme 16-8) [47c].

Under transfer hydrogenation conditions employing a ruthenium catalyst, hydrogen exchange between primary alcohols and 2-butyne triggers the formation of aldehyde–vinylruthenium pairs, which combine to form products of carbonyl vinylation [48a]. Under more forcing conditions, the initially formed allylic alcohols dehydrogenate to form conjugated enones, thus enabling intermolecular alkyne hydroacylation from the alcohol oxidation level [48b]. Under related conditions employing formic acid as terminal reductant, nonsymmetric alkynes reductively couple to aldehydes with high levels of regiocontrol (Scheme 16-9).

Using nickel catalysts, nonsymmetric disubstituted alkynes engage in reductive coupling to paraformaldehyde in the absence of exogenous reductant [48d]. Here, paraformaldehyde serves as electrophile and reductant, as evidenced by isolation of the reaction products as their formate esters. Under the conditions of ruthenium catalysis, formic acid-mediated reductive coupling of nonsymmetric disubstituted alkynes to paraformaldehyde exhibits complementary regioselectivity [48d]. Thus, both regioisomeric primary trisubstituted allylic alcohols are available with complete levels of stereocontrol in the absence of stoichiometric metallic reagents (Scheme 16-10).

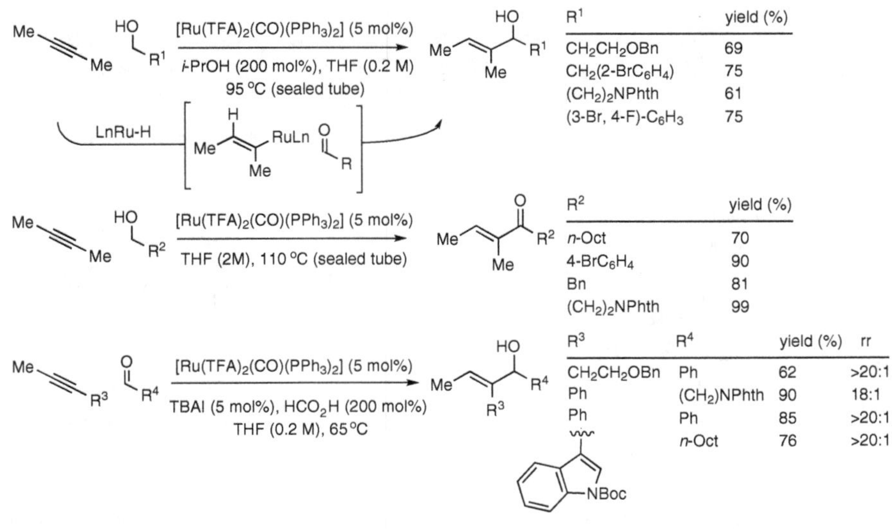

SCHEME 16-9. Ru-catalyzed alkyne–carbonyl C–C coupling from the alcohol or aldehyde oxidation level.

16.3 CARBONYL ALLYLATION AND PROPARGYLATION

Enantioselective carbonyl allylation is one of the principal protocols used for polyketide construction [49]. As illustrated in seminal work by Herold and Hoffmann [50a], enantioselective carbonyl allylation may be achieved using chirally modified allyl metal reagents [50]. Subsequently, to circumvent the use of stoichiometric chiral modifiers, enantioselective chiral Lewis acid- and chiral Lewis base-catalyzed additions of allylmetal reagents were developed [51]. Other catalytic methods for carbonyl allylation include the reduction of metallo-π-allyls derived from allylic alcohols and allylic carboxylates [52,53] and asymmetric variants of Fürstner's modification [54] of the Nozaki–Hiyama reaction

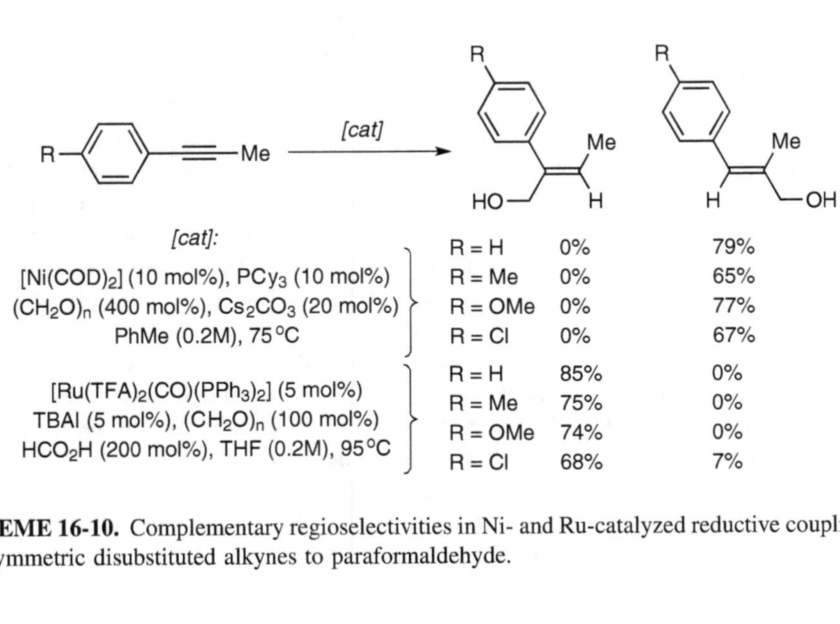

SCHEME 16-10. Complementary regioselectivities in Ni- and Ru-catalyzed reductive couplings of nonsymmetric disubstituted alkynes to paraformaldehyde.

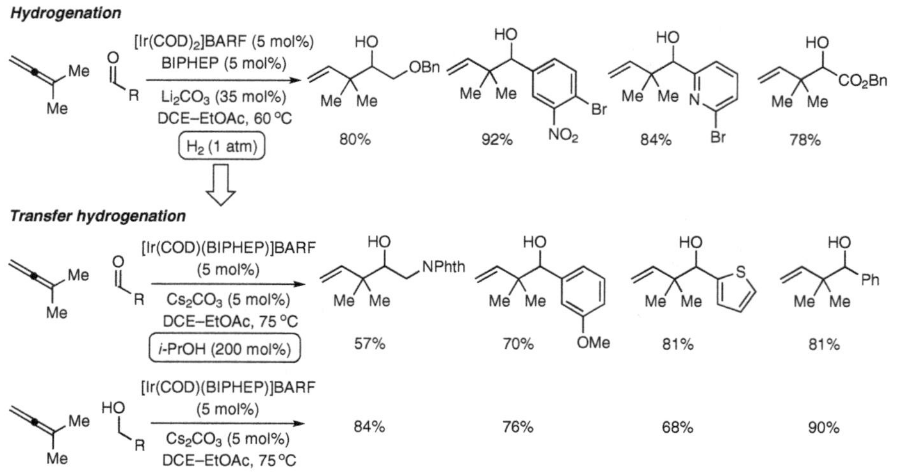

SCHEME 16-11. Catalytic carbonyl reverse prenylation: the transition from hydrogenative to transfer hydrogenative C–C coupling.

[55,56]. While these processes have had a positive impact on organic synthesis, they typically employ terminal reductants that are metallic, pyrophoric, or highly mass intensive, which impedes implementation on scale.

To achieve carbonyl allylation in the absence of stoichiometric by-products, the hydrogenation of allenes in the presence of carbonyl electrophiles was explored. It was found that iridium-catalyzed hydrogenation of dimethylallene in the presence of activated aldehydes or ketones delivers products of reverse prenylation [57a]. Remarkably, under nearly identical transfer hydrogenation conditions employing isopropanol as the terminal reductant, equivalent transformations are achieved [57b]. The ability to perform isopropanol-mediated allene–aldehyde reductive coupling via transfer hydrogenation ultimately led to related processes wherein primary alcohol reactants serve dually as hydrogen carriers and aldehyde precursors, thus enabling by-product-free carbonyl reverse prenylation from the alcohol oxidation level (Scheme 16-11) [57b].

Using ruthenium-based transfer hydrogenation catalysts, allene–aldehyde reductive coupling is also possible. For example, ruthenium-catalyzed reductive coupling of 1,1-disubstituted allenes to formaldehyde and higher aldehydes occurs efficiently using isopropanol as the hydrogen donor [58a]. Notably, reactions performed from the alcohol oxidation level are highly diastereoselective, whereas reactions conducted from the aldehyde oxidation level are not. The collective data suggest that in reactions conducted from the alcohol oxidation level, high diastereoselectivities arise in response to Curtin–Hammett effects associated with turnover limiting carbonyl addition of the allylruthenium intermediate to the transient aldehyde (Scheme 16-12) [58b].

A more general means of directing diastereoselectivity involves the use of allenes that hydrometallate to form geometrically defined σ-allylruthenium intermediates. For the indicated allenamide, which is conveniently prepared through base-catalyzed isomerization of the propargyl amine, products of *anti*-aminoallylation are formed from the aldehyde [59b] or alcohol [59a] oxidation level as single diastereomers. The collective data suggest intervention of an (*E*)-σ-allylruthenium intermediate that participates in stereospecific carbonyl addition through a closed, chair-like transition structure (Scheme 16-13).

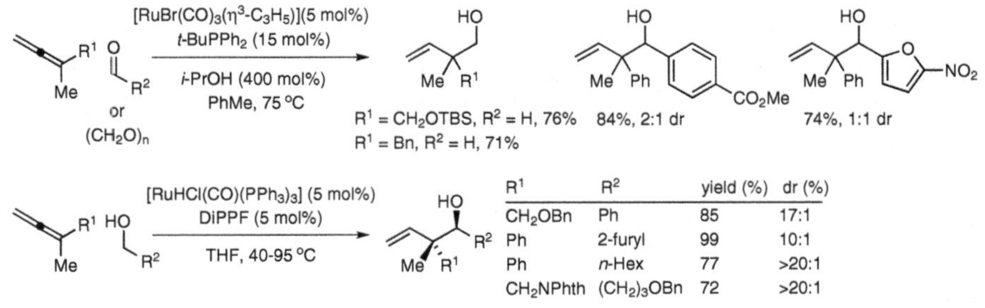

SCHEME 16-12. Ru-catalyzed hydrohydroxyalkylation of 1,1-disubstituted allenes from the alcohol or aldehyde oxidation level (DiPPF = 1,1′-bis(diisopropylphosphino)ferrocene).

Motivated by the prospect of engaging alcohols and π-unsaturated reactants as partners for C−C coupling [60], the transfer hydrogenative coupling of conjugated dienes was explored. Under the conditions of iridium-catalyzed transfer hydrogenation employing isopropanol as terminal reductant, 1,3-cyclohexadiene reductively couples to aldehydes in a highly diastereoselective fashion [61]. Again, an equivalent set of products is generated from the corresponding alcohols under nearly identical conditions (Scheme 16-14).

Under the conditions of ruthenium-catalyzed transfer hydrogenation employing $[RuHCl(CO)(PPh_3)_3]$ as precatalyst, simple acyclic dienes (butadiene, isoprene, and 2,3-dimethylbutadiene) reductively couple to aldehydes using isopropanol or formic acid as terminal reductant [62a]. Analogous products of diene hydrohydroxyalkylation are formed from the alcohol oxidation level. Notably, although alcohol reactants readily dehydrogenate under these conditions, the alcohol-containing products resist oxidation, as the homoallylic olefin of the product occupies a coordination site on ruthenium at the stage of the homoallylic alkoxide to suppress β-hydride elimination. However, through the use of a more highly coordinatively unsaturated ruthenium catalyst, $[Ru(TFA)(H)(CO)(PPh_3)_2]$, which is generated *in situ* through the acid–base reaction of $[Ru(H)_2(CO)(PPh_3)_3]$ and TFA, further oxidation of the initially formed homoallylic alcohol occurs to furnish β,γ-unsaturated ketones. In this way, products of diene hydroacylation are formed from the alcohol or aldehyde oxidation level (Scheme 16-15) [62b].

For ruthenium catalysts that embody cationic character, the hydrometallation of 2-substituted dienes is reversible, ultimately resulting in generation of isomeric π-allyls that couple at the 2-position to furnish products of carbonyl addition that incorporate all

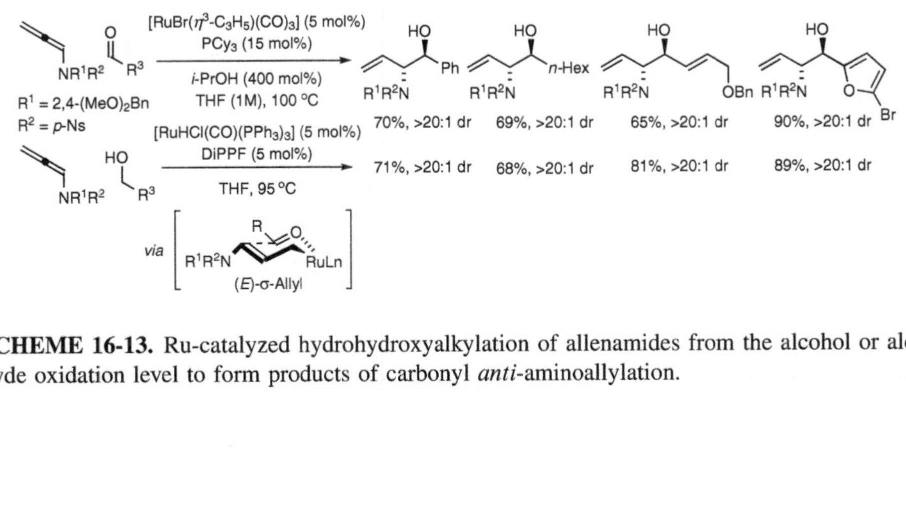

SCHEME 16-13. Ru-catalyzed hydrohydroxyalkylation of allenamides from the alcohol or aldehyde oxidation level to form products of carbonyl *anti*-aminoallylation.

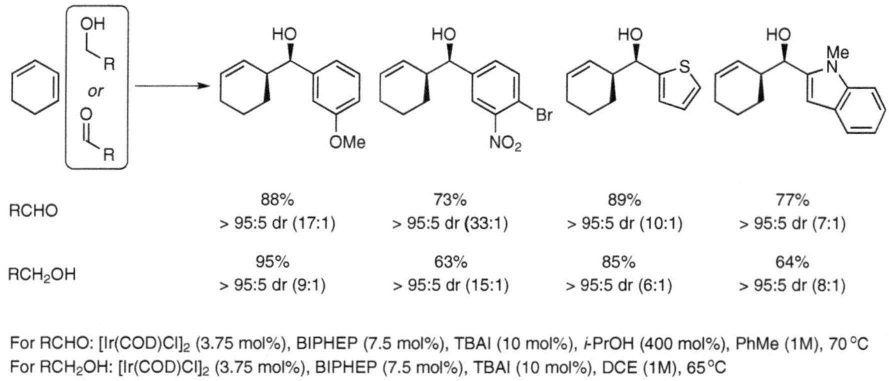

SCHEME 16-14. Coupling of dienes to alcohols or aldehydes via Ir-catalyzed transfer hydrogenation (the ratio of the indicated homoallylic alcohol and the isomeric bishomoallylic alcohol is indicated in parentheses).

carbon quaternary centers. For example, exposure of 2-substituted dienes to paraformaldehyde in the presence of a cationic ruthenium catalyst derived upon the acid–base reaction of $[RuH_2(CO)(PPh_3)_3]$ and $HO_2CC_7F_{15}$ results in hydrohydroxymethylation to form primary neopentyl alcohols [63a]. Similarly, exposure of 2-substituted dienes to ethanol results in diastereoselective hydrohydroxyethylation to form secondary neopentyl alcohols, albeit with incomplete levels of regioselectivity (Scheme 16-16) [63b].

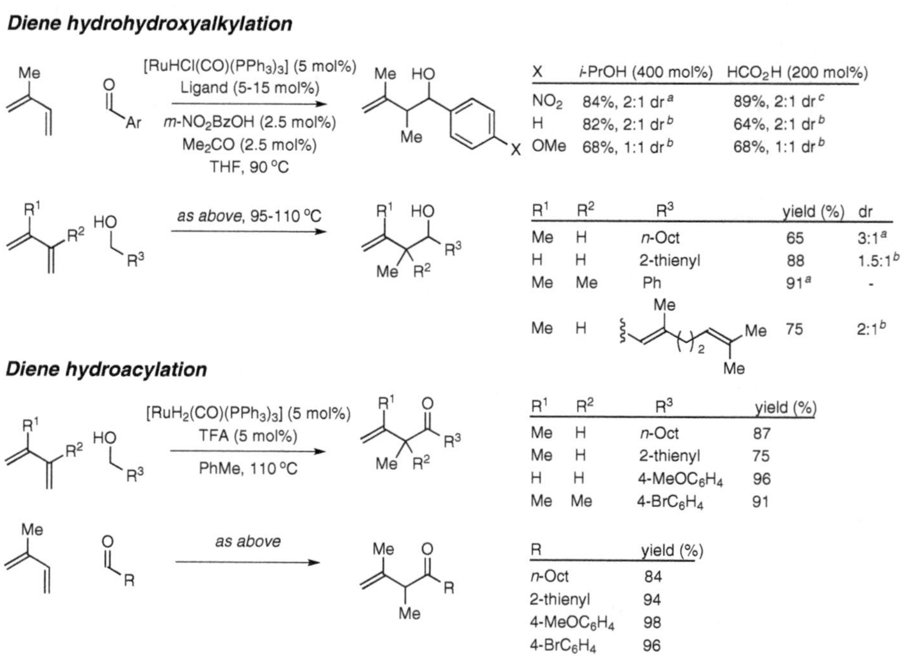

SCHEME 16-15. Hydrohydroxyalkylation and hydroacylation of acyclic dienes from the alcohol or aldehyde oxidation level. [a]Ligand = $(p\text{-MeOPh})_3P$. [b]Ligand = *rac*-BINAP. [c]No ligand added.

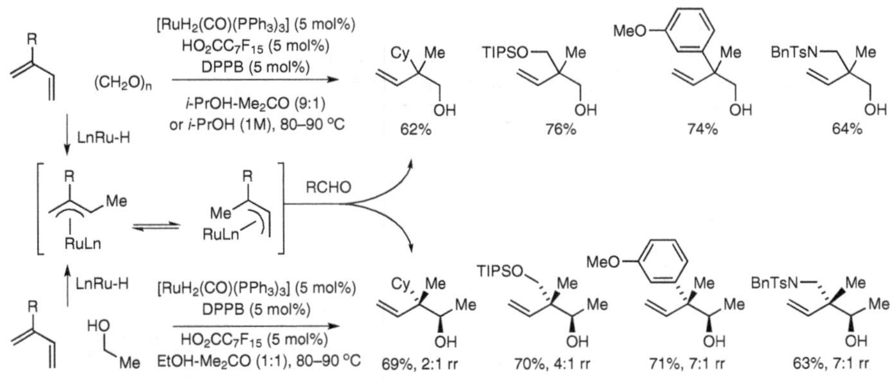

SCHEME 16-16. Hydrohydroxyalkylation of 2-substituted dienes to form all-carbon quaternary centers.

Diastereo- and enantioselective hydrohydroxyalkylations of butadiene would represent a powerful, by-product-free method for asymmetric carbonyl crotylation. However, to address the issue of relative stereocontrol, geometrically defined σ-allylruthenium intermediates are again required. It is known that 2-trialkylsilyl-substituted crotylmetal species adopt a "*pseudo*-(*Z*)-σ-crotylruthenium" geometry by virtue of allylic strain [64]. Use of the indicated silyl-butadiene, prepared from chloroprene, in combination with chiral ruthenium catalysts modified by (*R*)-SEGPHOS or (*R*)-DM-SEGPHOS, enables *syn*-diastereo- and enantioselective carbonyl crotylation from the alcohol or aldehyde oxidation level. The C–C coupling products participate in 9-BBN-mediated diastereoselective hydroboration–Suzuki cross-coupling to furnish adducts, which upon Tamao–Fleming oxidation of the C–Si bond deliver *anti,syn*-stereotriads found in numerous polyketide natural products (Scheme 16-17) [62c].

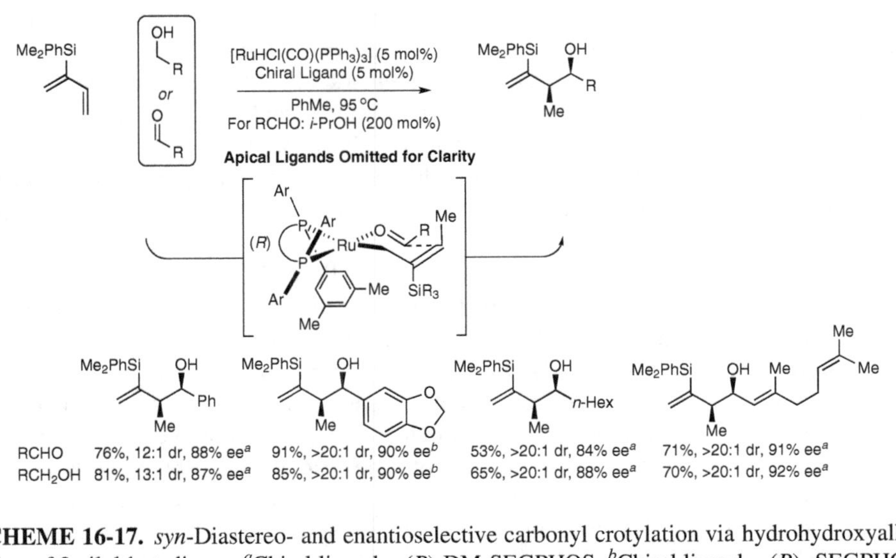

SCHEME 16-17. *syn*-Diastereo- and enantioselective carbonyl crotylation via hydrohydroxyalkylation of 2-silyl-butadienes. [a]Chiral ligand = (*R*)-DM-SEGPHOS. [b]Chiral ligand = (*R*)- SEGPHOS.

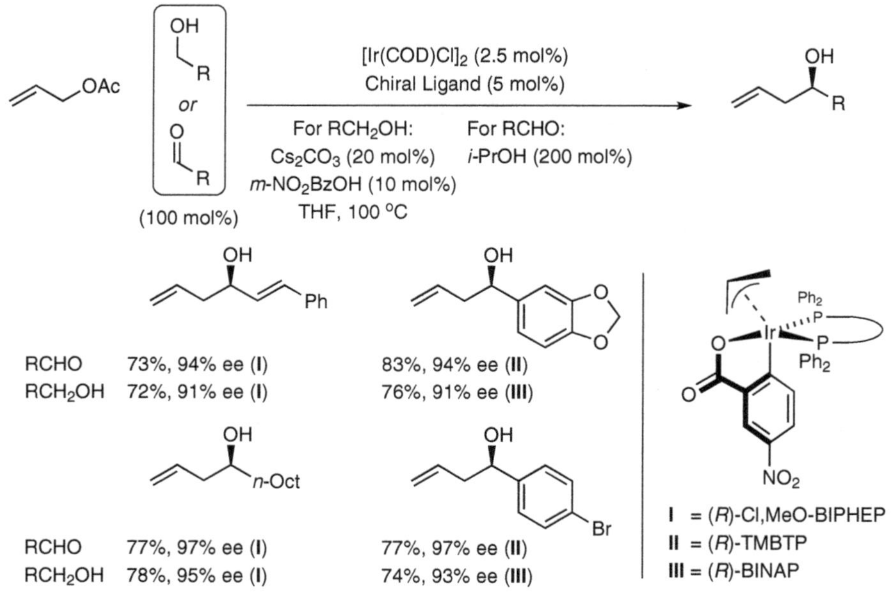

SCHEME 16-18. Enantioselective carbonyl allylation from the alcohol or aldehyde oxidation level using allyl acetate as the terminal allyl donor.

Using *ortho*-cyclometalated iridium catalysts generated *in situ* from allyl acetate, 3-nitrobenzoic acid, and a chiral bis(phosphine) ligand, enantioselective carbonyl allylation is achieved from the alcohol or aldehyde oxidation level using allyl acetate as the terminal allyl donor [65]. Aliphatic, allylic, and benzylic alcohols are converted to the corresponding homoallylic alcohols with uniformly high levels of enantioselectivity. In the presence of isopropanol, but under otherwise identical conditions, aldehydes are converted to an equivalent set of homoallylic alcohols (Scheme 16-18).

The active cyclometalated iridium π-allyl *C,O*-benzoate complexes have been characterized by single-crystal X-ray diffraction. Such complexes are sufficiently robust that they may be purified chromatographically. Even more remarkably, these same complexes can be recovered chromatographically from reaction mixtures and recycled multiple times without any significant decrease in the isolated yield of product or level of enantioselectivity. This capability is established in highly enantioselective reactions of aliphatic, allylic, and benzylic alcohols under microwave conditions in aqueous organic media (Scheme 16-19) [66].

The indicated catalytic mechanism is consistent with the collective data (Scheme 16-20). In reactions conducted from the alcohol oxidation level, entry into the catalytic cycle occurs upon protonolysis of the iridium π-allyl complex by the reactant alcohol to furnish a pentacoordinate iridium alkoxide. β-Hydride elimination produces the aldehyde and an iridium hydride, which suffers deprotonation to generate an anionic iridium(I) species. Oxidative addition of the allylic acetate provides the π-allyl complex. The ability to chromatographically recover this complex from the reaction mixture suggests that the π-allyl complex is the catalyst resting state, and that carbonyl addition to the aldehyde is the turnover-limiting event. The iridium alkoxide that results upon carbonyl addition is a hexacoordinate 18-electron complex. This intermediate resists β-hydride elimination due to

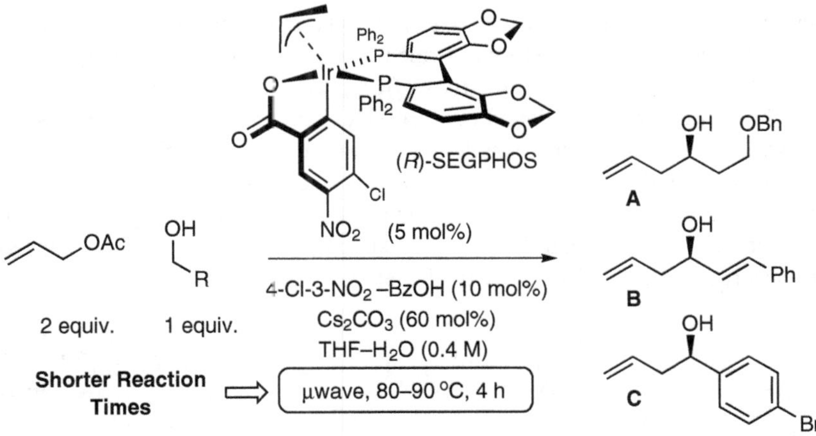

Chromatographic Recovery and Recycling of Catalyst

	First Run	Second Run	Third Run
A	75%, 93% ee 84% cat. recovery	71%, 94% ee 82% cat. recovery	70%, 94% ee 85% cat. recovery
B	81%, 97% ee 86% cat. recovery	78%, 98% ee 84% cat. recovery	75%, 97% ee 89% cat. recovery
C	81%, 94% ee 92% cat. recovery	77%, 96% ee 88% cat. recovery	77%, 96% ee 85% cat. recovery

SCHEME 16-19. Chromatographic recovery and recycling of the cyclometalated iridium catalyst in enantioselective carbonyl allylations from the alcohol oxidation level.

the lack of an open coordination site. Exchange of the homoallylic iridium alkoxide with reactant alcohol releases the product of carbonyl allylation and regenerates pentacoordinate iridium alkoxide to close the catalytic cycle.

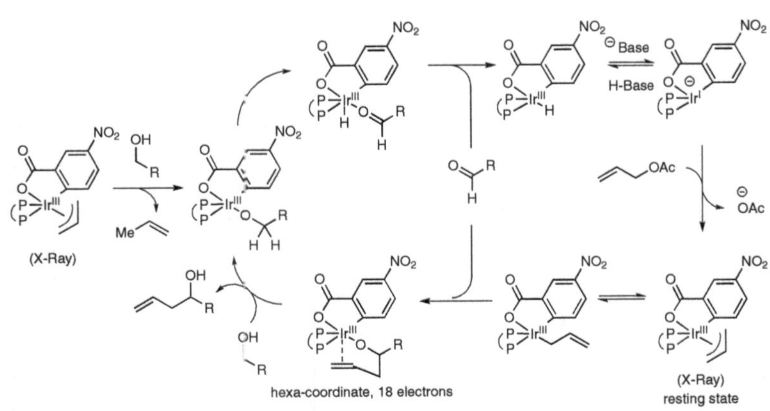

SCHEME 16-20. Mechanism of carbonyl allylation from the alcohol or aldehyde oxidation level using allyl acetate as the terminal allyl donor.

Initial investigations into stereoselective carbonyl crotylation employing α-methyl allyl acetate as the crotyl donor were performed using the *ortho*-cyclometalated iridium *C,O*-benzoate prepared *in situ* from $[Ir(COD)Cl]_2$, allyl acetate, 4-cyano-3-nitrobenzoic acid, and the chiral phosphine ligand (*S*)-SEGPHOS [67a]. Although *in situ* assembly of the catalyst is convenient and excellent enantioselectivities typically were observed (>95% ee), only modest levels of *anti*-diastereoselectivity were evident (5:1–11:1 dr). Conventional silica gel flash chromatographic purification of the catalyst permits alcohol-mediated carbonyl crotylation at lower temperatures (60 °C), resulting in enhanced levels of *anti*-diastereo- and enantioselectivity (Scheme 16-21) [67b].

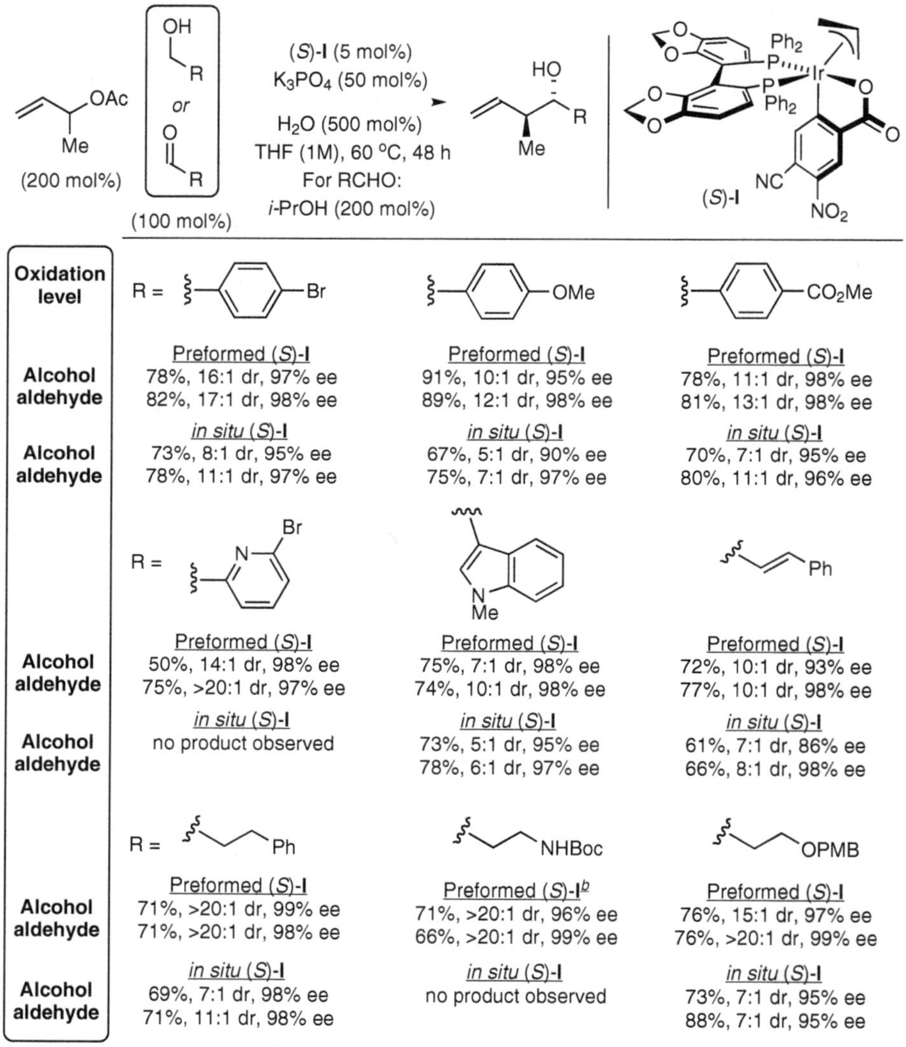

SCHEME 16-21. Enhanced levels of *anti*-diastereo- and enantioselectivity in alcohol-mediated carbonyl crotylations using a chromatographically purified single component iridium catalyst. [a]*In situ* method [67a]: $[Ir(COD)Cl]_2$ (2.5 mol%), (*S*)-SEGPHOS (5 mol%), 4-CN-3-NO_2BzOH (10 mol%), Cs_2CO_3 (20 mol%), THF (2.0 M), 90 °C, 48 h, α-methyl allyl acetate (200 mol%). For aldehyde substrates: *i*-PrOH (200 mol%).

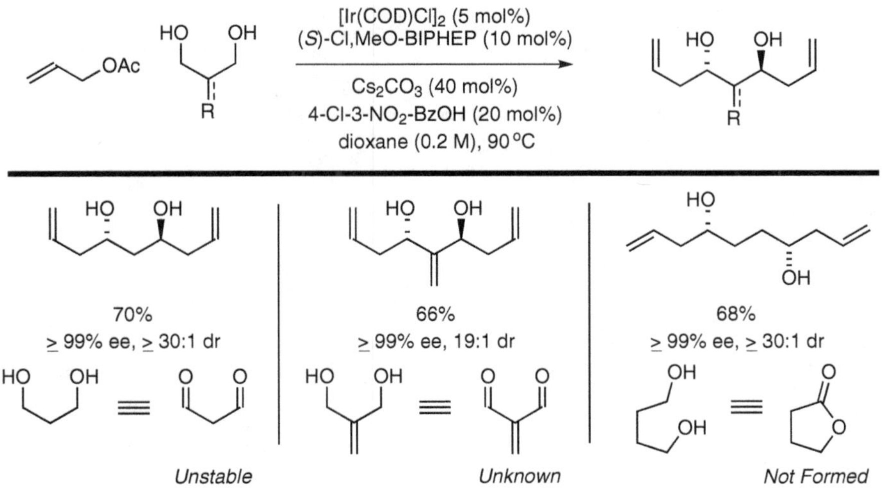

SCHEME 16-22. 1,*n*-Glycols as dialdehyde equivalents in iridium-catalyzed enantioselective carbonyl allylation from the alcohol oxidation level.

Enantioselective carbonyl allylation from the alcohol oxidation level enables carbonyl allylation processes that are not possible using conventional allylmetal reagents. For example, whereas 1,3-dialdehydes are highly intractable and cannot be used in direct two-directional chain elongation via double allylation [68], 1,3-propanediols and higher glycols are stable and abundant. Remarkably, such diols participate in bidirectional allylation through the sequential generation and capture of transient *mono*-aldehydes to provide C_2-symmetric adducts in good yield [69]. In these processes, the minor enantiomer of the *mono*-allylated intermediate is transformed to the *meso*-stereoisomer of the product, thus amplifying the level of enantiomeric enrichment (Scheme 16-22) [70].

Through iterative two-directional chain elongation of 1,3-propanediol, the total synthesis of the oxo-polyene macrolide (+)-roxaticin is achieved in 20 steps [71]. In this approach, nine of ten C–C bonds formed in the longest linear sequence were made via metal catalysis, including seven C–C bonds formed using iridium-catalyzed alcohol C–C coupling. This total synthesis represents the most concise preparation of any oxo-polyene macrolide reported, to date, and is achieved in the absence of chiral reagents, chiral auxiliaries, and with minimal use of premetalated *C*-nucleophiles (Scheme 16-23).

Double crotylations of 2-methyl-1,3-propanediol result in the direct generation of pseudo-C_2-symmetric polypropionate stereoquintets, which appear as substructures in diverse polyketide natural products. Using a chromatographically purified iridium precatalyst modified by (*R*)-SEGPHOS, the desired polypropionate stereoquintet is formed predominantly as 1 of 16 possible stereoisomers [72]. Differentiation of the hydroxyl and olefin moieties of the adduct, and definition of the latent stereocenter residing on the C_2-plane, are achieved in a single operation through iodoetherification. This two-directional approach to polypropionate construction avails a simplified approach to previously prepared polypropionate substructures, including the C19–C27 substructure of rifamycin S (Scheme 16-24).

Diverse processes beyond allylation and crotylation are possible using *ortho*-cyclometalated iridium catalysts. For example, diastereo- and enantioselective α-(trimethylsilyl)allylation [73a], α-(hydroxymethyl)allylation [73b], and α-(trifluoromethyl)

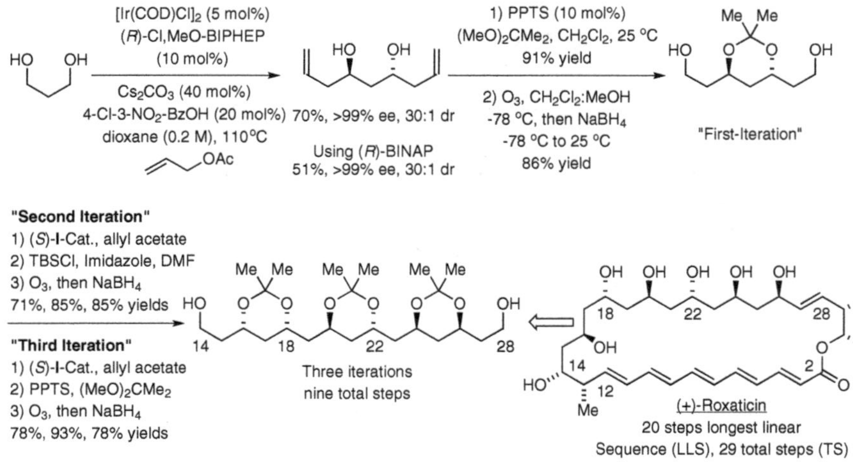

SCHEME 16-23. Elongation of 1,3-polyols via iterative catalyst-directed carbonyl allylation from the alcohol oxidation level.

allylation [73c] have been performed from the alcohol or aldehyde oxidation level. Finally, as illustrated in the reaction of the allylic *gem*-dibenzoate derived from acrolein, α-(hydroxy)allylation is achieved under the conditions of iridium-catalyzed transfer hydrogenation to provide *anti*-1-ene-2,3-diols with excellent control of relative and absolute stereochemistry [73d]. In the latter case, additions from the alcohol oxidation level remain undeveloped due to competing transesterification from the *gem*-dibenzoate to the reactant alcohol (Scheme 16-25). Finally, in previously described reactions of 1,

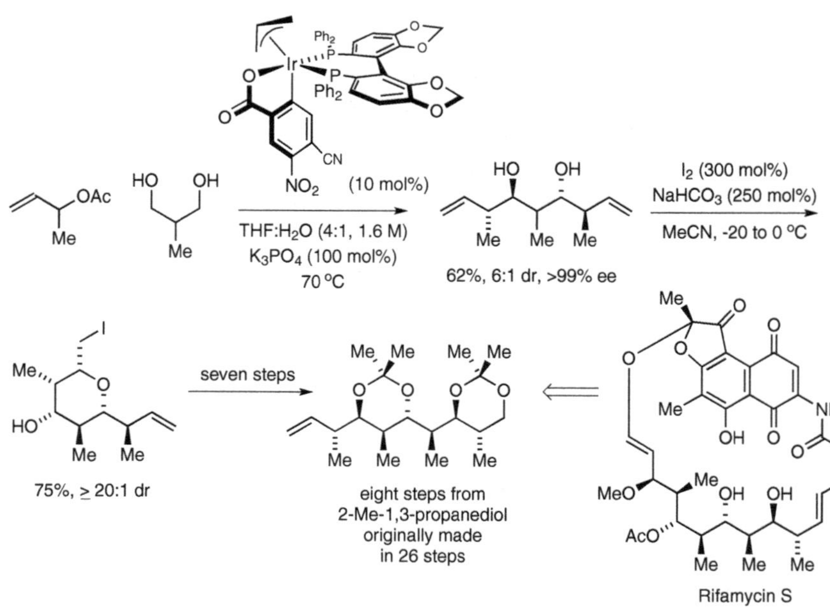

SCHEME 16-24. Diastereo- and enantioselective double crotylation of 2-methyl-1,3-propanediol and formal synthesis of rifamycin S through construction of the C19–C27 stereoheptad.

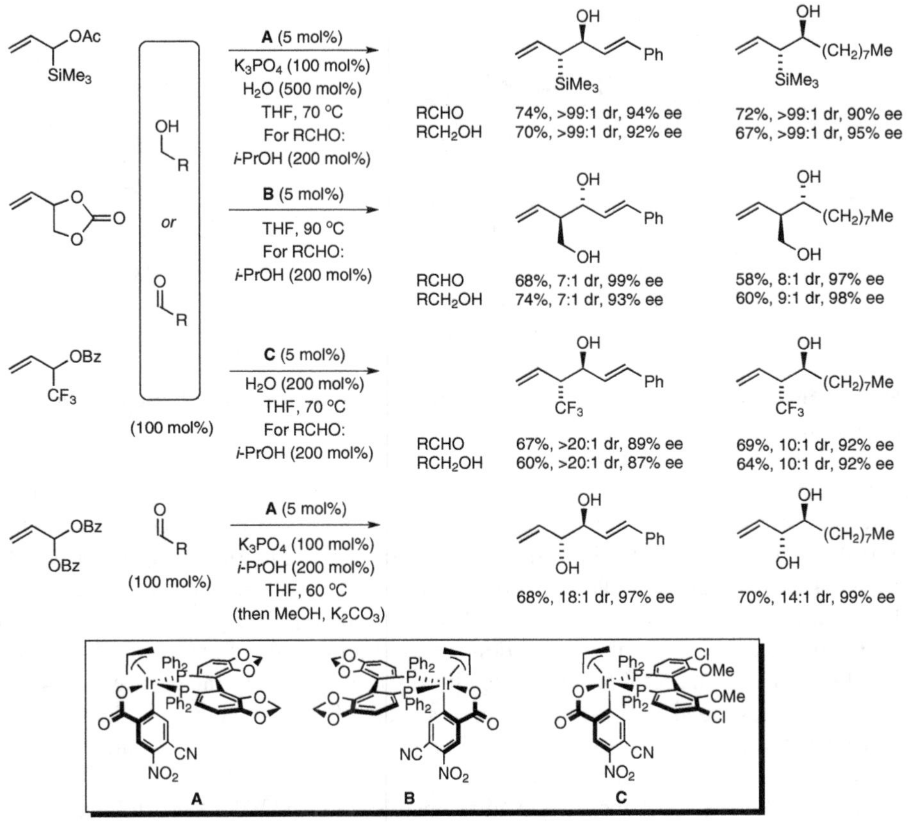

SCHEME 16-25. Diverse diastereo- and enantioselective allylation methods availed through use of ortho-cyclometalated iridium catalysts.

1-dimethylallene, products of carbonyl *tert*-prenylation were obtained as racemates (Scheme 16-11) [57]. Through the use of a cyclometalated iridium *C,O*-benzoate modified by (*S*)-SEGPHOS, enantioselective carbonyl *tert*-prenylation is achieved from the alcohol or aldehyde oxidation level under exceptionally mild conditions (30–50 °C) in the absence of stoichiometric metallic reagents [73e]. Indeed, for reactions conducted from the alcohol oxidation level, stoichiometric by-products are completely absent. Interestingly, the absolute stereochemistry of *tert*-prenylation is opposite to that observed in the aforementioned allylations (Scheme 16-26).

The complete levels of branched regioselectivity and high levels of *anti*-diastereoselectivity observed in the previously described allylations suggest carbonyl addition occurs with allylic inversion from the primary (*E*)-σ-allyl haptomer. Linear carbonyl addition products are accessible upon stabilization of the secondary σ-allyl haptomer through the introduction of a carbonyl moiety, in which case the secondary σ-allyl haptomer is equivalent to an η^1-*C*-bound iridium enolate. Upon exposure of the indicated carboxy-substituted allylic carbonate to primary alcohols in the presence of the chiral *ortho*-cyclometalated iridium *C,O*-benzoate complex modified by (*R*)-Cl,MeO-BIPHEP, a highly enantioselective vinylogous Reformatsky-aldol addition is achieved [74]. The degree of linear regioselectivity increases in response to increasing steric demand of the transient

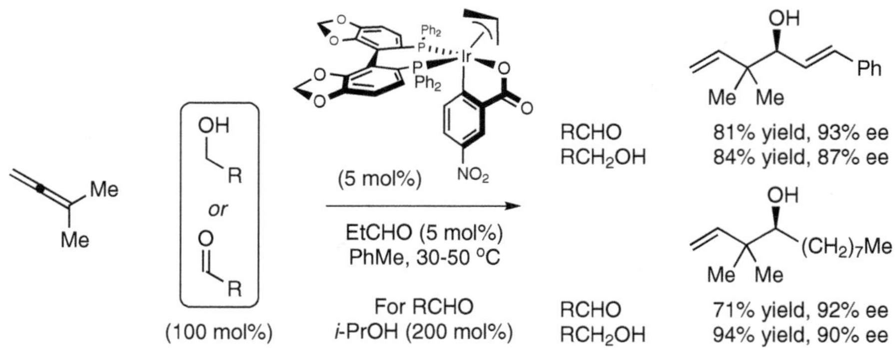

SCHEME 16-26. Enantioselective carbonyl *tert*-prenylation availed through use of *ortho*-cyclometalated iridium catalysts.

aldehyde, suggesting a Curtin–Hammett scenario wherein carbonyl addition occurs selectively from an equilibrating mixture of primary and secondary σ-allyl haptomers. Identical products are accessible from the aldehyde oxidation level under transfer hydrogenation conditions employing isopropanol as terminal reductant (Scheme 16-27).

Diastereo- and enantioselective carbonyl crotylation employing 1,3-butadiene, a basic petrochemical feedstock, as the crotyl donor would represent a significant advance over established crotylation protocols involving chirally modified crotylmetal reagents [49,50]. As previously described, ruthenium complexes promote 1,3-butadiene-mediated crotylation from the alcohol or aldehyde oxidation level, but these transformations are not stereoselective (Scheme 16-15) [62a]. Using the cyclometalated iridium *C,O*-benzoate derived from allyl acetate, 4-methoxy-3-nitrobenzoic acid, and BIPHEP, catalytic carbonyl crotylation employing 1,3-butadiene from the aldehyde or alcohol oxidation was achieved under transfer hydrogenation conditions [75]. Although carbonyl addition occurs with roughly equal facility from the alcohol or aldehyde oxidation level, stereoselective variants of these processes remain elusive (Scheme 16-28).

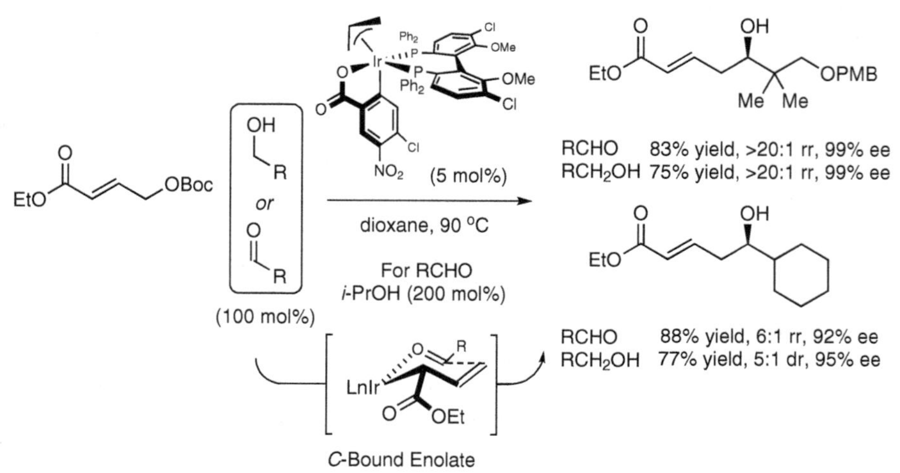

SCHEME 16-27. Enantioselective vinylogous Reformatsky-aldol additions from the alcohol or aldehyde oxidation level.

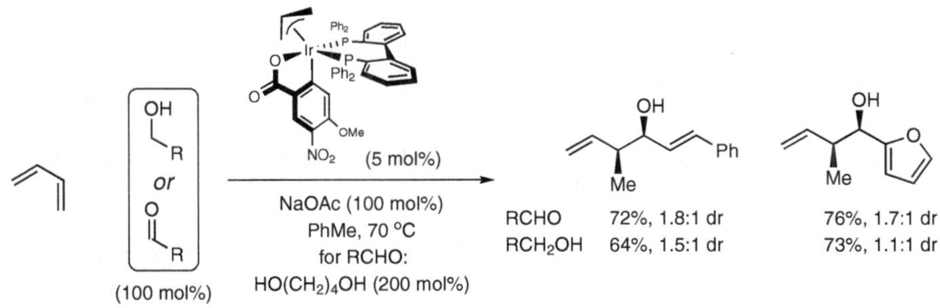

SCHEME 16-28. Ir-catalyzed C–C couplings of 1,3-butadiene under transfer hydrogenation conditions.

To explore the applicability of this emergent allylation methodology in the context of ketone addition, the allylation, crotylation, and *tert*-prenylation of substituted isatins were examined using the cyclometalated iridium *C,O*-benzoate generated *in situ* from $[Ir(COD)Cl]_2$, 4-cyano-3-nitrobenzoic acid and CTH-(*R*)-P-PHOS under the conditions of isopropanol-mediated transfer hydrogenation [76]. Using allyl acetate, α-methyl allyl acetate, and 1,1,-dimethylallene pronucleophiles, highly enantioselective allylation, crotylation, and *tert*-prenylation were observed, respectively (Scheme 16-29).

In a recent work by Obora and Ishii, alkynes were established as allyl donors in carbonyl arylallylations from the alcohol oxidation level [77]. Specifically, upon exposure to an iridium catalyst generated *in situ* from $[Ir(COD)(OH)]_2$ and $P(n\text{-Oct})_3$, 1-aryl-2-methyl-alkynes couple to primary alcohols to furnish homoallylic alcohols with complete branched regioselectivity and excellent levels of *anti*-diastereoselectivity (Scheme 16-30).

The propargylation of carbonyl compounds is typically accomplished using allenyl-metal reagents [78,79]. Under the conditions of ruthenium-catalyzed transfer hydrogenation, 1,3-enynes hydrometallate to form transient allenylruthenium intermediates, which

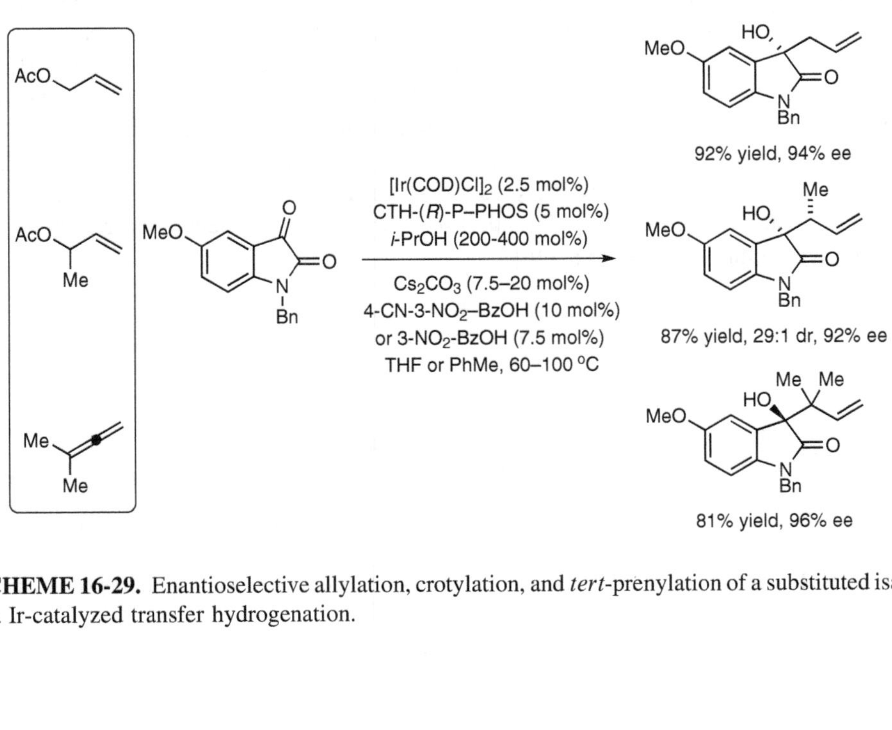

SCHEME 16-29. Enantioselective allylation, crotylation, and *tert*-prenylation of a substituted isatin via Ir-catalyzed transfer hydrogenation.

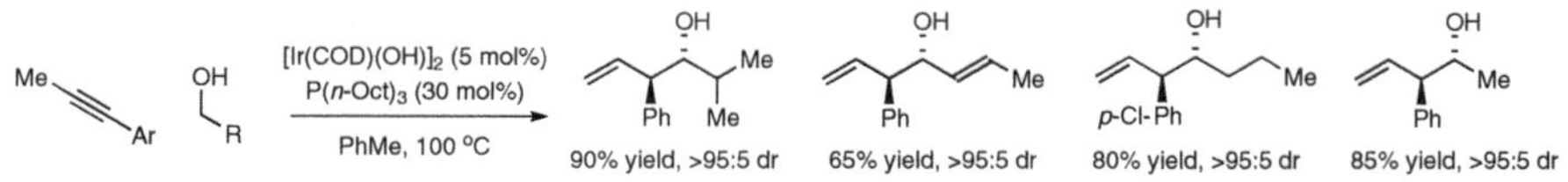

SCHEME 16-30. Carbonyl arylallylation from the alcohol oxidation level via Ir-catalyzed transfer hydrogenation employing alkynes as allyl donors.

propargylates benzylic, allylic, and aliphatic aldehydes directly from the alcohol oxidation level [80]. Using isopropanol as the terminal reductant, enyne–aldehyde reductive coupling provides identical propargylation products. Stereocontrolled variants of these enyne-mediated propargylations are currently under development (Scheme 16-31).

16.4 ALDOL, MANNICH, AND RELATED PROCESSES

Discovered nearly 150 years ago [81], the aldol reaction represents a major gateway to polyketide natural products [82]. Accordingly, enormous effort has been devoted to the development of diastereo- and enantioselective aldol addition protocols [83]. A particularly significant advance resides in the identification of metallic [84] and organic [85] catalysts for the direct asymmetric aldol addition of unmodified ketones, as these processes herald a departure from the use of chiral auxiliaries and preformed enol(ate) derivatives. However, a significant limitation associated with these emerging technologies resides in the issue of regiocontrolled enolization. For example, as illustrated by the direct enantioselective aldol coupling of 2-butanone catalyzed by L-proline [86,87] or the heterobimetallic catalyst $LaLi_3$-tris(binaphthoxide) (LLB), branched aldol adducts are inaccessible (Scheme 16-32) [88].

As illustrated in pioneering studies by Stork and coworkers [89], dissolving metal reduction of enones enables regiospecific generation and capture of enolate isomers that cannot be prepared exclusively from saturated ketones. Subsequently, catalytic reductive couplings of enones to aldehydes emerged [90]. To date, myriad metallic catalysts for "reductive aldol coupling" have been devised, including those based on rhodium [91,92], cobalt [93], iridium [94], ruthenium [95], palladium [96], copper [97,98], nickel [99], and indium [100,101]. These protocols invariably employ metallic terminal reductants, such as stannanes, silanes, and organozinc reagents, which mandate the generation of stoichiometric by-products. Inspired by the prospect of developing completely by-product-free processes, catalytic reductive aldol additions employing elemental hydrogen as the terminal reductant were investigated [92].

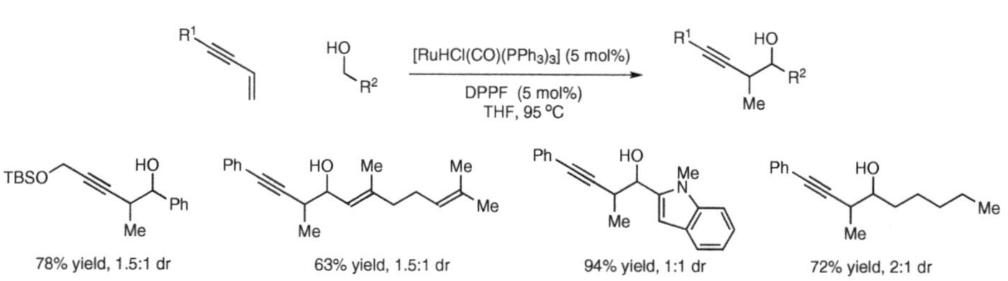

SCHEME 16-31. Ru-catalyzed carbonyl propargylation employing 1,3-enynes as allenylmetal donors.

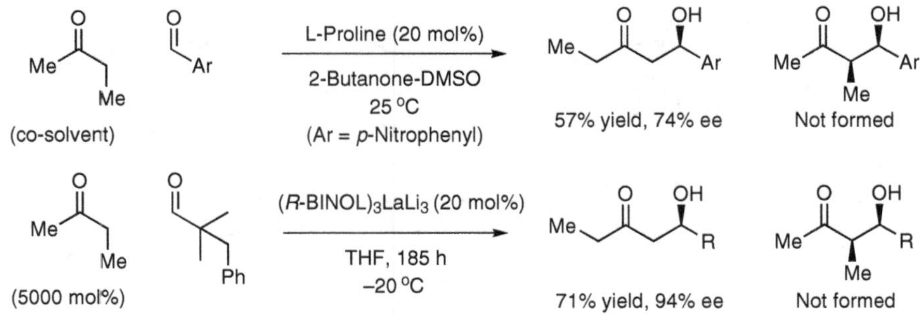

SCHEME 16-32. Regiocontrol in the direct asymmetric aldol addition of unmodified ketones.

Initial efforts focused on the intramolecular hydrogen-mediated reductive aldol addition of tethered enone-aldehydes [92a]. Notably, whereas neutral rhodium complexes, such as Wilkinson's catalyst, $[RhCl(PPh_3)_3]$, promote conventional conjugate enone hydrogenation, cationic rhodium precatalysts along with substoichiometric quantities of mild basic additives promote reductive C–C bond formation. Mechanistically, the difference in behavior for the cationic and neutral rhodium precatalysts may be explained as follows. Neutral rhodium(I) complexes may induce homolytic hydrogen activation to furnish dihydride intermediates [102]. Enone hydrometallation from the rhodium dihydride would deliver a rhodium enolate that retains a hydride ligand and, hence, can engage in enolate–hydrogen reductive elimination to furnish the product of conventional hydrogenation. In contrast, cationic rhodium dihydrides, as a consequence of their enhanced acidity [103], are subject to deprotonation in the presence of basic additives, promoting entry into monohydride-based catalytic cycles for which direct enolate–hydrogen reductive elimination is no longer possible [104]. The latter process represents a formal heterolytic activation of hydrogen ($H_2 + M\text{-}X \rightarrow M\text{-}H + HX$) [105]. An alternative mechanistic interpretation, which cannot be excluded on the basis of deuterium labeling, involves enone–carbonyl oxidative coupling. For cationic rhodium complexes, hydrogen activation is typically turnover limiting [37,38]. Additionally, cationic d^8 complexes possess a labile coordinate site on the square planar metal center, which enables simultaneous binding of two reactant molecules. The synergistic influence of these effects may induce oxidative coupling en route to high valent rhodacycles, which upon hydrogenolysis deliver products of C–C coupling. The intervention of oxidative coupling pathways may account for the fact that easily reduced enones participate in the hydrogen-mediated reductive aldol coupling, while α,β-unsaturated esters do not.

It was found that diverse enone-aldehydes engage in highly diastereoselective reductive cyclization to deliver five- and six-membered ring products upon exposure to cationic rhodium catalysts under ambient pressures of hydrogen [92a]. In a similar fashion, enone-ketones and enone-diones cyclize under nearly identical conditions to furnish *syn*-aldol adducts as single diastereomers [92b]. However, in these cases, the diminished electrophilicity of the ketone leads to substantial quantities of simple enone reduction product. The addition of aldehyde enolates to ketones, for which a single stoichiometric variant is known [106], represents a highly challenging class of aldol addition. Under hydrogenation conditions, enal-ketones cyclize with a high degree of efficiency to provide products of aldehyde enolate-ketone addition, although competitive 1,4-reduction is evident (Scheme 16-33) [92c].

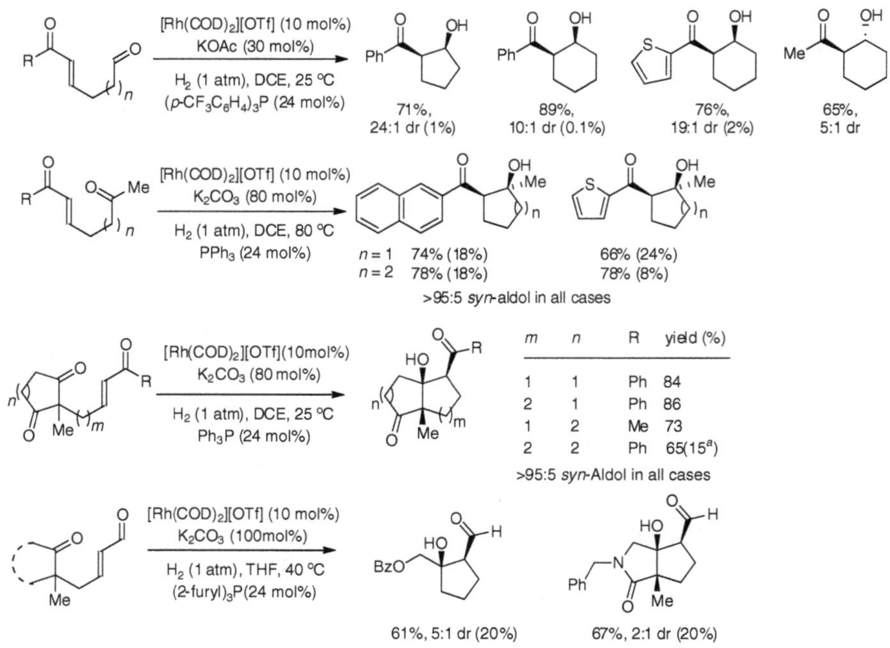

SCHEME 16-33. Reductive aldol cyclization via Rh-catalyzed hydrogenation. [a]Yield of conjugate reduction product.

Intermolecular reductive aldol addition under hydrogenation conditions provides the desired aldol adducts; however, diastereomeric mixtures initially were observed. Using tri (2-furyl)phosphine-modified rhodium catalysts, enone-aldehyde reductive coupling occurs with high levels of *syn*-diastereocontrol [92e,107]. Additionally, reactants that incorporate functional groups that are potentially liable to conventional hydrogenation, such as alkynes, alkenes, benzylic ethers, nitroarenes, and aryl bromides, remain intact. Indeed, even crotyl vinyl ketone, which incorporates a second enone moiety, engages in aldol addition with preservation of the crotyl moiety [92f]. High levels of substrate-directed asymmetric induction are achieved in hydrogen-mediated reductive aldol additions to *N*-Boc-α-amino aldehydes. Specifically, *syn*-diastereoselectivity is accompanied by exceptional levels of chelation control due to the intervention of hydrogen-bonded chelates, which are stabilized by the low dielectric reaction media. Remarkably, although such aldehydes are configurationally labile, racemization is not observed under the neutral conditions of hydrogenation (Scheme 16-34) [92g].

The ability to direct *syn*-diastereoselectivity through the use of tri(2-furyl)phosphine-modified rhodium catalysts inspired efforts toward enantioselective variants. Although π-acidic monodentate phosphine ligands are required to enforce diastereocontrol, commercially available phosphines of this type (e.g., phosphoramidites and BINOL-derived phosphites) give rise to inactive rhodium complexes, suggesting a very narrow window in terms of ligand π-acidity. Consequently, the design of an effective chiral monodentate *P*-based ligand was undertaken. Using TADDOL-like phosphonites, key structure-selectivity trends were determined, ultimately leading to the design of an effective ligand. Thus, by simply exposing methyl or ethyl vinyl ketone to aldehydes under an atmosphere of gaseous hydrogen in the presence of the rhodium phosphonite complex, aldol addition

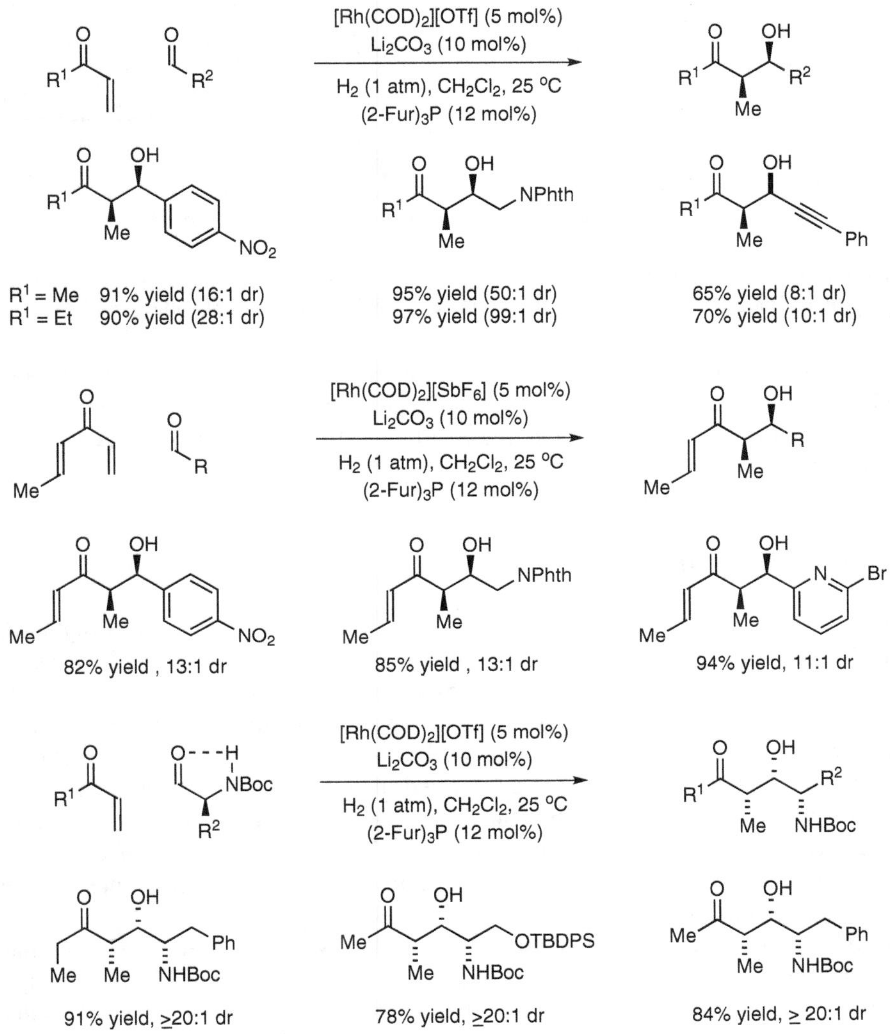

SCHEME 16-34. *syn*-Diastereoselective hydrogen-mediated reductive aldol addition employing cationic Rh catalysts ligated by tri(2-furyl)phosphine.

occurs with high levels of *syn*-diastereo- and enantioselectivity. This method generates optically enriched polyketide substructures in the absence of stoichiometric by-products (Scheme 16-35) [92h].

Based on the preceding results, reductive Mannich additions of vinyl ketones were explored. Reductive Mannich additions employing silane [108], the Hantzsch ester [109], or diethyl zinc [110] as the terminal reductant are known. Under the conditions of hydrogenation employing a tri(2-furyl)phosphine-ligated rhodium catalyst, vinyl ketones couple to *N*-(*o*-nitrophenyl)sulfonyl aldimines to provide the Mannich addition products with good levels of *syn*-diastereoselectivity (Scheme 16-36) [111].

The reactivity embodied by the hydrogen-mediated reductive aldol and Mannich addition provides a basis for other C–C bond forming hydrogenations. For example,

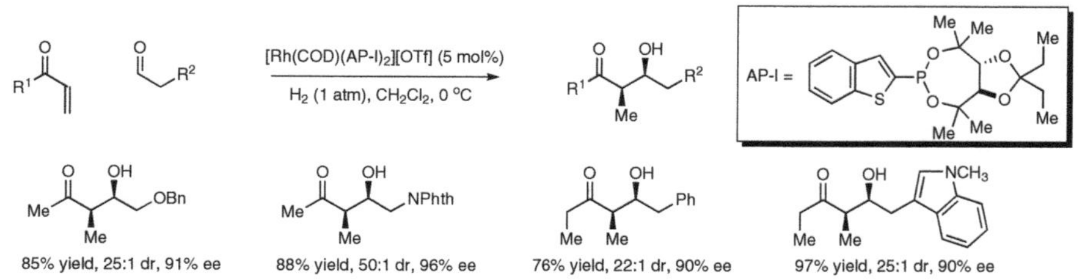

SCHEME 16-35. *syn*-Diastereoselective and enantioselective hydrogen-mediated aldol coupling employing cationic Rh catalysts ligated by a novel TADDOL-like phosphonite ligand.

appreciating the structural homology of enones and vinylazines, it was found that rhodium-catalyzed hydrogenation of 2-vinylazines in the presence of *N*-arylsulfonyl imines results in regio- and diastereoselective reductive coupling to furnish branched products of imine addition (Scheme 16-37) [112].

16.5 FUTURE DIRECTIONS

Carbonyl addition and catalytic hydrogenation are two of the most broadly utilized reductive transformations in Organic Chemistry. As perhaps foreshadowed by the 1912 Nobel Prize shared between Grignard and Sabatier, the past 10 years has witnessed the productive merger of these areas, resulting in a broad new family of highly stereoselective and atom economic catalytic C–C bond forming processes. Yet many unmet practical and fundamental challenges remain. For established transformations to be brought to scale, process optimization aimed at achieving more favorable catalyst loadings and turnover frequencies is required. Additionally, alcohol-mediated C–C couplings of ethylene and α-olefins remain elusive, as are corresponding transfer hydrogenative couplings of amines, that is, hydroaminoalkylations (Scheme 16-38) [113,114].

Another important future direction relates to the by-product-free C–C coupling of renewable alcohols and π-unsaturated reactants. Toward this end, the transfer hydrogenative C–C coupling of the prototypical, but notoriously unreactive, alcohol methanol was achieved recently (Scheme 16-39) [115]. Related couplings of glycerol, a by-product

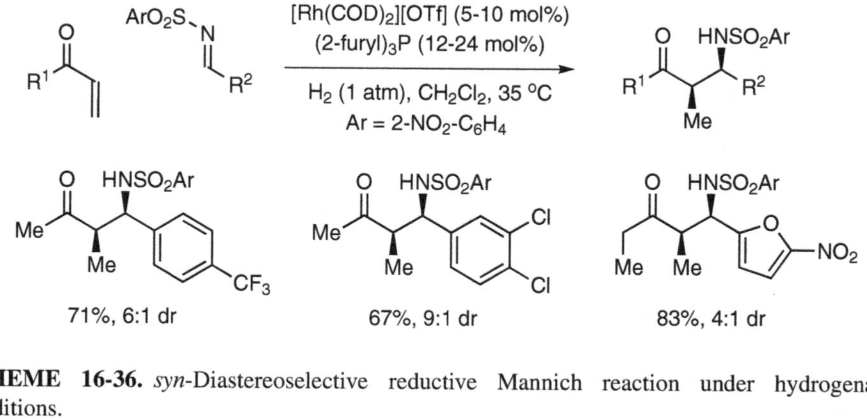

SCHEME 16-36. *syn*-Diastereoselective reductive Mannich reaction under hydrogenation conditions.

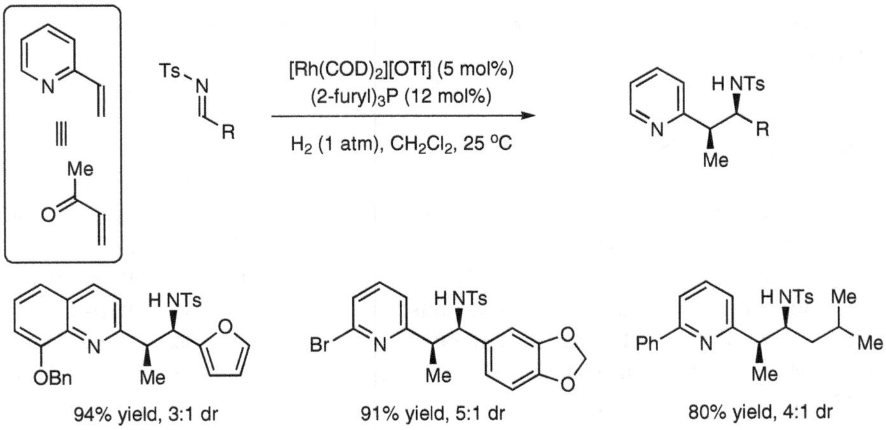

SCHEME 16-37. *syn*-Diastereoselective reductive Mannich reaction under hydrogenation conditions.

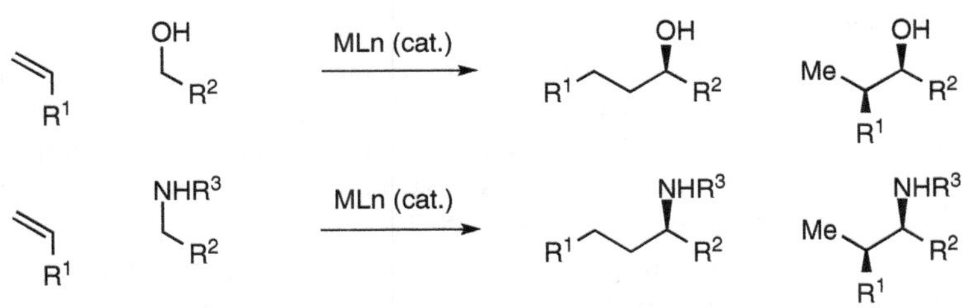

SCHEME 16-38. Aspirational transfer hydrogenative C–C bond formations: byproduct-free coupling of α-olefins with alcohols and amines.

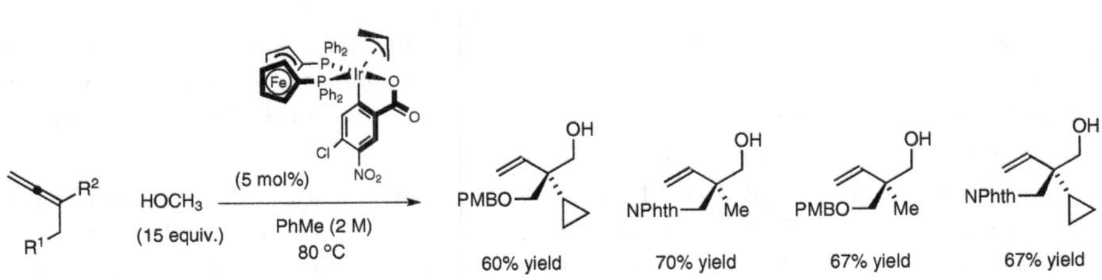

SCHEME 16-39. Ir-catalyzed direct C–C coupling of methanol—a renewable feedstock.

of biofuel production, and other site-selective C–C couplings of polyhydroxy compounds are underway.

ACKNOWLEDGMENTS

Acknowledgment is made to the Robert A. Welch Foundation, Johnson & Johnson, Eli Lilly, Merck, the NIH-NIGMS (RO1-GM69445), and the ACS-GCI, for partial support of the research described in this account. Dr. Oliver Briel of Umicore is thanked for the generous donation of rhodium and iridium salts. Dr. Yasunori Ino, Dr. Wataru Kuriyama, and Dr. Taichiro Touge of Takasago are thanked for the generous donation of SEGPHOS ligands. J.M. thanks the National Sciences and Engineering Research Council of Canada (NSERC) for a Postdoctoral Fellowship.

REFERENCES

1. Roelen O. inventor; Chemische Verwertungsgesellschaft, mbH Oberhausen, assignee. German Patent DE 849 548. September 20, 1938.
2. (a) Frohning CD, Kohlpaintner CW. In: Cornils B, Herrmann WA, editors. *Applied Homogeneous Catalysis with Organometallic Compounds*. Vol. 1, Wiley-VCH, Weinheim;1996, pp 29–104. (b) van Leeuwen, PWNM *Homogeneous Catalysis: Understanding the Art*. Kluwer, Dordrecht; 2004.
3. Hydroformylation is an outgrowth of the Fischer–Tropsch process, which represents another key hydrogen-mediated C–C bond formation: (a) Fischer F, Tropsch H (1923). The preparation of synthetic oil mixtures (synthol) from carbon monoxide and hydrogen. *Brennstoff-Chemie* 4, 276–285. (b) Fischer F, Tropsch H (1923). Synthesis of higher members of the aliphatic series from carbon monoxide. *Chem. Ber.* 56B, 2428–2443.
4. For reviews on C–C bond forming hydrogenation and transfer hydrogenation, see (a) Ngai MY, Kong JR, Krische MJ (2007). Hydrogen-mediated C–C bond formation: a broad new concept in catalytic C–C coupling. *J. Org. Chem.* 72, 1063–1072. (b) Skucas E, Ngai MY, Komanduri V, Krische MJ (2007). Enantiomerically enriched allylic alcohols and allylic amines via C–C bond-forming hydrogenation: asymmetric carbonyl and imine vinylation. *Acc. Chem. Res.* 40, 1394–1401. (c) Iida H, Krische MJ (2007). Catalytic reductive coupling of alkenes and alkynes to carbonyl compounds and imines mediated by hydrogen. *Top. Curr. Chem.* 279, 77–104. (d) Shibahara F, Krische MJ (2008). Formation of C–C bonds via ruthenium-catalyzed transfer hydrogenation: carbonyl addition from the alcohol or aldehyde oxidation level. *Chem. Lett.* 37, 1102–1107. (e) Bower JF, Kim IS, Patman RL, Krische MJ (2009). Catalytic carbonyl addition through transfer hydrogenation: a departure from preformed organometallic reagents. *Angew. Chem., Int. Ed.* 48, 34–46. (f) Han SB, Kim IS, Krische MJ (2009). Enantioselective iridium-catalyzed carbonyl allylation from the alcohol oxidation level via transfer hydrogenation: minimizing pre-activation for synthetic efficiency. *Chem. Commun.* 7278–7287. (g) Bower JF, Krische MJ (2011). Formation of C–C bonds via iridium-catalyzed hydrogenation and transfer hydrogenation. *Top. Organomet. Chem.* 34, 107–138.
5. For reviews, see (a) Trost BM (1991). The atom economy – a search for synthetic efficiency. *Science* 254, 1471–1477. (b) Trost BM (1995). Atom economy – a challenge for organic synthesis – homogeneous catalysis leads the way. *Angew. Chem., Int. Ed.* 34, 259–281.
6. "The ideal synthesis creates a complex skeleton . . . in a sequence only of successive construction reactions involving no intermediary refunctionalizations, and leading directly

to the structure of the target, not only *its skeleton but also its correctly placed functionality*." Hendrickson JB (1975). Systematic synthesis design. 4. Numerical codification of construction reactions. *J. Am. Chem. Soc.* 97, 5784–5800.

7. For a review on the topic of "redox economy" in organic synthesis, see Burns NZ, Baran PS, Hoffmann RW (2009). Redox economy in organic synthesis. *Angew. Chem., Int. Ed.* 48, 2854–2867.
8. For reviews addressing issues of oxidation on scale, see (a) Dugger RW, Ragan JA, Brown Ripin, D. H (2005). Survey of GMP bulk reactions run in a research facility between 1985 and 2002. *Org. Process Res. Dev.* 9, 253–258. (b) Carey JS, Laffan D, Thomson C, Williams MT (2006). Analysis of the reactions used for the preparation of drug candidate molecules. *Org. Biomol. Chem.* 4, 2337–2347. (c) Caron S, Dugger RW, Gut Ruggeri S, Ragan JA, Brown Ripin DH (2006). Large-scale oxidations in the pharmaceutical industry. *Chem. Rev.* 106, 2943–2989.
9. For enantioselective catalytic addition of vinylzinc reagents to aldehydes, see (a) Oppolzer W, Radinov RN (1992). Catalytic asymmetric synthesis of secondary (*E*)-allyl alcohols from acetylenes and aldehydes via (1-alkenyl)zinc intermediates. *Helv. Chim. Acta* 75, 170–173. (b) Oppolzer W, Radinov RN (1993). Synthesis of (*R*)-(−)-muscone by an asymmetrically catalyzed macrocyclization of an ω-alkynal. *J. Am. Chem. Soc.* 115, 1593–1594. (c) Soai K, Takahashi K (1994). Asymmetric alkenylation of chiral and prochiral aldehydes catalyzed by chiral or achiral amino alcohols: catalytic diastereoselective synthesis of protected *erythro*-sphingosine and enantioselective synthesis of chiral diallyl alcohols. *J. Chem. Soc., Perkin Trans. 1* 1257–1258. (d) Wipf P, Xu W (1994). Preparation of allylic alcohols by alkene transfer from zirconium to zinc. *Tetrahedron Lett.* 35, 5197–5200. (e) Oppolzer W, Radinov RN, De Brabander J (1995). Total synthesis of the macrolide (+)-aspicilin by an asymmetrically catalyzed macrocyclization of an ω-alkynal ester. *Tetrahedron Lett.* 36, 2607–2610. (f) Wipf P, Ribe S (1998). Zirconocene–zinc transmetalation and *in situ* catalytic asymmetric addition to aldehydes. *J. Org. Chem.* 63, 6454–6455. (g) Oppolzer W, Radinov RN, El-Sayed E (2001). Catalytic asymmetric synthesis of macrocyclic (*E*)-allylic alcohols from ω-alkynals via intramolecular 1-alkenylzinc/aldehyde additions. *J. Org. Chem.* 66, 4766–4770. (h) Dahmen S, Bräse S (2001). [2,2]Paracyclophane-based *N,O*-ligands in alkenylzinc additions to aldehydes. *Org. Lett.* 3, 4119–4122. (i) Chen YK, Lurain AE, Walsh PJ (2002). A general, highly enantioselective method for the synthesis of D- and L-α-amino acids and allylic amines. *J. Am. Chem. Soc.* 124, 12225–12231. (j) Ji JX, Qiu LQ, Yip CW, Chan ASC (2003). A convenient, one-step synthesis of optically active tertiary aminonaphthol and its applications in the highly enantioselective alkenylations of aldehydes. *J. Org. Chem.* 68, 1589–1590. (k) Lurain AE, Walsh PJ (2003). A catalytic asymmetric method for the synthesis of γ-unsaturated α-amino acid derivatives. *J. Am. Chem. Soc.* 125, 10677–10683. (l) Ko DH, Kang SW, Kim KH, Chung Y, Ha DC (2004). Asymmetric alkenylzinc additions to aldehydes catalyzed by a binaphthyl based *N,O*-ligand. *Bull. Korean Chem. Soc.* 25, 35–36. (m) Sprout CM, Richmond ML, Seto CT (2004). Solid-phase synthesis of chiral *N*-acylethylenediamines and their use as ligands for the asymmetric addition of alkylzinc and alkenylzinc reagents to aldehydes. *J. Org. Chem.* 69, 6666–6673. (n) Jeon SJ, Chen YK, Walsh PJ (2005). Catalytic asymmetric synthesis of hydroxy enol ethers: approach to a two-carbon homologation of aldehydes. *Org. Lett.* 7, 1729–1732. (o) Lauterwasser F, Gall J, Höfener S, Bräse S (2006). Second-generation *N,O*-[2.2]paracyclophane ketimine ligands for the alkenylzinc addition to aliphatic and aromatic aldehydes: scope and limitations. *Adv. Synth. Catal.* 348, 2068–2074. (p) Jeon SJ, Fisher EL, Carroll PJ, Walsh PJ (2006). Direct, stereospecific generation of (Z)-disubstituted allylic alcohols. *J. Am. Chem. Soc.* 128, 9618–9619. (q) Schmidt F, Rudolph J, Bolm C (2006). Catalyzed enantioselective synthesis of allyl alcohols from aldehydes and alkenylboronic acids. *Synthesis* 3625–3630. (r) Salvi L, Jeon SJ, Fisher EL, Carroll PJ, Walsh PJ (2007). Catalytic asymmetric generation of (Z)-disubstituted allylic alcohols. *J. Am. Chem. Soc.* 129, 16119–16125. (s) Wu HL, Wu PY, Uang BJ (2007). Highly enantioselective synthesis of (*E*)-allylic alcohols. *J. Org. Chem.* 72, 5935–5937.

10. For reviews on catalytic enantioselective aldehyde vinylation using organozinc reagents, see (a) Wipf P, Kendall C (2002). Novel applications of alkenyl zirconocenes. *Chem. – Eur. J.* 8, 1778–1784. (b) Wipf P, Nunes RL (2004). Selective carbon–carbon bond formations with alkenylzirconocenes. *Tetrahedron* 60, 1269–1279.

11. (a) Brown HC, Bhat KS (1986). Enantiomeric (*Z*)-crotyldiisopinocampheylborane and (*E*)-crotyldiisopinocampheylborane—synthesis in high optical purity of all 4 possible stereoisomers of β-methylhomoallyl alcohols. *J. Am. Chem. Soc.* 108, 293–294. (b) Brown HC, Bhat KS (1986). Chiral synthesis via organoboranes. 7. Diastereoselective and enantioselective synthesis of erythro-β-methylhomoallyl and threo-β-methylhomoallyl alcohols via enantiomeric (*Z*)-crotylborane and (*E*)-crotylborane. *J. Am. Chem. Soc.* 108, 5919–5923.

12. (a) Schlosser M, Rauchschwalbe G (1978). 2-Methylpentadienyl-potassium and 2,4-dimethylpentadienylpotassium—1st examples of U-shaped, though open-chain, organometallics. *J. Am. Chem. Soc.* 100, 3258–3260. (b) Schlosser M, Stähle M (1980). Allylic compounds of magnesium, lithium, and potassium—σ-structures or π-structures. *Angew. Chem., Int. Ed.* 19, 487–489.

13. As documented below, the stoichiometric generation of isopinocampheol frequently complicates product isolation in the Brown allylation protocol: (a) Ireland RE, Armstrong, III, JD, Lebreton J, Meissner RS, Rizzacasa MA (1993). Convergent synthesis of polyether ionophore antibiotics—synthesis of the spiroketal and tricyclic glycal segments of monensin. *J. Am. Chem. Soc.* 115, 7152–7165. (b) Burova SA, McDonald, FE (2004). Total synthesis of the polyene-polyol macrolide RK-397, featuring cross-couplings of alkynylepoxide modules. *J. Am. Chem. Soc.* 126, 2495–2500. (c) Ramachandran PV, Prabhudas B, Chandra JS, Reddy MVR (2004). Diastereoselective dihydroxylation and regioselective deoxygenation of dihydropyranones: a novel protocol for the stereoselective synthesis of C-1-C-8 and C-15-C-21 subunits of (+)-discodermolide. *J. Org. Chem.* 69, 6294–6304. (d) White JD, Hansen JD (2005). Total synthesis of (−)-7-epicylindrospermopsin, a toxic metabolite of the freshwater cyanobacterium *Aphanizomenon ovalisporum*, and assignment of its absolute configuration. *J. Org. Chem.* 70, 1963–1977. (e) Gao D, O'Doherty GA (2005). Enantioselective synthesis of 10-epi-anamarine via an iterative dihydroxylation sequence. *Org. Lett.* 7, 1069–1072. (f) Gao D, O'Doherty GA (2005). De novo asymmetric synthesis of anamarine and its analogues. *J. Org. Chem.* 70, 9932–9939. (g) Liu D, Xue J, Xie Z, Wei L, Zhang X, Li Y (2008). Formal total synthesis of benzylpedamide: the right half of (+)-pederin. *Synlett* 1526–1528.

14. (a) Sheldon RA (1997) Catalysis and pollution prevention. *Chem. Ind.* 12–15. (b) Sheldon RA (2007) The E factor: fifteen years on. *Green Chem.* 9, 1273–1283.

15. Prior to our systematic studies, two isolated reports of hydrogenative C–C coupling were reported in the literature: (a) Molander GA, Hoberg JO (1992). Organoyttrium-catalyzed cyclization of substituted 1,5- and 1,6-dienes. *J. Am. Chem. Soc.* 114, 3123–3125. (b) Kokubo K, Miura M, Nomura M (1995). Rhodium-catalyzed reaction of benzoic anhydride with styrene under molecular hydrogen. *Organometallics* 14, 4521–4524.

16. Side products of reductive C–C bond formation have been observed in catalytic hydrogenation on rare occasions: (a) Moyes RB, Walker DW, Wells PB, Whan DA, Irvine EA. In: Dines TJ, Rochester CH, Thomson J, editors. *Catalysis and Surface Characterisation (Special Publication).* Vol. 114, Royal Society of Chemistry, Cambridge;1992, pp 207–212. (b) Bianchini C, Meli A, Peruzzini M, Vizza F, Zanobini F, Frediani P (1989). A homogeneous iron(II) system capable of selectively catalyzing the reduction of terminal alkynes to alkenes and buta-1,3-dienes. *Organometallics* 8, 2080–2082.

17. For reviews on the synthesis of allylic alcohols, see (a) Nagato N. In: Seidel A, editor. *Kirk-Othmer's Encyclopedia of Chemical Technology*. Vol. 2, 5th ed. Wiley, Hoboken; 2004, pp 234–249. (b) Banerjee AK, Poon PS, Laya MS, Vera WJ (2004). Synthesis of allylic alcohols from alkenes and their derivatives. *Russ. Chem. Rev.* 73, 621–636. (c) Hodgson DM, Humphreys PG. Product class 5: allylic alcohols. In: Clayden JP, editor. *Science of*

Synthesis: Houben-Weyl Methods of Molecular Transformations. Vol. 36, Thieme, Stuttgart; 2007, pp 583–665.

18. For reviews on the synthesis of allylic amines, see (a) Bencheikh R, Chaabouni R, Laurent A, Mison P, Nafti A (1983). Synthesis of primary allylic amines. *Synthesis* 685–700. (b) Laurent A, Mison P, Nafti A, Cheikh RB, Chaabouni R (1984). Thermal rearrangement of *N*-ethoxycrbonylaziridines: a route to primary allylic amines. *J. Chem. Res.* 354–355. (c) Johannsen M, Jørgensen KA (1998). Allylic amination. *Chem. Rev.* 98, 1689–1708.
19. For reviews on metal-catalyzed allylic amination and alkoxylation, see (a) Acemoglu L, Williams JMJ, Negishi E, de Meijere A editors. *Handbook of Organopalladium Chemistry for Organic Synthesis*. Vol. 2, Wiley, New York; 2002, pp 1689–1705. (b) Trost BM, Crawley ML (2003). Asymmetric transition-metal-catalyzed allylic alkylations: applications in total synthesis. *Chem. Rev.* 103, 2921–2943. (c) Trost BM (2004). Asymmetric allylic alkylation, an enabling methodology. *J. Org. Chem.* 69, 5813–5837. (d) Miyabe H, Takemoto Y (2005). Regio- and stereocontrolled palladium- or iridium-catalyzed allylation. *Synlett* 1641–1655. (e) Lu Z, Ma SM (2008). Metal-catalyzed enantioselective allylation in asymmetric synthesis. *Angew. Chem., Int. Ed.* 47, 258–297. (f) Hartwig JF, Stanley LM (2010). Mechanistically driven development of iridium catalysts for asymmetric allylic substitution. *Acc. Chem. Res.* 43, 1461–1475. (g) Trost BM, Zhang T, Sieber JD (2010). Catalytic asymmetric allylic alkylation employing heteroatom nucleophiles: a powerful method for C−X bond formation. *Chem. Sci.* 1, 427–440.
20. For catalytic enantioselective ketone vinylation using organozinc reagents, see (a) Li H, Walsh PJ (2004). Catalytic asymmetric vinylation of ketones. *J. Am. Chem. Soc.* 126, 6538–6539. (b) Li H, Walsh PJ (2005). Catalytic asymmetric vinylation and dienylation of ketones. *J. Am. Chem. Soc.* 127, 8355–8361. (c) Jeon SJ, Li H, García C, LaRochelle LK, Walsh PJ (2005). Catalytic asymmetric addition of alkylzinc and functionalized alkylzinc reagents to ketones. *J. Org. Chem.* 70, 448–455.
21. Rhodium-catalyzed addition of vinylmetal reagents to aldehydes has been reported. References (a) and (f) report asymmetric transformations; however, only modest enantioselectivities are observed: (a) Sakai M, Ueda M, Miyaura N (1998). Rhodium-catalyzed addition of organoboronic acids to aldehydes. *Angew. Chem., Int. Ed.* 37, 3279–3281. (b) Batey RA, Thadani AN, Smil DV (1999). Potassium alkenyl- and aryltrifluoroborates: stable and efficient agents for rhodium-catalyzed addition to aldehydes and enones. *Org. Lett.* 1, 1683–1686. (c) Fürstner A, Krause H (2001). Practical method for the rhodium-catalyzed addition of aryl- and alkenylboronic acids to aldehydes. *Adv. Synth. Catal.* 343, 343–350. (d) Fujii T, Koike T, Mori A, Osakada K (2002). Rhodium-catalyzed addition of aryl- and alkenylsilanediols to aldehydes. *Synlett* 298–300. (e) Takezawa A, Yamaguchi K, Ohmura T, Yamamoto Y, Miyaura N (2002). Inter- and intramolecular additions of 1-alkenylboronic acids or esters to aldehydes and ketones catalyzed by rhodium(I) complexes in basic, aqueous solutions. *Synlett* 1733–1735. (f) Duan HF, Xie JH, Shi WJ, Zhang Q, Zhou QL (2006). Enantioselective rhodium-catalyzed addition of arylboronic acids to aldehydes using chiral spiro monophosphite ligands. *Org. Lett.* 8, 1479–1481. (g) Xing CH, Liu TP, Zheng JR, Ng J, Eposito M, Hu QS (2009). Rh(I)/diene-catalyzed addition reactions of aryl/alkenylboronic acids with aldehydes. *Tetrahedron Lett.* 50, 4953–4957.
22. For enantioselective copper-catalyzed addition of vinylmetal reagents to aldehydes, see Tomita D, Kanai M, Shibasaki M (2006). Nucleophilic activation of alkenyl and aryl boronates by a chiral CuIF complex: catalytic enantioselective alkenylation and arylation of aldehydes. *Chem. Asian J.* 1, 161–166.
23. For enantioselective organocatalyzed aldehyde vinylation employing vinylboron reagents, see Shono T, Harada T (2010). Catalytic enantioselective synthesis of secondary allylic alcohols from terminal alkynes and aldehydes via 1-alkenylboron reagents. *Org. Lett.* 12, 5270–5273.
24. Imine vinylation using vinylzinc reagents is known, but enantioselective variants have not been developed: (a) Wipf P, Kendall C, Stephenson CRJ (2003). Dimethylzinc-mediated additions of

alkenylzirconocenes to aldimines. New methodologies for allylic amine and *C*-cyclopropylalkylamine syntheses. *J. Am. Chem. Soc.* 125, 761–768. (b) Wipf P, Janjic J, Stephenson CRJ (2004). Microwave-assisted synthesis of allylic amines: considerable rate acceleration in the hydrozirconation-transmetalation-aldimine addition sequence. *Org. Biomol. Chem.* 2, 443–445.

25. For rhodium-catalyzed imine vinylation employing vinylmetal reagents, see (a) Oi S, Moro M, Fukuhara H, Kawanishi T, Inoue Y (1999). Rhodium-catalyzed arylation and alkenylation of imines with organostannanes. *Tetrahedron Lett.* 40, 9259–9262. (b) Kakuuchi A, Taguchi T, Hanzawa Y (2003). Rh(I)-catalyzed addition of alkenylzirconocene chlorides to aldimine derivatives. *Tetrahedron Lett.* 44, 923–926. (c) Nakao Y, Takeda M, Chen J, Hiyama T, Ichikawa Y, Shintani R, Hayashi T (2008). Rhodium-catalyzed addition of organo[2-(hydroxymethyl)phenyl]dimethylsilanes to arenesulfonylimines. *Chem. Lett.* 290–291. (d) Brak K, Ellman JA (2009). Asymmetric synthesis of α-branched allylic amines by the Rh(I)-catalyzed addition of alkenyltrifluoroborates to *N-tert*-butanesulfinyl aldimines. *J. Am. Chem. Soc.* 131, 3850–3851. (e) Brak K, Ellman JA (2010). Asymmetric Rh(I)-catalyzed addition of MIDA boronates to *N-tert*-butanesulfinyl aldimines: development and comparison to trifluoroborates. *J. Org. Chem.* 75, 3147–3150.

26. For enantioselective organocatalyzed imine vinylation employing vinylboron reagents, see Bishop JA, Lou S, Schaus SE (2009). Enantioselective addition of boronates to acyl imines catalyzed by chiral biphenols. *Angew. Chem., Int. Ed.* 48, 4337–4340.

27. Ojima I, Tzamarioudaki M, Tsai CY (1994). Extremely chemoselective silylformylation and silylcarbocyclization of alkynals. *J. Am. Chem. Soc.* 116, 3643–3644.

28. (a) Crowe WE, Rachita MJ (1995). Titanium-catalyzed reductive cyclization of δ,ε-unsaturated ketones and aldehydes. *J. Am. Chem. Soc.* 117, 6787–6788. For an aligned study, see (b) Kablaoui NM, Buchwald SL (1995). Reductive cyclization of enones by a titanium catalyst. *J. Am. Chem. Soc.* 117, 6785–6786.

29. (a) Oblinger E, Montgomery J (1997). A new stereoselective method for the preparation of allylic alcohols. *J. Am. Chem. Soc.* 119, 9065–9066. (b) Tang XQ, Montgomery J (1999). Nickel catalysis in the stereoselective preparation of quinolizidine, pyrrolizidine, and indolizidine alkaloids: total synthesis of (+)-allopumiliotoxin 267A. *J. Am. Chem. Soc.* 121, 6098–6099. (c) Tang XQ, Montgomery J (2000). Nickel-catalyzed preparation of bicyclic heterocycles: total synthesis of (+)-allopumiliotoxin 267A, (+)-allopumiliotoxin 339A, and (+)-allopumiliotoxin 339B. *J. Am. Chem. Soc.* 122, 6950–6954. (d) Mahandru GM, Liu G, Montgomery J (2004). Ligand-dependent scope and divergent mechanistic behavior in nickel-catalyzed reductive couplings of aldehydes and alkynes. *J. Am. Chem. Soc.* 126, 3698–3699. (e) Knapp-Reed B, Mahandru GM, Montgomery J (2005). Access to macrocyclic endocyclic and exocyclic allylic alcohols by nickel-catalyzed reductive cyclization of ynals. *J. Am. Chem. Soc.* 127, 13156–13157. (f) Chaulagain MR, Sormunen GJ, Montgomery J (2007). New *N*-heterocyclic carbene ligand and its application in asymmetric nickel-catalyzed aldehyde/alkyne reductive couplings. *J. Am. Chem. Soc.* 129, 9568–9569. (g) Malik HA, Sormunen GJ, Montgomery J (2010). A general strategy for regiocontrol in nickel-catalyzed reductive couplings of aldehydes and alkynes. *J. Am. Chem. Soc.* 132, 6304–6305.

30. (a) Huang WS, Chan J, Jamison TF (2000). Highly selective catalytic intermolecular reductive coupling of alkynes and aldehydes. *Org. Lett.* 2, 4221–4223. (b) Miller KM, Huang WS, Jamison TF (2003). Catalytic asymmetric reductive coupling of alkynes and aldehydes: enantioselective synthesis of allylic alcohols and α-hydroxy ketones. *J. Am. Chem. Soc.* 125, 3442–3443. (c) Miller KM, Jamison TF (2005). Highly regioselective catalytic asymmetric reductive coupling of 1,3-enynes and ketones. *Org. Lett.* 7, 3077–3080.

31. Takai K, Sakamoto S, Isshiki T (2003). Regioselective reductive coupling of alkynes and aldehydes leading to allylic alcohols. *Org. Lett.* 5, 653–655.

32. Zhou CY, Zhu SF, Wang LX, Zhou QL (2010). Enantioselective nickel-catalyzed reductive coupling of alkynes and imines. *J. Am. Chem. Soc.* 132, 10955–10957. Enantioselective Ni-catalyzed alkyne, imine, triethylborane three-component coupling has been reported, but modest selectivities (51–89% ee) are observed. For this method, vinylation is accompanied by ethyl transfer: Patel SJ, Jamison TF (2004). Asymmetric catalytic coupling of organoboranes, alkynes, and imines with a removable (trialkylsilyloxy)ethyl group—direct access to enantiomerically pure primary allylic amines. *Angew. Chem., Int. Ed.* 43, 3941–3944.

33. (a) Kong JR, Ngai MY, Krische MJ (2006). Highly enantioselective direct reductive coupling of conjugated alkynes and α-ketoesters via rhodium-catalyzed asymmetric hydrogenation. *J. Am. Chem. Soc.* 128, 718–719. (b) Cho CW, Krische MJ (2006). α-Hydroxy esters via enantioselective hydrogen-mediated C–C coupling: regiocontrolled reactions of silyl-substituted 1,3-diynes. *Org. Lett.* 8, 3873–3876. (c) Hong YT, Cho CW, Skucas E, Krische MJ (2007). Enantioselective reductive coupling of 1,3-enynes to glyoxalates mediated by hydrogen: asymmetric synthesis of β,γ-unsaturated α-hydroxy esters. *Org. Lett.* 9, 3745–3748. (d) Komanduri V, Krische MJ (2006). Enantioselective reductive coupling of 1,3-enynes to heterocyclic aromatic aldehydes and ketones via rhodium-catalyzed asymmetric hydrogenation: mechanistic insight into the role of brønsted acid additives. *J. Am. Chem. Soc.* 128, 16448–16449.

34. Kong JR, Cho CW, Krische MJ (2005). Hydrogen-mediated reductive coupling of conjugated alkynes with ethyl (*N*-sulfinyl)iminoacetates: synthesis of unnatural α-amino acids via rhodium-catalyzed C–C bond forming hydrogenation. *J. Am. Chem. Soc.* 127, 11269–11276.

35. For reviews on the use of *N*-tert-butanesulfinyl imines in asymmetric synthesis, see (a) Davis F, Chen BC (1998). Asymmetric synthesis of amino acids using sulfinimines (thiooxime S-oxides). *Chem. Soc. Rev.* 27, 13–18. (b) Ellman JA, Owens TD, Tang TP (2002). *N-tert*-Butanesulfinyl imines: versatile intermediates for the asymmetric synthesis of amines. *Acc. Chem. Res.* 35, 984–995. (c) Zhou P, Chen BC, Davis F (2004). Recent advances in asymmetric reactions using sulfinimines (*N*-sulfinyl imines). *Tetrahedron* 60, 8003–8030.

36. (a) Liu P, Krische MJ, Houk KN (2011). Mechanism and origins of regio- and enantioselectivities in Rh-I-catalyzed hydrogenative couplings of 1,3-diynes and activated carbonyl partners: intervention of a cumulene intermediate. *Chem. – Eur. J.* 17, 4021–4029. (b) Musashi Y, Sakaki S (2002). Theoretical study of rhodium(III)-catalyzed hydrogenation of carbon dioxide into formic acid. Significant differences in reactivity among rhodium(III), rhodium(I), and ruthenium(II) complexes. *J. Am. Chem. Soc.* 124, 7588–7603. (c) Williams VM, Kong JR, Ko BJ, Mantri Y, Brodbelt JS, Baik MH, Krische MJ (2009). ESI-MS, DFT, and synthetic studies on the H2-mediated coupling of acetylene: insertion of C=X bonds into rhodacyclopentadienes and Brønsted acid cocatalyzed hydrogenolysis of organorhodium intermediates. *J. Am. Chem. Soc.* 131, 16054.

37. For mechanistic studies on the hydrogenation of dehydroamino acids catalyzed by cationic Rh(I) complexes, see (a) Landis CR, Halpern J (1987). Asymmetric hydrogenation of methyl-(Z)-α-acetamidocinnamate catalyzed by (1,2-bis((phenyl-ortho-anisoyl)phosphino)ethane) rhodium(I) – kinetics, mechanism, and origin of enantioselection. *J. Am. Chem. Soc.* 109, 1746–1754. (b) Chan ASC, Halpern J (1980). Interception and characterization of a hydridoalkylrhodium intermediate in a homogeneous catalytic-hydrogenation reaction. *J. Am. Chem. Soc.* 102, 838–840. (c) Halpern J, Riley DP, Chan ASC, Pluth JJ (1977). Novel coordination chemistry and catalytic properties of cationic 1,2-bis(diphenylphosphino)ethanerhodium(I) complexes. *J. Am. Chem. Soc.* 99, 8055–8057. (d) Halpern J (1982). Mechanism and stereoselectivity of asymmetric hydrogenation. *Science* 217, 401–407.

38. For reviews encompassing the mechanism of asymmetric hydrogenation catalyzed by cationic rhodium complexes, see (a) Halpern J (1985). Asymmetric catalytic hydrogenation: mechanism and origin of enantioselection. *Asymmetric Synth.* 5, 41–69. (b) Landis CR, Brauch TW (1998). Probing the nature of H_2 activation in catalytic asymmetric hydrogenation. *Inorg. Chim. Acta* 270, 285–297.

39. Rhee JU, Krische MJ (2006). Highly enantioselective reductive cyclization of acetylenic aldehydes via rhodium catalyzed asymmetric hydrogenation. *J. Am. Chem. Soc.* 128, 10674–10675.

40. (a) Vaska L, Peone JJr. (1971). Simple synthetic route to d^8 complexes of rhodium and iridium. Electronegativity scale for anionic ligands. *Chem. Commun.* 418–419. (b) Haynes A, McNish J, Pearson JM (1998). *Cis–trans* isomerism in $[M(CO)_2I_4]^-$ (M = Rh, Ir): kinetic, mechanistic and spectroscopic studies. *J. Organomet. Chem.* 551, 339–347. (c) Grotjohn DB, Collins LSB, Wolpert M, Bikzhanova GA, Lo HC, Combs D, Hubbard JL (2001). First direct structural comparison of complexes of the same metal fragment to ketenes in both C,C- and C,O-bonding mode. *J. Am. Chem. Soc.* 123, 8260–8270.

41. Li J, Schreckenbach G, Ziegler T (1995). A reassessment of the first metal–carbonyl dissociation energy in M(CO)4 (M = Ni, Pd, Pt), M(CO)5 (M = Fe, Ru, Os), and $M(CO)_6$ (M = Cr, Mo, W) by a quasirelativistic density functional method. *J. Am. Chem. Soc.* 117, 486–494, and references cited therein.

42. (a) Dewar MJS (1951). A review of the π-complex theory. *Bull. Soc. Chim. Fr.* C71–C79. (b) Chatt J, Duncanson LA (1953). Olefin co-ordination compounds. III. Infrared spectra and structure: attempted preparation of acetylene compounds. *J. Chem. Soc.* 2939–2947. (c) Dewar MJS, Ford GP (1979). Relationship between olefinic π-complexes and three-membered rings. *J. Am. Chem. Soc.* 101, 783–791.

43. Alkyne complexation by iridium(I) results in substantial deviation from linearity, as revealed by single crystal X-ray diffraction analysis: (a) Kirchner RM, Ibers JA (1973). A (π-acetylene)(σ-vinyl) transition-metal complex. The structure of (σ-dicyanovinyl)(carbonyl)(π-dicyanoacetylene)-bis(triphenylphosphine)iridium(I). *J. Am. Chem. Soc.* 95, 1095–1101. (b) Calabrese JC, Roe DC, Thorn DL, Tulip TH (1984). Conversion of a (methoxymethyl)iridium(I) acetylene complex to a metallacyclobutene iridium(III) complex – crystal and molecular-structures of Ir(CH_2OCH_3)(*p*-tol-C≡C-p-tol)$(PMe_3)_3$ and *fac*-Ir[CH_2C(p-tol)≡C(p-tol)]Br$(PMe_3)_3$. *Organometallics* 3, 1223–1230. (c) Rappoli BJ, Churchill MR, Janik TS, Rees WM, Atwood JD (1987). Crystal-structure of the quasitetrahedral iridium(I) complex, Ir$(COCH_2CMe_3)$[P(*p*-tolyl)$_3]_2[C_2(CO_2Me)_2]$ – an intermediate in cyclotrimerization of activated alkynes by 16-electron alkyl complexes of iridium, *trans*-RIr(CO)L_2 (R = Me, CH_2CMe_3, L = PPh_3, P(p-tolyl)$_3$). *J. Am. Chem. Soc.* 109, 5145–5149. (d) Rees WM, Churchill MR, Fettinger JC, Atwood JD (1987). 1. Addition of dimethyl acetylenedicarboxylate $[C_2(CO_2Me)_2]$ to *trans*-MeIr(CO)$(PPh_3)_2$. Formation and isomerization of MeIr(CO)$(PPh_3)_2[C_2(CO_2Me)_2]$, and crystal structure of the thermodynamic isomer. *J. Organomet. Chem.* 319, 411–422. (e) Marinelli G, Streib WE, Huffman JC, Caulton KG, Gagne MR, Takats J, Dartiguenave M, Chardon C, Jackson SA, Eisenstein O (1990). The origin of structural variety of alkyne complexes of d^8 metals - an example of structural isomerism. *Polyhedron* 9, 1867–1881.

44. (a) Barchuk A, Ngai MY, Krische MJ (2007). Allylic amines via iridium-catalyzed C–C bond forming hydrogenation: imine vinylation in the absence of stoichiometric by-products or metallic reagents. *J. Am. Chem. Soc.* 129, 8432–8433. (b) Ngai MY, Barchuk A, Krische MJ (2007). Enantioselective iridium-catalyzed imine vinylation: optically enriched allylic amines via alkyne-imine reductive coupling mediated by hydrogen. *J. Am. Chem. Soc.* 129, 12644–12645.

45. Seidel A, editor. *Kirk-Othmer's Encyclopedia of Chemical Technology*, 5th ed. Vol. 1, Wiley, Hoboken; 2004, pp 216–217.

46. Bianchini C, Caulton KG, Chardon C, Eisenstein O, Folting K, Johnson TJ, Meli A, Peruzzini M, Rauscher DJ, Streib WE, Vizza F (1991). An η4-benzene species mediates acetylene cyclotrimerization. *J. Am. Chem. Soc.* 113, 5127–5129, and references cited therein.

47. (a) Kong JR, Krische MJ (2006). Catalytic carbonyl *Z*-dienylation via multicomponent reductive coupling of acetylene to aldehydes and α-ketoesters mediated by hydrogen: carbonyl insertion into cationic rhodacyclopentadienes. *J. Am. Chem. Soc.* 128, 16040–16041.

(b) Skucas E, Kong JR, Krische MJ (2007). Enantioselective reductive coupling of acetylene to *N*-arylsulfonyl imines via rhodium catalyzed C−C bond-forming hydrogenation: (Z)-dienyl allylic amines. *J. Am. Chem. Soc.* 129, 7242–7243. (c) Han SB, Kong JR, Krische MJ (2008). Catalyst-directed diastereoselectivity in hydrogenative couplings of acetylene to α-chiral aldehydes: formal synthesis of all eight L-hexoses. *Org. Lett.* 10, 4133–4135.

48. (a) Patman RL, Chaulagain MR, Williams VM, Krische MJ (2009). Direct vinylation of alcohols or aldehydes employing alkynes as vinyl donors: a ruthenium catalyzed C−C bond-forming transfer hydrogenation. *J. Am. Chem. Soc.* 131, 2066–2067. (b) Williams VM, Leung JC, Patman RL, Krische MJ (2009). Hydroacylation of 2-butyne from the alcohol or aldehyde oxidation level via ruthenium catalyzed C−C bond forming transfer hydrogenation. *Tetrahedron* 65, 5024–5029. (c) Bausch CC, Patman RL, Breit B, Krische MJ (2011). Divergent regioselectivity in the synthesis of trisubstituted allylic alcohols by nickel- and ruthenium-catalyzed alkyne hydrohydroxymethylation with formaldehyde. *Angew. Chem., Int. Ed.* 50, 5686–5689. (d) Leung JC, Patman RL, Sam B, Krische MJ (2011) Alkyne-aldehyde reductive C−C coupling via ruthenium catalyzed transfer hydrogenation: regio- and stereoselective carbonyl vinylation to form trisubstituted allylic alcohols in the absence of vinylmetal reagents. *Chem. Eur. J.* 17, 12437–12443.

49. For reviews on enantioselective carbonyl allylation, see (a) Ramachandran PV (2002). Pinane based versatile allylboranes. *Aldrichimica Acta* 35, 23–35. (b) Denmark SE, Fu J (2003). Catalytic enantioselective addition of allylic organometallic reagents to aldehydes and ketones. *Chem. Rev.* 103, 2763–2794. (c) Yu CM, Youn J, Jung HK (2006). Regulation of stereoselectivity and reactivity in the inter- and intramolecular allylic transfer reactions. *Bull. Korean Chem. Soc.* 27, 463–472. (d) Marek I, Sklute G (2007). Creation of quaternary stereocenters in carbonyl allylation reactions. *Chem. Commun.* 1683–1691. (e) Hall DG (2007). Lewis and Bronsted acid catalyzed allylboration of carbonyl compounds: from discovery to mechanism and applications. *Synlett* 1644–1655. (f) Lachance H, Hall DG (2008). Allylboration of carbonyl compounds. *Org. React.* 73, 1–573.

50. For selected examples of chirally modified allyl metal reagents, see (a) Herold T, Hoffmann RW (1978). Enantioselective synthesis of homoallyl alcohols via chiral allylboronic esters. *Angew. Chem., Int. Ed.* 17, 768–769. (b) Hoffmann RW, Herold T (1981). Stereoselective synthesis of alcohols. 7. Optically-active homoallyl alcohols via addition of chiral boronates to aldehydes. *Chem. Ber.* 114, 375–383. (c) Hayashi T, Konishi M, Kumada M (1982). Optically-active allylsilanes. 2. High stereoselectivity in asymmetric reaction with aldehydes producing homoallylic alcohols. *J. Am. Chem. Soc.* 104, 4963–4965. (d) Brown HC, Jadhav PK (1983). Asymmetric carbon–carbon bond formation via β-allyldiisopinocampheylborane – simple synthesis of secondary homoallylic alcohols with excellent enantiomeric purities. *J. Am. Chem. Soc.* 105, 2092–2093. (e) Roush WR, Walts AE, Hoong, LK (1985). Diastereoselective and enantioselective aldehyde addition-reactions of 2-allyl-1,3,2-dioxaborolane-4,5-dicarboxylic esters, a useful class of tartrate ester modified allylboronates. *J. Am. Chem. Soc.* 107, 8186–8190. (f) Reetz MT (1988). Asymmetric C−C bond formation using organometallic chemistry. *Pure Appl. Chem.* 60, 1607–1614. (g) Short RP, Masamune S (1989). Asymmetric allylboration with B-allyl-2-(trimethylsilyl)borolane. *J. Am. Chem. Soc.* 111, 1892–1894. (h) Corey EJ, Yu CM, Kim SS (1989). A practical and efficient method for enantioselective allylation of aldehydes. *J. Am. Chem. Soc.* 111, 5495–5496. (i) Seebach D, Beck AK, Imwinkeiried R, Roggo S, Wonnacott A (1987). Chiral alkoxytitanium(IV) complexes for enantioselective nucleophilic additions to aldehydes and as Lewis-acids in Diels-Alder reactions. *Helv. Chim. Acta* 70, 954–974. (j) Riediker M, Duthaler RO (1989). Enantioselective syntheses with titanium-carbohydrate complexes. 1. Enantioselective allylation of carbonyl-compounds with titanium-carbohydrate complexes. *Angew. Chem., Int. Ed.* 28, 494–496. (k) Panek JS, Yang M (1991). Diastereofacial selectivity with optically-active α-substituted β-silyl-(E)-hexenoates – enantioselective construction of homoallylic ethers via

reaction with aryl acetals. *J. Am. Chem. Soc.* 113, 6594–6600. (l) Kinnaird JWA, Ng PY, Kubota K, Wang X, Leighton JL (2002). Strained silacycles in organic synthesis: a new reagent for the enantioselective allylation of aldehydes. *J. Am. Chem. Soc.* 124, 7920–7921. (m) Hackman BM, Lombardi PJ, Leighton JL (2004). Highly diastereo- and enantioselective reagents for aldehyde crotylation. *Org. Lett.* 6, 4375–4377. (n) Lachance H, Lu X, Gravel M, Hall DG (2003). Scandium-catalyzed allylboration of aldehydes as a practical method for highly diastereo- and enantioselective construction of homoallylic alcohols. *J. Am. Chem. Soc.* 125, 10160–10161. (o) Burgos CH, Canales E, Matos K, Soderquist JA (2005). Asymmetric allyl- and crotylboration with the robust, versatile, and recyclable 10-TMS-9-borabicyclo [3.3.2]decanes. *J. Am. Chem. Soc.* 127, 8044–8049.

51. For seminal examples of catalytic asymmetric carbonyl allylation, see (a) Furuta K, Mouri M, Yamamoto H (1991). Chiral (acyloxy) borane catalyzed asymmetric allylation of aldehydes. *Synlett* 561–562. (b) Costa AL, Piazza MG, Tagliavini E, Trombini C, Umani-Ronchi A (1993). Catalytic asymmetric-synthesis of homoallylic alcohols. *J. Am. Chem. Soc.* 115, 7001–7002. (c) Keck GE, Tarbet KH, Geraci LS (1993). Catalytic asymmetric allylation of aldehydes. *J. Am. Chem. Soc.* 115, 8467–8468. (d) Denmark SE, Coe DM, Pratt NE, Griedel BD (1994). Asymmetric allylation of aldehydes with chiral Lewis-bases. *J. Org. Chem.* 59, 6161–6163. (e) Denmark SE, Fu J (2001). Catalytic, enantioselective addition of substituted allylic trichlorosilanes using a rationally-designed 2,2′-bispyrrolidine-based bisphosphoramide. *J. Am. Chem. Soc.* 123, 9488–9489.

52. For selected examples of carbonyl allylation via umpolung of transition metal π-allyls, see (a) Palladium: Tabuchi T, Inanaga J, Yamaguchi M (1986). Reductive coupling of allylic acetates with carbonyl-compounds by using Pd(0)-SmI2 system. *Tetrahedron Lett.* 27, 1195–1196. (b) Takahara JP, Masuyama Y, Kurusu Y (1992). Palladium-catalyzed carbonyl allylation by allylic alcohols with $SnCl_2$. *J. Am. Chem. Soc.* 114, 2577–2586. (c) Kimura M, Ogawa Y, Shimizu M, Sueishi M, Tanaka S, Tamaru Y (1998). Novel carbonyl-dependent regioselective allylation via diethylzinc-mediated umpolung of π-allylpalladium. *Tetrahedron Lett.* 39, 6903–6906. (d) Kimura M, Shimizu M, Shibata K, Tazoe M, Tamaru Y (2003). Pd-catalyzed nucleophilic alkylation of aliphatic aldehydes with allyl alcohols: allyl, 2-tetrahydrofuryl, and 2-tetrahydropyranyl ethers as useful C-3, C-4, and C-5 sources. *Angew. Chem., Int. Ed.* 42, 3392–3395. (e) Zanoni G, Gladiali S, Marchetti A, Piccinini P, Tredici I, Vidari G (2004). Enantioselective catalytic allylation of carbonyl groups by umpolung of π-allyl palladium complexes. *Angew. Chem., Int. Ed.* 43, 846–849. Rhodium: (f) Masuyama Y, Kaneko Y, Kurusu Y (2004). Rhodium-catalyzed carbonyl allylations by allylic alcohols with tin(II) chloride. *Tetrahedron Lett.* 45, 8969–8971. (g) Ruthenium: Tsuji Y, Mukai T, Kondo T, Watanabe Y (1989). Ruthenium complex catalyzed allylation of aldehydes with allylic acetates. *J. Organomet. Chem.* 369, C51–C53. (h) Kondo T, Ono H, Satake N, Mitsudo T, Watanabe Y (1995). Nucleophilic and electrophilic allylation reactions – synthesis, structure, and ambiphilic reactivity of (η3-allyl)ruthenium(II) complexes. *Organometallics* 14, 1945–1953.

53. For selected reviews of carbonyl allylation based on the reductive coupling of metallo-π-allyls derived from allylic alcohols, ethers or carboxylates, see (a) Masuyama Y. Palladium-catalyzed carbonyl allylation *via* π-allylpalladium complexes. In: Liebeskind LS, editor. *Advances in Metal-Organic Chemistry*, Vol. 3, JAI Press, Greenwich;1994, pp 255–303. (b) Tamaru Y. Palladium-catalyzed reactions of allyl and related derivatives with organoelectrophiles. In: Negishi E, de Meijere A, editors. *Handbook of Organopalladium Chemistry for Organic Synthesis*. Vol. 2, Wiley, New York; 2002, pp 1917–1943. (c) Tamaru Y (1999). Novel catalytic reactions involving π-allylpalladium and -nickel as the key intermediates: umpolung and β-decarbopalladation of π-allylpalladium and nickel-catalyzed homoallylation of carbonyl compounds with 1,3-dienes. *J. Organomet. Chem.* 576, 215. (d) Kondo T, Mitsudo T (2002). Advances in ruthenium catalysis. Novel catalytic reactions via η3-allylruthenium intermediates. *Curr. Org. Chem.* 6, 1163–1179. (e) Tamaru Y (2005). Activation of allyl alcohols as allyl

cations, allyl anions, and amphiphilic allylic species by palladium. *Eur. J. Org. Chem.* 13, 2647–2656. (f) Zanoni G, Pontiroli A, Marchetti A, Vidari G (2007). Stereoselective carbonyl allylation by umpolung of allylpalladium(II) complexes. *Eur. J. Org. Chem.* 22, 3599–3611.

54. Fürstner A, Shi N (1996). A multicomponent redox system accounts for the first Nozaki–Hiyama–Kishi reactions catalytic in chromium. *J. Am. Chem. Soc.* 118, 2533–2534.

55. For selected examples of enantioselective carbonyl allylation and crotylation via catalytic Nozaki–Hiyama coupling, see (a) Bandini M, Cozzi PG, Umani-Ronchi A (2000). The first catalytic enantioselective Nozaki–Hiyama–Kishi reaction. *Polyhedron* 19, 537–539. (b) Bandini M, Cozzi PG, Umani-Ronchi A (2001). Enantioselective catalytic addition of allyl organometallic reagents to aldehydes promoted by [Cr(Salen)]: the hidden role played by weak Lewis acids in metallo-Salen promoted reactions. *Tetrahedron* 57, 835–843. (c) Bandini M, Cozzi PG, Umani-Ronchi A (2000). Salen as a chiral activator: anti versus syn switchable diastereoselection in the enantioselective addition of crotyl bromide to aromatic aldehydes. *Angew. Chem., Int. Ed.* 39, 2327–2330. (d) Inoue M, Suzuki T, Nakada M (2003). Asymmetric catalysis of Nozaki–Hiyama allylation and methallylation with a new tridentate bis(oxazolinyl) carbazole ligand. *J. Am. Chem. Soc.* 125, 1140–1141. (e) Lee JY, Miller JJ, Hamilton SS, Sigman MS (2005). Stereochemical diversity in chiral ligand design: discovery and optimization of catalysts for the enantioselective addition of allylic halides to aldehydes. *Org. Lett.* 7, 1837–1839. (f) McManus HA, Cozzi PG, Guiry PJ (2006). Application of tridentate bis (oxazoline) ligands in catalytic asymmetric Nozaki–Hiyama allylation and crotylation: an example of high enantioselection with a non-symmetric bis(oxazoline) ligand. *Adv. Synth. Catal.* 348, 551–558. (g) Xia G, Yamamoto H (2006). Catalytic enantioselective Nozaki–Hiyama allylation reaction with tethered bis(8-quinolinolato) (TBOx) chromium complex. *J. Am. Chem. Soc.* 128, 2554–2555. (h) Hargaden GC, Müller-Bunz H, Guiry PJ (2007). New proline-oxazoline ligands and their application in the asymmetric Nozaki–Hiyama–Kishi reaction. *Eur. J. Org. Chem.* 4235–4243. (i) Hargaden GC, O'Sullivan TP, Guiry PJ (2008). Synthesis of non-symmetric bis(oxazoline)-containing ligands and their application in the catalytic enantioselective Nozaki–Hiyama–Kishi allylation of benzaldehyde. *Org. Biomol. Chem.* 6, 562–566. (j) Liu X, Henderson JA, Sasaki T, Kishi Y (2009). Dramatic improvement in catalyst loadings and molar ratios of coupling partners for Ni/Cr-mediated coupling reactions: heterobimetallic catalysts. *J. Am. Chem. Soc.* 131, 16678–16680. (k) White JD, Shaw S (2011). *cis*-2,5-Diaminobicyclo[2.2.2]octane, a new scaffold for asymmetric catalysis via Salen-metal complexes. *Org. Lett.* 13, 2488–2491.

56. For selected reviews on catalytic Nozaki–Hiyama coupling, see (a) Avalos M, Babiano R, Cintas P, Jiménez JL, Palacios JC (1999). Synthetic variations based on low-valent chromium: new developments. *Chem. Soc. Rev.* 28, 169–177. (b) Bandini M, Cozzi PG, Umani-Ronchi A (2002). [Cr(Salen)] as a 'bridge' between asymmetric catalysis, Lewis acids and redox processes. *Chem. Commun.* 919–927. (c) Hargaden GC, Guiry PJ (2007). The development of the asymmetric Nozaki–Hiyama–Kishi reaction. *Adv. Synth. Catal.* 349, 2407–2424. (d) Inoue M, Suzuki T, Kinoshita A, Nakada M (2008). Catalytic asymmetric Nozaki–Hiyama reactions with a tridentate bis(oxazolinyl)carbazole ligand. *Chem. Rec.* 8, 169–181.

57. (a) Skucas E. Bower J.F. Krische MJ (2007). Carbonyl allylation in the absence of preformed allyl metal reagents: reverse prenylation via iridium-catalyzed hydrogenative coupling of dimethylallene. *J. Am. Chem. Soc.* 129, 12678–12679 (b) Bower J.F. Skucas E. Patman R.L. Krische MJ (2007). Catalytic C–C coupling via transfer hydrogenation: reverse prenylation, crotylation, and allylation from the alcohol or aldehyde oxidation level. *J. Am. Chem. Soc.* 129, 15134–15135

58. (a) Ngai MY Skucas E Krische MJ (2008). Ruthenium catalyzed C–C bond formation via transfer hydrogenation: branch-selective reductive coupling of allenes to paraformaldehyde and higher aldehydes. *Org. Lett.* 10, 2705–2708. (b) Zbieg J.R. McInturff E.L. Leung J.C. Krische MJ

(2011). Amplification of anti-diastereoselectivity via Curtin–Hammett effects in ruthenium-catalyzed hydrohydroxyalkylation of 1,1-disubstituted allenes: diastereoselective formation of all-carbon quaternary centers. *J. Am. Chem. Soc.* 133, 1141–1144.

59. (a) Zbieg JR McInturff EL Krische MJ (2010). Allenamide hydro-hydroxyalkylation: 1,2-amino alcohols via ruthenium-catalyzed carbonyl anti-aminoallylation. *Org. Lett.* 12, 2514–2516. (b) Skucas E. Zbieg J.R. Krische MJ (2009). Anti-aminoallylation of aldehydes *via* ruthenium-catalyzed transfer hydrogenative coupling of sulfonamido allenes: 1,2-aminoalcohols. *J. Am. Chem. Soc.* 131, 5054–5055.

60. In related "hydrogen auto-transfer" reactions, alcohol dehydrogenation and nucleophile generation occur independently. Hence, preactivated nucleophiles are required. Such processes promote formal alcohol substitution rather than carbonyl addition. For selected reviews, see (a) Nixon TD, Whittlesey MK, Williams JMJ (2009). Transition metal catalyzed reactions of alcohols using borrowing hydrogen methodology. *Dalton Trans.* 753–762. (b) Dobereiner GE Crabtree RH (2010). Dehydrogenation as a substrate-activating strategy in homogeneous transition-metal catalysis. *Chem. Rev.* 110, 681–703. (c) Guillena G Ramón DJ Yus M (2010). Hydrogen autotransfer in the *N*-alkylation of amines and related compounds using alcohols and amines as electrophiles. *Chem. Rev.* 110, 1611–1641.

61. Bower JF, Patman RL, Krische MJ (2008). Iridium-catalyzed C–C coupling via transfer hydrogenation: carbonyl addition from the alcohol or aldehyde oxidation level employing 1,3-cyclohexadiene. *Org. Lett.* 10, 1033–1035.

62. (a) Shibahara F, Bower JF, Krische MJ (2008). Ruthenium-catalyzed C–C bond forming transfer hydrogenation: carbonyl allylation from the alcohol or aldehyde oxidation level employing acyclic 1,3-dienes as surrogates to preformed allyl metal reagents. *J. Am. Chem. Soc.* 130, 6338–6339. (b) Shibahara F, Bower JF, Krische MJ (2008). Diene hydro-acylation from the alcohol or aldehyde oxidation level via ruthenium-catalyzed C–C bond-forming transfer hydrogenation: synthesis of β, γ-unsaturated ketones. *J. Am. Chem. Soc.* 130, 14120–14122. (c) Zbieg JR, Moran J, Krische MJ (2011). Diastereo- and enantioselective ruthenium-catalyzed hydrohydroxyalkylation of 2-silyl-butadienes: carbonyl syn-crotylation from the alcohol oxidation level. *J. Am. Chem. Soc.* 133, 10582–10586.

63. (a) Smejkal T, Han H, Breit B, Krische MJ (2009). All carbon quaternary centers *via* ruthenium catalyzed hydroxymethylation of 2-substituted butadienes mediated by formaldehyde: beyond hydroformylation. *J. Am. Chem. Soc.* 131, 10366–10367. (b) Han H, Krische MJ (2009). Direct ruthenium catalyzed C–C coupling of ethanol: diene hydro-hydroxyethylation to form all carbon quaternary centers. *Org. Lett.* 12, 2844–2846.

64. Crotylmagnesium and crotylzinc reagents with Me3Si-substituents at the 2-position react with aldehydes to give racemic syn-adducts: (a) Sato F, Kusakabe M, Kobayashi Y (1984). Acyclic stereoselection using relative 1,2-asymmetric induction. 4. An effective method for controlling 3,4-stereochemistry of aldols - direct preparation of optically-active 2,3-anti-3-hydroxy-2-methyl carbonyl-compounds. *J. Chem. Soc., Chem. Commun.* 1131–1132. (b) Helm MD, Mayer P, Knochel P (2008). Preparation of silyl substituted crotylzinc reagents and their highly diastereoselective addition to carbonyl compounds. *Chem. Commun.* 1916–1917.

65. (a) Kim IS, Ngai MY, Krische MJ (2008). Enantioselective iridium-catalyzed carbonyl allylation from the alcohol or aldehyde oxidation level using allyl acetate as an allyl metal surrogate. *J. Am. Chem. Soc.* 130, 6340–6341. (b) Kim IS, Ngai MY, Krische MJ (2008). Enantioselective iridium-catalyzed carbonyl allylation from the alcohol or aldehyde oxidation level via transfer hydrogenative coupling of allyl acetate: departure from chirally modified allyl metal reagents in carbonyl addition. *J. Am. Chem. Soc.* 130, 14891–14899.

66. Gao X, Krische MJ, Manuscript in preparation.

67. (a) Kim IS, Han SB, Krische MJ (2009). Anti-diastereo- and enantioselective carbonyl crotylation from the alcohol or aldehyde oxidation level employing a cyclometallated iridium catalyst: alpha-methyl allyl acetate as a surrogate to preformed crotylmetal reagents. *J. Am. Chem. Soc.* 131, 2514–2520. (b) Gao X, Townsend IA, Krische MJ (2011). Enhanced anti-diastereo- and enantioselectivity in alcohol-mediated carbonyl crotylation using an isolable single component iridium catalyst. *J. Org. Chem.* 76, 2350–2354.

68. For a review on two-directional chain synthesis, see Poss CS, Schreiber SL (1994). 2-Directional chain synthesis and terminus differentiation. *Acc. Chem. Res.* 27, 9–17.

69. Lu Y, Kim IS, Hassan A, Del Valle DJ, Krische MJ (2009). 1,n-Glycols as dialdehyde equivalents in iridium-catalyzed enantioselective carbonyl allylation and iterative two-directional assembly of 1,3-polyols. *Angew. Chem., Int. Ed.* 48, 5018–5021.

70. This mechanism for enantiomeric enrichment is documented by Eliel and Midland: (a) Kogure T, Eliel EL (1984). A convergent asymmetric-synthesis of (−)-malyngolide and its 3 stereo-isomers. *J. Org. Chem.* 49, 576–578. (b) Midland MM, Gabriel J (1985). Synthesis of (−)-talaromycin-A. *J. Org. Chem.* 50, 1143–1144.

71. Han SB, Hassan A, Kim IS, Krische MJ (2010). Total synthesis of (+)-roxaticin via C−C bond forming transfer hydrogenation: a departure from stoichiometric chiral reagents, auxiliaries, and premetalated nucleophiles in polyketide construction. *J. Am. Chem. Soc.* 132, 15559–15561.

72. Gao X, Han H, Krische MJ (2011). Direct generation of acyclic polypropionate stereopolyads via double diastereo- and enantioselective iridium catalyzed crotylation of 1,3-diols: beyond stepwise carbonyl addition in polyketide construction. *J. Am. Chem. Soc.* 133, 12795–12800.

73. Han SB, Gao X, Krische MJ (2010). Iridium-catalyzed anti-diastereo- and enantioselective carbonyl (trimethylsilyl)allylation from the alcohol or aldehyde oxidation level. *J. Am. Chem. Soc.* 132, 9153–9156. (b) Zhang YJ, Yang JH, Kim SH, Krische MJ (2010). Anti-diastereo- and enantioselective carbonyl (hydroxymethyl)allylation from the alcohol or aldehyde oxidation level: allyl carbonates as allylmetal surrogates. *J. Am. Chem. Soc.* 132, 4562–4563. (c) Gao X, Zhang YJ, Krische MJ (2011). Iridium-catalyzed anti-diastereo- and enantioselective carbonyl (alpha-trifluoromethyl)allylation from the alcohol or aldehyde oxidation level. *Angew. Chem., Int. Ed.* 50, 4173–4175. (d) Han SB, Han H, Krische MJ (2010). Diastereo- and enantioselective anti-alkoxyallylation employing allylic gem-dicarboxylates as allyl donors via iridium-catalyzed transfer hydrogenation. *J. Am. Chem. Soc.* 132, 4562–4563. (e) Han SB, Kim IS, Han H, Krische MJ (2009). Enantioselective carbonyl reverse prenylation from the alcohol or aldehyde oxidation level employing 1,1-dimethylallene as the prenyl donor. *J. Am. Chem. Soc.* 131, 6916–6917.

74. Hassan A, Zbieg JR, Krische MJ (2011). Enantioselective iridium-catalyzed vinylogous Reformatsky-aldol reaction from the alcohol oxidation level: linear regioselectivity by way of carbon-bound enolates. *Angew. Chem., Int. Ed.* 50, 3493–3496.

75. Zbieg JR, Fukuzumi T, Krische MJ (2010). Iridium-catalyzed hydrohydroxyalkylation of butadiene: carbonyl crotylation. *Adv. Synth. Catal.* 352, 2416–2420.

76. Itoh J, Han SB, Krische MJ (2009). Enantioselective allylation, crotylation, and reverse prenylation of substituted isatins: iridium-catalyzed C−C bond-forming transfer hydrogenation. *Angew. Chem., Int. Ed.* 48, 6313–6316.

77. Obora Y, Hatanaka S, Ishii Y (2009). Iridium-catalyzed coupling reaction of primary alcohols with 1-aryl-1-propynes leading to secondary homoallylic alcohols. *Org. Lett.* 11, 3510–3513.

78. For reviews encompassing carbonyl propargylation employing allenyl metal reagents, see (a) Moreau JL In: Patai S, editor. *The Chemistry of Ketenes, Allenes and Related Compounds.* Wiley, New York;1980, pp 363–413. (b) Marshall JA (1996). Chiral allylic and allenic stannanes as reagents for asymmetric synthesis. *Chem. Rev.* 96, 31–48. (c) Gung BW (2004). Additions of allyl, allenyl, and propargylstannanes to aldehydes and imines. *Org. React.* 64, 1–113. (d) Marshall JA, Gung BW, Grachan ML In: Krause N, Hashmi ASK, editors. *Modern Allene Chemistry.*

Wiley-VCH, Weinheim; 2004, pp 493–592. (e) Marshall JA (2007). Chiral allylic and allenic metal reagents for organic synthesis. *J. Org. Chem.* 72, 8153–8166.

79. For selected milestones in carbonyl propargylation, see (a) Prévost C, Gaudemar M, Honigberg J (1950). The propargylic rearrangement. *Comptes Rendus* 230, 1186–1188. (b) Wotiz JH (1950). The propargylic rearrangement. I. The carbonation products of Grignard reagents from primary propargylic bromides. *J. Am. Chem. Soc.* 72, 1639–1642. (c) Karila M, Capmau ML, Chodkiewicz W (1969). Stereochemistry of addition of asymmetrical allene organometals on carbonyl. *Comptes Rendus* 269, 342–345. (d) Lequan M, Guillerm G (1973). Condensation reaction of tin propargylic derivatives and tin allenic derivatives on chloral – kinetic study. *J. Organomet. Chem.* 54, 153–164. (e) Mukaiyama T, Harada T (1981). A convenient synthesis of α-hydroxyallenes by the reaction of propargyl iodides with aldehydes in the presence of stannous halide. *Chem. Lett.* 621–624. (f) Favre E, Gaudemar M (1966). Sur les comportements compares du bore et de l'aluminium en synthèse organometallique. *Comptes Rendus* 263, 1543–1545. (g) Danheiser RL, Carini DJ (1980). (Trimethylsilyl)allenes as propargylic anion equivalents - synthesis of homopropargylic alcohols and ethers. *J. Org. Chem.* 45, 3925–3927. (h) Haruta R, Ishiguro, M, Ikeda N, Yamamoto H (1982). Chiral allenylboronic esters: a practical reagent for enantioselective carbon-carbon bond formation. *J. Am. Chem. Soc.* 104, 7667–7669. (i) Corey EJ, Yu CM, Lee DH (1990). A practical and general enantioselective synthesis of chiral propa-1,2-dienyl and propargyl carbinols. *J. Am. Chem. Soc.* 112, 878–879. (j) Minowa N, Mukaiyama T (1987). Asymmetric allylation with a new chiral allylating agent prepared from tin(II) triflate, chiral diamine, and allylaluminum. *Bull. Chem. Soc. Jpn.* 60, 3697–3704. (k) Marshall JA, Wang XJ (1991). Highly diastereoselective SE′ additions of enantioenriched allenylstannanes to *(S)*-2-(benzyloxy)propanal. *J. Org. Chem.* 56, 3211–3213. (l) Marshall JA, Maxson K (2000). Stereoselective synthesis of stereotriad subunits of polyketides through additions of nonracemic allenylsilanes to (*R*)- and (*S*)-2-methyl-3-oxygenated propanals. *J. Org. Chem.* 65, 630–633. (m) Matsumoto Y, Naito M, Uozumi Y, Hayashi T (1993). Axially chiral allenylboranes – catalytic asymmetric-synthesis by palladium-catalyzed hydroboration of but-1-en-3-ynes and their reaction with an aldehyde. *J. Chem. Soc., Chem. Commun.* 1468–1469. (n) Keck GE, Krishnamurthy D, Chen X (1994). Asymmetric-synthesis of homopropargylic alcohols from aldehydes and allenyltri-*n*-butylstannane. *Tetrahedron Lett.* 35, 8323–8324. (o) Denmark SE, Wynn T (2001). Lewis base activation of Lewis acids: catalytic enantioselective allylation and propargylation of aldehydes. *J. Am. Chem. Soc.* 123, 6199–6200.

80. Patman RL, Williams VM, Bower JF, Krische MJ (2008). Carbonyl propargylation from the alcohol or aldehyde oxidation level employing 1,3-enynes as surrogates to preformed allenylmetal reagents: a ruthenium-catalyzed C–C bond-forming transfer hydrogenation. *Angew. Chem., Int. Ed.* 47, 5220–5223.

81. Though largely attributed to Würtz, the aldol reaction was reported first by Borodin: (a) von Richter V (1869). *Ber. Dtsch. Chem. Ges.* 2, 552–554. (Borodin's earliest results are cited in this article). (b) Würtz CA (1872). *Bull. Soc. Chim. Fr.* 17, 436–442. (c) Borodin A (1873). Ueber einen neuen abkömmling des valerals. *Ber. Dtsch. Chem. Ges.* 6, 982–985. See also: (d) Kane R (1838). Ueber eine aus dem essiggeist reihe von verbindungen. *Ann. Phys. Chem.* 120, 473–494.

82. For selected reviews on synthetic methods for polyketide constructions, see (a) Paterson I, Doughty VA, Florence G, Gerlach K, McLeod MD, Scott JP, Trieselmann T (2001). Asymmetric aldol reactions using boron enolates: applications to polyketide synthesis. *ACS Symp. Ser.* 783, 195–206. (b) Koskinen AMP, Karisalmi K (2005). Polyketide stereotetrads in natural products. *Chem. Soc. Rev.* 34, 677–690. (c) Yeung KS, Paterson I (2005). Advances in the total synthesis of biologically important marine macrolides. *Chem. Rev.* 105, 4237–4313. (d) Schetter B, Mahrwald R (2006). Modern aldol methods for the total synthesis of polyketides. *Angew. Chem., Int. Ed.* 45, 7506–7525. (e) Morris JC, Nicholas GM, Phillips AJ

(2007). Marine natural products: synthetic aspects. *Nat. Prod. Rep.* 24, 87–108. (f) Paterson I. Total synthesis of polyketides using asymmetric aldol reactions. In: Christmann M, Bräse S, editors. *Asymmetric Synthesis*, 2nd edition, Wiley-VCH, Weinheim; 2008, pp 293–298. (g) Paterson I, Findlay AD (2009). Recent advances in the total synthesis of polyketide natural products as promising anticancer agents. *Aust. J. Chem.* 62, 624–638.

83. For selected reviews on stereoselective aldol additions, see (a) Heathcock CH (1981). Acyclic stereocontrol through the aldol condensation. *Science* 214, 395–400. (b) Heathcock CH (1982). Acyclic stereoselection via the aldol condensation. *ACS Symp. Ser.* 185, 55–72. (c) Evans DA, Nelson JV, Taber TR (1982). Stereoselective aldol condensations. *Top. Stereochem.* 13, 1–115. (d) Machajewski TD, Wong CH (2000). The catalytic asymmetric aldol reaction. *Angew. Chem., Int. Ed.* 39, 1352–1374. (e) Palomo C, Oiarbide M, García JM (2004). Current progress in the asymmetric aldol addition reaction. *Chem. Soc. Rev.* 33, 65–75.
84. For a recent review on the use of metallic catalysts for direct enantioselective aldol additions, see Shibasaki M, Matsunaga S, Kumagai N In: Mahrwald R, editor. *Modern Aldol Reactions.* Vol. 2, Wiley-VCH, Weinheim; 2004, pp 197–227.
85. For recent reviews on the use of organic catalysts for direct enantioselective aldol additions, see (a) List B. In: Mahrwald R, editor. *Modern Aldol Reactions.* Vol. 1, Wiley-VCH, Weinheim; 2004, pp 161–200. (b) Notz W, Tanaka F, Barbas, III, CF (2004). Enamine-based organocatalysis with proline and diamines: the development of direct catalytic asymmetric aldol, Mannich, Michael, and Diels–Alder reactions. *Acc. Chem. Res.* 37, 580–591. (c) Guillena G, Najera C, Ramon DJ (2007). Enantioselective direct aldol reaction: the blossoming of modern organocatalysis. *Tetrahedron Asymmetry* 18, 2249–2293. (d) Zlotin SG, Kucherenko AS, Beletskya IP (2009). Organocatalysis of asymmetric aldol reaction. Catalysts and reagents. *Russ. Chem. Rev.* 78, 737–784.
86. Sakthivel K, Notz W, Bui T, Barbas, III, CF (2001). Amino acid catalyzed direct asymmetric aldol reactions: a bioorganic approach to catalytic asymmetric carbon-carbon bond-forming reactions. *J. Am. Chem. Soc.* 123, 5260–5267.
87. Using amides of L-proline, direct catalyzed aldol coupling of 2-butanone to p-nitrobenzaldehyde affords mixtures of regioisomeric products: (a) Tang Z, Yang ZH, Chen XH, Cun LF, Mi AQ, Jiang YZ, Gong LZ (2005). A highly efficient organocatalyst for direct aldol reactions of ketones with aldehydes. *J. Am. Chem. Soc.* 127, 9285–9289. (b) Luo S, Lu H, Li J, Zhang L, Cheng JP (2007). A simple primary-tertiary diamine-Bronsted acid catalyst for asymmetric direct aldol reactions of linear aliphatic ketones. *J. Am. Chem. Soc.* 129, 3074–3075.
88. Yoshikawa N, Yamada YMA, Das J, Sasai H, Shibasaki M (1999). Direct catalytic asymmetric aldol reaction. *J. Am. Chem. Soc.* 121, 4168–4178.
89. (a) Stork G, Rosen P, Goldman NL (1961). The α-alkylation of enolates from the lithium-ammonia reduction of α,β-unsaturated ketones. *J. Am. Chem. Soc.* 83, 2965–2966. (b) Stork G, Rosen P, Goldman NL, Coombs RV, Tsuji J (1965). Alkylation and carbonation of ketones by trapping the enolates from the reduction of α,β-unsaturated ketones. *J. Am. Chem. Soc.* 87, 275–286.
90. For recent reviews on the reductive aldol reaction, see (a) Nishiyama H, Shiomi T (2007). Reductive aldol, Michael, and Mannich reactions. *Top. Curr. Chem.* 279, 105–137. (b) Garner SA, Han SB, Krische MJ In: Andersson P, Munslow I, editors. *Modern Reductions.* Wiley-VCH, Weinheim;2008, pp 387–408.
91. For rhodium-catalyzed reductive aldol reactions mediated by silane, see (a) Revis A, Hilty TK (1987). Novel synthesis of β-siloxy esters by condensation of carbonyls and trimethylsilane with α,β-unsaturated esters catalyzed by $RhCl_3$. *Tetrahedron Lett.* 28, 4809–4812. (b) Matsuda I, Takahashi K, Sato S (1990). Rhodium catalyzed direct coupling of α,β-unsaturated ketone, aldehyde, and trialkylsilane: an easy access to regio-defined aldol derivatives. *Tetrahedron Lett.* 31, 5331–5334. (c) Emiabata-Smith D, McKillop A, Mills C, Motherwell WB, Whitehead

AJ (2001). Some preliminary studies on a novel rhodium(I)-catalyzed tandem hydrosilylation-intramolecular aldol reaction. *Synlett* 1302–1304. (d) Freiría M, Whitehead AJ, Tocher DA, Motherwell WB (2004). Further observations on the rhodium (I)-catalyzed tandem hydrosilylation-intramolecular aldol reaction. *Tetrahedron* 60, 2673–2692. (e) Taylor SJ, Morken JP (1999). Catalytic diastereoselective reductive aldol reaction: optimization of interdependent reaction variables by arrayed catalyst evaluation. *J. Am. Chem. Soc.* 121, 12202–12203. (f) Taylor SJ, Duffey MO, Morken JP (2000). Rhodium-catalyzed enantioselective reductive aldol reaction. *J. Am. Chem. Soc.* 122, 4528–4529. (g) Zhao CX, Bass J, Morken JP (2001). Generation of (E)-silylketene acetals in a rhodium-DuPhos catalyzed two-step reductive aldol reaction. *Org. Lett.* 3, 2839–2842. (h) Russell AE, Fuller NO, Taylor SJ, Aurriset P, Morken JP (2004). Investigation of the Rh-catalyzed asymmetric reductive aldol reaction. Expanded scope based on reaction analysis. *Org. Lett.* 6, 2309–2312. (i) Fuller NO, Morken JP (2005). Direct formation of synthetically useful silyl-protected aldol adducts via the asymmetric reductive aldol reaction. *Synlett* 1459–1461. (j) Nishiyama H, Shiomi T, Tsuchiya Y, Matsuda I (2005). High performance of Rh(Phebox) catalysts in asymmetric reductive aldol reaction: high anti-selectivity. *J. Am. Chem. Soc.* 127, 6972–6973. (k) Willis MC, Woodward RL (2005). Rhodium-catalyzed reductive aldol reactions using aldehydes as the stoichiometric reductants. *J. Am. Chem. Soc.* 127, 18012–18013. (l) Ito JI, Shiomi T, Nishiyama H (2006). Efficient preparation of new rhodium- and iridium-[bis(oxazolinyl)-3,5-dimethylphenyl] complexes by C–H bond activation: applications in asymmetric synthesis. *Adv. Synth. Catal.* 348, 1235–1240. (m) Shiomi T, Ito JI, Yamamoto Y, Nishiyama H (2006). 4-Substituted-phenyl(bisoxazoline)-rhodium complexes: efficiency in the catalytic asymmetric reductive aldol reaction. *Eur. J. Org. Chem.* 5594–5600. (n) Shiomi T, Nishiyama H (2007). Intermolecular asymmetric reductive aldol reaction of ketones as acceptors promoted by chiral Rh(phebox) catalyst. *Org. Lett.* 9, 1651–1654. (o) Osborne JD, Willis MC (2008). Rhodium-catalyzed hydroacylation or reductive aldol reactions: a ligand dependent switch of reactivity. *Chem. Commun.* 5025–5027. (p) Hashimoto T, Ito J, Nishiyama H (2008). Felkin-Anh selectivity in Rh(bisoxazolinyl-phenyl)-catalyzed reductive aldol coupling reaction: asymmetric synthesis of stereotriads. *Tetrahedron* 64, 9408–9412. (q) Shiomi T, Adachi T, Nishiyama H (2009). Intermolecular antiselective and enantioselective reductive coupling of enones and aromatic aldehydes with chiral Rh(Phebox) catalysts. *Org. Lett.* 11, 1011–1014.

92. For rhodium-catalyzed reductive aldol reaction mediated by hydrogen, see (a) Jang HY, Huddleston RR, Krische MJ (2002). Reductive generation of enolates from enones using elemental hydrogen: catalytic C–C bond formation under hydrogenative conditions. *J. Am. Chem. Soc.* 124, 15156–15157. (b) Huddleston RR, Krische MJ (2003). Enolate generation under hydrogenation conditions: catalytic aldol cycloreduction of keto-enones. *Org. Lett.* 5, 1143–1146. (c) Koech PK, Krische MJ (2004). Catalytic addition of metallo-aldehyde enolates to ketones: a new C–C bond-forming hydrogenation. *Org. Lett.* 6, 691–694. (d) Marriner GA, Garner SA, Jang HY, Krische MJ (2004). Metallo-aldehyde enolates via enal hydrogenation: catalytic cross aldolization with glyoxal partners as applied to the synthesis of 3,5-disubstituted pyridazines. *J. Org. Chem.* 69, 1380–1382. (e) Jung CK, Garner SA, Krische MJ (2006). Hydrogen-mediated aldol reductive coupling of vinyl ketones catalyzed by rhodium: high syn-selectivity through the effect of tri-2-furylphosphine. *Org. Lett.* 8, 519–522. (f) Han SB, Krische MJ (2006). Reductive aldol coupling of divinyl ketones via rhodium-catalyzed hydrogenation: syn-diastereoselective construction of beta-hydroxyenones. *Org. Lett.* 8, 5657–5660. (g) Jung CK, Krische MJ (2006). Asymmetric induction in hydrogen-mediated reductive aldol additions to α-amino aldehydes catalyzed by rhodium: selective formation of syn-stereotriads directed by intramolecular hydrogen-bonding. *J. Am. Chem. Soc.* 128, 17051–17056. (h) Bee C, Han SB, Hassan A, Iida H, Krische MJ (2008). Diastereo- and enantio-selective hydrogenative aldol coupling of vinyl ketones: design of effective monodentate TADDOL-like phosphonite ligands. *J. Am. Chem. Soc.* 130, 2746–2747.

93. For cobalt-catalyzed reductive aldol reaction, see (a) Isayama S, Mukaiyama T (1989). Cobalt (II) catalyzed coupling reaction of α,β-unsaturated compounds with aldehydes by the use of phenylsilane - new method for preparation of β-hydroxy nitriles, amides, and esters. *Chem. Lett.* 2005–2008. (b) Baik TG, Luis AL, Wang LC, Krische MJ (2001). Diastereoselective cobalt-catalyzed aldol and Michael cycloreductions. *J. Am. Chem. Soc.* 123, 5112–5113. (c) Wang LC, Jang HY, Roh Y, Lynch V, Schultz AJ, Wang X, Krische MJ (2002). Diastereoselective cycloreductions and cycloadditions catalyzed by Co(dpm)2-silane (dpm = 2,2,6,6-tetramethylheptane-3,5-dionate): mechanism and partitioning of hydrometallative versus anion radical pathways. *J. Am. Chem. Soc.* 124, 9448–9453. (d) Lam HW, Joensuu PM, Murray GJ, Fordyce EAF, Prieto O, Luebbers T (2006). Diastereoselective cobalt-catalyzed reductive aldol cyclizations using diethylzinc as the stoichiometric reductant. *Org. Lett.* 8, 3729–3732. (e) Lumby RJR, Joensuu PM, Lam HW (2007). Diastereoselective intermolecular cobalt-catalyzed reductive aldol reactions of α,β-unsaturated amides with ketones. *Org. Lett.* 9, 4367–4370. (f) Lumby RJR, Joensuu PM, Lam HW (2008). Racemic and asymmetric cobalt-catalyzed reductive aldol couplings of α,β-unsaturated amides with ketones. *Tetrahedron* 64, 7729–7740.

94. For iridium-catalyzed reductive aldol reaction, see Zhao CX, Duffey MO, Taylor SJ, Morken JP (2001). Enantio- and diastereoselective reductive aldol reactions with iridium-pybox catalysts. *Org. Lett.* 3, 1829–1831.

95. For ruthenium-catalyzed reductive aldol reaction, see (a) Doi T, Fukuyama T, Minamino S, Ryu I (2006). $RuHCl(CO)(PPh_3)_3$-catalyzed reductive dimerization of α,β-unsaturated aldehydes leading to α-hydroxymethyl ketones. *Synlett* 3013–3016. (b) For a related process, see Denichoux A, Fukuyama T, Doi T, Horiguchi J, Ryu I (2010). Synthesis of 2-hydroxymethyl ketones by ruthenium hydride-catalyzed cross-coupling reaction of α,β-unsaturated aldehydes with primary alcohols. *Org. Lett.* 12, 1–3.

96. For palladium-catalyzed reductive aldol reaction, see Kiyooka SI, Shimizu A, Torii S (1998). A mild aldol reaction of aryl aldehydes through palladium-catalyzed hydrosilation of α,β-unsaturated carbonyl compounds with trichlorosilane. *Tetrahedron Lett.* 39, 5237–5238.

97. For copper-promoted reductive aldol reactions, see (a) Chiu P., Chen B, Cheng KF (1998). A conjugate reduction-intramolecular aldol strategy toward the synthesis of pseudolaric acid A. *Tetrahedron Lett.* 39, 9229–9232. (b) Chiu P (2004). Organosilanes in copper-mediated conjugate reductions and reductive aldol reactions. *Synthesis* 13, 2210–2215. For copper-promoted reductive intramolecular Henry reaction, see (c) Chung WK, Chiu P (2005). Reductive intramolecular Henry reactions induced by Stryker's reagent. *Synlett* 55–58. For copper-promoted and -catalyzed reductive cyclizations of α,β-acetylenic ketones tethered to ketones, see (d) Chiu P, Leung SK (2004). Stoichiometric and catalytic reductive aldol cyclizations of alkynediones induced by Stryker's reagent. *Chem. Commun.* 2308–2309.

98. For copper-catalyzed reductive aldol reactions, see (a) Ooi T, Doda K, Sakai D, Maruoka K (1999). Unique property of copper(I) chloride as a radical initiator as well as a Lewis acid: application to CuCl-catalyzed aldol reaction of α,β-unsaturated ketones with Bu_3SnH. *Tetrahedron Lett.* 40, 2133–2136. (b) Lam HW, Joensuu PM (2005). Cu(I)-catalyzed reductive aldol cyclizations: diastereo- and enantioselective synthesis of β-hydroxylactones. *Org. Lett.* 7, 4225–4228. (c) Lam HW, Murray GJ, Firth JD (2005). Diastereoselective synthesis of 4-hydroxypiperidin-2-ones via Cu(I)-catalyzed reductive aldol cyclization. *Org. Lett.* 7, 5743–5746. (d) Deschamp J, Chuzel O, Hannedouche J, Riant O (2006). Highly diastereo- and enantioselective copper-catalyzed domino reduction/aldol reaction of ketones with methyl acrylate. *Angew. Chem., Int. Ed.* 45, 1292–1297. (e) Chuzel O, Deschamp J, Chauster C, Riant O (2006). Copper(I)-catalyzed enantio- and diastereoselective tandem reductive aldol reaction. *Org. Lett.* 8, 5943–5946. (f) Zhao D, Oisaki K, Kanai M, Shibasaki M (2006). Catalytic enantioselective intermolecular reductive aldol reaction to ketones. *Tetrahedron Lett.* 47, 1403–1407. (g) Zhao D, Oisaki K, Kanai M, Shibasaki M (2006). Dramatic ligand effect in

catalytic asymmetric reductive aldol reaction of allenic esters to ketones. *J. Am. Chem. Soc.* 128, 14440–14441. (h) Welle A, Diez-Gonzalez S, Tinant B, Nolan SP, Riant O (2006). A three-component tandem reductive aldol reaction catalyzed by *N*-heterocyclic carbene–copper complexes. *Org. Lett.* 8, 6059–6062. (i) Lipshutz BH, Amorelli B, Unger JB (2008). CuH-catalyzed enantioselective intramolecular reductive aldol reactions generating three new contiguous asymmetric stereocenters. *J. Am. Chem. Soc.* 130, 14378–14379. (j) Deschamp J, Riant O (2009). Efficient construction of polycyclic derivatives via a highly selective Cu-I-catalyzed domino reductive-aldol cyclization. *Org. Lett.* 11, 1217–1220. (k) Kato M, Oki H, Ogata K, Fukuzawa S (2009). Copper–ClickFerrophos-complex-catalyzed enantioselective reductive aldol reaction. *Synlett* 1299–1302.

99. For nickel-catalyzed reductive aldol reactions, see (a) Chrovian CC, Montgomery J (2007). Surprising role of aryl halides in nickel-catalyzed reductive aldol reactions. *Org. Lett.* 9, 537–540. (b) Joensuu PM, Murray GJ, Fordyce EAF, Luebbers T, Lam HW (2008). Diastereoselective nickel-catalyzed reductive aldol cyclizations using diethylzinc as the stoichiometric reductant: scope and mechanistic insight. *J. Am. Chem. Soc.* 130, 7328–7338. (c) Margalef IV, Rupnicki L, Lam HW (2008). Formal synthesis of salinosporamide A using a nickel-catalyzed reductive aldol cyclization-lactonization as a key step. *Tetrahedron* 64, 7896–7901.

100. For a reductive aldol coupling employing stoichiometric quantities of indium reagent, see Inoue K, Ishida T, Shibata I, Baba A (2002). Remarkable dependence of diastereoselectivity on anhydrous or aqueous solvent in the indium hydride promoted reductive aldol reaction of α,β-unsaturated ketones. *Adv. Synth. Catal.* 344, 283–287.

101. For indium-catalyzed reductive aldol reaction, see (a) Shibata I, Kato H, Ishida T, Yasuda M, Baba A (2004). Catalytic generation of indium hydride in a highly diastereoselective reductive aldol reaction. *Angew. Chem., Int. Ed.* 43, 711–714. (b) Miura K, Yamada Y, Tomita M, Hosomi A (2004). Indium(III) acetate-catalyzed 1,4-reduction and reductive aldol reactions of α-enones with phenylsilane. *Synlett* 1985–1989.

102. For mechanistic studies on alkene hydrogenation catalyzed by neutral Rh(I) complexes, see (a) Tolman CA, Meakin PZ, Lindner DL, Jesson JP (1974). Triarylphosphine, hydride, and ethylene complexes of rhodium(I) chloride. *J. Am. Chem. Soc.* 96, 2762–2774. (b) Halpern J, Wong CS (1973). Hydrogenation of *tris*(triphenylphosphine)chlororhodium(I). *J. Chem. Soc., Chem. Commun.* 629–630. (c) Halpern J, Okamoto T, Zakhariev A (1977). Mechanism of the chloro*tris*(triphenylphosphine) rhodium(I)-catalyzed hydrogenation of alkenes. The reaction of chlorodihydridotris(triphenyl-phosphine)rhodium(III) with cyclohexene. *J. Mol. Catal.* 2, 65–68.

103. For a review of the acidity of metal hydrides, see Kristjánsdóttir SS, Norton JR Acidity of hydrido transition metal complexes in solution. In: Dedieu A, editior. *Transition Metal Hydrides*. VCH, New York; 1992, pp 309–357.

104. Monohydride catalytic cycles initiated via deprotonation of cationic rhodium dihydrides have been postulated and their veracity further established through the stoichiometric conversion of cationic rhodium(I) complexes to the corresponding rhodium(I) monohydrides via exposure to hydrogen in the presence of a tertiary amine base: (a) Schrock RR, Osborn JA (1976). Catalytic hydrogenation using cationic rhodium complexes. I. Evolution of the catalytic system and the hydrogenation of olefins. *J. Am. Chem. Soc.* 98, 2134–2143. (b) Schrock RR, Osborn JA (1976). Catalytic hydrogenation using cationic rhodium complexes. II. The selective hydrogenation of alkynes to cis olefins. *J. Am. Chem. Soc.* 98, 2143–2147. (c) Schrock RR, Osborn JA (1976). Catalytic hydrogenation using cationic rhodium complexes. 3. The selective hydrogenation of dienes to monoenes. *J. Am. Chem. Soc.* 98, 4450–4455.

105. For heterolytic activation of elemental hydrogen, see (a) Brothers PJ (1981). Heterolytic activation of hydrogen by transition metal complexes. *Progr. Inorg. Chem.* 28, 1–61. See also: (b) Jeske G, Lauk, H, Mauermann H, Schumann H, Marks TJ (1985). Highly reactive

organolanthanides. A mechanistic study of catalytic olefin hydrogenation by bis(pentamethyl-cyclopentadienyl) and related 4f complexes. *J. Am. Chem. Soc.* 107, 8111–8118.

106. Yachi K, Shinokubo H, Oshima K (1999). Reaction of titanate-type aldehyde enolate with ketones to provide 3-hydroxyaldehydes. *J. Am. Chem. Soc.* 121, 9465–9466.

107. For tri-2-furylphosphine and triphenylarsine effects in metal catalyzed reactions, see (a) Farina V, Krishnan B (1991). Large rate accelerations in the Stille reaction with tri-2-furylphosphine and triphenylarsine as palladium ligands: mechanistic and synthetic implications. *J. Am. Chem. Soc.* 113, 9585–9595. (b) Farina V (1996). New perspectives in the cross-coupling reactions of organostannanes. *Pure Appl. Chem.* 68, 73–78. (c) Andersen NG, Keay BA (2001). 2-Furyl phosphines as ligands for transition-metal-mediated organic synthesis. *Chem. Rev.* 101, 997–1030.

108. For metal-catalyzed reductive Mannich couplings mediated by silane, see (a) Muraoka T, Kamiya SI, Matsuda I, Itoh K (2002). Rhodium-catalyzed approach to Mannich-type products using aldimine, α,β-unsaturated ester, and hydrosilane. *Chem. Commun.* 1284–1285. (b) Townes JA, Evans MA, Queffelec J, Taylor SJ, Morken JP (2002). Stereoselective synthesis of *trans* β-lactams through iridium-catalyzed reductive coupling of imines and acrylates. *Org. Lett.* 4, 2537–2540.

109. For secondary amine-catalyzed reductive Mannich coupling of enal to imines mediated by Hantzsch ester, see Zhao GL, Cordova A (2006). Direct organocatalytic asymmetric reductive Mannich-type reactions. *Tetrahedron Lett.* 47, 7417–7421.

110. Priet O, Lam HW (2008). Cobalt-catalyzed reductive Mannich reactions of 4-acryloylmorpholine with *N*-tosyl aldimines. *Org. Biomol. Chem.* 6, 55–57.

111. Garner SA, Krische MJ (2007). Rhodium-catalyzed reductive Mannich coupling of vinyl ketones to *N*-sulfonylimines mediated by hydrogen. *J. Org. Chem.* 72, 5843–5846.

112. Komanduri V, Grant CD, Krische MJ (2008). Branch-selective reductive coupling of 2-vinyl pyridines and imines via rhodium catalyzed C–C bond forming hydrogenation. *J. Am. Chem. Soc.* 130, 12592–12593.

113. For early transition metal-catalyzed hydroaminoalkylation of α-olefins, see (a) Clerici MG, Maspero F (1980). Catalytic *C*-alkylation of secondary amines with alkenes. *Synthesis* 305–306. (b) Nugent WA, Ovenall DW, Holmes SJ (1983). Catalytic C–H activation in early transition-metal dialkylamides and alkoxides. *Organometallics* 2, 161–162. (c) Herzon SB, Hartwig JF (2007). Direct, catalytic hydroaminoalkylation of unactivated olefins with *N*-alkyl arylamines. *J. Am. Chem. Soc.* 129, 6690–6691. (d) Herzon SB, Hartwig JF (2008). Hydroaminoalkylation of unactivated olefins with dialkylamines. *J. Am. Chem. Soc.* 130, 14940–14941. (e) Kubiak R, Prochnow I, Doye S (2009). Titanium-catalyzed hydroaminoalkylation of alkenes by C–H bond activation at sp^3 centers in the α-position to a nitrogen atom. *Angew. Chem., Int. Ed.* 48, 1153–1156. (f) Bexrud JA, Eisenberger P, Leitch DC, Payne PR, Schafer LL (2009). Selective C–H activation α to primary amines. Bridging metallaaziridines for catalytic, intramolecular α-alkylation. *J. Am. Chem. Soc.* 131, 2116–2118. (g) Prochnow I, Kubiak R, Frey ON, Beckhaus R, Doye S (2009). Tetrabenzyltitanium: an improved catalyst for the activation of sp^3 C–H bonds adjacent to nitrogen atoms. *ChemCatChem* 1, 162–172. (h) Kubiak R, Prochnow I, Doye S (2010). [Ind_2TiMe_2]: a catalyst for the hydroaminomethylation of alkenes and styrenes. *Angew. Chem., Int. Ed.* 49, 2626–2629. (i) Zi G, Zhang F, Song H (2010). Highly enantioselective hydroaminoalkylation of secondary amines catalyzed by group 5 metal amides with chiral biarylamidate ligands. *Chem. Commun.* 6296–6298. (j) Reznichenko AL, Emge TJ, Audörsch S, Klauber EG, Hultzsch KC, Schmidt B (2011). Group 5 metal binaphtholate complexes for catalytic asymmetric hydroaminoalkylation and hydroamination/cyclization. *Organometallics* 30, 921–924. (k) Prochnow I, Zark P, Müller T, Doye S (2011). The mechanism of the titanium-catalyzed hydroaminoalkylation of alkenes. *Angew. Chem., Int. Ed.* 50, 6401–6405. (l) Zhang F, Song H, Zi G (2011). Synthesis and catalytic activity of group 5 metal amides with chiral biaryldiamine-based ligands. *Dalton Trans.* 1547–1566.

114. For late transition metal-catalyzed hydroaminoalkylation of α-olefins, see (a) Jun CH, Hwang DC, Na SJ (1998). Chelation-assisted alkylation of benzylamine derivatives by Ru(0) catalyst. *Chem. Commun.* 1405–1406. (b) Chatani N, Asaumi T, Yorimitsu S, Ikeda T, Kakiuchi F, Murai S (2001). $Ru_3(CO)_{12}$-catalyzed coupling reaction of sp^3 C–H bonds adjacent to a nitrogen atom in alkyamines with alkenes. *J. Am. Chem. Soc.* 123, 10935–10941. (c) Sakaguchi S, Kubo T, Ishii Y (2001). A three-component coupling reaction of aldehydes, amines and alkynes. *Angew. Chem., Int. Ed.* 40, 2534–2536. (d) Yi CS, Yun SY (2004). Intermolecular coupling reaction of cyclic amines and alkenes catalyzed by a ruthenium-hydride complex $(PCy_3)_2(CO)RuHCl$. *Organometallics* 23, 5392–5395. (e) Tsuschikama K, Kasagawa M, Endo K, Shibata T (2009). Cationic Ir(I)-catalyzed sp^3 C–H bond alkenylation of amides with alkynes. *Org. Lett.* 11, 1821–1823.

115. Moran J, Preetz A, Mesch RA, Krische MJ (2011). Iridium-catalyzed direct C–C coupling of methanol and allenes. *Nat. Chem.* 3, 287–290.

INDEX

Sustainable Catalysis: Challenges and Practices for the Pharmaceutical and Fine Chemical Industries, First Edition. Edited by Peter J. Dunn, K. K. (Mimi) Hii, Michael J. Krische, and Michael T. Williams.

www.ingramcontent.com/pod-product-compliance
Lightning Source LLC
LaVergne TN
LVHW062214200926
848484LV00006B/1394